Animal Signals and Communi

Volume 2

Series editors

Vincent M. Janik
School of Biology
University of St Andrews
Fife, UK

Peter McGregor
Centre for Applied Zoology
Cornwall College
Newquay, UK

For further volumes:
http://www.springer.com/series/8824

Henrik Brumm
Editor

Animal Communication and Noise

 Springer

Editor
Henrik Brumm
Max Planck Institute for Ornithology
Seewiesen
Germany

ISSN 2197-7305   ISSN 2197-7313 (electronic)
DOI 10.1007/978-3-642-41494-7
Springer Heidelberg New York Dordrecht London

Printed on acid-free paper

Springer is part of Springer Science+Business Media (www.springer.com/mycopy)

Preface

Animal communication is not only intriguing, but the scientific study of it has made important contributions to areas such as neurobiology, sensory physiology, ethology, behavioural ecology, and evolutionary biology. Many studies on animal communication have investigated which information is encoded in a given signal and how this information is used by receivers. However, information coding is only one of two crucial steps in communication: before a signal can be detected and recognized it must first be transmitted successfully. Signal transmission is not a trivial task, because the exchange of information between sender and receiver can be crucially constrained by noise, which will lead to errors in the receiver.

Interference of communication by noise may be most obvious in acoustic signalling (and most research has been done in this area) but the problem of signal detection is a general one that applies to all signal modalities. In line with this notion, this book not only considers acoustic communication but also visual, chemical, and electric signals. Within the chapters of this volume you will find reviews of the literature on communication in many different groups of animals, including insects, fish, amphibians, lizards, birds and mammals.

Noise pollution is an issue of growing concern, and this book also addresses the implications of anthropogenic noise for conservation. There are several books that deal with the impact of noise on animal behaviour, and this is with no doubt an important area that needs more research in the future. However, the main thrust of this book is a conceptual one. We advocate that the mitigation of noise is a very fundamental process in communication, and that we need to consider the effects of noise if we want to understand how animal communication systems operate.

It was a great pleasure to work with the 24 biologists who joined their expert forces to write this book, and I would like to thank all of them for their contributions. Without the series editors Vincent Janik and Peter McGregor this book would not exist, not only because they kindly invited me to edit this volume, but they were also most helpful during all stages of the project. Also, I would like to warmly thank Andrea Schlitzberger from Springer Publishing for her patience and for making it all happen.

I owe a great deal of gratitude to many colleagues and mentors who helped shaping ideas about communication and noise. Henrike Hultsch started it all 15 years ago when she suggested that I try playing noise to nightingales. Dietmar Todt taught me independence and encouraged me not to follow the mainstream.

Peter Slater has been the most wonderful mentor one can think of and he truly is a role model for me—as rigorous scientist, enthusiastic naturalist, and all-around decent fellow. Finally, I would also like to thank Sue Anne Zollinger and all other members past and present of my research group in Seewiesen who shared countless discussions about communication in noise during the last years

Seewiesen, 2013   Henrik Brumm

Contents

Chapter 1
Introduction

Henrik Brumm

Abstract The study of animal communication has led to significant progress in our general understanding of motor and sensory systems, evolution, and speciation. However, one aspect that is often neglected is that signal exchange in every modality is constrained by noise. In this introduction to the volume, I give an overview of the organisation of the book and the contents of each of the chapters. I highlight that the widespread problem of communication in noise has led to similar solutions across taxa and signal modalities. In addition, special features that have evolved in only a few taxa are considered, such as particular forms of signal plasticity or derived sensory mechanisms.

Communication is a key area of animal behaviour because all social interactions between individuals are based on the exchange of information. Hence, animals have evolved the most astounding ways to pass on messages, by using, for example, optical, acoustic, electric, or chemical signals. In many species, sexual reproduction relies on the exchange of signals between mating partners in one way or another. This means that the advertisement signals of sexually displaying animals play a particularly important role in sexual selection and speciation.

For communication to occur, a sender has to encode information in a signal, which is then transmitted to a receiver (Shannon and Weaver 1949). Our own experience tells us that acoustic signals, such as speech, can be impaired by noise. However, as this book demonstrates, noise is not only a hindrance for acoustic communication but is a basic problem for all forms of signal exchange. Moreover, in addition to extrinsic noise in the transmission channel, communication is also disturbed by intrinsic noise within the nervous system of the receiver. Although these two forms of noise are fundamentally different in nature, they are similar in that they decrease the contrast between signal and background, which increases the probability of errors in the receiver. However, animals possess a huge arsenal of

H. Brumm (✉)
Communication and Social Behaviour Group, Max Planck Institute for Ornithology,
82319 Seewiesen, Germany
e-mail: brumm@orn.mpg.de

H. Brumm (ed.), *Animal Communication and Noise*,
Animal Signals and Communication 2, DOI: 10.1007/978-3-642-41494-7_1,
© Springer-Verlag Berlin Heidelberg 2013

mechanisms that allow them to deal with noise. A review of the literature on these, sometimes stunning, capacities is the core of this volume, in which several authors investigate microevolutionary adaptations and individual signal plasticity that enhance communication in noise, and the sophisticated sensory and cognitive mechanisms for signal detection and recognition by the receivers.

An introudctory text by Haven Wiley sets the stage for the chapters on assorted taxa and signalling modalities which follw. This introductory text reviews that noise in communication is equivalent to errors by receivers and that receivers' errors have fundamental consequences for optimal behaviour of both receivers and signallers. By applying signal detection theory to animal communication, Wiley shows that exaggeration of signals should evolve to improve the detectability of signals by receivers, and that noise drives the evolution of signals to a signal detection balance, in which signals reach optimal but not ideal detectability and receivers reach optimal but not ideal performance. These theoretical considerations are then picked up by the chapters that follow in Parts II and III, which focus on specific forms of animal communication. Several of these chapters investigate the effects of noise on both signal production and perception in a given group of animals or a given modality. However, in cases where the literature is very extensive, production and perception are treated in separate chapters. In Chap. 3, Heinrich Römer explores the effects of noise on acoustic communication in insects. Friedrich Ladich reviews the consequences of noise for acoustic signalling in fish in Chap. 4. The following two chapters deal with acoustic signal masking in anurans: Chap. 5, by Joshua Schwartz and Mark Bee, investigates acoustic signal production in noise. Chapter 6, by the same authors together with Alejandro Veléz, addresses the receivers' side. In Chap. 7, Sue Anne Zollinger and I review the effects of noise on vocal production in birds. This is followed by an analysis of avian sound perception in noise by Bob Dooling and Sandra Blumenrath in Chap. 8. Chapter 9, written by Peter Tyack and Vincent Janik, explores the influence of noise on vocal production in marine mammals. How receivers deal with acoustic signal masking in this group of animals is treated in Chap. 10 by James Finneran and Brian Branstetter. In Part III of the book, the concept of noise is expanded to modalities other than the acoustic channel: Richard Peters addresses the role of noise in visual communication in Chap. 11, in particular he investigates the effects of motion noise from windblown plants on movement-based signalling in lizards. Chapter 12 by Jan Benda, Jan Grewe and Rüdiger Krahe, explores the effects of intrinsic noise on electric signal detection in fish. In Chap. 13, Volker Nehring, Tristram Wyatt and Parizia d'Ettorre venture into unchartered waters by exploring signal masking in chemical communication, providing one of the first reviews on noise in chemical signal transmission in animals.

Taken together, the chapters of this volume suggest that the widespread problem of communication in noise has led to similar solutions across taxa and modalities, including individual adjustments of signal properties, environmental selection for particularly contrasting signals, and perceptual mechanisms of receivers for signal detection in noise. However, there are also special features that have evolved in only a few taxa, such as particular forms of signal plasticity or derived sensory mechanisms. In acoustic communication, the Lombard effect appears to be a general mechanism of vocal plasticity used by birds and mammals to make themselves heard in noise. The evidence for a lack of the Lombard effect in anurans suggests that it has probably evolved in amniotes, either as a synapomorphy of sauropsids and mammals or independently in the two clades. However, the picture is still patchy. First, not many anuran species have been tested and much more work is needed to confirm the absence of the Lombard effect in this clade. Second, to establish the Lombard effect as a derived trait in birds and mammals it would be necessary to study it in marsupials and monotremes, as well as testudines, squamates and crocodilians. Some tortoise, lizard, and crocodile species are vocally very active, and future research should target these species to further elucidate the phylogenetic origin of the Lombard effect.

Other widespread solutions to the noise problem that have been observed across a wide range of taxa include, e.g. the frequency tuning of neurons. This perceptual feature is based mainly on a sharpening of stimulus filtering by the peripheral or central nervous system, which means that energy outside the sensitivity range of the filter does not lead to signal masking. Frequency tuning has been demonstrated for the hearing systems of invertebrates and vertebrates, as well as for electroreception in fishes. In this book, a similar phenomenon is also suggested for chemoreception.

Another recurrent theme found in several chapters is the role of noise on the evolution of multicomponent and multimodal signals. If noise terms differ in different sensory channels, signal transmission gains can be considerably increased by the use of multiple modalities (Higham and Hebets 2013). Indeed, the use of multimodal displays for communication in noise is suggested for a number of insect and vertebrate species (Chaps. 2, 3, 5, 7, 13). However, the current evidence is sparse and hence interpretations are often stretched in a way that is disproportional to the small amount of data available. Therefore, we need much more research on this exciting topic, including both work on the effects of noise on the evolution and performance of multimodal signals and studies investigating how receivers integrate different sensory modalities to decode signals in noise.

Finally, a particular kind of acoustic noise is raising growing concerns—the sounds produced by humans. Anthropogenic noise is a severe form of pollution that can have massive impacts on the health of humans and probably also animals. The World Health Organisation estimates that in the European Union alone more than 200,000 people die every year because of noise-induced illnesses (WHO 2011). In Chap. 14, Peter McGregor, Andrew Horn, Marty Leonard, and Frank Thomsen review how anthropogenic noise is a critical cause for concern in conservation, particularly through effects on animal communication. As it is such a pressing issue, the topic of acoustic noise pollution is also addressed in most chapters on acoustic signals in Part II. The review of the recent literature shows that we are just beginning to understand the effects of anthropogenic noise on animal communication, but whether impairments of signal exchange result in changes in population dynamics is still unknown at this time.

The organisation of the book into different chapters on certain taxa is somewhat artificial and reflects the history of research rather than biological necessity.

However, this taxonomic assembly is also an advantage because the chapters are authored by experts who have studied the respective groups of animals for many years. Moreover, by connecting the different chapters, I hope that this book will help in bridging this historical gap, at least partly. Whichever species you may look at, the study of the effects of noise on communication provides opportunities for research of proximate mechanisms as well as evolutionary processes. With this book, we want to advocate the integration of the knowledge gained by the two approaches and to highlight particularly interesting fields of current and future research.

References

Higham JP, Hebets PA (2013) An introduction to multimodal communication. Behav Ecol Sociobiol 67:1381–1388

Shannon CE, Weaver W (1949) The mathematical theory of communication. Illinois University Press, Urbana

WHO (2011) Burden of disease from environmental noise—quantification of healthy life years lost in Europe. World Health Organization, Regional Office for Europe

Part I
Signal Detection Theory

Chapter 2
Signal Detection, Noise, and the Evolution of Communication

R. Haven Wiley

Abstract Signal detection theory has had limited application in studies of animal communication. Yet by specifying constraints placed by noise on a receiver's performance, it provides a way to investigate optimal performance and thus the evolution of communication. Noise in this case is anything influencing a receiver's receptors other than a signal of interest. The essential features of signal detection theory are (1) a distinction between the detectability of a signal in noise and the criterion or threshold for a receiver's response and (2) a realization that any decision by a receiver to respond has four possible outcomes, not all of which are independent. Although presented here in terms of a receiver's threshold for response to one kind of signal, signal detection theory applies also to more complex criteria for response as well as complex discriminations among multiple signals. A receiver's optimal performance always depends on the payoffs of the four possible outcomes of a decision to respond and on the detectability of a signal. By incorporating detectability, signal detection theory can provide a complete explanation for the evolution of exaggerated signals. An alternative explanation, based only on sexual selection and necessary costs of signals, does not do so. In particular, signal detection theory shows that exaggeration of signals should evolve so as to improve the detectability of signals by receivers. By shifting the emphasis from a receiver's preferences and to its performance, this theory also clarifies the co-evolution of signalers and receivers. The result is a signal-detection balance, in which signals reach optimal but not ideal detectability and receivers reach optimal but not ideal performance. The crucial importance of the detectability of signals by receivers means that noise in natural situations, just as much as costs and benefits for the participants, determines the features of communication.

R. H. Wiley (✉)
Department of Biology, University of North Carolina, Chapel Hill,
NC 27599-3280, USA
e-mail: rhwiley@email.unc.edu

H. Brumm (ed.), *Animal Communication and Noise*,
Animal Signals and Communication 2, DOI: 10.1007/978-3-642-41494-7_2,
© Springer-Verlag Berlin Heidelberg 2013

2.1 Introduction

In recent decades, the study of animal communication has been transformed by steadily expanding research on the effects of noise on communication, as this volume demonstrates. Initially, interest focused on the attenuation and degradation of acoustic signals as they propagated from the signaler to a receiver. This work quickly led to questions about adaptations of signals to minimize these effects in different habitats. It also became clear that receivers could often use attenuation and degradation of signals to judge the distance to signalers. There was also some early interest in ways that animals can avoid masking of signals by environmental noise, but this possibility has recently received much more attention. It is now clear, as this volume shows, that animals counteract the effects of environmental noise on communication in several ways. The discovery that animals make these adjustments to anthropogenic noise, which presumably presents a novel challenge from an evolutionary perspective, suggests that these adaptations can result from behavioral plasticity in addition to or instead of evolution. With this diversity of interests in the effects of noise on animal communication, it seems appropriate to take a broad view of the role of noise in communication. This chapter reviews the argument that noise in communication is equivalent to errors by receivers and that receivers' errors have fundamental consequences for optimal behavior of both receivers and signalers.

People have always recognized that noise, as commonly conceived, is a problem for communication. People have also always known that communication is prone to errors. It was Shannon (1948), however, who first realized that noise in communication is nothing more or less than a receiver's errors (see Shannon and Weaver 1963). This insight provided the start for his mathematical analysis of the limitations on the rate of communication, now known as information theory. Shannon formulated communication as the process of reproducing, at one point, signals generated at another point. In addition to correct reproduction, he recognized that there are also the possibilities of two kinds of error, ambiguity, and equivocation (reproduction without signal and signal without reproduction). There is no need to pursue the mathematical details of Shannon's analysis to cast this situation into one familiar in the study of animal communication. Whenever a receiver samples the input of its receptors, in all but ideal conditions, and makes a decision to respond or not, there are four mutually exclusive outcomes possible. These four outcomes are a consequence of a combination of two possible situations (the presence of a signal or not) and a binary decision by the receiver (respond or not). Two outcomes are correct (correct detection and correct rejection) and two are errors (false alarm and missed detection). Just as Shannon first emphasized, when signals of interest to a receiver cannot be completely separated from other coincident events, the receiver inevitably makes errors in deciding whether or not a signal has occurred.

We return shortly to what constitutes a signal, but for the moment we need only recognize that examples of signals include the presence of an optimal mate or a rival, the presence of a predator or parasite, the nutritional state of offspring, the identity of a nest mate, and so forth. An example of a correct detection is a response to an optimal mate or to a territorial intrusion by a rival, and a missed detection is a failure to respond despite signals from an optimal mate or a rival. A correct rejection is an absence of response to a suboptimal mate or to an individual that does not represent a threat such as a territorial neighbor still inside its own territory. A false alarm is a response to signals from such individuals. Experiments that present signals to territorial individuals or females seeking mates routinely elicit all four of these kinds of responses. Even "successful" experiments, in which the subjects respond with statistically significant probability to the "correct" signals and not to the "false" signals, nearly always include instances of false alarms and missed detections as well as correct detections and correct rejections.

Decision theory provides a method to determine the optimal decision when the outcomes of decisions are uncertain. It computes the expected utility of any decision from the payoffs (positive or negative) and probabilities of each of its possible outcomes. Von Neumann introduced a measure of the expected utility of a decision,

$$E(U) = \Sigma(i)\,U(i)p(i),$$

the sum of the utility, $U(i)$, times the probability, $p(i)$, of each of i mutually exclusive outcomes (von Neumann and Morgenstern 1944). Decision theory has since been widely applied in economics and has become familiar to behavioral ecologists, who routinely consider costs, benefits, and probabilities of alternatives in order to predict optimal behavior.

The optimal performance of a receiver facing four possible outcomes of any decision to respond or not is easily formulated in terms of decision theory. This fusion of decision theory and information theory occurred over a period of a decade or so and resulted in a general theory of a receiver's performance known as signal detection theory (Green and Swets 1966). Perhaps because this theory was first introduced by psychophysicists interested in studying the sensory capabilities of humans and other animals, its application has not diffused widely. Nevertheless, for over half a century, signal detection theory has provided the foundation for psychophysics and cognitive psychology. In particular, it has provided a way to separate the motivation of subjects from the inherent detectability of signals. Although it has remained peripheral in studies of animal communication, this chapter will suggest that signal detection theory, by addressing the fundamental problems of noise, can explain some basic adaptations for communication and reveal some unrecognized problems.

The application of signal detection theory to animal communication has been presented in some detail elsewhere (Wiley 1994, 2006), so this chapter focuses on the generality of this theory for understanding communication and then develops ways it can help to think about the evolution of communication, including the co-evolution of an equilibrium in the performance of signalers and receivers.

2.2 Signal Detection Theory as a General Model for Communication

To justify the general application of signal detection theory, we must address three issues: what constitutes a signal, what constitutes a receiver, and what constitutes an error by a receiver. In considering these issues, it will become apparent that many fundamental features of communication apply to interactions between machines or between humans and machines, as well as those between animals including those between humans. Indeed these features apply to interactions within organisms, among organs and cells, and even to those between molecules. So there arises a fourth issue, what special features apply to communication among living organisms? Although the following discussion of these four issues concentrates on animal communication, a wider scope is sometimes appropriate.

First, to qualify as a signal, an event must affect some receiver's behavior. In other words, signals are associated with responses, a point frequently emphasized. A response might be overt but it could also be covert. We often think of responses as actions quickly following a signal, but they could also be changes in a receiver's state that alter the probabilities of further actions. Beyond this basic condition for a signal, there have been proposals to separate signals from signs (characterized by representation), cues (characterized by a lack of intention or evolutionary specialization), or indices (characterized by an invariant relation with some property of interest to a receiver) (see for instance, Markl 1985; Maynard Smith and Harper 2003). These distinctions have inevitably proven difficult to characterize operationally.

The present perspective can ignore these distinctions. A signal is any event that influences a receiver's behavior, immediately or subsequently, without providing all of the power for that behavior (Wiley 1994). At least some of the power necessary for the receiver's response must come from the signal, because there must be enough to alter the receiver's sensory receptors. For most familiar kinds of signals, including human language and animal displays, however, it is clear that most of the energy for responses comes from the receiver. According to this definition, moving out of the way of approaching danger as a result of a push is not an example of communication, but jumping aside in response to a shout, or even responding to the sound of approaching danger, is. The essential feature of any signal, in this view, is its limited power, insufficient to produce the response. As a result, the receiver itself has a crucial role in determining the response. The receiver therefore is in a position to get what it wants, as Grafen (1990) has emphasized, although, as we see below, only within some limits.

Second, this definition of a signal leads to a conclusion that any receiver has three essential components. A receiver must acquire a signal, must differentiate it from other events, and then must generate the power and arrange the coordination for a response. Electrical engineers have distinct terms for these components: a transducer, a switch (or gate), and an amplifier. If the receiver is an animal, these three basic components are often neural: sensory receptors, associative neurons, and motor effectors (in combination with a musculo-skeletal apparatus). These three components are sometimes not obvious, however. Acquiring a signal often involves transducing it from its original form of energy or matter to one appropriate for the receiver's nervous system (for instance, from sound waves to action potentials in sensory neurons). Differentiating between a signal and irrelevant events might involve no more than a filter, a simple physical, electrical, or chemical connection between an input and an output, but it could involve extremely complex connections, such as human cognition. Generating the power for a response might be a process that nearly consumes a receiver but in many cases it is nearly trivial. Even if it involves no more than cleaving one molecule of ATP, a response requires some energy from a receiver to amplify the direct effects of a signal. As already mentioned, an amplified response need not be an overt action. It could instead be an altered internal state, such as a memory, neural association, or other physiological state, that can affect future actions. The possibility of such covert responses recurs in all forms of communication. For instance, in electrical apparatus, capacitors and computer memories provide this possibility. Altered molecular states of a cell do too. In each case, receivers have the three fundamental components just mentioned. For living organisms, we might call the three components a sensor, an associator, and an effector. A crucial factor is the second one: all receivers must make associations between signals and responses.

Third, the insufficient power of a signal and the necessity of association by a receiver together impose a special state of affairs on any receiver. Receivers are, fundamentally, decision makers susceptible to error. This inescapable conclusion arises from the possibility that receivers cannot in every instance separate the occurrence of signals from other events impinging on them. It might be possible to arrange a situation in which a particular receiver can almost always differentiate correctly between particular signals and irrelevant events. Living organisms including humans might try their best to attain such situations, and they might evolve to maximize the possibility of these ideal situations, but it seems unlikely that they often achieve them in the real world. Later, in this chapter, it will become apparent that approaching this ideal of error-free communication has diminishing returns. Consequently, communication among living organisms is not likely to evolve, nor is communication among machines likely to be designed, in a way that reaches this ideal. At best, we can expect an occasional close approach to the ideal. In all but ideal circumstances, receivers make errors, more or less frequently.

What constitutes an error by a receiver? To recognize an error, one must have a goal. If our goal is to understand the evolution of communication, then our concern is the relative rates of spread of alleles associated with receivers that differ in their mechanisms for response. In this view, those responses that make a receiver less likely to survive or reproduce are errors and those responses that do otherwise are correct. For a living organism, this ultimate goal might be less prominent at any particular moment than a more proximate one of maintaining homeostasis and of managing its relationships with other individuals. Nevertheless, the goals of homeostasis and behavior are themselves ultimately subject to the goal of propagating alleles. Because there is a single ultimate goal, the ultimate costs of errors are continuous with the ultimate benefits of correct responses. These costs and benefits are measured by decrements and increments on the same scale. In a more proximal view, scales for measurements of costs and benefits might coincide but they do not have to. For instance, the costs of errors and benefits of responses might both be measured by probabilities of obtaining a mate. On the other hand, they might be measured, respectively, by probabilities of attracting a parasite and attracting a mate. The ultimate costs and benefits would remain the same: differences in the spread of alleles associated with receivers' mechanisms for response.

This concept of error in communication includes the normal human concept of error. The human view becomes a special case of this general view. We think of error as an opinion or action that tends to thwart a person's own objectives or that fails to conform to the opinions or actions of other people. Error often seems to require a goal set by human judgment (or attributed by humans to divine judgment). For our purposes here, these cases all represent proximate mechanisms of human behavior subsumed in the ultimate one of evolution.

The common human approach also recognizes that error has two inevitable aspects, errors of omission and commission, although it seems to take some effort for humans to keep these possibilities routinely in mind. Nevertheless, the approach here emphasizes that these two aspects of error are a fundamental aspect of any decision. Because decisions are a fundamental part of any receiver, so are these two forms of error. Whenever an animal samples its sensors and decides to respond or not, it faces four possible outcomes, two of which are correct and two of which are errors.

So far, this chapter has argued that the basic definition of a signal, as an event that evokes a response from a receiver but lacks sufficient power to produce the response, leads to the important conclusion that a receiver must have three components, one of which makes decisions prone to errors. The following section explores the nature of a receiver's decisions further and leaves us with a conclusion that all receivers face a double bind. Furthermore, we can see more clearly the relationship between errors and noise.

2.3 A Receiver's Double Bind

A receiver's dilemma results from the convergence of signal and noise. A simple example, in line with our focus on animal communication, is a sensory neuron tuned to a particular frequency of sound. In this case, a signal is a tone of this frequency emitted by an appropriate signaler, and any other sound with this frequency is noise. We must imagine that different occurrences of a signal have some random variation around a mean intensity, because the conditions under which the signal is produced and then received are never exactly the same. Likewise, the activation of a receiver's receptors by a signal varies. Nevertheless, we expect that the activity in a receiver's receptors, provided they are well matched to features of the signal, is often greater during the occurrence of a signal than during its absence. Taking the variation into account, we find that the probability density functions for the activity of a receptor in the presence and absence of a signal often overlap (Fig. 2.1). If they overlap at all, then the receiver cannot completely avoid errors. As we have seen, receivers in the real world must usually, if not always, face such situations.

The possibility of error is thus the inevitable result of a decision by a receptor. A mechanism that makes a decision to respond or not requires a criterion for response. The simplest criterion is a threshold: if activity in the input reaches a predetermined level, then respond, otherwise do not. Of course, a criterion for response, even one based on just one receptor, can be more complex, and decisions can be based on the inputs from many receptors. The basic conundrum for a receiver, however, is not affected by the complexity of criteria or the number of inputs, a point discussed in more detail elsewhere (Wiley 1994, 2006). All the basic features of a receiver's conundrum are evident in the case of a simple threshold for activity in a single neuron (Fig. 2.1).

The receiver can adjust its threshold upward or downward. The location of the threshold is its decision. Such a decision might change from time to time depending on the receiver's physiology or development, and it might differ from individual to individual as a result of their genetic or epigenetic differences. Nevertheless, in any situation a receiver confronts, its threshold for response fixes four probabilities, one for each of the four mutually exclusive and exhaustive possible outcomes when the receiver samples its sensors.

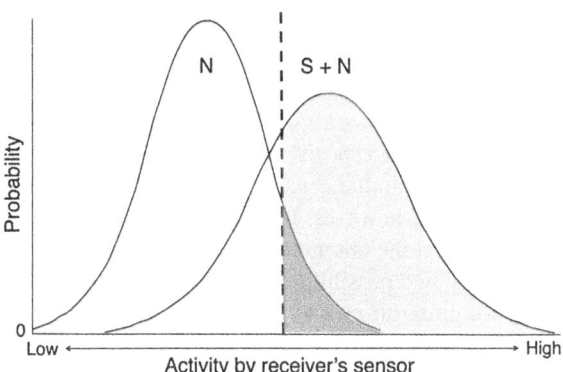

Fig. 2.1 Receiver performance depends on the activity of its sensors, the signal-to-noise ratio, and its threshold for response. Horizontal axis, the level of activity of the receiver's sensors. Vertical axis, probability that activity reaches any level when only noise is present (N) and when a signal is present with noise ($S \mid N$). The latter distribution would often have greater variance as a result of variation in the signal. *Dashed line*, an example of a threshold for response set by the receiver. *Light shading*, the cumulative probability of a correct detection when a signal is present and the receiver's threshold is at the indicated level. *Dark shading*, the cumulative probability of a false alarm when only noise is present. There are corresponding probabilities of a missed detection, when a signal is present, and a correct rejection, when it is not. A receiver can adjust its threshold for response in order to maximize the utility of its threshold and thus to optimize its performance

Furthermore, the four possible outcomes are not independent of each other. By raising its threshold, for instance, a receiver might reduce the chance of a false alarm, but it would concomitantly raise the chance of a missed detection. By lowering its threshold, a receiver might reduce the chance of a missed detection, but it would raise the chance of a false alarm. Receivers thus face an inevitable trade-off between the consequences of false alarms and missed detections (coinciding with this trade-off there is also one between correct detections and correct rejections). This trade-off is equivalent to the well-known trade-off in electronic receivers between sensitivity and selectivity. Only by accepting more false alarms (less selectivity) can a receiver reduce missed detections (more sensitivity). Evolution should thus result in receivers that optimize the expected utility, $E(U)$, of their criteria for response (Wiley 1994).

A receiver's criteria for response can vary in complexity. As described above, a simple case is a threshold on a single dimension of a signal, such as frequency or intensity. Other cases can include multidimensional criteria for responses to complex patterns of stimulation. Experimental demonstrations that a species' own vocalizations are easier to detect in background noise, for instance, indicate that channels for filtering and decision-making have evolved complex filters for detection of these signals (Okanoya and Dooling 1991; Dooling et al. 1992; Benney and Braaten 2000). Regardless of the complexity of a receiver's criteria for response, it faces the same inevitable trade-off in minimizing errors (Wiley 1994).

Although beyond the scope of our discussion here, it is also important to realize that a receiver only "knows" two possible states of the world prior to its decision to respond or not: input-above-criterion or input-below-criterion. The view presented here suggests that there are actually four possible states of the world, depending both on whether or not the receiver's input is above or below its criterion but also on whether or not a signal has actually occurred. We can imagine a privileged observer, one with a special vantage or special equipment for studying both signals and receivers simultaneously, who might realize these four states of the world. For the receiver, however, the world has only two states. And, going one step farther, we see that the observer, in deciding any "fact" about signals and responses, also sees only two possibilities: the evidence at hand is either sufficient or not. But we are not going to pursue this point here.

2.4 Applications of Signal Detection Theory to Animal Communication

The application of signal detection theory to human psychophysics has provided quantitative confirmation of many of its predictions. Controversies in this field have focused on the validity of assumptions for mathematical convenience, rather than on the underlying trade-off any receiver must face (reviewed by Wiley 2006).

Experiments in psychophysics have repeatedly demonstrated that a receiver's performance increases with higher signal-to-noise ratios, in other words higher contrast between signals and noise. Furthermore, performance improves under any conditions that allow a receiver to predict the timing and features of a signal. Identifying intervals when a signal might occur by means of alerting signals, using signals with features known in advance, and including redundancy (predictable temporal or spatial structure) all increase performance (reviewed by Wiley 2006). Other aspects of receiver psychology (Guilford and Dawkins 1991, 1993), including the "peak shift" so frequent in discrimination learning (Enquist and Arak 1998; Lynn et al. 2005), also follow from signal detection theory.

These results from experimental psychophysics have analogies with communication in natural circumstances (Wiley 2006). Adaptations that increase contrast between signal and noise, for instance, are widespread. Birds and mammals, including humans, increase the intensity of their vocalizations in the presence of background sound (Brumm and Todt 2002; Pytte et al. 2003; Brumm and Slabbekoorn 2005; Leonard and Horn 2005; Brumm and Zollinger 2011). In another case, two closely related populations of birds differ in the dominant frequencies in their songs, so that each minimizes overlap of its songs with background noise (Slabbekoorn and Smith 2002a). Many long-range acoustic signals of birds and mammals have attributes that reduce attenuation and degradation during transmission through their natural habitats and thus increase contrast between signal and noise for a receiver (Morton 1975; Wiley and Richards 1982; Wiley 1991; Brown et al. 1995; Mitani and Stuht 1998; Boncoraglio and Saino 2007; Brumm and Naguib 2009; Ey and Fischer 2009). Visual signals also provide evidence for adaptations that improve a receiver's signal/noise ratio. For instance, the movements in territorial displays of *Anolis* lizards are faster than the movement of vegetation in the background (Fleishman 1988; 1992). To maintain this contrast, lizards increase the speed of movements in their displays in windier conditions (Ord et al. 2007). The contrast between a bird's coloration and its background depends on the spectral properties of ambient light (irradiance) as well as the reflectance of the bird's plumage and the nearby vegetation (Endler 1990). The colors of manakins and other lekking birds of neotropical forests contrast best with the background at the sites where they perform their displays (Endler and Théry 1996; Heindl and Winkler 2003; Doucet et al. 2007). Species that display on the forest floor clear away leaf litter in order to increase the contrast of their plumage with the ground (Uy and Endler 2004).

It has also become apparent that animals include alerting components in their signals—introductory components poor in information that serve to attract the attention of potential receivers to subsequent components rich in information (Richards 1981; Wiley and Richards 1982; Peters and Evans 2003; Mitchell et al. 2006; Ord and Stamps 2008). In at least one case this alerting component becomes longer and more conspicuous in noisy conditions (Peters et al. 2007).

Redundancy is also prominent in many animals' signals, sometimes producing spectacular patterns in time or space. Temporal patterns in signals include simple repetition of movements or acoustic elements, as well as complex arrangements.

Spatial patterns of coloration and movement are also prevalent. Nevertheless, there has been little attention to the possibility that levels of redundancy differ in different levels of noise. Humans, in the presence of noise, speak more distinctly (as well as more loudly, as mentioned above) and thus with greater redundancy in enunciation, a change that improves intelligibility (van Summers et al. 1988). Birds close to noisy waterfalls and torrents repeat their songs more often (Brumm and Slater 2006), and birds also increase their rates of signaling in the presence of artificial ambient noise (Potash 1972). The use of multiple "ornaments" for communication might also provide redundancy (Møller and Pomiankowski 1993; Johnstone 1996; Candolin 2003). If the expression of these ornaments were positively correlated, they might improve detection (provide "backup" for missed detections), as predicted for increased redundancy. In contrast, multiple ornaments with negative or no correlation might serve as separate signals for distinct sets of receivers or responses (Andersson et al. 2002). So far, theoretical treatment and experimental investigation of multiple ornaments have only tangentially considered the possibility that features of signals correlated in time or space might improve detection by increasing redundancy.

Receivers might evolve adaptations to background noise as well as signalers. The optimal frequency for detection of sound by great tits *Parus major* is higher in the presence of natural noise such as wind in a forest than it is in quiet conditions such as in a sound-attenuating chamber. The higher optimal frequency in natural conditions is a better match for the dominant frequencies in the species' vocalizations (Langemann and Klump 2001).

Among the more important consequences of background sound is the limit it sets for the active space of a signal (Brenowitz 1982; Römer and Bailey 1986; Janik 2000; Nemeth and Brumm 2010). A striking example of this limitation occurs in choruses of frogs. The phonotactic responses of female green treefrogs *Hyla cinerea* to calls of individual males differ in the presence and absence of sound from a chorus of these frogs (Gerhardt and Klump 1988). Females preferentially approach a male's calls only when the calls exceed the sound of the chorus by 3 dB. A male's call attenuates by spherical spreading alone to this level in a distance of about 1 m. Female frogs in such a chorus would thus respond to individual males only within this short distance. As a result of the spacing of calling males, even in the densest part of a chorus, a female is within this distance of only 2–3 males at a time. To sample more males, she would have to move around and thus risk exposure to predators such as snakes. A similar conclusion was reached by Wollerman (1999) for female *Hyla ebraccata* at a large chorus with eight species of frogs in a Costa Rican rainforest. Because of background noise, a female's choice of males is much more limited in a large aggregation than we might imagine.

The adaptations of receivers to noise might explain one of the striking features of sympatric animals' signals. Biologists have long realized that sympatric species usually have signals with distinctly different features, at least for communication with conspecifics (Marler 1957). Evidence suggests that signals of closely related species can diverge in sympatry in comparison to allopatry (reviewed by

Slabbekoorn and Smith 2002b; Pfennig and Pfennig 2009). Sympatric species' signals, however, are not only distinct but also disjunct. In other words, the signals of sympatric species are separated by gaps in signal space (the multidimensional space with axes defined by the features of signals). This disjunction of sympatric species' signals occurs even in highly diverse faunas, such as among birds in neotropical rainforests. In these situations, sympatric species' signals can diverge enough to produce significant overdispersion in signal space (Luther and Wiley 2009).

This disjunction of signals raises an evolutionary problem. Although it is clear that natural selection for character divergence might result in differences in the signals of two populations, the strength of this selection should fall to zero once the signals no longer overlap (distinct signals). So it is hard to see how natural selection for character divergence could routinely produce gaps between sympatric populations' signals (disjunct signals). A possible resolution of this problem comes from experiments in Neotropical forests on two bird species with similar (but disjunct) songs (Luther and Wiley 2009). The results showed that receivers have a broader scope for responses than do conspecific signalers for producing songs. When digitally synthesized songs were morphed to produce exemplars intermediate between the two species, playbacks revealed that individuals of each species responded to exemplars beyond the natural range of conspecific songs. The responses of the two species left no gaps in signal space. Receivers' responses were thus not disjunct and not even quite distinct. Presumably, the greater scope of receivers' responses allows them to compensate for variation in noisy signals. Receivers' scope for responses is thus wider than the scope for signals as produced by signalers and measured in clean recordings. If this result applies to animal communication broadly, the disjunct signals of sympatric species are not explained entirely by selection on signalers but also by selection on error-prone receivers—in other words, by noise.

2.5 Detection Versus Discrimination

The problem for a receiver becomes more complicated when the task is to classify two or more relevant signals as well as to detect the presence or absence of any one (Miller et al. 1951; Green and Birdsall 1978; Wiley 2006). This problem arises whenever a receiver must make appropriate, but different, responses to more than one signal. An animal that must respond in different ways to different types of prey or food might face this situation. Social situations that require recognition of several different individuals also fit this situation. Another occurs when appropriate responses must be given to signals warning about different kinds of predators (Owings and Leger 1980; Cheney and Seyfarth 1990; Blumstein and Armitage 1997).

These situations require classification (often called discrimination) of signals, in other words, different responses to each of several signals. In contrast, detection requires the same response to exemplars of one signal. A test for discrimination thus requires a comparison of responses to two different sets of signals, each mixed with noise, and to noise alone. A test for detection, as described above, requires only a comparison between responses to one set of signals, mixed with noise, and noise alone. A complete analysis of discrimination thus requires three situations, noise alone and with each of two sets of signals, whereas an analysis of detection only requires two situations.

In a complete analysis of discrimination between two signals, there are nine possible outcomes as viewed by a privileged observer, instead of the four possibilities for detection. There are three states of the world (noise with signal one, noise with signal two, or noise alone), and there are three possible responses of the subject (appropriate for signal one, for signal two, or none). The analysis of this situation is correspondingly complex, with more than twice the number of relevant probabilities and utilities.

Despite this complexity, applying signal detection theory to discrimination leads to an important prediction (Macmillan 2002; Wiley 2006): performance of receivers in tasks that require discrimination is lower than performance in tasks with detection only. Consequently, we expect to find that individuals can detect signals in higher levels of noise but can discriminate among them only in lower levels of noise. Humans, for instance, can detect occurrences of a single known word in higher levels of noise than they can discriminate between two or more words (Miller et al. 1951).

This difference between detection and discrimination applies to female frogs mating in a dense chorus. In a number of species of frogs, we know that females prefer conspecific male advertisement calls with lower dominant frequencies. This preference has, for instance, been confirmed for *H. ebraccata* in Costa Rica (Wollerman 1998). As already discussed, we also know that female frogs, including *H. ebraccata*, have difficulty detecting individual male's calls in large choruses. Do they have even greater difficulty discriminating males' dominant frequencies?

In a test of discrimination in natural levels of noise, Wollerman and Wiley (2002) presented gravid female *H. ebraccata* with males' calls mixed with the background sound of a chorus. One speaker presented calls with a dominant frequency at the population mean, while a second speaker presented calls with a dominant frequency two SD below the mean. With no added chorus noise ($S/N > 25$ dB), females reliably preferred the lower frequency. With added chorus noise ($S/N = 6$ or 9 dB), they no longer preferred the lower frequency, although they still detected (responded preferentially to) a single male's calls in chorus sounds. The discrimination made in relatively quiet conditions thus disappeared in conditions that still allowed detection of the signals. This result is thus in agreement with the prediction of signal detection theory: discrimination requires a higher S/N ratio than does detection.

2.6 Evolution of Receivers

By providing a method for analyzing the performance of a receiver, signal detection theory allows us to determine a receiver's optimal performance and thus the expected evolution of communication. The constraints on a receiver's performance can provide a sufficient explanation for such problematic features of communication as the prevalence of honesty, the persistence of deception, and the exaggeration of signals. In addition, signal detection theory suggests that the co-evolution of signalers and receivers (or the behavior of signaling and receiving) lead to a signal detection balance.

Because receivers provide the power necessary for a response, they evolve to optimize performance in the conditions they experience. The first step in understanding the evolution of communication is thus an explanation for a receiver's decisions to respond or not. As we have seen, this explanation requires optimization of the expected utility, $E(U)$, of the receiver's criterion for response. Procedures for calculating optimal thresholds for response have been presented elsewhere (Wiley 1994). Here we use some limiting cases to illustrate the main conclusions. Compare, for instance, situations in which missed detections have relatively high costs with those in which false alarms have relatively high costs.

Missed detections might be especially costly when an individual is listening for alarm calls. A missed detection (failing to respond to an alarm call) is likely to mean increased exposure to a predator. A false alarm (briefly fleeing when there is no alarm call) would often require only a little energy and a little time lost from other activities. If predators are a relatively frequent danger, the cost of a missed detection multiplied by its probability might well dominate other terms in the expected utility of any threshold for response. In this case, a low threshold is optimal. The result would be a receiver with "adaptive gullability" (Wiley 1994), one prone to false alarms but subject to few missed detections. Such an individual would be susceptible to frequent deception, for instance, when calls that mimic alarms allow a subordinate individual to take advantage of a dominant rival.

Examples of adaptive gullability include birds that respond to false alarms by subordinates that usurp food or by rivals that interrupt sexual activity (Munn 1986; Møller 1988, 1990). Monkeys are also manipulated by subordinates in this way (Cheney and Seyfarth 1990). Another example comes from species in which satellite males encroach upon matings by dominant males. In many cases, the subordinate males look like females. Dominants trying to detect cheating males thus run the risk of false alarms, with the consequence that they chase away some females. When missed detections are expensive, adaptive gullability should evolve and dominant males should fail to exclude all satellites from matings.

False alarms, on the other hand, might have especially negative consequences when individuals make infrequent but crucial choices. Mate choice might often fit this situation. In most species, a female chooses a mate infrequently and yet mistakenly mating with a low-quality male, a male with inadequate resources, or even another species could substantially reduce the spread of her genes. In this case, a high threshold is optimal. The result would be a receiver with "adaptive fastidiousness" (Wiley 1994), one liable to miss detections of suitable signals but subject to few false alarms. From a privileged observer's perspective, such a receiver would appear to be "choosy" or "coy," because they would often fail to respond to suitable signals.

This situation would apply whenever reproductive success of a female is limited by the number of eggs she matures, while reproductive success of a male is limited by the number of matings he gets. A mistake in mating in this case has greater consequences for a female than for a male. As Wiley and Poston (1996) have argued, females in many species have evolved choosiness in mating and males have not because the consequences of errors in mating differ for the two sexes.

Adding signal detection theory to an investigation of mating signals and preferences has advantages over the usual approach based exclusively on sexual selection. First, it emphasizes that the evolution of receivers is likely to depend on the probabilities and consequences of all four outcomes of an interaction. Second, it emphasizes the detectability of a signal, which in relation to the receiver's criterion for response, determines the probabilities of the possible outcomes. Overall, it stresses features of communication with noisy signals in natural situations, as opposed to communication with clean signals in expurgated situations.

2.7 Evolution of Signals

Once the performance of receivers begins to evolve toward its optimum, the evolution of signaling should adapt to the changing behavior of receivers. On one hand, the presence of receivers with "adaptive gullability" opens opportunities for signalers that can manipulate receivers with misleading signals, like the deceptive alarm calls mentioned above. In this case, the evolution of deceptive signals is limited by the payoffs and probabilities of the four outcomes for receivers and by the probabilities of honest and deceptive signals (Wiley 1983).

On the other hand, the presence of receivers with "adaptive fastidiousness" favors signalers that produce exaggerated signals that exceed the high thresholds or other stringent criteria set by these receivers. For instance, among oropendolas and caciques, males of species with only brief interactions with females have displays with high repetitiveness and complexity (Price 2013). Signal detection theory predicts that the evolution of exaggerated signals should result in increased detectability or discriminability of signals by intended receivers (potential receivers whose responses would have advantages for the signaler). Evidence for greater detectability of exaggerated signals comes from a study of nestling birds begging for food from their parents. When begging, nestlings often reveal bright colors in their mouths, particularly red gapes and yellow flanges. Heeb et al. (2003) showed that nestling great tits with gapes and flanges that were more detectable under natural light conditions (the dim light available in nest cavities) gained more weight than did other nestlings. The detectability of the markings was a better predictor of parental response than was their complexity or redness (which might indicate the nestling's nutritional state).

Studies of fish have revealed a connection between discriminability and the evolution of colorful signals for mate choice. In Lake Victoria, female preferences for the colors of males contribute to reproductive isolation between many coexisting species of cichlids. Sedimentation of the lake in areas with high agricultural runoff, however, has obscured colors and resulted in loss of reproductive isolation (Seehausen et al. 1997). Another case involves sticklebacks in lakes of coastal British Columbia. In some populations, males have bright red on their underparts and in others they lack red, differences that contribute to reproductive isolation between sympatric populations. In lakes with high concentrations of tannin, the tea-colored water masks red signals. In these lakes, males have lost their red markings, and females have lost not only their preferences for red males but also their sensitivity to red light (Boughman 2001 also see Fuller and Noa 2010). Colorful signals and receivers' responses to them thus persist only where the ambient light does not mask them. Between populations, lower thresholds for responses to red by females correlate with redder males. Within a population, on the other hand, females with higher thresholds for red should tend to mate with redder males.

2.8 Signal Detection in Relation to Previous Theories

This approach to the evolution of signals based on signal detection theory complements previous ones based on sexual selection and costs for the signaler and receiver. The effects of sexual selection on communication have attracted widespread attention, because the evolution of exaggerated signals, one of the most striking features of animal communication, is especially associated with mate choice. Not all mate choice is a result of communication, however (Wiley and Poston 1996). Mate choice, behavior that results in mating with some potential mates more than others, includes both direct choice (preferences for perceived traits of potential mates) and indirect choice (any other behavior that results in narrowing the set of potential mates). It is direct choice that requires communication between potential mates. Both forms of mate choice generate sexual selection, the evolution of alleles associated with the traits of mating individuals. The distinctive feature of sexual selection, as opposed to other forms of natural selection, is the genetic correlation that inevitably results from nonrandom mating between individuals with a preference and those with the corresponding trait. This genetic correlation produces the explosive evolution that makes sexual selection distinctive. If this genetic correlation becomes sufficiently strong, the evolution of a preference and a corresponding trait become self-reinforcing, and alleles for a preferred trait spread in association with alleles for the corresponding preference until the benefit of additional matings is balanced by the cost of the trait (Lande 1981; Kirkpatrick 1982). Subsequent analyses have emphasized that alleles for a preference can spread unless the direct costs of the preference (from searching for or interacting with males) completely compensate for the benefit (direct or indirect) of mating with a preferred male (Pomiankowski 1987, 1988). If a preference has no costs, then, a preference can spread even if it has no benefits. Matings of females with high thresholds and males with exaggerated signals produce the sort of genetic correlation that characterizes sexual selection.

At first it seemed that sexual selection could result in the evolution of arbitrary traits, those with no benefits for females and none other than multiple matings for males. This possibility provided an attractive explanation for many secondary sexual traits that seem exaggerated to an extreme of preposterousness. The expanded esophageal sacs of male greater Sage-Grouse *Centrocercus urophasianus*, so laboriously inflated during displays, provide an example (Wiley 1973). The selection on a male trait depends on the sum of direct selection as a result of its effect on the male's viability and selection as a result of females' preferences (reviewed by Heisler 1994). Taking both costs and benefits of male traits into account, sexual selection favors the evolution of preferences with the greatest net benefit for females and a corresponding trait with the greatest net benefits for males. Strictly arbitrary preferences (those with no costs for the choosy partner) and arbitrary traits (those with costs limited only by mating success of the chosen partner) seem unlikely to evolve.

Although sexual selection can explain the explosive rate of evolution of preferences and traits, it does not explain the direction of evolution. Sexual selection puts no constraints other than costs on the nature of the preference or the corresponding trait. Even when we consider the costs and benefits of the partners, sexual selection could in principle result in preferences for either augmented or diminished traits. Nevertheless, sexual selection has always been assumed to produce augmentation of signals. This gap between theory and preconception poses a dilemma. The explanation for the exaggeration of signals by sexual selection alone is incomplete. To complete the argument, it has been proposed that high costs of traits are necessary to insure reliable (or honest) signaling of mate quality, which in turn insures a net benefit for females' preferences (Zahavi 1975; Grafen 1990; Johnstone 1995, 1997; Zahavi and Zahavi 1997; Maynard Smith and Harper 2003).

This expanded argument has plausibility. Exaggeration of signals should normally increase the costs for signalers. These costs might include any of those previously identified for signals: additional time and energy, developmental compromises with other traits as a result of physiological interactions or genetic epistasis, and risks of interception by unwanted receivers, like predators, parasites, and conspecific rivals (McGregor 1993; Zuk and Kolluru 1998). In many cases, exaggeration of a signal at a cost could increase the discriminability of high-quality mates, those able to absorb the additional costs. For a graphic demonstration of how costs produce honesty, see Fig. 2.2, from Wiley (2000, 2013); more or less similar graphs are presented by Johnstone (1997) and Getty (1998, 2006).

This argument for costly exaggeration of signals nevertheless raises problems. It has been pointed out that some costs do not insure honesty (Hurd 1995; Getty 2006; also consider the final comment in the legend of Fig. 2.2), so the argument

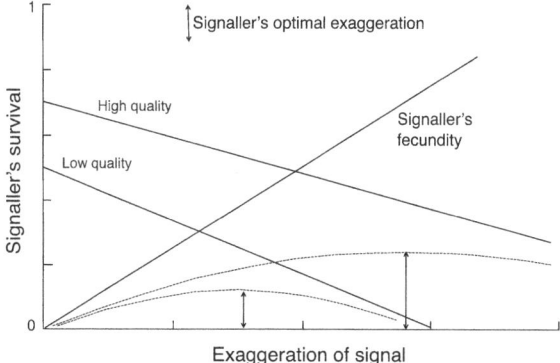

Fig. 2.2 Reliability of signals occurs when signalers of different quality adjust their levels of signaling to maximize fitness (the product of survival and fecundity). Signalers with higher quality have higher intrinsic survival than those with lower quality when no signal is produced and higher marginal survival when signaling. Females only respond to males' signals, so all males have the same fecundity for any level of signaling. A male's fitness as a function of his level of signaling (*dotted lines*) reaches a maximum at a higher level for males with higher quality than for those with lower quality. Signals of male quality are reliable (honest) unless the survival functions for males of different quality cross. In other words, reliability requires that quality correlate with intrinsic survival in the absence of signaling or marginal survival at any level of signaling. Otherwise the shapes of the curves do not matter. Notice that signals would still be reliable even if males had equal intrinsic survival provided their marginal survival correlates with quality—or if they had equal marginal survival provided their intrinsic survival correlates with quality (see Wiley 2013 for more discussion)

for exaggeration might then not apply. A more serious problem is that any level of cost can separate individuals with different capacities to bear those costs (Getty 1998; Wiley 2000). Formal arguments that costs are necessary for honest signals have shown only that signals must have some cost but not that a receiver's benefits must rise as a signal's cost rises (Grafen 1990; Maynard Smith 1991; Johnstone and Grafen 1992; Maynard Smith and Harper 2003). Because all signals presumably have some costs, these arguments do not explain why honest signals must have exaggerated costs.

There is now extensive evidence that preferred traits have costs. Less extensive, but still substantial, evidence shows that individuals with preferred traits also have high phenotypic quality, such as higher survival, lower resistance to disease, greater foraging abilities, or greater success in competition with conspecific. Some evidence indicates that females benefit from preferences for mating with these individuals, either directly as a result of greater survival or reproduction or indirectly as a result of genetic advantages for their offspring (Andersson 1994; Searcy and Nowicki 2005). Some of this evidence comes from comparisons of benefits for females mated to two categories of males, with higher or lower expression of a trait, and some comes from measurements made partly or entirely in laboratories, rather than in natural situations. Evidence that females' benefits correlate with the size of their partners' traits in nature is absent. In cases of extremely exaggerated traits, it has sometimes not been possible to find correlations between the size of the trait and a preference for the trait or the benefits of the preference (Poston 1997).

A corollary has also been proposed that the costs of signals should be structurally related to their "meaning." For instance, a signal must reduce foraging success in order to indicate a greater capability for foraging, or it must reduce survival in order to demonstrate a greater capability for survival (Zahavi 1975; Zahavi and Zahavi 1997). This corollary could also provide an explanation for exaggeration of signals. Some signals might have this feature, but it is possible to imagine cases in which they do not. An ability to fight could be indicated by a signal that imposes a cost on foraging, if reduced foraging reduced fighting ability. Or parental ability could be indicated reliably by a signal that imposes a cost on fighting, if success in fighting improved opportunities for parenting.

Despite a superficial plausibility, arguments that the reliability of signals is proportional to their costs so far have no formal proof and little if any confirmation in the field. Instead the theory of sexual selection indicates that, for any net benefit for receivers (as a result of the reliability of a signal), a signal should evolve to minimize costs. These arguments and the corresponding evidence do not produce a strong explanation for the widespread evolution of exaggerated signals.

2.9 Signal Detection Theory as an Explanation for Exaggeration

Signal detection theory, on the other hand, provides an unequivocal prediction that signals intended for choosy receivers should evolve exaggeration. Exaggerated signals evolve in response to high thresholds. High thresholds of receivers are a result of adaptive fastidiousness, which, as described above, occur when receivers face situations with low inherent detectability or discriminability of signals (low signal-to-noise ratio) and costly missed detections.

In this case, however, there is no requirement that exaggeration of signals should correlate with their costs, although as we have seen this possibility might often arise. Instead, exaggeration of signals should correlate with their discriminability in the intended receiver's local environment. Signals should evolve to reduce the possibility of confusion with irrelevant perturbations of the receiver's receptors (Endler 1992; Wiley and Richards 1982; Wiley 1994, 2006). It is thus not the cost of a signal that is the primary consideration in its evolution, but its impact on the receiver. Exaggerated signals should evolve to become spectacular. The cost is a secondary consideration.

Just as signal detection theory requires shifting our emphasis from preferences to an emphasis on thresholds (or to criteria for response in general), it also requires shifting our emphasis from a receiver's benefits to an emphasis on the expected utility of its threshold. In signal detection theory, the benefit of choosiness is a result of the difference in payoffs from mating with an optimal partner as opposed to a suboptimal one, in other words the difference in payoffs for a correct detection and a false alarm. Other payoffs affect the receiver's optimal threshold as well. The cost of additional search is the payoff for a missed detection, when an optimal mate is actually present, or for a correct rejection, when no optimal mate is present. The inevitable trade-offs between these possible outcomes are summarized in the expected utility, $E(U)$, of the receiver's threshold, which depends on the payoffs and probabilities of all four possible outcomes.

This approach also reinforces the improbability of arbitrary mating preferences and traits. Preferences could have equal benefits for receivers only when different thresholds for response have equal expected utilities, $E(U)$. This condition requires that the four possible outcomes have equal probabilities and equal consequences for different thresholds (or exactly compensating effects on their expected utilities). In other words, alternative signals would have equivalent consequences for a female only if they had exactly the same correlation with male quality and exactly the same detectability by females (or exactly compensating effects). Meeting these conditions seems so unlikely that arbitrary signals and preferences seem doubly implausible. As a consequence, optimizing a receivers' performance would nearly always oppose runaway evolution of arbitrary thresholds and signals.

By shifting our emphasis away from the costs of signals and the strengths of preferences, as the explanation for the exaggeration of signals, to new emphases on the performance of receivers, we find that the dominant influence on the evolution of exaggerated signals is the detectability or discriminability of signals in the receiver's natural environment. From the perspective of signal detection theory, the costs of signals are secondary. Costly signaling should evolve only when it increases the performance of receivers. The primary consideration is the detectability of signals from the perspective of receivers. The detectability of signals depends on the prevailing signal-to-noise ratio, the relationship between the properties of signals and properties of irrelevant events that alter activity in the receiver's sensors. Noise is thus an inescapable, if not dominant, consideration in explaining the evolution of exaggerated signals.

There is a further benefit from an application of signal detection theory to communication. Although the evolution of signalers and receivers must be mutually related, it has not been easy to formulate the nature of this relationship. It is easy to see that the evolution of signalers must depend on the evolution of receivers, and vice versa. Furthermore, it is routine to show that the properties of signals are related to the properties of corresponding detectors. If males have certain traits, we can test the expectation that females respond to these traits and that their sensory mechanisms have corresponding filters. Conversely, females' preferences often provide a match for male's traits. Perhaps in the course of evolution one side of this relationship drives the other. Perhaps, as in the theory of sensory exploitation, the mechanisms of females' responses set constraints for the evolution of males' traits. Although never previously suggested, one could conversely imagine that males' traits might drive the evolution of females' preferences.

A more likely result would be coevolution of both receivers and signalers to a signal detection balance. By providing an explicit measure of a receiver's performance, signal detection theory can provide to a way to think about the evolution of this balance. As before, it becomes apparent that noise is a predominant consideration.

2.10 Signal-Detection Balance

Although we have discussed exaggeration mostly in terms of its implications for the costs of signals and the increased probability of responses (correct detections) by receivers, signal detection theory identifies an additional consequence of exaggeration: diminishing returns for a signaler. As a signal becomes more detectable to the intended receivers, the probabilities of errors by receivers decrease asymptotically toward zero and the probability of correct detections increases towards one. In the later stages of this process, any further increase in a receiver's threshold would result in progressively fewer additional correct detections and more additional missed detections. As receivers' thresholds stabilized, further exaggeration of signals would yield little or no increase in benefits for them. Selection on receivers for increasing thresholds would thus progressively decrease. Even if further exaggeration of signals had little or no cost, selection on signalers for further exaggeration would also progressively decrease as a result of the diminishing returns from improved performance of receivers. Although high costs of false alarms and noisy discriminations could result in the evolution of highly fastidious receivers and extravagant exaggeration of signals, both receivers and signalers face diminishing returns.

Eventually, an equilibrium between diminishing benefits and augmenting costs of exaggeration would put an end to further exaggeration of a signal. Furthermore, these diminishing returns suggest that this equilibrium would be reached at a point short of perfect discriminability of signals by intended receivers (Wiley 2013). At this equilibrium, receivers would make some mistakes, and signals would sometimes fail to evoke the intended response. Receivers would have evolved optimal, not ideal, performance, and signals would have evolved optimal, not complete, efficacy. Both receivers and signalers would have adapted to the constraints of environmental noise on signal detection or discrimination. We should therefore avoid a naive expectation that evolution leads to signals that are always detectable by receivers or receivers that never make mistakes. At a signal-detection balance, ideal signals and ideal receivers would not exist.

It seems likely that most communication is poised in such a signal-detection balance. If so, the properties of communication would be difficult to understand without an investigation of all the constraints on optimal performance of receivers and on optimal detectability or discriminability of signals. Noise, as much as costs and benefits of signals or responses, would determine the properties of communication.

Acknowledgments My approach to understanding communication has developed over several decades in the course of many discussions and experiments with students and colleagues at Chapel Hill, many of whose papers are cited in my chapter. Those who took a particular interest in signal detection included Douglas Richards, Lori Wollerman, Marc Naguib, David Luther, and Jonathan Micancin. In addition, many ideas were vetted in a course in animal communication taught jointly with Steve Nowicki with students from both Chapel Hill and Durham. Continuation of my work has always been supported in many indispensible ways by my wife, Minna Wiley.

References

Andersson MB (1994) Sexual selection. Princeton University Press, Princeton

Andersson S, Pryke SR, Örnborg J, Lawes MJ, Andersson M (2002) Multiple receivers, multiple ornaments, and a trade-off between agonistic and epigamic signaling in a widowbird. Am Nat 160:683–691

Benney KS, Braaten RF (2000) Auditory scene analysis in estrildid finches (*Taeniopygia guttata* and *Lonchura striata domestica*): a species advantage for detection of conspecific song. J Comp Psych 114:174–182

Blumstein DT, Armitage KB (1997) Alarm calling in yellow-bellied marmots: I. The meaning of situationally variable alarm calls. Anim Behav 53:143–171

Boncoraglio G, Saino N (2007) Habitat structure and the evolution of bird song: a meta-analysis of the evidence for the acoustic adaptation hypothesis. Funct Ecol 21:134–142

Boughman JW (2001) Divergent sexual selection enhances reproductive isolation in sticklebacks. Nature 411:944–948

Brenowitz EA (1982) The active space of red-winged blackbird song. J Comp Physiol 147:511–522

Brown CH, Gomez R, Waser PM (1995) Old world monkey vocalizations: adaptation to the local habitat? Anim Beahv 50:945–961

Brumm H, Slabbekoorn H (2005) Acoustic communication in noise. Adv Study Behav 35:151–209

Brumm H, Todt D (2002) Noise-dependent song amplitude regulation in a territorial songbird. Anim Behav 63:891–897

Brumm H, Slater PJB (2006) Ambient noise, motor fatigue and serial redundancy in chaffinch song. Behav Ecol Sociobiol 60:475–481

Brumm H, Naguib M (2009) Environmental acoustics and the evolution of bird song. Adv Study Behav 40:1–33

Brumm H, Zollinger SA (2011) The evolution of the Lombard effect: 100 years of psychoacoustic research. Behaviour 148:1173–1198

Candolin U (2003) The use of multiple cues in mate choice. Biol Rev 78:575–595

Cheney DL, Seyfarth RM (1990) How monkeys see the world. University of Chicago Press, Chicago

Dooling RJ, Brown SD, Klump GM, Okanoya K (1992) Auditory perception of conspecific and heterospecific vocalizations in birds: evidence for special processes. J Comp Psych 106:20–28

Doucet SM, Mennill DJ, Hill GE (2007) The evolution of signal design in manakin plumage ornaments. Am Nat 169:S62–S80

Endler JA (1990) On the measurement and classification of colour in studies of animal colour patterns. Biol J Linn Soc 41:315–352

Endler JA (1992) Signals, signal donations, and the direction of evolution. Am Nat 139:S125

Endler JA, Thery M (1996) Interacting effects of lek placement, display behavior, ambient light, and color patterns in three neotropical forest-dwelling birds. Am Nat 148:421–452

Enquist M, Arak A (1998) Neural representation and the evolution of signal form. In: Dukas R (ed) Cognitive ecology. University of Chicago Press, Chicago, pp 21–87

Ey E, Fischer J (2009) The "Acoustic Adaptation Hypothesis"—a review of the evidence from birds, anurans and mammals. Bioacoustics 19:21–48

Fleishman LJ (1988) Sensory and environmental influences on display form in *Anolis auratus*, a grass anole from Panama. Behav Ecol Sociobiol 22:309–316

Fleishman LJ (1992) The influence of the sensory system and the environment on motion patterns in the visual displays of anoline lizards and other vertebrates. Am Nat 139:S36–S61

Fuller RC, Noa LA (2010) Female mating preferences, lighting environment, and a test of sensory bias in bluefin killifish. Anim Behav 80:23–35

Gerhardt H, Klump G (1988) Masking of acoustic signals by the chorus background noise in the green tree frog: a limitation on mate choice. Anim Behav 36:1247–1249

Getty T (1998) Reliable signalling need not be a handicap. Anim Behav 56:253–255

Getty T (2006) Sexually selected signals are not similar to sports handicaps. Trends Ecol Evol 21:83–88

Grafen A (1990) Biological signals as handicaps. J Theor Biol 144:517–546

Green DM, Swets JA (1966) Signal detection theory and psychophysics (reprinted with additions by Krieger, New York, 1974). Wiley, New York

Green DM, Birdsall TG (1978) Detection and recognition. Psychol Rev 85:192–206

Guilford T, Dawkins M (1991) Receiver psychology and the evolution of animal signals. Anim Behav 42:1–14

Guilford T, Dawkins MS (1993) Receiver psychology and the design of animal signals. Trends Neurosci 16:430–436

Heeb P, Schwander T, Faoro S (2003) Nestling detectability affects parental feeding preferences in a cavity-nesting bird. Anim Behav 66:637–642

Heindl M, Winkler H (2003) Vertical lek placement of forest-dwelling manakin species (Aves, Pipridae) is associated with vertical gradients of ambient light. Biol J Linn Soc 80:647–658

Heisler IL (1994) Quantitative genetic models of the evolution of mating behavior. In: Boake CRB (ed) Quantitative genetic studies of behavioral evolution. University of Chicago Press, Chicago, pp 101–125

Hurd P (1995) Communication in discrete action-response games. J Theor Biol 174:217–222

Janik VM (2000) Source levels and the estimated active space of bottlenose dolphin (Tursiops truncatus) whistles in the Moray Firth, Scotland. J Comp Physiol A 186:673–680

Johnstone RA, Grafen A (1992) The continuous Sir Philip Sydney game: a simple model of biological signalling. J Theor Biol 156:215–234

Johnstone RA (1995) Sexual selection, honest advertisement and the handicap principle: reviewing the evidence. Biol Rev 70:1–65

Johnstone RA (1996) Multiple displays in animal communication: 'backup signals' and 'multiple messages'. Phil Trans Roy Soc London B 351:329–338

Johnstone RA (1997) The evolution of animal signals. In: Krebs JR, Davies NB (eds) Behavioural ecology, 4th edn. Oxford University Press, Oxford, pp 157–178

Kirkpatrick M (1982) Sexual selection and the evolution of female choice. Evolution 36:1–12

Lande R (1981) Models of speciation by sexual selection on polygenic traits. Proc Natl Acad Sci USA 78:3721–3725

Langemann U, Klump GM (2001) Signal detection in amplitude-modulated maskers. I. Behavioural auditory thresholds in a songbird. Eur J Neurosci 13:1025–1032

Leonard ML, Horn AG (2005) Ambient noise and the design of begging signals. Proc Roy Soc B 272:651–656

Luther DA, Wiley RH (2009) Production and perception of communicatory signals in a noisy environment. Biol Lett 5:183–187

Lynn SK, Cnaani J, Papaj DR (2005) Peak shift discrimination learning as a mechanism of signal evolution. Evolution 59:1300–1305

Macmillan (2002) Signal detection theory. In: Pashler HE (ed.) Stevens' handbook of experimental psychology, 3rd ed, vol 4. Wiley, New York, pp 43–90

Markl H (1985) Manipulation, modulation, information, cognition: some of the riddles of communication. Fortschr Zool 31:163–194

Marler P (1957) Species distinctiveness in the communication signals of birds. Behaviour 2:13–39

Maynard Smith J (1991) Honest signalling: the Philip Sidney game. Anim Behav 42:1034–1035

Maynard Smith J, Harper DGC (2003) Animal signals. Oxford University Press, Oxford

McGregor PK (1993) Signalling in territorial systems: a context for individual identification, ranging and eavesdropping. Phil Trans Roy Soc B 340:237–244

Miller GA, Heise GA, Lichten W (1951) The intelligibility of speech as a function of the context of the test materials. J Exp Psychol 41:329–335

Mitani JC, Stuht J (1998) The evolution of nonhuman primate loud calls: acoustic adaptation for long-distance transmission. Primates 39:171–182

Mitchell BR, Makagon MM, Jaeger MM, Barrett RH (2006) Information content of coyote barks and howls. Bioacoustics 15:289–314

Møller AP (1988) False alarm calls as a means of resource usurpation in the great tit Parus major. Ethol 79:25–30

Møller AP (1990) Deceptive use of alarm calls by male swallows. Behav Ecol 1:1–6

Møller AP, Pomiankowski A (1993) Why have birds got multiple sexual ornaments? Behav Ecol Sociobiol 32:167–176

Morton ES (1975) Ecological sources of selection on avian sounds. Am Nat 109:17–34

Munn CA (1986) Birds that cry 'wolf'. Nature 319:143–145

Nemeth E, Brumm H (2010) Birds and anthropogenic noise: are urban songs adaptive? Am Nat 176:465–475

Okanoya K, Dooling RJ (1991) Perception of distance calls by budgerigars (Melopsittacus undulatus) and zebra finches (*Poephila guttata*): assessing species-specific advantages. J Comp Psych 105:60–72

Ord TJ, Peters RA, Clucas B, Stamps JA (2007) Lizards speed up visual displays in noisy motion habitats. Proc Roy Soc B 274:1057–1062

Ord TJ, Stamps JA (2008) Alert signals enhance animal communication in "noisy" environments. Proc Natl Acad Sci USA 105:18830–18835

Owings DH, Leger DW (1980) Chatter vocalizations of California ground squirrels: predator- and social-role specificity. Zeitschr Tierpsychol 54:163–184

Peters RA, Evans CS (2003) Design of the Jacky dragon visual display: signal and noise characteristics in a complex moving environment. J Comp Physiol A 189:447–459

Peters RA, Hemmi JM, Zeil J (2007) Signaling against the wind: modifying motion-signal structure in response to increased noise. Curr Biol 17:1231–1234

Pfennig KS, Pfennig DW (2009) Character displacement: ecological and reproductive responses to a common evolutionary problem. Quart Rev Biol 84:253–276

Pomiankowski A (1987) The costs of choice in sexual selection. J Theor Biol 128:195–218

Pomiankowski AN (1988) The evolution of female mate preferences for male genetic quality. Oxford Surv Evol Biol 5:136–184

Poston J (1997) Mate choice and competition for mates in the boat-tailed grackle. Anim Behav 54:525–534

Potash LM (1972) A signal detection problem and possible solution in Japanese quail (*Coturnix coturnix japonica*). Anim Behav 20:192–195

Price JJ (2013) Why is birdsong so repetitive? Signal detection and the evolution of avian singing modes. Behav 150:995–1014

Pytte C, Rusch KM, Ficken MS (2003) Regulation of vocal amplitude by the blue-throated hummingbird, *Lampornis clemenciae*. Anim Behav 66:703–710

Richards DG (1981) Alerting and message components in songs of rufous-sided towhees. Behaviour 76:223–249

Römer H, Bailey WJ (1986) Insect hearing in the field. II. Male spacing behaviour and correlated acoustic cues in the bush-cricket *Mygalopsis marki*. J Comp Physiol A 159:627–638

Searcy WA, Nowicki S (2005) The evolution of animal communication: reliability and deception in signalling systems. Princeton University Press, Princeton

Seehausen O, van Alphen JJM, Witte F (1997) Cichlid fish diversity threatened by eutrophication that curbs sexual selection. Science 277:1808–1811

Shannon CE (1948) The mathematical theory of communication, I and II. Bell Syst Tech J 27:379–423, 623–656

Shannon CE, Weaver W (1963) The mathematical theory of communication. University of Illinois Press, Urbana

Slabbekoorn H, Smith TB (2002a) Habitat-dependent song divergence in the little greenbul: an analysis of environmental selection pressures on acoustic signals. Evolution 56:1849–1858

Slabbekoorn H, Smith TB (2002b) Bird song, ecology and speciation. Phil Trans Roy Soc Lond B 357:493–503

van Summers W, Pisoni DB, Bernacki RH, Pedlow RI, Stokes MA (1988) Effects of noise on speech production: acoustic and perceptual analyses. J Acoust Soc Am 84:917–928

Von Neumann J, Morgenstern O (1944) Theory of Games and Economic Behavior. Princeton University Press, Princeton

Uy JAC, Endler JA (2004) Modification of the visual background increases the conspicuousness of golden-collared manakin displays. Behav Ecol 15:1003–1010

Wiley RH (1973) The strut display of sage grouse: a "fixed" action pattern. Behaviour 47:129–152

Wiley RH, Richards DG (1982) Adaptations for acoustic communication in birds: sound transmission and signal detection. In: Kroodsma DH, Miller EH (eds) Acoustic communication in birds, vol 1, Communication and behavior. Academic Press, New York, pp 131–181

Wiley RH (1983) The evolution of communication: information and manipulation. In: Halliday TR, Slater PJB (eds) Animal behaviour, vol 2, Communication. Blackwell Scientific Publications, Oxford, pp 156–189

Wiley RH (1991) Associations of song properties with habitats for territorial oscine birds of eastern North America. Am Nat 138:973–993

Wiley RH (1994) Errors, exaggeration, and deception in animal communication. In: Real L (ed) Behavioral mechanisms in evolutionary ecology. University of Chicago Press, Chicago, pp 157–189

Wiley RH, Poston J (1996) Perspective: indirect mate choice, competition for mates, and coevolution of the sexes. Evolution 50:1371–1381

Wiley RH (2000) Sexual selection and mate choice: trade-offs for males and females. In: Apollonio M, Festa-Bianchet M, Mainardi D (eds) Vertebrate mating systems. World Scientific Publishing Co, Singapore, pp 8–46

Wiley RH (2006) Signal detection and animal communication. Adv Study Behav 36:217–247

Wiley RH (2013) A receiver-signaler equilibrium in the evolution of communication in noise. Behav 150:957–993

Wollerman L (1998) Stabilizing and directional preferences of female *Hyla ebraccata* for calls differing in static properties. Anim Behav 55:1619–1630

Wollerman L (1999) Acoustic interference limits call detection in a neotropical frog *Hyla ebraccata*. Anim Behav 57:529–536

Wollerman L, Wiley RH (2002) Background noise from a natural chorus altos female discrimination of male calls in a neotropical frog. Anim Behav 63:15–22

Zahavi A (1975) Mate choice: a selection for a handicap. J Theor Biol 53:205–214

Zahavi A, Zahavi A (1997) The handicap principle. Oxford University Press, Oxford

Zuk M, Kolluru GR (1998) Exploitation of sexual signals by predators and parasitoids. Quart Rev Biol 73:415–438

Part II
Acoustic Signals

Chapter 3
Masking by Noise in Acoustic Insects: Problems and Solutions

Heiner Römer

Abstract In most environments, acoustic signals of insects are a source of high background noise levels for many birds and mammals, but at the same time, their own communication channel is noisy due to conspecific and heterospecific signalers as well. In this chapter, I first demonstrate how this situation influences communication and the evolution of related traits at the population level. Solutions for communicating under noise differ between insect taxa, because their hearing system evolved independently many times, and the signals vary strongly in the time and frequency domain. After describing some solutions from the senders' point of view the focus of the chapter is on properties of the sensory and central nervous system, and how these properties enable receivers to detect relevant acoustic events from irrelevant noise, and to discriminate between signal variants.

3.1 Introduction

This book is mainly on the impact of background noise on intraspecific communication. However, a chapter about noise and hearing in insects should consider the fact that in some taxa (Lepidoptera, Neuroptera, Dictyoptera, and Coleoptera), ears appear to have evolved primarily for the function to escape attacks of insectivorous bats, through the ability to detect their echolocation calls in flight (e.g., Roeder 1967; Miller and Olesen 1979). By contrast, in two groups of Orthoptera, the katydids and crickets, hearing evolved in the context of intraspecific communication, most likely long before the appearance of bats in the Miocene (Alexander 1962; review in Hoy 1992). The fact that many katydids and crickets adopted a nocturnal lifestyle made them potential prey for insectivorous bats,

H. Römer (✉)
Department of Zoology, Karl-Franzens-University, Universitätsplatz 2,
8010 Graz, Austria
e-mail: heinrich.roemer@uni-graz.at

H. Brumm (ed.), *Animal Communication and Noise*,
Animal Signals and Communication 2, DOI: 10.1007/978-3-642-41494-7_3,
© Springer-Verlag Berlin Heidelberg 2013

when on the wing, which most likely has been the selection pressure for the evolution of bat avoidance as a secondary function of their hearing system (Popov and Shuvalov 1977; Moiseff et al. 1978; Libersat and Hoy 1991; Yager 1999; Faure and Hoy 2000). In any case, hearing in insects includes the two behavioral contexts of identification and localization of mates or rivals (intraspecific communication), and the detection and localization of predators (or, in the case of parasitoids, the detection and localization of hosts). Given the dramatic consequences of not detecting a predator acoustically under masking conditions compared to those missing a mate, I will include some aspects of predator detection under noise as well.

3.2 The Problem

The information needed by an organism for shaping its behavior and for decision making is transmitted via afferent nerves and encoded in trains of action potentials. Sensory systems and the brain have to make adaptive assumptions about what had happened in the physical world, by decoding this information. In all sensory systems investigated, receptor cells or sensory interneurons always reveal short episodes of high-frequency firing of action potentials (bursts) in addition to single, spontaneous APs (Eggermont and Smith 1996; Metzner et al. 1998; Krahe and Gabbiani 2004). These bursts convey information about important stimulus features (Metzner et al. 1998; Marsat and Pollack 2006). In the past, behavioral ecologists had a tendency to study communication systems by looking at signal design and signaling behavior, but ignoring the sensory and brain mechanisms that enable receivers to make sense of signals in a noisy world. They simply assumed that natural or sexual selection would have provided individuals with the sensory and neuronal machinery to perform a given task sufficiently well (i.e., to increase their fitness). However, in recent years claims were made by more and more scientists that a comprehensive understanding of communication systems and sexual selection by female choice greatly benefits from considering the cognitive mechanisms underlying decisions where signal processing is involved (e.g., Guilford and Dawkins 1991; Römer 1992; Bateson and Healy 2005; Ryan et al. 2007; Castellano 2009; Miller and Bee 2012). Such a brain-based point of view is illustrated in Fig. 3.1.

Here, the action potential activity of a first-order sensory interneuron of a katydid was recorded at night in the insects' habitat, a tropical rainforest. A crucial task of the auditory pathway is the recognition and classification of acoustic objects important for survival and reproduction. If the brain has to rely on the bursting activity of the cell, how does the insect form object classes based on bursting activity? For example, how does the brain distinguish the calling activity of a conspecific male (burst marked by asterisk) from irrelevant events caused by heterospecific signalers (noise)? Another important acoustic object class would be represented by bursts induced by a predator (echolocating bat; repetitive bursts

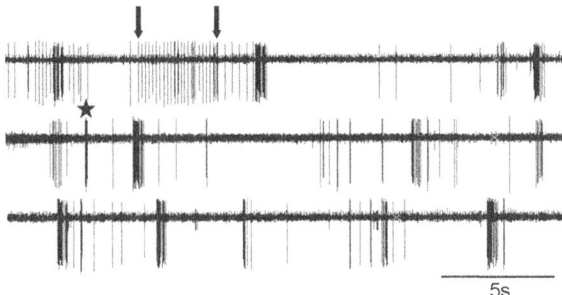

Fig. 3.1 Outdoor recording of the activity of an auditory interneuron (omega neuron) of a katydid at about 2 h after sunset in the tropical rainforest of Panama. Note the different bursting activity of the cell in response to sound events in the background. The task of the CNS of the receiver is to identify and classify acoustic objects based solely on this information. The short burst marked by the *asterisk* was elicited by a conspecific signal, the repetitive bursts between *arrows* by an echolocating bat. For further explanation see text

between arrows). How does the brain form one common "predator" object class from such repetitive bursts when different bat species vary in the rate of calls in their search phase? Moreover, and even more demanding: how to distinguish variations within one object class which carry important information (e.g., quality of a signaler or its distance to the receiver) from variations caused by the noisiness of the transmission channel or sensory processing? In the example given in Fig. 3.1, does variation in duration and spike count of the longer bursts carry information about distance of the same signaler, or differences in the signal structure of different signalers varying in quality? Ronacher et al. (2004) summarized the causes and consequences of spike train variability for processing temporal acoustic patterns in insects, and the interested reader is strongly referred to this comprehensive review. The authors list a number of factors contributing to this variability, including external noise caused by signal degradation on the transmission channel and masking signals from conspecific and heterospecific signalers, as well as intrinsic noise induced at various levels from signal transduction, spike generation, and synaptic transmission in the sensory system.

As a result of the unavoidable noisiness of spike trains in neurons of sensory pathways one should expect that mechanisms evolved which reduce the mistakes, that a nervous system falsely classifies noisy events as signals. On the other hand, minute variations in spike trains may well reflect differences between objects or object classes which are important for the receiver, such as small differences in the size of a sender, or the loudness or frequency composition in the sound signal of a mate. Such small differences, in contrast to those caused by noise, should be preserved during sensory processing, since they represent the neuronal basis for discrimination between mates or other decisions of importance for the fitness of receivers (Ronacher et al. 2004).

Signal detection theory represents a conceptual framework with the objective to assess the efficiency by which a given receptor/detector system can detect a single signal or a group of specified signals against a specified background noise, or to distinguish between signal variants. Signal detection theory can handle both behavioral and neuronal data and is therefore useful for many chapters in this book. Thus, for a more general consideration of SDT the reader is referred to Chap. 2 by H. Wiley, this volume.

3.2.1 Ecological Evidence for Masking: Acoustic Niche Partitioning

Some of the best evidence for the important role acoustic masking can have for the fitness of individuals would be to demonstrate spectral, temporal, and/or spatial niche partitioning as a result of calling activity of other species (see also Chap. 5 by Schwartz and Bee, and Chap. 7 by Brumm and Zollinger this volume). Of course, the role of competition for limited resources in natural selection has been known for a long time, and ecologists recognized that competition for resources may be an important factor affecting the abundance and the distribution of species. The competition for a communication channel has however, only rarely been considered in this context (but see Greenfield 1983; Greenfield and Karandinos 1979 for chemical communication, and Chap. 13), although it should be evident that as the number of species in an ecosystem using the same channel increases, the chances of successful communication will decrease.

Competition for a communication channel is particularly evident for the air-borne-sound channel, since for every additional species that vocalizes at the same time and location, the background noise level increases, the signal-to-noise level decreases and signal detection and/or discrimination is severely impaired. Impressive examples are known for insects (e.g., Gogala and Riede 1995). In particular, tropical rainforests are among the habitats with highest species diversity, and acoustic noise measurements in a Neotropical rainforest of Panama at night have demonstrated sound pressure levels as high as 70 dB (Lang et al. 2005). Spectral analysis revealed that a great proportion is due to the signaling activity of insects (Diwakar and Balakrishnan 2006; Ellinger and Hödl 2003; Lang et al. 2005; see sonogram of a recording in Fig. 3.2). Of these, the calling activity of crickets constitutes the main frequency band between about 2 and 9 kHz where most acoustic energy is concentrated. However, the frequency channel in the high audio and ultrasonic range is also occupied, mainly by the calling activity of katydids (Fig. 3.3), and of course the echolocation activity of bats. The potential for masking at these higher frequencies is often underestimated, because such recordings are usually made from the ground, and many of these signaling katydids broadcast from canopy or mid-canopy regions, and as a result their high frequency or ultrasonic-signals suffer from stronger excess attenuation compared to frequencies used by crickets below 10 kHz (Römer and Lewald 1992). Thus, the situation in a nocturnal tropical rainforest looks terribly complicated for any

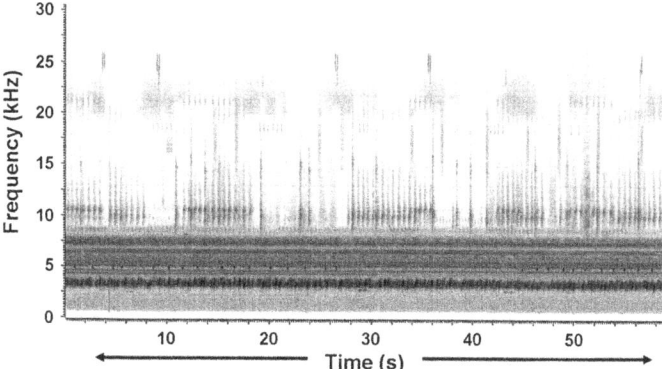

Fig. 3.2 Sonogram of a sound recording of 1 min in the tropical rainforest of Panama (Barro Colorado Island) at about 2 h after sunset. Note the strong audio component between 3 and 9 kHz mainly due to calling of crickets and frogs, and the various song patterns in the high sonic and ultrasonic frequency range (mainly katydid calls)

involved taxon. The fact, however, that so many species still communicate under these conditions means that they can deal with these environmental conditions to an extent that the use of acoustic signals in fitness-related tasks is still a likely evolutionary outcome of natural selection (Brumm and Slabberkoorn 2005).

3.2.2 Different Solutions for Different Taxa

Depending on the kind of signals used by the different taxa of acoustic insects, one would expect that niche partitioning in the acoustic communication channel should be different. Male crickets produce calling songs which usually have a pure-tone character limited to frequencies from 2 to 9 kHz (Bennett-Clark 1998; for few exceptions of ultrasonic signaling in crickets see Robillard et al. 2007). Most grasshoppers and katydids, however, produce broadband signals with a frequency spectrum that may extend far into the ultrasonic range; some include only ultrasonic frequencies, in some cases of tropical species up to more than 100 kHz (Heller 1988; Morris et al. 1994). Again, exceptions have been reported for a group of nine tettigoniid species, the calls of four were narrow band and in the audible range similar to those of gryllids (Diwakar and Balakrishnan 2006). Thus, theoretically, crickets could solve the problem by partitioning their signals in the frequency domain, simply by using a narrow, but different carrier frequency within the available range from 2 to 9 kHz. By contrast, grasshoppers and katydids would suffer from strong frequency overlap in heterospecific interactions, and should rely on partitioning in time and space. We can further predict that in such interactions the species with the higher duty cycle will gain an advantage over the other species, since the detection/recognition of a high duty cycle signal will be favored under these conditions (but see other solutions below).

Fig. 3.3 Fifteen of about 70 species of katydid in the rainforest of Panama which cause most of the high-frequency or ultrasonic noise at night, as seen in the sonogram of Fig. 3.2. More than 50 species of cricket add to the sonic background noise with their calling songs between 2 and 10 kHz. The *upper* nine species are Phaneropterine katydids, the six *below* are Pseudophyllines. Photographs by Alexander Lang

Evidence for the latter hypothesis comes from two ecological studies on katydids (Greenfield 1988; Römer et al. 1989) which demonstrate that when only two species use a spectrally similar signal, this can result in complete suppression of calling activity of one species by the other, or a shift in the diurnal calling activity of one species. In one study, the katydid *Hemisaga denticulata* (a species with a low duty cycle call) was acoustically active over the afternoon, but showed a strong decline of signaling after sunset, when another katydid *Mygalopsis marki* started singing, which increased the noise level from 48 dB SPL to 60 dB SPL (Römer et al. 1989). Experimental manipulation of song interference by removal of all singing *M. marki* males resulted in a recovery of singing activity of *Hemisaga* males after only 15 min. Furthermore, in another habitat without the interfering species and an overall noise level of only 40–45 dB SPL the number of singing *H. denticulata* remained constant even after sunset. A similar removal experiment was performed by Greenfield (1988) with two species of Neocono-cephalus (*N. spiza* and *N. nebrascensis*). The species with the low duty cycle song shifted its singing activity to the day as a result of the masking sound of the other species, but after removal of the competing species became nocturnally active again (see also Sect. 4.5 for a further argument why in these interactions the signal with a low duty cycle is at a disadvantage). Of course, the argument of strong song interference does also hold for narrow-band acoustic signals such as in crickets. For two species of rainforest crickets (*Paroecanthus podagrosus* and *Diatrypa spec.*) with their carrier frequency of calling songs at 3.9 and 4.0 kHz, we found in more than 100 h of sound recordings not a single case where both species were calling at the same time and space (Schmidt and Römer unpublished). This is true even though both species have more selective frequency filters compared to European field crickets (see below). Finally, frequency overlap can result in het-erospecific interference between insect and vertebrate taxa, as has been suggested by Ryan and Brenowitz (1985). Cicada choruses have been shown to inhibit the evoked territorial calling of male frogs, because the call frequencies of both species overlap between 3 and 7 kHz (Paez et al. 1993).

Although the majority of behavioral and neurophysiological studies agree with the hypothesis of a strong advantage of signals with a high duty cycle over low ones, other solutions have been discovered as well, based on the habituation/dishabituation properties of nerve cells. If a highly repetitive signal of one species results in strong habituation of a sensory neuron, but the less redundant signal evokes dishabituation in the same neuron due to some novel property in its signal, then the less redundant signal could still be detected in the noise of the competing species. Schul and Sheridan (2006) provided an example for such a "novelty detector," where the "noise" constitutes the conspecific signal with pulse repetition rates of 140 Hz, under which the echolocation pulses of bats have to be detected. Given that carrier frequencies between these two signals are different (e.g., 15 kHz vs. 40 kHz), the dishabituation described above resulted in almost 100 % response probability to the bat signal. Future behavioral studies need to demonstrate, however, that such physiological properties of identified nerve cells are also found in the behavioral ability to respond to the less redundant signal.

In addition to the frequency domain, the above examples indicate the second possibility of niche partitioning in the time domain. Another example comes from a Bornean mixed dipterocarp forest, where the "dusk community" consists of a well-defined ensemble of cicada, cricket, and frog species, in which the first half-hour is dominated by cicadas and the second half-hour by crickets and frogs. Furthermore, the signaling activity of a given cicada species exhibits a surprisingly narrow temporal segregation in the range of minutes (Gogala and Riede 1995). Furthermore, two studies on cicadas (Sueur 2002) and crickets and katydids (Diwakar and Balakrishnan 2007) included other parameters for acoustic niche partitioning in their analysis. Sueur (2002) found a set of properties that facilitated niche partitioning among a cicada community, which also included (apart from call frequency) calling height and timing, as well as behavioral categories such as the tendency to aggregate, and the calling strategy ('call-fly' vs. 'call-stay'). In the rainforest study by Diwakar and Balakrishnan (2007) calling heights of both gryllid and katydid species ranged from the ground to the canopy, with more gryllid than katydid species occupying the ground and herb layer. Their study revealed vertical stratification of calling heights, with three main layers corresponding to the canopy, understorey, and the ground layer. Importantly, these clusters emerged from the raw data of calling heights of individuals of each species without a priori distinction of layers.

Although this chapter is on acoustic insects where communication happens in the acoustic far field, it should be evident that acoustic masking may also happen in the acoustic near field (i.e., the range close to a sound source where the energy component due to particle displacement is greater than the sound pressure component). Samarra et al. (2009) reported the masking of courtship song in *Drosophila montana* by background noise at frequencies overlapping with those in the song, based on female behavioral responses. This happened at a signal-to-noise-ratio of −6 dB, and it is highly unlikely that natural habitat noise levels can account for the observed masking in the acoustic near field. The authors therefore speculate that when a female is courted by several males it might create the relevant biotic noise. This remains to be tested in future experiments, in which the near-field acoustic environment of females is determined with appropriate microphones.

3.2.3 Background Noise and Signal Synchrony or Alternation

Some of the most impressive interactions occur in insect and anuran choruses when the signaling of individuals is influenced by the precise timing of signaling of other individuals (see also Chap. 5 by Schwartz and Bee). If individuals are able to signal in silent gaps of masking noise (Zelick and Narins 1985) this may be a solution to the problem of masking interference. Extreme forms of fine-scale signal timing occur when neighboring individuals either synchronize or alternate their signals with those of neighbors, i.e., when the phase angles approximate either 0° or 180°, respectively (Walker 1969; Sismondo 1990; Greenfield 1994; Greenfield and Roizen 1993; Hartbauer et al. 2005). Although the models explaining synchrony and alternation at the proximate level differ to some extent, it is obvious that the phase response curve of the underlying song oscillator should be sensitive to background noise, because of two possible reasons: On the one hand, the signal of the neighbor could simply be masked by the background and thus cannot influence the oscillator any more. On the other hand, noisy events in the background could reset the oscillator if they occur within a certain phase of the phase response curve.

Figure 3.4 shows one result of the breakdown of synchrony due to background noise (Hartbauer et al. 2012). In the undisturbed situation, a male katydid (*Mecopoda elongata*) was entrained to conspecific chirps and established a very regular synchronous interaction, with his signal being delayed relative to the broadcast one (follower role; upper trace). Under masking noise conditions, the fixed temporal relationship broke down at a SNR of −1 dB (lower trace). It remains to be examined whether the breakdown is a simple masking effect and/or due to noise resetting the oscillator.

3.3 Solutions by the Sender

In the following section, I discuss some solutions to the problem of communicating under noise from the senders' point of view, which are by no means restricted to insects. We would expect that rather different taxa dealing with the same problem might have evolved similar, though not identical, solutions. Thus,

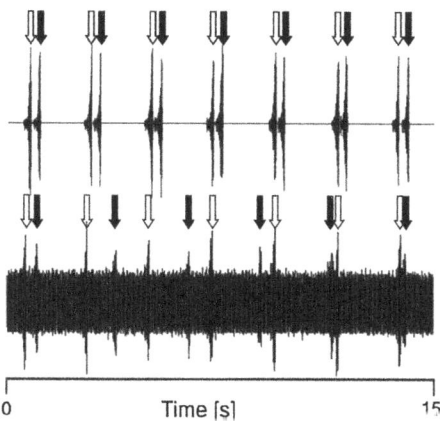

Fig. 3.4 Breakdown of call synchrony under background noise. In the upper panel, a male *Mecopoda elongata* (*filled arrow*) synchronized his chirp with a regularly repeated conspecific stimulus (*open arrow*) and establishes a constant follower relationship with the stimulus. Under background noise at a SNR of −1 dB synchrony breaks down and the males' chirp is produced at different phases of the stimulus period. For further explanation see text (modified from Hartbauer et al. 2012)

Chap. 5 by Schwarz and Bee on frogs and Chap. 7 by Brumm and Zollinger on birds deal with the same issues discussed in this section.

3.3.1 No Lombard Effect, But Strong Selection for Increased Loudness

The Lombard effect describes one of the most obvious mechanisms to overcome masking noise problems, namely an increase in the amplitude of the signal under noise. The effect appears to be very common in birds and mammals, and Chaps. 7 and 9 describe in detail findings related to the Lombard effect, and for a failure of a demonstration of the effect in anurans see Chap. 5. Similarly, there is no such report for insects. This is somewhat surprising given the fact that the decrease in SPL of a male calling song at higher distances is rather flat, so that a small increase in loudness (or efficiency in transmission) of only a few decibels may result in a better SNR, and thus a relatively large increase in the active range of the signal. In fact, the greatest advantage one would expect for those species where the signal suffers little or no excess attenuation, so that the decrease in loudness over distance follows the 6 dB per doubling of distance rule. In these cases, the decrease is exponential and thus rather flat at greater distances. The bladder grasshopper *Bullacris membracioides* (Pneumoridae; Orthoptera) is a striking case for this phenomenon, and achieves hearing distances between 1.5 and 2 km (van Staaden and Römer 1997), due to the use of a resonator for sound production and favorable atmospheric conditions for sound transmission after sunset. Theoretically, the active range of the male signal being just at the masked threshold for a receiver at, e.g., 500 m would increase with a small increase in loudness of 5 dB to 800 m, with a corresponding impressive effect for the broadcast area (from 0.78 to 2.01 km^2).

Differential attraction of females to louder calling songs is known for insects (see Forrest and Green 1991 for a field study); comparable results have been obtained in many laboratory-based choice experiments, where 2–3 dB have been sufficient for females choosing the more intense song. Fruitflies, mole crickets, crickets, and katydids preferentially approach the louder of two conspecific signals of different intensity, and selection has favored the use of resonators, amplifying burrows, and baffles to achieve an increased sound output (review in Römer 1998). However, although many acoustic insects suffer from masking by conspecific and heterospecific signalers, in order to argue for a Lombard-like effect in this taxon it has to be demonstrated that either (i) populations differing in the amount of masking also differ in their sound output, or (ii) individuals regulate their sound output depending on the SPL of background noise. So far, no empirical evidence has been reported which would support these ideas.

Importantly, the lack of empirical evidence for a Lombard effect in insects does not mean that there is no plasticity in modifying the SPL depending on ecological conditions. This can, curiously enough, best be demonstrated in cases where insects *down-regulate* their sound output. Again males of the bladder grasshopper *B. membracioides* represent one example: they produce their 100 dB SPL call until they receive an acoustic female reply, which is a low intensity call at 60 dB SPL. The female response then induces a duet and male phonotaxis, until she is finally contacted. Interestingly, observations in the field indicate that the SPL of the male call is not always at the maximum close to 100 dB, in particular when the male has established reliable duetting with the female (which means that both are within the active range of the signals of the opposite sex). In these cases males often down-regulate the SPL of their call. The most likely explanation for this behavior is competition by "unintended receivers" from conspecific males, particularly alternate male morphs, which cannot call and fly at all, but intercept the acoustic duet of calling males with the female (Alexander and van Staaden 1989; Donelson and van Staaden 2005).

3.3.2 Use of Multimodal or Alternative Signals Under Masking Noise Conditions

In recent years it has become evident that many animal displays may be rather complex, including more than one signal component in different sensory modalities. Several hypotheses have been proposed why such complexity exists, in particular in mating signals (reviewed by Candolin 2003). Of these, the back-up signal (or redundant signal) hypothesis proposes that multiple signals allow a better assessment of mate quality as each signal reflects the same quality with some error (Møller and Pominakowski 1993; Johnstone 1997). Under masking noise conditions, where errors in the detection of differences among mates will increase, mate choice would become more reliable with multiple back-up cues. In Acridid grasshoppers, for example, the act of sound production (stridulation) involves the strong up-and-down movement of the hindlegs rubbing against a stridulatory file on the hindwings. In some species the hind legs or joints between femur and tibia are brightly colored, so that at close range the acoustic display also includes a striking visual display (Riede 1986). Some Acrid families have given up acoustic signaling altogether (Leptysminae, Rhytidochrotinae, Ommatolampinae, Melanoplinae, Proctolabinae, and Bactrophorinae), but show conspicuous movements of hindlegs ("knee-waving") and antennae (Riede 1987). Katydids with their elytral stridulation broadcast airborne-sound signals, but the same act of stridulation produces vibrations on the substrate where signaling takes place (Keuper and Kühne 1983). The additional information via the vibrational channel improves the localization of singing males by females (Latimer and Schatral 1983).

Neither in grasshoppers nor katydids is there conclusive evidence that the use of such additional modalities is correlated with the amount of background noise in the airborne-sound channel. However, in a neotropical katydid which uses airborne sound for long distance communication, but also an alternative form of private signaling through substrate vibration, the various trade-offs when communicating in these two modalities have been studied in more detail (Römer et al. 2010). As demonstrated earlier, the background noise level for the airborne-sound channel can be quite high in the nocturnal rainforest, whereas it is low in the vibration channel in the low frequency range of the vibration signal (carrier frequency 13 Hz) and in the plant in the understory where the insect lives (but see Cocroft and Rodriguez 2005 for arguments for a noisy vibratory channel). Indeed, in a comparison of signal perception using neurophysiological methods under outdoor conditions, the detection of the signal in the vibratory channel was more reliable than the detection of the short, airborne-sound signal, in particular with respect to the false alarm rate. One should keep in mind, that the benefit of using such an alternative signaling modality is reduced or offset due to a reduced active space of the vibration signal.

3.3.3 Signal Duration and Redundancy to Counteract Masking by Noise

The difference in the duration and/or rate of acoustic signals in different insect species is striking: from single clicks lasting less than a millisecond (see Sect. 4.4) to stereotyped repetition of single song elements for many minutes and hours (Gerhardt and Huber 2002). It appears evident that stereotyped repetitions could support both the detection and recognition by receivers when amplitude fluctuations, reverberations, or masking noise in the transmission channel are superimposed on the signal at some distance from the source (see also Chap. 7 for experimental evidence in birds). Indeed, when the effect of amplitude fluctuations and reverberations on the perception of conspecific song patterns was studied in a katydid outdoors, the temporal song pattern was represented in the central nervous system of a receiver with remarkable accuracy at distances well beyond the nearest neighbor distance (Rheinlaender and Römer 1986; Römer and Lewald 1992).

However, in a series of behavioral studies on the grasshopper *Chorthippus biguttulus* the stereotyped repetition of song elements did not improve the ability to detect and recognize the conspecific signal substantially (Ronacher and Krahe 1998; Ronacher et al. 2000; Ronacher and Hoffmann 2003). In this species, the song of females to which the male performs phonotaxis is composed of a series of identical subunits each characterized by their species-specific amplitude modulation. Although the natural female song lasts for more than one second, males responded behaviorally to a shortened song containing only three subunits (corresponding to 250 ms duration). Ronacher et al. (2000) conducted similar experiments under unmodulated noise, which decreases the depth of the AM-pattern. The expectation in these experiments was that with female songs containing more subunits higher noise levels would be tolerated. Surprisingly, however, even under high noise levels the results indicated an upper limit for temporal integration in the order of 450 ms, since the performance of males did not improve with more than five subunits. When these experiments were conducted under amplitude-modulated noise, modulation frequencies >15 Hz were the most efficient in masking the AM-pattern of the song. Thus, their results indicate that a chorus-like, temporally structured noise does more efficiently mask the signal than unmodulated noise with the same carrier frequency spectrum as the signal.

Altogether, the authors concluded that in the case of *Ch. biguttulus* the insect does not seem to rely very much on the serial redundancy of the signal for recognition under masking noise conditions. They point out however, that such redundancy, or longer duration signals may be most relevant in the context of sexual selection, via both female choice and male–male competition (see also Römer 1998). And as outlined in Sect. 4.5, the gain control mechanism observed in katydid and cricket receivers is most effective with intense and long duration signals, so that signalers with these signal properties may better be able to outcompete other signalers from being represented in the sensory system of receivers.

At the other end of the continuum of signal duration and redundancy in insect signals are those species with extremely short signals, repeated at a very low rate. Acoustically orienting predators may represent one selection pressure for their evolution (Zuk and Kolluru 1998). Except for duetting Phaneropterine katydids with their use of temporal windows we are lacking empirical data on the behavioral performance of receivers concerning detection/recognition of these signals. We might predict that species lacking redundant signals have to maintain smaller interindividual distances to achieve better SNRs. Based on an informal survey of the genus *Neoconocephalus* and other katydids Greenfield (1990) concluded that indeed discontinuously (less redundant) singing species experience high-density populations more frequently than do continuously singing species. Using unsupervised clustering as a tool to analyze the bursting activity of an auditory interneuron recorded under noisy conditions of the rainforest, Pfeiffer et al. (2012) demonstrated that small modifications of a stimulus (e.g., a double syllable compared to a single syllable) strongly enhanced the ability of the algorithm to separate bursts resulting from a stimulus from those resulting from noise.

3.4 Solutions by the Receiver

Similar to Chap. 4, I will now discuss solutions to the problem of communicating under noise from the receivers' point of view, which are again not restricted to insects. The section is particularly connected with Chaps. 6 and 8 on the same topics in frogs and birds, respectively.

3.4.1 Frequency Tuning: Increasing the Selectivity of Filters

In contrast to katydids with their broad range of frequencies in the calling songs another solution does only work for taxa such as crickets, where the sender concentrates acoustic energy within a small frequency range. One of the potential solutions to cope with a complex noisy acoustic environment is an improvement (mostly sharpening) in stimulus filtering by the peripheral or central nervous systems, which is found in other taxa as well (see Chaps. 6, 8, 10, and 12). Thus, any sound outside the sensitivity range of the filter would play a reduced role in masking of the signals, depending on the sharpness of the tuning (the matched filter hypothesis, Capranica and Moffat 1983; Wehner 1989).

Schmidt et al. (2011) studied the frequency tuning of an auditory neuron (AN1-neuron) mediating phonotaxis in a rainforest cricket (*P. podagrosus*; carrier frequency of calling song at 3.7 kHz) which suffers from strong acoustic competition, in comparison with the same, homologous neuron in European field crickets where such competition does not exist. As predicted, the neuron in the rainforest species exhibited a more selective tuning compared to the one in its European counterparts (Fig. 3.5). Remarkably, a comparison of the filters indicates that the increased filter performance of the *Paroecanthus* AN1 (best frequency at 3.9 kHz) is mainly due to the increased steepness of the slope toward higher frequencies. If the filter has been shaped by natural selection to avoid masking interference, this is exactly what we would expect to happen, because in the crickets' habitat there is more masking potential in the noise spectrum at higher compared to lower frequencies (see sonogram in Fig. 3.2). A rather similar situation has been reported for the two sympatric cricket species *Teleogryllus oceanicus* and *T. commodus* with calling song frequencies of 4.8 and 4.0 kHz, respectively, where the AN1 filter of

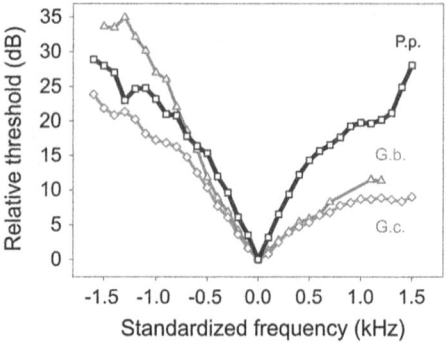

Fig. 3.5 Comparison of the standardized average sensitivity tuning of the AN1-neuron in *P. podagrosus* (P.p.), a rainforest cricket under strong acoustic competition from other crickets, with the tuning of the same homologous neuron in two species of field crickets *G. bimaculatus* (G.b.) and *G. campestris* (G.c.) where acoustic competition in neighboring frequency bands does not exist. Data for the *Gryllus* species are taken from Kostarakos et al. (2009) (modified from Schmidt et al. 2011)

T. commodus exhibits a steeper slope toward higher frequencies compared to other field crickets, which could aid in separating the frequency of its own calling song from that of the sympatric species (Kostarakos et al. 2009).

The performance of the filter of the rainforest cricket has been quantified by Schmidt and colleagues in two ways: first, by examining the representation of the species-specific amplitude modulation of the male calling song in the sound signal, when embedded in background noise. The filter of the rainforest cricket performed significantly better in representing this important signal parameter (Fig. 3.6). Second, the neuronal representation of the song pattern within receivers was maintained for a wide range of signal-to-noise ratios, up to −6 to −9 dB.

Although the above study appears to be conclusive concerning the hypothesis of environmental selection on a frequency filter to avoid masking, it cannot exclude

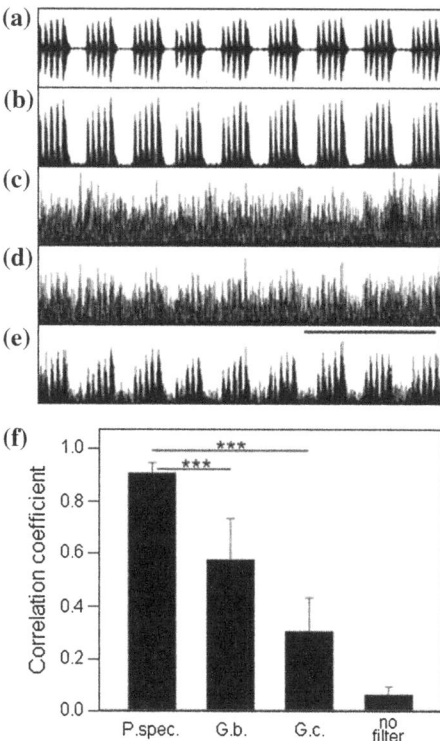

Fig. 3.6 The effect of the different filter functions in detecting the specific amplitude modulation (AM) of the *Paroecanthus* calling song embedded in background noise. **a** Oscillogram and **b** AM of *Paroecanthus* calling song. **c** AM of *Paroecanthus* calling song embedded in background noise without any filtering, **d** with the filter of *G. campestris*, and **e** with the filter of *P. podagrosus*. Note the increase in the quality of representation of the AM of conspecific song by using the more selective filter. Time bar 120 ms. **f** Correlation of the AM of *Paroecanthus* calling song with the AM of the same calling song embedded in background noise ($N = 9$) under the filter regimes of the three cricket species, and without any filter (Schmidt et al. 2011)

the possibility that the differences in filter characteristics between the European and rainforest cricket species are the outcome of phylogenetic constraints rather than adaptations to environmental conditions. Future studies on more species (of different subfamilies) would be helpful, or a comparison of different populations of a single species communicating under different noise levels and/or noise spectra (e.g., Amézquita et al. 2005, 2006). Such studies could also give us additional information about the possible impact such changes in receivers might have for signalers. More selectively, tuned receivers could impose strong selection on signalers to call exactly at the carrier frequency where they are tuned, because otherwise males would be unable to stimulate the females' hearing system adequately. Given this bias in female selectivity, we would expect to find in future studies a reduction in the variance of the male carrier frequency compared to those species where the selectivity in tuning is reduced. Interestingly, a reduced variance in this important song trait for female preference would in turn reduce the potential for female preference for the trait, pointing to the close interrelationship between signals, environmental conditions and the sensory and central nervous system of receivers (the "sensory drive hypothesis"; Endler 1992, 1993).

3.4.2 Frequency Tuning: Changing the Best Frequency of Filters

A fundamental assumption for the solution presented in the above section is a match between the carrier frequency of the signal and the hearing sensitivity of the receiver (Ryan and Keddy-Hector 1992). Although there are a number of exceptions to this general rule among the insects (e.g., cicadas: Huber et al. 1990; katydids: Bailey and Römer 1991; for a sex-specific mismatch see Dobler et al. 1994; haglids: Mason et al. 1999), it can be assumed that such matching has been arrived at by co-evolution between signalers and receivers (Endler 1992). The case of mismatch between the CF of the male call at 5 kHz and the best hearing sensitivity between 10 and 20 kHz in the katydid *Sciarasaga quadrata* (Austrosaginae: Tettigoniidae) is therefore surprising, given the fact that masking calls of up to 16 sympatric katydid species are in this frequency range of hearing. The solution to the problem is a mechanism that allows peripheral control of a sound guide to the ear, thereby shifting the sensitivity of the ear to 5 kHz, i.e., the CF of the call (Römer and Bailey 1998). The advantage of signaling at a lower frequency is in fact threefold: (1) to escape the masking noise conditions of heterospecific katydid species, (2) to achieve a better transmission of the conspecific call, which is close to the 6 dB/doubling of distance (geometric spreading) for the 5 kHz signal in the insect's habitat, and (3) finally, as *S. quadrata* is heavily parasitized by an acoustically orienting parasitoid fly, *Homotrixa alleni,* the call of the katydid may be under strong selection to be outside the best frequency range of the flies hearing system, which is most sensitive to frequencies >10 kHz (Stumpner et al. 2007).

Thus, as this example shows, it may be difficult to disentangle unequivocally the specific impact of one particular selection pressure (background noise or escape from parasitoids) for the evolution of a specific trait.

3.4.3 Noisy Conspecifics and Solutions to Cocktail Party-Like Problems

Although masking noise is most often associated with the sound production of heterospecific signalers, insects may also communicate in dense populations of conspecific individuals. If more than one signaler is within hearing range of a receiver, the temporal overlap of several songs arriving from different directions may result in a severe masking of the individual species-specific temporal song pattern at the position of the receiver. Insect choruses are therefore noisy social environments for acoustic communication, a situation quite common among humans (known as the cocktail party problem) and nonhuman animals (see Chap. 6, this volume). Bee and Micheyl (2008) therefore strongly argued for studies of the cocktail party problem in the context of animal acoustic communication because many of the sensory solutions to the human cocktail party problem may also represent potentially important mechanisms underlying acoustic communication in nonhuman animals.

3.4.3.1 Spatial Release from Masking

One of the mechanisms discussed by Bee and Micheyl is spatial release from masking, which refers to the improved detection of a sound signal when the masker is spatially separated to some degree from the signal (Klump 1996). However, surprisingly little is known for insects on this mechanism. Ronacher and Hoffmann (2003) investigated the influence of amplitude-modulated noise on the recognition of species-specific communication signals in a grasshopper behaviorally, and found little evidence for spatial release from masking. They explained their negative finding with the particular mode of processing signals for pattern recognition in grasshoppers (summation of signals from both auditory sides; von Helversen 1984). However, this is not the case in crickets and katydids (Pollack 1988; von Helversen and von Helversen 1995; Schul et al. 1998; Römer and Krusch 2000), and although spatial release from masking was not addressed directly in these studies, they nevertheless suggest that the mechanism works effectively in these taxa. In particular, katydids, with their known high peripheral directionality and contrast enhancement through lateral inhibition along the longitudinal body axis appear to possess the proximate basis for spatial release from masking (review in Gerhardt and Huber 2002; Hedwig and Pollack 2008).

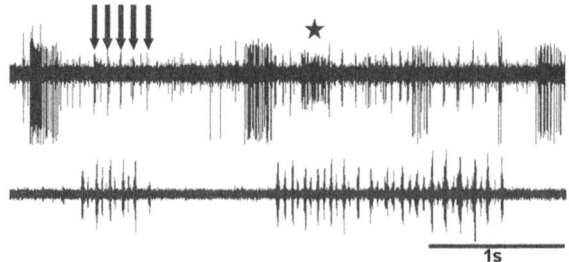

Fig. 3.7 An extreme case of spatial release from masking in a katydid. Simultaneous AP-recording of both omega cells (large spikes contralateral, small spikes ipsilateral cell, *upper line*) and the output of a bat detector, recording HF-sound ipsilaterally (*lower line*). Note that both auditory sides represent completely different "acoustic worlds" in their spike discharge: the ipsilateral cell responds to a bat (*arrows*) and to some other HF-background, not detected by the bat detector (*asterisk*), whereas the contralateral cell does not respond to these events at all. One would expect similar effects in crickets and other acoustic insects, depending on their degree of directional hearing

In one study designed to simulate the complex chorus situation in the katydid *Tettigonia viridissima*, Römer and Krusch (2000) investigated the representation of up to three acoustic signals, presented without a fixed temporal relationship, in the responses of a pair of local interneurons (omega cells), while varying the intensity and direction of these signals. The results suggest that the auditory world of the katydid is rather sharply divided into two azimuthal hemispheres, with signals arriving from any direction within one hemisphere being predominantly represented in the discharge of neurons of this side of the auditory pathway (see also for an extreme case of spatial release from masking in Fig. 3.7). Similar results were reported for crickets (Pollack 1986) where the homologous omega neuron did respond selectively to an ipsilateral stimulus when an equally intense stimulus was presented from the opposite side.

3.4.3.2 Do Results of Conventional Masking Experiments Tell us the Truth About Real-World Situations?

Spatial release from masking is usually tested in behavioral and neurophysiological experiments by presenting the signal and masker (the noise) both from the ipsilateral side, examining the masked threshold, and then by moving the masker spatially away from the location of the signal to test the threshold again. If we do this experiment with the rainforest cricket *P. podogrosus* introduced earlier, using the ambient nocturnal noise as masker (at realistic real-world SPLs between 55 and 60 dB), the signal-to-noise-ratio at the masked threshold is between −6 and −9 dB with masker and signal on the same, ipsilateral side, owing to the excellent filter performance (Schmidt et al. 2011; see Fig. 3.5). Shifting the masker to contralateral improves this value on average by further 8 dB, due to the directionality of the system. However, if we then place the same preparation directly in the rainforest with a background noise level between 56 and 58 dB SPL and repeat the threshold measurement, we never measured masked thresholds as high as those measured in the laboratory. Rather, masked thresholds were close to the *unmasked* threshold in the lab (range 32–35 dB SPL; Schmidt and Römer 2011). Thus, under natural conditions where the masking noise acts on the receiver from all directions, the signal-to-noise-ratio at threshold can amount to −23 dB. Such findings are consistent with the warning by Bee and Micheyl (2008) that "an approach using one or a limited number of masking noise sources in highly controlled laboratory studies of spatial unmasking does not wholly reflect the real-world listening conditions that many animals face." Furthermore, as pointed out by Brumm and Slabberkoorn (2005) in most studies the critical bandwidth of the signal for a perceptually relevant ratio is not known at all, because we only rarely know the filter properties of receivers (either from behavior or from physiological approaches), and in these cases overall SPL measurements of the noise do not tell us very much about the limits of hearing outdoors. Of course, signal-to-noise-ratio measures using "spectrally shaped" noise that has the spectrum of naturalistic acoustic scenes are more useful.

3.4.3.3 No Evidence for Comodulation Masking Release in Insects

Comodulation masking release describes the finding of improved detection of signals in masking noise as a result of coherent patterns of amplitude modulations in the noise across different frequency channels (Klump 1996; Buus 1998). Ronacher and Hoffmann (2003), and Ronacher et al. (2004) discuss in detail their results on signal detection in the grasshopper *Ch. biguttulus* with respect to co-modulation masking release. Since the ear of a grasshopper does not provide the basis for much frequency resolution (Römer 1976; Jacobs et al. 1999), and the spectrum of the noise and the signal was rather similar in their experiments, comodulation masking release was not likely to happen in these grasshoppers.

3.4.4 Listening for a Signal in a Short Time Window

As illustrated in Fig. 3.1, the masking problem for a receiver in most communication systems is mainly due to the fact that he does not know exactly, when the signaler(s) produced a signal, so that the afferent activity has to be evaluated continuously for relevant information. If, however, the receiver would know the timing of the signal, noisy events before and after this time could be completely ignored, which would make the task of signal detection/identification much easier. Such a system exists in most phaneropterine katydids, where both sexes produce sound, and pair formation is achieved by duetting (Zhantiev and Dubrovin 1977; Heller and von Helversen 1986; Robinson et al. 1986; Bailey 2003). Here, the male calling song elicits an acoustic reply in the female and the male then responds by phonotaxis. In order to elicit phonotaxis by the male, the time delay of the female response must occur within a rather narrow time window, which is species-specific and matches the species-specific female delay time (Heller and von Helversen 1986; Robinson et al. 1986). In the katydid *Leptophyes punctatissima*, for example, the actual width of the time window for accepting the reply is only 30 ms. Because the female reply is only 0.3 ms in duration and therefore unable to transmit information about species identity via its amplitude modulation, the temporal window could be used by the male as a feature for recognition. However, the extremely short female reply carries the problem that it will induce only a short, unspecific burst in afferent neurons of the male (Kostarakos et al. 2007), which might be confused with bursts of action potentials caused by noisy events in the transmission channel and create false alarms (and thus misdirected phonotaxis) in the male. Can the small time window for listening for females reduce or eliminate such a detrimental effect of noise?

In a neurophysiological study in the insect's habitat we recorded bursts of action potentials in an afferent interneuron in response to female replies and background noise (Ofner and Römer, unpublished). Based on responses of the neuron to the female signal in the undisturbed situation we could determine the amount of hits and false alarms using a given bursting criterion of the cell. With a call rate of the male of 1/3 s, and assuming that the male will listen to an acoustic reply of the female for the next 3 s, we measured an average of 1.5 false alarms over this time. If the male would only listen to the female reply within the species-specific time window of 30 ms, however, the rate of false alarms would be reduced 100 times to 0.015. Thus, temporal windows in these duetting species may not only solve the problem of species recognition with a signal that otherwise offers little chance of identification, but may at the same time reduce the effect of masking noise considerably.

3.4.5 Noise Reduction Due to Automatic Gain Control in the Afferent Auditory Pathway of Receivers

In the natural chorus situation of acoustic insects outlined above, the mechanism of spatial release from masking would be quite helpful to separate signalers on opposite sides of the receiver. However, the distribution of signaling males in populations of crickets and katydids (Thiele and Bailey 1980; Forrest and Green 1991; Arak and Eiriksson 1992) would suggest that the acoustic situation for receivers in such populations may be far more complex, since more than one signaler can broadcast from one auditory side, and the intensity between signalers at the position of the receiver (distances to receiver), and their differences, can vary strongly.

For crickets (Pollack 1988, 2000) and katydids (Römer and Krusch 2000) a neuronal mechanism has been described that can cope with these chorus situations. Although each auditory pathway is selectively listening to, and encodes the temporal pattern of predominantly ipsilateral sounds (see above), each pathway also selects for the most intense of several alternative sounds. The underlying synaptic mechanism, first described for crickets by Pollack (1988), is based on a dual mode of synaptic activity. In addition to a fast excitatory depolarization, a signal also causes an inhibition which can be seen as a hyperpolarization with a slow build-up and decay time. The latter component is most likely a calcium-activated potassium current (Sobel and Tank 1994; Baden and Hedwig 2007). The inhibition prevents suprathreshold depolarization of the membrane in response to softer signals, thus representing a gain control effectively filtering out the less intense of several competing signals. The information transmitted to the brain is thus not confounded in its amplitude modulation (Fig. 3.8).

Fig. 3.8 Schematic illustration, of how the gain control mechanism in an auditory neuron can create a selective response to only one signal in a chorus. A female receiver (R) is confronted with calling songs of three males (S1–3) from different distances. The SPL of their signal at the receiver differs due to distance, and a microphone at the receiver's position would record the combined, masked signal S1–3. Due to the long lasting inhibition, the EPSPs elicited by the more distant songs remain subthreshold, and the spike response of the neuron represents the temporal pattern of only one signaler (modified from Pollack 2000)

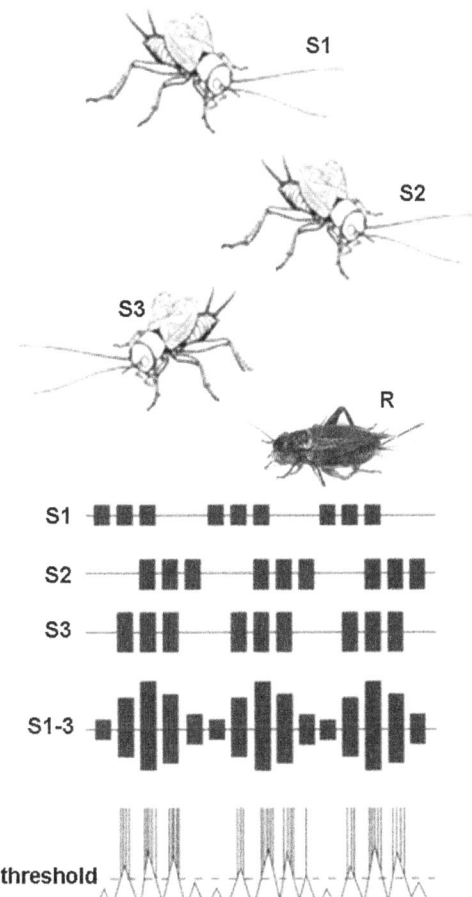

Both, the strong directionality and the proximate mechanism of gain control in crickets and katydids constitute properties of what Guilford and Dawkins (1991) called the receiver's psychology, and in sensory drive models of sexual selection (Endler and Basolo 1998) such characteristics (most likely their bias type 3 and 4) will bias the direction of evolution by affecting which new courtship signals will be most successful. For example, the properties of the gain control mechanism described above should result in selection acting on males to produce more intense signals because the active range of these signals is greater, and they inhibit the representation of competitive signals if the difference in amplitude is 2–5 dB. Indeed, this view is supported by virtually all studies on acoustic insects (reviews by Ryan and Keddy-Hector 1992; Forrest 1994; Römer 1998).

The second property of the gain control mechanism, i.e., the unusually long time-constant of the membrane hyperpolarization, may have a similarly strong impact for the evolution of acoustic signals and for intra- and heterospecific interactions in these two taxa. Short duration signals, or singing bouts of low duty cycle, would have little or no effect in eliciting the suppression of competitive signals in the receiver. In a population of males competing for phonotactically responding females, long duration signals or singing bouts would therefore be favored over short ones, since only the former would be able to reliably initiate the inhibitory effect. The choice of females for signals with longer duration, such as in female crickets (Hedrick 1986), may thus be explained at the proximate level by passive attraction only (Parker 1983) where females merely move to the male producing a signal that activates its sensory pathway most strongly.

The gain control mechanism could also play an important role for the structuring of mixed species choruses. If the mechanism is common to all crickets and katydids and represents an evolutionary conservative feature in their auditory system, this would result in a disadvantage for species with lower duty cycles/short duration signals: in the case of sympatry with higher duty cycle species the representation of their own song in conspecific receivers would be suppressed, due to the properties of the underlying membrane hyperpolarization. In fact, the behavioral observation in two sympatric pairs of katydid species discussed in Sect. 2.1 (Greenfield 1988; Römer et al. 1989), where noise produced by one species inhibited the calling activity of the other species would represent exactly such cases, because the species with the low duty cycle song was the one being suppressed in both studies. The examination of these and other sensory biases represents an extremely interesting field for future research for the study of acoustic communication, since we can expect that basic properties of the sensory and/or central nervous system will be modified by selection if the disadvantage is too high. The possibility of addressing this issue in insects using comparative studies of single, identified neurons presents many advantages over similar studies in vertebrates.

3.5 Acoustic Predator Detection and Decision Making for Evasive Responses in Noise

It is intuitively clear that the fitness loss of a receiver falsely interpreting the quality in the courtship signals of two males are quite moderate compared to a receiver missing the information about a nearby predator. Natural selection should thus favor appropriate solutions for predator detection and discrimination from noise. In this review I therefore include some information about what is currently known about the effect of noise in an acoustic predator-related detection system, using the now classical bat echolocation—insect prey as an example. Almost any nocturnally active insect on the wings will be under the risk of predation by insectivorous bats. The difficult task of the afferent auditory system of potential prey is therefore to provide reliable information about the presence and vicinity of the predator, so that higher brain centers are enabled to perform a decision regarding the type of behavioral response, as well as the best timing and direction (Altes and Anderson 1980). The task is difficult because the prey has to distinguish the echolocation calls of a bat from high frequency acoustic noise within the environment, from potential acoustic noise generated by its own movements (Waters and Jones 1994; Poulet and Hedwig 2002) and intrinsic noise within its own auditory system (Fullard 1987; Waters 1996). The solution to the task is also quite different in the three major taxa where this problem has been investigated, namely noctuid moth, crickets, and katydids.

For example, the evolution of bat-evasive behavior in crickets and katydids is shaped by different constraints. Cricket ears are most sensitive to the carrier of their calling songs but also to frequencies far into the ultrasonic range up to 100 kHz (review in Pollack 1998), the latter indicating the second major function in cricket audition, namely predator detection and avoidance (Fullard 1998; Hoy 1992). Behaviorally, Wyttenbach et al. (1996) demonstrated categorical perception of frequency, i.e., based on sound frequency crickets discriminate "good = mate" (<15 kHz) from "bad = bat" (>15 kHz). The extremely broad tuning of the HF - channel bears some cost, however, such as the susceptibility to any high-frequency noise produced by other insects, which could elicit unnecessary bat avoidance behavior.

Katydids cannot discriminate conspecific signals from bats simply based on spectral information, because their own calling songs are usually broadband signals with a frequency spectrum that extends far into the ultrasonic range (Heller 1988; Morris et al. 1994), also used by most aerial-hawking bats (Fenton et al. 1998). Thus the only reliable information for discrimination between "good" and "bad" should be based on temporal cues (Schul et al. 2000; but see Schul and Sheridan 2006 and Sect. 2.1).

3.5.1 External Noise and Predator Detection by the Repetitive Nature of Bat Search Calls

Katydid receivers face the problem of detecting cues from echolocating bats in the time domain, but afferent spike trains can be very noisy if katydids are active in the nocturnal rainforest (see Fig. 3.1). Hartbauer et al. (2010) recorded AP activity of the omega neuron first in the laboratory when stimulated with sequences of bat calls at different repetition rates typical for the guild of insectivorous bats, in the presence of background noise. The spike activity typical for responses to bat echolocation contrasts with responses to background noise, producing different distributions of inter-spike intervals. These interval distributions allowed the development of a 'neuronal bat detector' algorithm, optimized to detect responses to bats in afferent spike trains. Application of the algorithm to more than 24 h of outdoor recordings of the same cell demonstrated a remarkably reliable detection rate: in 95 % of cases, the algorithm detected a bat reliably, even under high background noise, and correctly rejected responses due to background noise when an electronic bat detector showed no response.

3.5.2 Internal Noise, Coding Inaccuracy, and Predator Detection

When a noctuid moth in flight has to make a decision about the presence of an echolocating bat, it has to rely on information from only two sensory cells in each ear, the A1 and A2 cell, the latter being about 20 dB less sensitive (Roeder 1964). Waters (1996) investigated two kinds of noise in the A1 cell for its ability to encode information about a nearby bat: the first was spontaneous discharge, which may produce an incorrect decision that a bat is present. Spontaneous APs are the main problem in distinguishing bat from non-bat at low intensity levels; they occur at a median rate of 7.4 Hz. Depending on the recognition criterion (number of APs within a given time period) the author could calculate that false alarms (i.e., the moth mistakenly identifies a bat as being present on the basis of spontaneous APs) would occur every few seconds. Ecological data based on bat—moth encounters by Roeder and Treat (1962) demonstrate that the selective advantage of reacting over nonreacting moths in encounters with bats is 44 % per encounter, indicating that the costs of a missed detection (the moth mistakenly deciding a bat is not present) are high. Ironically, the high directionality of the moth hearing system (interaural intensity differences of about 20–30 dB; Payne et al. 1966; Madsen and Miller 1987) does prevent an easy solution to the problem: simultaneous APs in A1 cells of both ears would represent a reliable information of a real source, but due to the high directionality of the system this does not happen.

Can the afferent information be processed in a way to increase information transfer to decision-making centers in the brain? Boyan and Fullard (1988) described an interneurone (IN 501) which they suggested represents a "noise filter", because it shows a 1:1 spiking relationship with the A1 afferent only at very high A1 discharge rates. However, Waters (1996) calculated that with such decision criteria the moth would strongly reduce its sensitivity and maximum detection range. The second kind of intrinsic noise is inaccurate encoding of stimulus intensity due to response variability, in particular to the short duration calls of bats. Altogether, these results led Waters suggest that, for bats using short duration calls, the moth would only be able to recognize an approaching bat from the repetitive nature of the incoming signal. This is consistent with results on katydid receivers and detection of bat-like sound described above.

3.6 Conclusions

Compared with vertebrates, insects explore a wide range of signal carrier frequencies, from below 1 kHz to more than 100 kHz, with strong differences in the amount of excess attenuation, and thus the range of communication. Because many insect species communicate in choruses of both conspecific and heterospecific signalers, the problem of masking interference is severe. Despite the broad range of possible frequency channels insects are often forced into certain frequency ranges due to phylogenetic or biophysical constraints, and clearly suffer from cocktail party-like hearing problems similar to vertebrates. They solve some of the problems by exploiting microhabitats with favorable communication conditions, via the evolution of novel traits in signaling and/or hearing, or via individual plasticity in signaling behavior.

In my review, I emphasized the particular advantage that insects offer compared to vertebrates for an organismic approach to acoustic communication systems, in which behavioral and environmental approaches are combined with neurophysiological approaches on the receiver side under laboratory and field conditions. In this way the role of the often ignored "receiver psychology" for insect signaling and hearing can be examined and placed in the context of the possible selection pressures that may have shaped the character of the signals and the sensory systems necessary for signal detection, discrimination and final decision making. Results achieved with this approach are promising for future research on the sensory ecology of hearing, which is still in its infancy.

Acknowledgments I acknowledge the comments of Mark Bee, Henrik Brumm and an anonymous reviewer, which greatly improved the chapter. My own research on the sensory ecology of acoustic communication in insects was strongly influenced by collaboration with Jürgen Rheinlaender and Win Bailey, who focused my interest on field work and evolutionary aspects of acoustic signaling and hearing. I also thank M. van Staaden, G.K. Morris, I. Dadour, and D. Gwynne for numerous discussions during our field work. Own research for this article was supported by grants from the Austrian Science Foundation FWF, projects P09523-BIO and P20882-B09.

References

Amézquita A, Castellanos L, Hödl W (2005) Auditory masking of male *Epipedobates femoralis* (Anura: Dendrobatidae) under field conditions. Anim Behav 70:1377–1386

Amézquita A, Hödl W, Lima AP, Castellanos L, Erdtmann L, De Araújo MC (2006) Masking interference and the evolution of the acoustic communication system in the Amazonian dendrobatid frog *Allobates femoralis*. Evolution 60:1874–1887

Alexander RD (1962) Evolutionary change in cricket acoustical communication. Evolution 16:443–467

Alexander AJ, van Staaden MJ (1989) Alternative sexual tactics in male bladder grasshoppers (Orthoptera, Pneumoridae). In: Bruton MN (ed) Alternative life-history styles of animals. Kluwer Academic Publishers, Dordrecht, pp 261–277

Altes RA, Anderson GM (1980) Binaural estimation of cross-range velocity and optimum escape manoeuvres by moths. In: Busnel RG, JF Fish (eds) Animal sonar systems. Plenum, New York, pp 851–852

Arak A, Eiriksson T (1992) Choice of singing sites by male bushcrickets (*Tettigonia viridissima*) in relation to signal propagation. Behav Ecol Sociobiol 30:365–372

Baden T, Hedwig B (2007) Neurite specific Ca2+-dynamics underlying sound processing in an auditory interneurone. J Neurobiol 67:68–80

Bailey WJ (2003) Insect duets: underlying mechanisms and their evolution. Physiol Entomol 28:157–174

Bailey WJ, Römer H (1991) Sexual differences in auditory sensitivity: mismatch of hearing threshold and call frequency in a tettigoniid (Orthoptera, Tettigoniidae: Zaprochilinae). J Comp Physiol A 169:349–353

Bateson M, Healy SD (2005) Comparative evaluation and its implication for mate choice. Trends Ecol Evol 20:659–664

Bee MA, Micheyl C (2008) The cocktail party problem: what is it? How can it be solved? And why should animal behaviourists study it? J Comp Psychol 122:235–251

Bennet-Clark HC (1998) Size and scale effects as constraints in insect sound communication. Phil Trans Roy Soc B 353:407–419

Boyan GS, Fullard JH (1988) Information processing at a central synapse suggests a noise filter in the auditory pathway of the noctuid moth. J Comp Physiol A 164:251–258

Brumm H, Slabberkoorn H (2005) Acoustic communication in noise. Adv Study Behav 35:151–209

Buus S (1998) Auditory masking. In: Crocker MJ (ed) Handbook of acoustics. Wiley, New York, pp 1147–1165

Candolin U (2003) The use of multiple cues in mate choice. Biol Rev 78:575–595

Capranica RR, Moffat AJM (1983) Neurobehavioral correlates of sound communication in anurans. In: Ewert JP, Capranica RR, Ingle D (eds) Advances in vertebrate neuroethology. Plenum, New York, pp 701–730

Castellano S (2009) Towards an information-processing theory of mate choice. Anim Behav 78:1493–1497

Cocroft RB, Rodriguez RL (2005) The behavioral ecology of insect vibrational communication. Bioscience 55(4):323–334

Diwakar S, Balakrishnan R (2006) The assemblage of acoustically communicating crickets of a tropical evergreen forest in southern India: call diversity and diel calling patterns. Int J Anim Sound Record 16:1–23

Diwakar S, Balakrishnan R (2007) Vertical stratification in an acoustically communicating ensiferan assemblage of a tropical evergreen forest in southern India. J Tropical Ecol 23:479–486

Dobler S, Stumpner A, Heller K-G (1994) Sex-specific spectral tuning for the partner's song in the duetting bushcricket *Ancistrura nigrovittata* (Orhtoptera: Phaneropterinae). J Comp Physiol A 175:303–310

Donelson NC, van Staaden MJ (2005) Alternate tactics in male bladder grasshoppers *Bullacris membracioides* (Orgtoptera: Pneumoridae). Behaviour 142:761–778

Eggermont JJ, Smith GM (1996) Burst-firing sharpens frequency tuning in primary auditory cortex. NeuroReport 7:753–757

Ellinger N, Hödl W (2003) Habitat acoustics of a neotropical lowland forest. Bioacoustics 13:297–321

Endler JA (1992) Signals, signal conditions, and the direction of evolution. Am Nat 139:125–153

Endler JA (1993) Some general comments on the evolution and design of animal communication systems. Philos Trans R Soc B 340:215–225

Endler JA, Basolo AL (1998) Sensory ecology, receiver biases and sexual selection. TREE 13:415–420

Faure PA, Hoy RR (2000) Neuroethology of the katydid T-cell. I. Tuning and responses to pure tones. J Exp Biol 203:3225–3242

Fenton MB, Portfors CV, Rautenbach IL, Waterman JM (1998) Compromises: sound frequencies used in echolocation by aerial-feeding bats. Can J Zool 76:1174–1182

Forrest TG (1994) From sender to receiver: propagation and environmental effects on acoustic signals. Am Zool 34:644–654

Forrest TG, Green DM (1991) Sexual selection and female choice in mole crickets (Scapteriscus: Gryllotalpidae): modelling the effects of intensity and male spacing. Bioacoustics 3:93–109

Fullard JH (1987) Sensory ecology and neuroethology of moths and bats: interactions in a global perspective. In: Fenton MB, Racey PA, Rayner JMV (eds) Recent advances in the study of bats. Cambridge University Press, Cambridge, pp 244–272

Fullard JH (1998) The sensory co evolution of moths and bats. In: Hoy RR, Popper AN, Fay RR (eds) Comparative hearing: insects. Springer, Heidelberg, pp 279–326

Gerhardt HC, Huber F (2002) Acoustic communication in insects and anurans: common problems and diverse solutions. The University of Chicago Press, Chicago

Gogala M, Riede K (1995) Time sharing of song activity by cicadas in Temengor Forest Reserve, Hulu Perak, and in Sabah, Malaysia. Malay Nat J 48:297–305

Greenfield MD (1983) Reproductive isolation in clearwing moths (Lepidoptera: Sesiidae): a tropical-temperate comparison. Ecology 64(2):362–375

Greenfield MD (1988) Interspecific acoustic interactions among katydids *Neoconocephalus*: inhibition-induced shifts in diel periodicity. Anim Behav 36:684–695

Greenfield MD (1990) Evolution of acoustic communication in the genus *Neoconocephalus*: discontinuous songs, synchrony, and interspecific interactions. In: Bailey WJ, Rentz DCF (eds) The Tettigoniidae: biology, systematics and evolution. Crawford House Press, Bathurst, pp 71–98

Greenfield MD (1994) Synchronous and alternating choruses in insects and anurans: common mechanisms and diverse functions. Annu Rev Ecol Syst 25:97–126

Greenfield MD, Karandinos MG (1979) Resource partitioning of the sex communication channel in clearwing moths (Lepidoptera: Sesiidae) of Wisconsin. Ecol Monogr 49(4):403–426

Greenfield MD, Roizen I (1993) Katydid synchronous chorusing is an evolutionary stable outcome of female choice. Nature 364:618–620

Guilford T, Dawkins MS (1991) Receiver psychology and the evolution of animals signals. Anim Behav 42:1–14

Hartbauer M, Kratzer S, Steiner K, Römer H (2005) Mechanisms for synchrony and alternation in song interactions of the bushcricket *Mecopoda elongata* (Tettigoniidae: Orthoptera). J Comp Physiol 191:175–188

Hartbauer M, Radspieler G, Römer H (2010) Reliable detection of predator cues in afferent spike trains of a katydid under high background noise levels. J Exp Biol 213:3036–3046

Hartbauer M, Siegert ME, Fertschai I, Römer H (2012) Acoustic signal perception in a noisy habitat: lessons from synchronising insects. J Comp Physiol A. doi:10.1007/s00359-012-0718-1

Hedrick AV (1986) Female preference for male calling bout duration in a field cricket. Behav Ecol Sociobiol 19:73–77

Hedwig B, Pollack GS (2008) Invertebrate auditory pathways. In: Basbaum AI, Akimichi K, Shepard GM, Westheiner G (eds) Invertebrate auditory pathways. The senses: a comprehensive reference. Dallos P, Oertel D. Academic Press, San Diego, pp 525–564

Heller KG (1988) Zur Bioakustik der Europäischen Laubheuschrecken. Josef Margraf, Weikersheim

Heller KG, von Helversen D (1986) Acoustic communication in phaneropterid bushcrickets: species-specific delay of female stridulatory response and matching male sensory time window. Behav Ecol Sociobiol 18:189–198

von Helversen D (1984) Parallel processing in auditory pattern recognition and directional analysis by the grasshopper *Chorthippus biguttulus* L. (Acrididae). J Comp Physiol A 154:837–846

von Helversen D, von Helversen O (1995) Acoustic pattern recognition and orientation in orthopteran insects: parallel or serial processing. J Comp Physiol 177:767–774

Hoy RR (1992) The evolution of hearing in insects as an adaptation to predation from bats. In: Webster DB, Fay RR, Popper AN (eds) The evolutionary biology of hearing. Springer, New York, pp 115–129

Huber F, Kleindienst H-U, Moore TE, Schildberger K, Weber T (1990) Acoustic communication in periodical cicadas: neuronal responses to songs of sympatric species. In: Gribakin FG, Wiese K, Popov AV (eds) Sensory systems and communication in arthropods. Birkhäuser, Basel, pp 217–228

Jacobs K, Otte B, Lakes-Harlan R (1999) Tympanal receptor cells of *Schistocerca gregaria*: correlation of soma positions and dendrite attachment sites, central projections and physiologies. J Exp Zool 283:270–285

Johnstone RA (1997) The evolution of animal signals. In: Krebs JR, Davies NB (eds) Behavioural ecology. An evolutionary approach. Blackwell Science, Oxford, pp 155–178

Klump GM (1996) Bird communication in a noisy world. In: Kroodsma DE, Miller EH (eds) Ecology and evolution of acoustic communication in birds. University Press, Ithaca, pp 321–338

Kostarakos K, Rheinlaender J, Römer H (2007) Spatial orientation in the bushcricket *Leptophyes punctatissima* (Phaneropterinae; Orthoptera). III. Peripheral directionality and central nervous spatial cues. J Comp Physiol A 193:1115–1123

Kostarakos K, Hennig M, Römer H (2009) Two matched filters and the evolution of mating signals in four species of cricket. Frontiers Zool 6:22

Krahe R, Gabbiani F (2004) Burst firing in sensory systems. Nat Rev Neurosci 24:10731–10740

Kcuper Λ, Kühne R (1983) The acoustic behaviour of the bushcricket *Tettigonia cantans*. II. Transmission of air-borne sound and vibration signals in the biotope. Behav Proc 5:55–74

Lang A, Teppner I, Hartbauer M, Römer H (2005) Predation and noise in communication networks of neotropical katydids. In: McGregor P (ed) Animal communication networks. Cambridge University Press, Cambridge, pp 152–169

Latimer W, Schatral A (1983) The acoustic behaviour of the bushcricket *Tettigonia cantans*. I. Behavioural responses to sound and vibration. Behav Proc 8:113–124

Libersat F, Hoy RR (1991) Ultrasonic startle behavior in bushcrickets (Orthoptera; Tettigoniidae). J Comp Physiol A 169:507–514

Mason AC, Morris GK, Hoy RR (1999) Peripheral frequency mismatch in the primitive ensiferan *Cyphoderris monstrosa* (Orthopterea: Haglidae). J Comp Physiol 184:543–551

Madsen BM, Miller LA (1987) Auditory input to motor neurones of the dorsal longitudinal flight muscles of the noctuid moth (*Barathra brassicae* L.). J Comp Physiol A 160:23–31

Marsat G, Pollack G (2006) A behavioral role for feature detection by sensory bursts. J Neurosci 26:10542–10547

Metzner W, Koch C, Wessel R, Gabbiani F (1998) Feature extraction by burst-like spike patterns in multiple sensory maps. J Neurosci 18:2283–2300

Miller LA, Olesen J (1979) Avoidance behaviour in green lacewings. I. Behaviour of free flying green lacewings to hunting bats and ultrasound. J Comp Physiol A 131:113–120

Miller CT, Bee MA (2012) Receiver psychology turns 20: is it time for a broader approach? Anim Behav 83:331–343

Moiseff A, Pollack GS, Hoy RR (1978) Steering response of flying crickets to sound and ultrasound: mate attraction and predator avoidance. Proc Natl Acad Sci USA 75:4052–4056

Morris GK, Mason AC, Wall P (1994) High ultrasonic and tremulation signals in neotropical katydids (Orthoptera: Tettigoniidae). J Zool Lond 233:129–163

Møller AP, Pominakowski A (1993) Why have birds got multiple sexual ornaments? Behav Ecol Sociobiol 32:167–176

Paez VP, Bock BC, Rand AS (1993) Inhibition of evoked calling of *Dendrobates pumilio* due to acoustic interference from cicada calling. Biotropica 25(2):242–245

Parker GA (1983) Mate quality and mating decisions. In: Bateson P (ed) Mate choice. Cambridge University Press, Cambridge, pp 141–164

Payne RS, Roeder KD, Wallman J (1966) Directional sensitivity of the ears of noctuid moths. J Exp Biol 44:17–31

Pfeiffer M, Hartbauer M, Lang AB, Maass W, Römer H (2012) Probing real sensory worlds with unsupervised clustering. PLoS ONE 7:e37354

Pollack GS (1986) Discrimination of calling song models by the cricket, *Teleogryllus oceanicus*: the influence of sound direction on neural coding of the stimulus temporal pattern and on phonotactic behaviour. J Comp Physiol 158:549–561

Pollack GS (1988) Selective attention in an insect auditory neuron. J Neurosci 8:2635–2639

Pollack GS (1998) Neural processing of acoustic signals. In: Hoy RR, Popper AN, Fay RR (eds) Comparative hearing: insects. Springer, New York, pp 139–196

Pollack GS (2000) Who, what, where? Recognition and localization of acoustic signals by insects. Curr Opin Neurobiol 10:763–767

Popov AV, Shuvalov VF (1977) Phonotactic behaviour of crickets. J Comp Physiol A 119:111–126

Poulet JFA, Hedwig B (2002) A corollary discharge maintains auditory sensitivity during sound production. Nature 418:872–876

Rheinlaender J, Römer H (1986) Insect hearing in the field. I. The use of identified nerve cells as "biological microphones". J Comp Physiol 158:647–651

Riede K (1986) Modification of the courtship song by visual stimuli in the grasshopper *Gomphocerus rufus* (Acrididae). Physiol Entomol 11:61–74

Riede K (1987) A comparative study of mating behaviour in some neotropical grasshoppers (Acridoidea). Ethology 76:265–296

Robillard T, Grandcolas P, Desutter-Grandcolas L (2007) A shift toward harmonics for high-frequency calling shown with phylogenetic study of frequency spectra in Eneopterinae crickets (Orthoptera, Grylloidea, Eneopteridae). Can J Zool 85:1264–1275

Robinson D, Rheinlaender J, Hartley JC (1986) Temporal parameters of male-female sound communication in *Leptophyes punctatissirna*. Physiol Entomol 11:317–323

Roeder KD (1964) Aspects of the noctuid tympanic nerve response having significance in the avoidance of bats. J Insect Physiol 10:529–546

Roeder KD (1967) Nerve cells and insect behavior. Harvard University Press, Cambridge

Roeder KD, Treat AE (1962) The acoustic detection of bats by moths. In: Proceedings of the 11th international entomological conference, Wien, vol 3, pp 7–11

Römer H (1976) Die Informationsverarbeitung tympanaler Rezeptorelemente von *Locusta migratoria* (Acrididae, Orthoptera). J Comp Physiol 109:101–122

Römer H (1992) Ecological constraints for the evolution of hearing and sound communication in insects. In: Webster DB, Fay RR, Popper AN (eds) The evolutionary biology of hearing. Springer, New York, pp 79–93

Römer H (1998) The sensory ecology of acoustic communication in insects. In: Hoy RR, Popper AN, Fay RR (eds) Comparative hearing: insects. Springer handbook of auditory research. Springer, New York, pp 63–96

Römer H, Lewald J (1992) High-frequency sound transmission in natural habitats: implications for the evolution of insect acoustic communication. Behav Ecol Sociobiol 29:437–444

Römer H, Bailey W (1998) Strategies for hearing in noise: peripheral control over auditory sensitivity in the bushcricket *Sciarasaga quadrata* (Austrosaginae: Tettigoniidae). J Exp Biol 201:1023–1033

Römer H, Bailey WJ, Dadour I (1989) Insect hearing in the field. III. Masking by noise. J Comp Physiol 164:609–620

Römer H, Krusch M (2000) A gain-control mechanism for processing of chorus sounds in the afferent auditory pathway of the bushcricket *Tettigonia viridissima* (Orthoptera; Tettigoniidae). J Comp Physiol A 186:181–191

Römer H, Lang A, Hartbauer M (2010) The signaller's dilemma: a cost–benefit analysis of public and private communication. PLoS ONE e13325

Ronacher B, Krahe R (1998) Song recognition in the grasshopper *Chorthippus biguttulus* is not impaired by shortening song signals: implications for neural encoding. J Comp Physiol A 183:729–735

Ronacher B, Hoffmann C (2003) Influence of amplitude-modulated noise on the recognition of communication signals in the grasshopper *Chorthippus biguttulus*. J Comp Physiol A 189:419–425

Ronacher B, Krahe R, Hennig RM (2000) Effects of signal duration on the recognition of masked communication signals by the grasshopper *Chorthippus biguttulus*. J Comp Physiol A 186:1065–1072

Ronacher B, Franz A, Wohlgemuth S, Hennig R (2004) Variability of spike trains and the processing of temporal patterns of acoustic signals—problems, constraints, and solutions. J Comp Physiol A 190:257–277

Ryan MJ, Brenowitz EA (1985) The role of body size, phylogeny and ambient noise in the evolution of bird song. Am Nat 126:87–100

Ryan MJ, Keddy-Hector A (1992) Directional pattern of female mate choice and the role of sensory biases. Am Nat 139:S4–S35

Ryan MJ, Akre KI, Kirkpatrik M (2007) Mate choice (primer). Curr Biol 17:313–316

Samarra FIP, Klappert K, Brumm H, Miller PJO (2009) Background noise constrains communication: acoustic masking of courtship signals in the fruit fly *Drosophila montana*. Bahaviour 146:1635–1648

Schmidt AKD, Riede K, Römer H (2011) High background noise shapes selective auditory filters in a tropical cricket. J Exp Biol 214:1754–1762

Schmidt AKD, Römer H (2011) Solutions to the cocktail party problem in insects: selective filters, spatial release from masking and gain control in tropical crickets. PLoS ONE 6(12):c28593. doi:10.1371/journal.pone.0028593

Schul J, von Helversen D, Weber T (1998) Selective phonotaxis in *Tettigonia cantans* and *T. viridissima* in song recognition and discrimination. J Comp Physiol A 182:687–694

Schul J, Matt F, von Helversen O (2000) Listening for bats: the hearing range of the bushcricket *Phaneroptera falcata* for bat echolocation calls measured in the field. Proc Roy Soc Lond 267:1711–1715

Schul J, Sheridan RA (2006) Auditory stream segregation in an insect. Neuroscience 138:1–4

Sismondo E (1990) Synchronous, alternating, and phase-locked stridulation by a tropical katydid. Science 249:55–58

Sobel EC, Tank DW (1994) In vivo Ca2+ dynamics in a cricket auditory neuron: an example of chemical computation. Science 263:823–826

Sueur J (2002) Cicada acoustic communication: potential sound partitioning in a multispecies community from Mexico (Hemiptera: Cicadomorpha: Cicadidae). Biol J Linn Soc 75:379–394

van Staaden MJ, Römer H (1997) Sexual signalling in bladder grashoppers: tactical design for maximizing calling range. J Exp Biol 200:2597–2608

Stumpner A, Allen GR, Lakes-Harlan R (2007) Hearing and frequency dependence of auditory interneurons in the parasitoid fly *Homotrixa alleni* (Tachinidae: Ormiini). J Comp Physiol A 193:1113–1125

Thiele DR, Bailey WJ (1980) The function of sound in male spacing behavior of buschcrickets (Tettigoniidae: Orthoptera). Aust J Ecol 5:275–286

Walker TJ (1969) Acoustic synchrony: two mechanisms in the snowy tree cricket. Science 166:891–894

Waters DA (1996) The peripheral auditory characteristics of noctuid moths: information encoding and endogenous noise. J Exp Biol 199:857–868

Waters DA, Jones G (1994) Wingbeat-generated ultrasound in noctuid moths increases the discharge rate of the bat-detecting A1 cell. Proc R Soc Lond B 258:41–46

Wehner R (1989) "Matched filters"—neural models of the external world. J Comp Physiol A 161:511–531

Wyttenbach RA, May ML, Hoy RR (1996) Categorical perception of sound frequency by crickets. Science 273:1542–1544

Yager DD (1999) Structure, development, and evolution of insect auditory systems. Microsc Res Tech 47:380–400

Zelick R, Narins PM (1985) Characterization of the advertisement call oscillator in the frog *Eleutherodacytylus coqui*. J Comp Physiol 156:223–229

Zhantiev RD, Dubrovin NN (1977) Sound communication in the genus *Isophya* (Orthoptera, Tettigoniidae) (in Russian). Zool Zurnal 56:40–51

Zuk M, Kolluru GR (1998) Exploitation of sexual signals by predators and parasitoids. Quart Rev Biol 73:415–438

Chapter 4
Effects of Noise on Sound Detection and Acoustic Communication in Fishes

Friedrich Ladich

Abstract The ambient noise in aquatic habitats is characterized by a large variety of noise levels and spectral profiles due to various abiotic and biotic factors such as running water, wind, tides, and vocalizing animals. Fish hearing sensitivity declines when exposed to high noise levels or in the presence of masking noise, in particular, in taxa possessing hearing enhancements. Most vocal fishes communicate over short distances (<0.5 m), probably because of low sound levels produced, low sound frequencies and the ambient noise conditions. Some species exploit 'quiet windows' of low spectral noise levels for acoustic communication. Human-made noise such as ship noise masks the hearing abilities of fishes and hinders acoustic communication. Whether fishes are able to cope with anthropogenic noise by increasing sound amplitude, shifting dominant frequencies of sounds, or by other mechanisms remains unknown.

4.1 Introduction

Fishes rely on their auditory sense for collecting acoustic information of biotic or abiotic origin (acoustic orientation). In order to fulfill this task, fishes possess inner ears consisting of three semicircular canals and three otolithic end organs, the utricle, saccule, and lagena. In contrast to tetrapods, they lack external or middle ears and, to our knowledge, sensory structures solely devoted to hearing (Ladich 2010). Most fishes apparently utilize the saccule for sound perception. Interestingly, while all fishes are sensitive to particle motion at low frequencies, at least one-third of all species have developed accessory morphological structures, termed 'hearing specializations' to detect sound pressure and extend their hearing abilities

F. Ladich (✉)
Department of Behavioural Biology, University of Vienna, Althanstrasse 14,
1090 Vienna, Austria
e-mail: friedrich.ladich@univie.ac.at

H. Brumm (ed.), *Animal Communication and Noise*,
Animal Signals and Communication 2, DOI: 10.1007/978-3-642-41494-7_4,
© Springer-Verlag Berlin Heidelberg 2013

to lower sound levels and higher frequencies (Ladich and Popper 2004; Braun and Grande 2008). Popper and Fay (2011) propose to assign species onto a continuum of pressure detection mechanisms. At one end of the scale are fish with no air-filled structures such as shark and sculpins that only detect particle motion, on the other end are fish with hearing specializations such as swim bladders with an extensive use of sound pressure such as carps and catfishes. Fishes lacking hearing specializations have previously been termed 'hearing non-specialists or generalists' those possessing such structures were called 'hearing specialists'. The importance of hearing improvements is illustrated by the fact that one of the most successful bony fish groups, namely otophysines (carps, catfishes, tetras, and knifefishes, 8,000 species), are characterized by accessory hearing structures (Weberian ossicles) connecting the inner ear (greek: otos) to the swim bladder (physa).

The acoustic sense is a prerequisite for sound communication. Fishes evolved a unique diversity of sound-generating mechanisms among vertebrates. The main group of sound-producing mechanisms (sonic organs) is based on swim bladders. These can be vibrated by intrinsic drumming muscles located in the wall of the swim bladder (toadfishes, searobins), or by extrinsic drumming muscles originating on structures such as the skull or vertebral processes. Pectoral sound-producing mechanisms include vibration of the pectoral girdle (sculpins), rubbing of the enhanced pectoral spine in a groove of the shoulder girdle (catfishes), and plucking of enhanced fin tendons (croaking gouramis, genus *Trichopsis*). In addition, sounds can be produced by other mechanisms such as teeth grating (clownfish) but in many cases the exact process is still unknown (for reviews see Ladich and Fine 2006; Ladich and Bass 2011).

Sound production and acoustic communication usually do not take place in sound-proof chambers but in habitats with a certain amount of natural ambient and possibly anthropogenic noise. Thus, any acoustic process such as sound detection, sound transmission, and communication will be limited by noise and fishes as well as other animal that use sounds to communicate have to cope with this situation. Noise in the communication system may lead to errors by receivers in various contexts such as foraging, predator avoidance, agonistic, and reproductive behavior and these errors can have fundamental consequences for the optimal behavior of both receivers and signalers (see Chap. 2). The effects of underwater noise on fishes have been studied from quite different points of view. Most studies focussed on the effects of noise on inner ears and hearing, a few on sound production and transmission in the environment, and almost none on communication. The influence of anthropogenic noise on fishes in general has been reviewed recently (Popper 2003; Popper and Hastings 2009a, b; Slabbekoorn et al. 2010).

Studies on the effects of noise on hearing may be subdivided into three categories depending on the noise types involved: those applying artificial noise such as white (Gaussian) noise, ambient noise, and anthropogenic noise. Artificial sounds such as white noise are used to understand basic auditory capacities, e.g., the temporary hearing loss as a function of noise level and time of noise exposure, and the degree of masking in relationship to the noise level. Natural ambient noise consists of biotic (animal and plant sounds) and abiotic components (e.g., running waters, surf, rain) and is an integral part of each fish's life. On the other hand, anthropogenic sound primarily derives from ships and boats, from construction sites (pile driving), from geological surveys (airs guns), from military operations (LFA sonar), and fishing operations (sonar, bottom trawls).

This chapter first examines the effects of different kinds of noise on sound detection, then focuses on the potential influence of ambient noise on transmission of fish vocalizations, and finally concentrates on studies investigating the influence of ambient and anthropogenic noise on the detectability of communication sounds. Noise-related changes on behavior, on morphology or on fish kept in aquaculture are not treated specifically. Note that all the areas outlined are characterized by a major lack of field experiments. Accordingly, we need to know more about acoustic communication distances of fishes in the field to estimate negative effects of human-made noise on communication (see Chap. 14). Our knowledge on the effects of noise on acoustic orientation and communication in fishes and the responses of fishes is quite limited (Ladich 2008) and often based on assumptions derived from other animal taxa (Slabbekoorn et al. 2010).

4.2 Effects of White Noise on Hearing

4.2.1 Noise Exposure

Several studies have investigated the potential effects of high levels of white noise on sound detection in fishes. Animals were usually exposed to white (Gaussian) noise for several hours (or days) at different noise levels in order to study the decline as well as the recovery of hearing sensitivities. Typically, exposure to high noise levels resulted in a temporal shift in thresholds (TTS) for a particular time period depending on the absolute auditory sensitivity of the species, the exposure time, and the exposure level. Due to a lack of appropriate miniature particle motion sensors for lab purposes hearing thresholds in noise exposure and masking studies have been described in sound pressure units independently of the ability of species to detect sound pressure.

Scholik and Yan (2001, 2002a) exposed fathead minnows *Pimephales promelas* (family Cyprinidae) and bluegill sunfish *Lepomis macrochirus* (family Centrarchidae) for 24 h to white noise at 142 or 148 dB re 1 μPa. They observed a significant decline in hearing thresholds in the best hearing range by about 10–18 dB in the minnow but not in the sunfish. Recovery to baseline thresholds took more than two weeks in the minnow. Amoser and Ladich (2003) exposed goldfish *Carassius auratus* (family Cyprinidae) and the Amazonian catfish *Pimelodus pictus* (family Pimelodidae) to white noise at 158 dB for 12 and 24 h and found a threshold shift of up to 26 dB in the goldfish and up to 32 dB in the catfish. The recovery took much longer in the catfish than in the goldfish (>14 vs. 3 d). The higher TTS and longer recovery time in the catfish was explained by its higher baseline auditory sensitivity. Smith et al. (2003, 2004) exposed goldfish to white noise levels of 160–170 dB between 10 min and 21 d and found that recovery took up to 2 weeks when exposed for 3 weeks. In contrast to goldfish, the cichlid *Oreochromis niloticus* showed little or no hearing loss. The observation that sunfish and cichlid are not affected by noise at considerable levels can be explained by the fact that they lack hearing specialization; accordingly, they have rather low hearing sensitivities and are thus less affected by noise.

Wysocki and Ladich (2005b) investigated the effects of white noise exposure (158 dB) on the temporal resolution ability of the goldfish's auditory system. Fish communication sounds generally consist of series of pulses that differ mainly in pulse periods. Temporal patterns of pulses within sounds are important carriers of information in fish (Myrberg et al. 1978). Immediately after noise exposure, hearing sensitivity to click pulses was reduced on average by 21 dB and recovered within 1 week. Analysis of the response to double clicks showed that the minimum click period resolvable by the auditory system increased from 1.25 to 2.08 ms immediately after noise exposure and recovered within 3 days. Thus, environmental noise potentially impairs the detection of temporal patterns of sounds and subsequently gathering of information important for acoustic orientation and communication.

Other investigators examined the physiological and morphological effects of exposure to pure tones. Popper and Clark (1976) investigated the TTS after 4 h exposure to pure tones of 300, 500, 600, and 1,000 Hz at 149 dB. The TTS lasted for 2–4 h in goldfish. Recovery was complete. Exposure to very high sound pressure levels ($\sim$ 175–200 dB) resulted in morphological damage to sensory hair cells in various regions of the otolithic endorgans in the cod *Gadus morhua,* the cichlid *Astronotus ocellatus,* and the goldfish (Enger 1981; Hastings et al. 1996; Smith et al. 2011). In summary, fish lose their hearing abilities at least partly when exposed to levels of more than 100 dB above hearing thresholds.

4.2.2 Masking

Exposure to high levels of white noise deteriorates hearing sensitivity for hours or even days by shifting hearing thresholds and thus decreasing hearing sensitivities. Even much lower noise levels (below 110 dB), however, can decrease hearing abilities when the noise is present during sound detection, a phenomenon termed masking. Numerous authors investigated the masking effects from various points of views.

Fay (1974) showed that masking by broadband white noise increases the hearing thresholds of goldfish by a certain degree. The masking effect was lowest at 100 Hz with a signal-to-noise ratio (or critical ratio; critical ratio is defined as the sound pressure level of the tonal signal at thresholds minus the spectrum level of the noise) of 13 dB and highest at higher frequencies. Every increase in the spectrum level of the masking noise by 10 dB increases the hearing curve by

10 dB. Without elaborating on the critical ratio in more detail, the above experiments clearly showed that hearing abilities are limited by the level of noise and that any increase in the noise level decreases the sensitivity linearly in the goldfish, a well-studied species possessing hearing specializations.

Wysocki and Ladich (2005a) extended these investigations to fish taxa with differing hearing abilities and that use vocalizations for communication. They compared data from the goldfish to representatives of Amazonian doradid catfishes (family Doradidae) and North American centrarchids. The hearing abilities of the lined Raphael catfish *Platydoras armatulus* were determined between 200 and 4,000 Hz and of the pumpkinseed sunfish *Lepomis gibbosus* (no hearing specialization known) between 100 and 800 Hz in the presence of white noise. Noise levels of 110 dB RMS elevated the thresholds by 15–20 dB in *C. auratus* and by 4–22 dB in *P. costatus*. White noise of 130 dB RMS elevated overall hearing thresholds significantly in the otophysines by 23–44 dB, whereas the sunfish's sensitivity declined only at the higher noise level by 7–11 dB. Wysocki and Ladich (2005a) illustrate that the occurrence and degree of the threshold shift (masking) depend on the hearing sensitivity of fishes (with pressure sensitive fish showing a higher degree of masking), on the frequency, and on the noise levels tested. Ramcharitar and Popper (2004) observed differences in TTS within drums (family Sciaenidae). The black drum *Pogonias chromis* showed significantly greater shifts in auditory thresholds than the Atlantic croaker *Micropogonias undulates*, particularly in the frequency range of 300–600 Hz.

Noise exposure and masking studies that applied white noise indicate that noise affects sound detection and subsequently limits the abilities of fish to analyze the acoustic scene (or soundscape; Fay 2009). This might affect acoustic communication and orientation of fishes, in particular of species having enhanced hearing abilities.

4.3 Effects of Ambient Noise on Hearing

Studies showing the negative effects of white noise on auditory sensitivity raise the question if and to which degree fish may be masked under natural ambient noise conditions. In this chapter the term ambient noise refers to natural nonhuman noise sources. Theoretically, we postulate that the auditory (and sound-producing) system of fish are well adapted to their environment under calm conditions and that signal detection will be occasionally masked by short noise pulses. If this hypothesis is correct, then we furthermore assume that the large diversity in hearing sensitivities—based on a large number of accessory hearing structures—evolved as an adaptation to varying ambient noise levels and spectra (Ladich and Popper 2004). In order to test this hypothesis, ambient noise levels (RMS levels) and spectra need to be measured and analyzed in various habitats, and the auditory sensitivities of fish need to be measured in the presence of the habitat noise.

4.3.1 Natural Ambient Noise

Wenz (1962) and Urick (1983) estimated and described ambient noise spectra in the ocean in dependence of sea states, wind speeds, depth, and oceanic traffic. The general conclusion from these largely theoretical descriptions is that the noise spectrum level increases with sea state, wind speed, precipitation, and decreases at higher frequencies. Nonetheless, a single set of noise curves for all oceanic habitats is much too general for a meaningful assessment of the noise situation in the habitat of a particular fish species.

Recently, the ambient noise of several freshwater and marine habitats have been compared with regards to the particular fish species that live in these environments.

Wysocki et al. (2007) described a broad range of aquatic habitats in Central Europe, including running waters such as creeks and rivers and stagnant waters such as lakes and backwaters. They found considerable differences in noise levels and spectral profiles between the twelve habitats investigated. Stagnant habitats are quiet, with overall noise levels below 100 dB re 1 μPa ($L_{\text{Leq,1min}}$, RMS) under no-wind conditions. Noise levels in fast-flowing waters were typically above 110 dB and peaked at 135 dB in a free-flowing section of the Danube River. Noise levels ($L_{\text{Leq,1min}}$, RMS) differed by more than 50 dB between habitats, making it necessary to consider each habitat separately when looking for masking effects in the field. Note that RMS noise levels merely provide a rough estimate of the overall noise situation in a habitat. It is important to examine spectral levels in order to determine how fish might be affected in their particular hearing range and how well sounds may propagate. Low levels of spectral noise energy in a limited frequency range, sometimes termed 'noise windows', are far more suitable for sound propagation and sound detection than high levels over a wider range of frequencies (Lugli and Fine 2003). Wysocki et al. (2007) showed that most environmental noise in stagnant habitats is concentrated in the lower frequency range below 500 Hz. In fast-flowing waters, high amounts of sound energy were present in the frequency range above 1 kHz, leaving a low energy "noise window" below 1 kHz (Fig. 4.1).

The soundscape of aquatic habitats can be quite diverse even for closely related species. Lugli (2010) investigated the ambient noise at the typical breeding sites of northern Italian and Mediterranean gobies (family Gobiidae) that inhabit stony streams, vegetated springs, brackish lagoons, and sandy as well as rocky shores. Noise spectral levels in the 50–500 Hz band differed by more than 40 dB; they were much lower in the vegetated spring (60–70 dB re 1 Pa^2/Hz) and the stream (70–80 dB) than in the brackish/marine habitats (80–110 dB). The author concluded that lagoon and coastal gobies are exposed to higher levels of low-frequency masking noise than freshwater gobies (Fig. 4.2).

Studies by Lugli and Fine (2003) and Speares et al. (2011) showed that the ambient noise spectrum not only differs considerably between but also within habitats. Lugli and Fine (2003) measured quiet areas and areas adjacent to

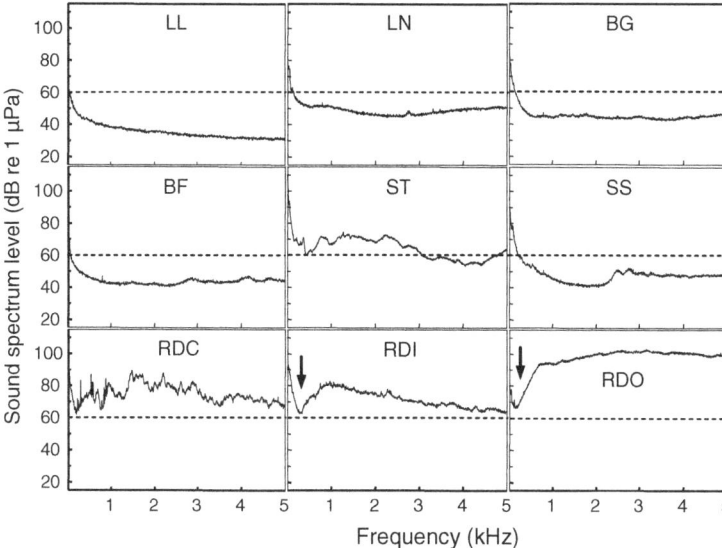

Fig. 4.1 Absolute amplitude spectra of the nine habitats within Central Europe illustrating the large diversity in ambient noise conditions. *BG* Backwater Gänsehaufen Traverse; *BS* Backwater Schönauer Traverse; *LL* Lake Lunz; *LN* Lake Neusiedl; *RDC* Danube Channel; *RDI* Danube River at Danube Island; *RDO* Danube River at Orth; *SS* Schwarza stream; and *ST* Stream Triesting. *Arrows* indicate potential noise windows. Modified with permission from Wysocki et al. (2007). Copyright 2007, Acoustical Society of America

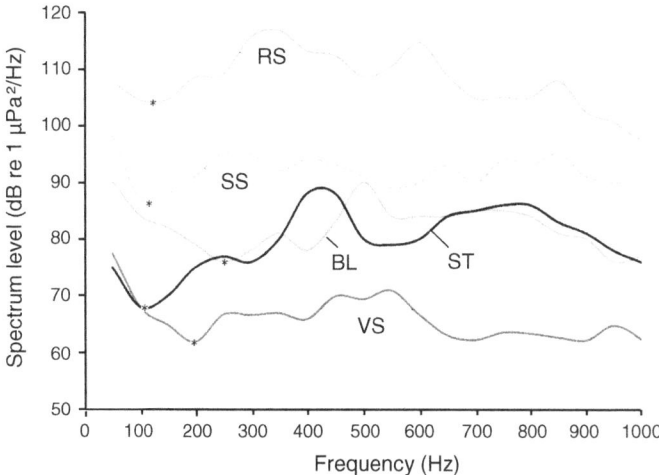

Fig. 4.2 Low-frequency spectra (0.05–1 kHz) of the ambient noise from five habitats inhabited by northern Italian and the Mediterranean goby species. Asterisks indicate the center frequency of the quiet window of the ambient noise. *BL* Brackish lagoon; *RS* Rocky shore; *SS* Sandy shore; *ST* Stony stream; *VS* Vegetated spring. Modified with kind permission from Springer Science + Business Media from Lugli (2010)

waterfalls and rapids in two shallow streams, the Stream Stirone in northern Italy and in the Serchio River in Tuscany in central Italy. Noise spectral levels differed by up to 50 dB in the frequency range below 1 kHz between quiet pools and locations close to rapids and waterfalls. Ambient noise from a waterfall attenuates as much as 30 dB between 1 and 2 m. Speares et al. (2011) investigated two creeks in Alabama, which are inhabited by small darter species (family Percidae), and reported that spectral levels differed between three microhabitats—a run, a riffle, and a pool—depending on the water flow velocity. The noisiest microhabitat in both streams was the fast-moving riffle. Spectral levels of the riffle were approximately 40–60 dB higher than levels of the other microhabitats in these creeks in the frequency range below 1 kHz. The observations that large water movements (running water, coastal surf) result in high noise levels (Wysocki et al. 2007; Lugli 2010) were corroborated by laboratory experiments. Flume experiments carried out by Tonolla et al. (2009) showed that increases in water velocity resulted in increased sound levels over a wide range of frequencies.

Changes in levels and spectral composition were not only found between and within habitats, but were also observed throughout the year in Central Europe. Amoser and Ladich (2010) determined that changes in sound pressure level (SPL) were smallest in the river (maximum: 10 dB), whereas higher changes were measured in stagnant habitats and streams (maximum: 31 dB). The spectral compositions of the ambient noise determined at different times of the year were similar at the river sites (mean cross-correlation coefficients: 0.85 and 0.94) and were weaker or not correlated at the other study sites (means: 0.24–0.76). Cross-correlation coefficients are measures of the similarity between the shapes of the amplitude spectra within each habitat. The mean cross-correlation coefficients of the ambient noise spectra were negatively correlated to changes in SPL, indicating that small changes in spectral composition (high coefficients) were accompanied by small changes in SPLs (RMS) and vice versa. These local and seasonal changes make the soundscape for fish rather complex, especially when fish migrate between habitats.

Besides large differences in noise conditions in freshwater habitats and microhabitats, pronounced differences were also described in marine habitats due to weather conditions and in coastal habitats due to tides. Chapman and Hawkins (1973) mentioned that the level of sea noise in Upper Loch Torridon on the west coast of Scotland, a typical habitat of cods (family Gadidae), was directly related to weather conditions. Any increase in wind speed, and hence surface motion, was accompanied by a proportional increase in the noise level. Heavy rain also considerably increased noise levels at higher frequencies.

Measurements in tidal zones were conducted by Coers et al. (2008). She and her colleagues investigated the ambient noise in the tidal zone of Fayal Island, the Azores, the preferred habitat of the rock-pool blenny *Hypsoblennius gilberti* (family Bleniidae). The researchers observed that the ambient noise revealed major spatial and temporal variation in levels throughout the tidal cycles. Overall levels (RMS) of ambient noise could increase up to 40 dB during high tide and up to 16 dB in spectral levels in the frequency range from 50 to 300 Hz.

Several studies described the ambient noise profiles of coral reefs in different contexts, such as to investigate guidance mechanisms for larvae, juveniles, and adult fish (Tolimieri et al. 2004; Kennedy et al. 2010; Radford et al. 2011). Reef noise is a combination of the sounds produced by reef-associated animals and various abiotic sources. Depending on the reef investigated, different high-energy peaks are found caused by vocal fish, crustaceans, and other marine invertebrates. Kennedy et al. (2010), for example, recorded the ambient noise at 40 reef sites of the Las Perlas archipelago in the Gulf of Panama and compared these sites to offshore sites. Acoustic recordings were taken at each site while the sea was calm. Each reef had a different spectral profile but a similar spectral peak at around 3 kHz, which was attributable to snapping shrimps. In contrast, offshore recordings were rather quiet and of lower levels, possessed a more flat spectrum and dropped off above 3 kHz (Fig. 4.3). Tolimeiri et al. (2004), in contrast, recorded the sea noise at the Feather reef in Northern Australia and found two energy peaks, one attributed to a fish chorus with energies below 1 kHz and a second to shrimps above 10 kHz.

In summary, several recent studies have described the acoustic environment of fishes in freshwater as well as marine habitats in much more detail than previously. These studies reveal that the noise situation is quite diverse, depending on a large

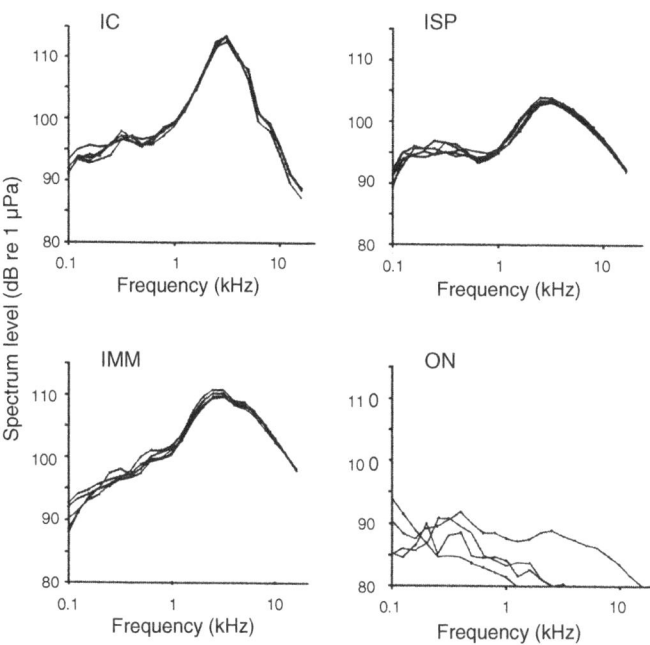

Fig. 4.3 Third octave spectra measured at several coral reef sites and at offshore sites measured in the Las Perlas Archipelago in the Gulf of Panama. Each plot shows five 'typical' reef acoustic profiles during the 2 min recording period. *IC* Isla Contadora; *IMM* Isla Mongo Mongo; *ISP* Isla San Pablo; *ON* Offshore Noise. Modified from Kennedy et al. (2010). Copyright 2010, with permission from Elsevier

degree on the movement of waters caused by natural water flow (e.g., rivers), by winds, by tides, etc., but also on the acoustic activity of various vocalizing animals. The conclusion was that standing waters are much quieter than moving or densely populated waters. Much more work is required to get a comprehensive picture of the degree soundscapes facilitate or limit sound detection and acoustic communication in fish when living in habitats from a depth of a few centimeters down to the deepest oceans.

4.3.2 Masking by Ambient Noise

Studies examining the effects of ambient noise on hearing in fishes are sparse. They were either carried out in the field or by recording the ambient noise in the field and playing it back in the lab. Chapman and Hawkins (1973) and Chapman (1973) measured hearing in the cod *Gadus morhua* and other representatives of the family Gadidae in the field. Fish were tested in a Scottish Loch 15 m below the sea surface and 5 m above the sea bed. Only in calm sea conditions where unmasked thresholds obtained. The authors clearly showed that any change in the sea noise level was accompanied by corresponding shifts in the hearing threshold in gadids. This hearing threshold to spectral level of the sea noise ratio at a particular frequency was constant and independent of the sea noise level. The ratio increased from 18 dB at 50 Hz to 24 dB at 380 Hz in the cod. This masking effect of the sea noise was confirmed when the noise level was raised artificially by transmitting random noise through underwater speakers. These findings were corroborated by laboratory experiments using white noise at different levels (Wysocki and Ladich 2005a).

Based on this knowledge, Amoser and Ladich (2005) attempted to determine the degree to which fish are masked under ambient noise conditions in various European freshwater habitats and what this masking effect looks like in species possessing different hearing abilities. They recorded ambient noise in four different habitats (Danube River, Triesting stream, Lake Neusiedl, backwaters of the Danube River), and played it back to native fish species while simultaneously measuring their auditory thresholds using the auditory evoked potential (AEP) recording technique. The results showed that the carp *Cyprinus carpio*, a pressure sensitive species, is only moderately masked by the quiet habitat noise level of standing waters (mean threshold shift 9 dB) but is heavily affected by stream and river noise by up to 49 dB in its best hearing range (0.5–1.0 kHz) (Fig. 4.4). In contrast, the hearing thresholds of the European perch *Perca fluviatilis*, a species lacking hearing specializations, were only slightly affected (mean up to 12 dB at 0.1 kHz) by the highest noise levels presented. Their results indicated that hearing abilities of otophysines are well adapted to the lowest noise levels encountered in freshwater habitats and that their hearing is considerably masked in some parts of their distribution range. A parallel study on the topmouth minnow *Pseudorasbora parva*, a common Eurasian cyprinid, supports these conclusions (Scholz and

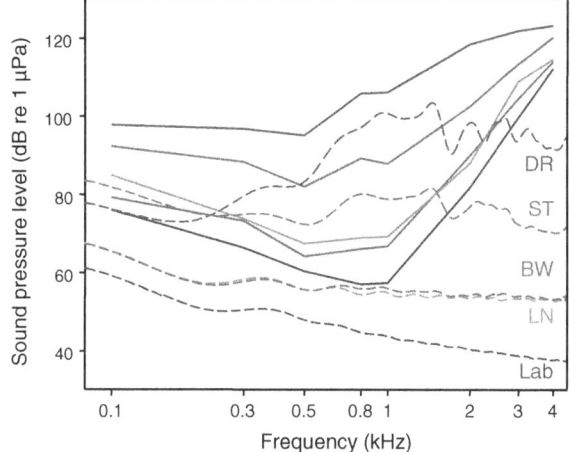

Fig. 4.4 Mean hearing thresholds of the carp *Cyprinus carpio* (*solid lines*) under laboratory conditions and in the presence of the different freshwater noise types (see also Fig. 4.1). *Broken lines* show the cepstrum-smoothed sound power spectra of the corresponding noise types. *BW* Backwater; *DR* Danube River; *Lab* Laboratory noise; *LN* Lake Neusiedl; *ST* Stream Triesting. After Amoser and Ladich (2005)

Ladich 2006). Their hearing sensitivity is slightly masked under ambient noise conditions recorded in their habitat. Their best hearing sensitivities were between 300 and 800 Hz at 57 dB re 1 μPa under quiet laboratory conditions and at 72 dB in the presence of lake noise.

Hearing in species lacking accessory hearing structures is minimally impaired by the typical noise in natural habitats. Belanger et al. (2010) examined the hearing sensitivity of the round goby *Neogobius melanostomus* (family Gobiidae) at ambient noise conditions encountered in the Detroit River. This species has been (most likely accidentally) introduced from the Black and Caspian Sea region of Eurasia to the Great Lakes region and thus is not native to the Great Lakes. At natural noise levels (135 dB RMS), the authors did not observe any shift in auditory thresholds. Slight shifts of up to 10 dB were found at much higher noise levels, which according the authors might occur under severe weather conditions.

To what degree are the hearing abilities of marine fish adapted to ambient noise? More recent studies on nonrelated taxa revealed that fish are well adapted to the ambient noise found during calm sea conditions. In addition to the study on cods by Chapman (1973), investigations on toadfish, on sciaenids or drums, damselfish, and gobies (family Gobiidae) revealed that the hearing sensitivities were only slightly masked. Vasconcelos et al. (2007) showed that ambient noise from the Tagus River estuary in Portugal affected the auditory sensitivity only at low frequencies (50–100 Hz) in the Lusitanian toadfish *Halobatrachus didactylus* compared to quiet lab conditions. Codarin et al. (2009) observed that the hearing sensitivity in the red-mouthed goby *Gobius cruentatus*, the Brown meagre *Sciaena umbra* (family Sciaenidae) and the Mediterranean damselfish *Chromis chromis* (family Pomacentridae) changed by less than 3 dB when exposed to the ambient noise recorded in their habitat, the Miramare Natural Marine Reserve in the Adriatic Sea (Fig. 4.5).

Lugli (2010) described a large diversity in noise profiles in goby habitats (Fig. 4.2). How does this large diversity affect hearing in representatives of this perciform family? The conclusion, based on the lack of threshold shifts in species lacking hearing specializations such as the European perch, the red-mouthed and

Fig. 4.5 Mean (±S.E.) hearing thresholds of (**a**) the sciaenid *Sciaena umbra,* (**b**) the Mediterranean pomacentrid *Chromis chromis* and (**c**) the gobiid *Gobius cruentatus* under laboratory conditions ("*baseline*"; *squares*) and during playback of the ambient (*circles*) and boat noise (*cabin-cruiser*; *triangles*) compared to sound spectra of ambient (*dotted line, blue*) and cabin-cruiser (*continuous line, red*) noises. ** = $p < 0.001$ (repeated measure ANOVA). Modified from Codarin et al. (2009). Copyright 2009, with permission from Elsevier

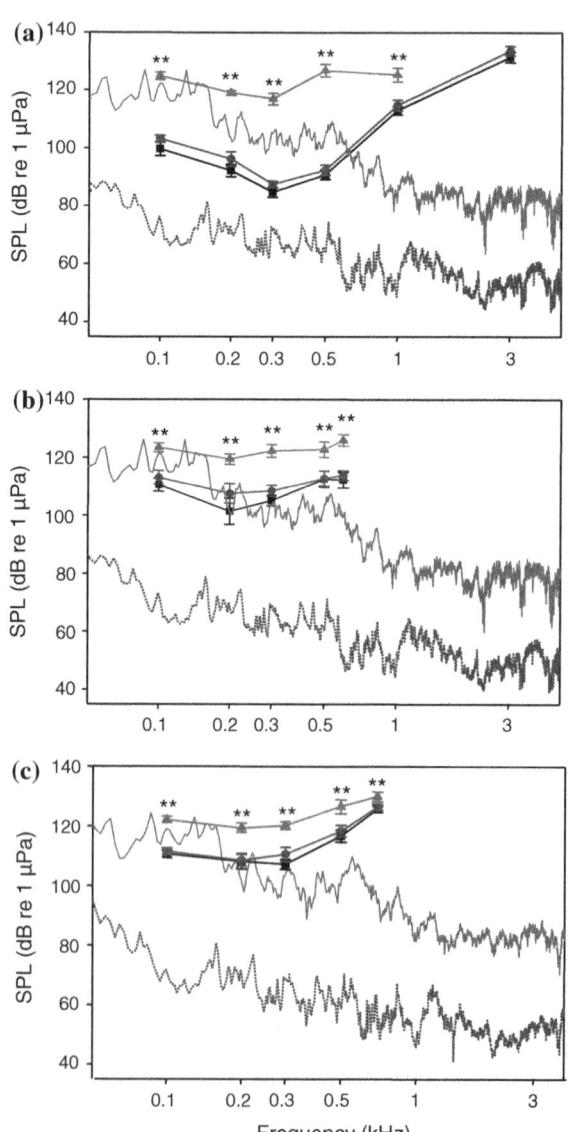

round goby, and the Mediterranean damselfish (Amoser and Ladich 2005; Codarin et al. 2009; Belanger et al. 2010), is that the hearing sensitivities of northern Italian gobies are minimally or not at all affected by the different noise levels in their habitats under calm conditions. This changes as we will see in the next chapter under anthropogenic noise and perhaps under severe weather conditions.

4.3.3 Anthropogenic Noise and Masking

The following section concentrates on how human-made noise changes the natural soundscape and how this affects sound detection (for the implications of anthropogenic noise for the conservation of fish and other animals see Chap. 14). The potential deterring or tissue-damaging effects of anthropogenic noise have been reviewed elsewhere (e.g., Popper and Hastings 2009a, b).

Boats and ships produce an increasing amount of noise, which could change the acoustic scene when a vessel passes, during certain seasons or even throughout a year. In the last decades, noise levels in many habitats have generally increased due to intense ship traffic close to coasts. Andrew et al. (2002, 2011) described an increase in noise levels at the North American west coast of approximately 10 dB at low frequencies. Shipping is the number one factor for this increase. Seasonal changes are pronounced in temperate zones due to human boating and recreational activities during the summertime. Such a seasonal change has been convincingly described by Samuel et al. (2005) in the Peconic Bay Estuary system in Long Island, New York. Between Independence Day and Labor Day the coastal habitats of New York waters are flooded with anthropogenic noise in the frequency range up to 1200 Hz, i.e., within the main hearing and communication range of fishes. During the period of highest human activity, average spectrum levels were about 26 dB higher than during the lowest period of human activity. Although the authors did not concentrate on fishes in particular, their study illustrates that human seasonal activity increases noise levels in coastal waters and that this is probably a worldwide phenomenon.

The main sound energy of surface vessels is almost always located at low frequencies and thus quite often within the hearing and communication range of fishes. The characteristic features of noise of ships and boats depend on propeller, engines, and load and may vary to a certain degree. Amoser et al. (2004) showed that the noise energies emitted by a Class 1 powerboat during a race at Lake Traunsee in Upper Austria peaked at 415 Hz, which is well within the most sensitive hearing range of cyprinids such as the carp *C. carpio* or the roach *Rutilus rutilus* inhabiting this lake. Cyprinids may be masked by this noise up to a distance of several hundred meters but fish lacking hearing specializations such as the coregonid *Coregonus lavaretus* (family Coregonidae) will be affected only at close distance.

The effects of anthropogenic noise from ships and boats on hearing sensitivity of fishes are similar to intense white noise described above in Sect. 4.2.2. Exposing fathead minnow *Pimephales promelas* (family Cyprinidae) to boat noise for 2 h elevated the hearing threshold in the minnow's most sensitive hearing range (Scholik and Yan 2002b). Masking effects have been demonstrated in representatives of several marine fish families. Vasconcelos et al. (2007) and Codarin et al. (2009) found that ferry and boat noise decrease the hearing sensitivities in the toadfish *H. didactylus*, the goby *G. cruentatus*, the sciaenid *S. umbra,* and the damselfish *C. chromis* between 10 dB and more than 30 dB (Figs. 4.5 and 4.8). The masking effect caused by ship noise as compared to ambient noise was more pronounced in the sciaenid than in representatives of the other families investigated because of its generally higher hearing sensitivities.

While ship traffic noise is the most ubiquitous anthropogenic noise source in aquatic habitats, other noise sources such as, e.g., construction sites or geological surveys can also affect hearing in fishes. Popper et al. (2005) reported 24 h threshold shifts in the northern pike *Esox lucius* (family Esocidae) and the lake chub *Couesius plumbeus* (family Cyprinidae) when exposed to airgun shots of a geological survey in the Mackenzie River Delta.

4.4 Sound Production and Transmission

Representatives of numerous bony fish families possess sound-generating mechanisms and vocalize in agonistic and reproductive contexts (Ladich and Fine 2006; Ladich and Myrberg 2006; Myrberg and Lugli 2006). The main energies of sounds are often concentrated at low frequencies of around 100 Hz or slightly above, based on the contraction rate of drumming muscles (100–200 Hz). In contrast, broadband high-frequency sounds with main energies at or above 1 kHz are found in gouramis, catfishes, and some cyprinids (Ladich 1988, 1997; Ladich et al. 1992) and are produced by sonic mechanisms other than swim bladders (Ladich and Bass 2011).

Are the main energies of sounds and thus sound-generating mechanisms adapted to ambient noise conditions? Do fish produce sound energies at low frequencies to optimize sound transmission and thus increase communication distances? Lugli and Fine (2003) suggest that vocal gobies utilize noise windows for communication. The authors found a quiet window or 'notch' around 100 Hz at noisy locations in shallow streams in northern Italy. The window lies between two noise sources, a low-frequency one attributed to turbulence, and a high-frequency one between 200 and 500 Hz attributed to bubble noise from water breaking the surface (Lugli and Fine 2007). Freshwater gobies such as *Padogobius martensii* and *Gobius nigricans* emit sounds with main frequencies in the 80–200 Hz band (Lugli et al. 2003). Therefore, both species utilize frequencies for sound communication that fall within the low-frequency quiet region of their habitats (Lugli et al. 2003).

In a subsequent paper, Lugli and Fine (2007) extended these observations by investigating particle motion of ambient noise and of vocalizations in addition to acoustic pressure. Gobies lack accessory hearing structures and will therefore only detect particle motion (particle velocity of particle acceleration) in a sound field but not the sound pressure components. So far the description of sound spectra and vocalizations in fish are almost exclusively based on sound pressure due to a lack of appropriate particle motion detectors. Lugli and Fine (2007) measuring both components with a new underwater acoustic pressure velocity probe found that the ambient noise spectrum is generally similar for sound pressure and particle velocity including the quiet window at noisy locations. The energy distribution of the velocity spectrum is shifted up by 50–100 Hz. The energy distribution of vocalizations was similar for sound pressure and particle velocity for the tonal sound, whereas the pulse-train sound exhibited larger differences. Transmission loss was high for both sound components and amplitudes declined by 6–10 dB/ 10 cm. The ratio between pressure and velocity did not change with distance from the sound source. The authors argued that SPL measurements, either for environmental noise or sounds emitted by a particle motion sensitive teleost are likely relevant for characterization of the dominant frequencies used for communication in the near field of a sound source.

Lugli (2010) investigated additional habitats such as rocky or sandy shores and found similar quiet windows at 100 Hz (stream, sandy/rocky sea shore) or at 200 Hz (spring, brackish lagoon) (asterisks in Fig. 4.2). The spectrum of the ambient noise showed that fish sound frequencies match the frequency band of the quiet window in several goby habitats (Fig. 4.8). In a further step, Lugli (2010) generalized this result by comparing the main frequencies of mating sounds of representatives of gobies, toadfishes (family Batrachoididae), sculpins (family Cottidae), minnows (family Cyprinidae), and darters (family Percidae) to the frequency band of the quiet window that he found in his study on goby habitats. Although this is only a rough comparison because the noise characteristics of each species' habitat need to be analyzed in detail, it indicates that fish other than gobies might utilize noise windows too (Fig. 4.6).

Crawford et al. (1997) and Speares et al. (2011) described acoustic or noise windows at higher frequencies than those observed by Lugli and coauthors. Crawford et al. (1997) investigated acoustic communication in the weakly electric mormyrid *Pollimyrus isidori* in shallow floodplains of the Niger River in Mali. The main energies of their vocalizations range from 300 Hz up to 2 kHz and fall within an acoustic window, thereby minimizing potential interference with sound sources from other abiotic and biotic sources. Strong high-frequency noise above 4 kHz was thought to emanate from stridulating aquatic insects. Speares et al. (2011) studied the aggressive vocalizations produced by two closely related species of darters, genus *Etheostoma* (family Percidae), and compared the spectrum to that of the ambient noise in their respective microhabitats, namely creeks in Alabama. Dominant frequencies of darters' aggressive drum sounds are concentrated between 100 and 400 Hz, thus avoiding high ambient noise levels at lower frequencies.

Nonetheless, this match of ambient noise windows and sound frequencies in gobies (and perhaps other vocal teleosts) should not conceal that the communication distances are quite short due to low sound levels and due to physical

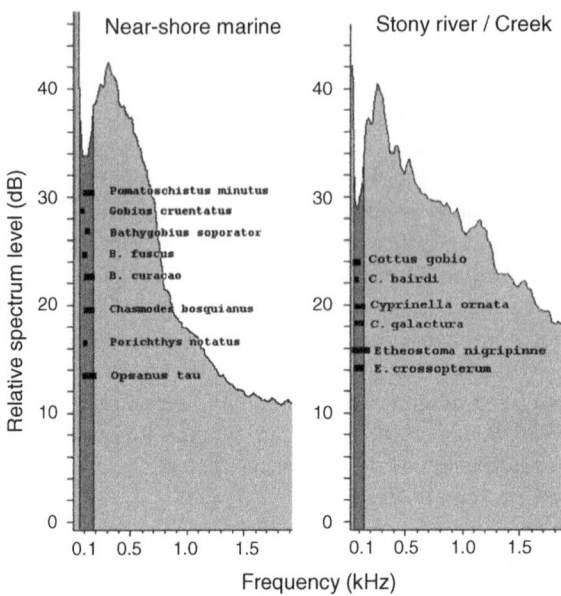

Fig. 4.6 Main frequencies of the mating sound (*dark horizontal bars*) emitted by vocal fish species breeding in near-shore marine habitats (*left graph*) or stony streams and creeks (*right graph*), superimposed on a generic ambient noise spectrum of an breaking wave (*left*), and a small waterfall (*right*), chosen among those available from the study (breaking wave noise spectrum: Rosolina beach, waterfall noise spectrum: Stream Stirone) by Lugli (2010). The low-frequency AN spectrum of both noise sources features a narrow quiet window at around 100 Hz (*dark gray area*). Modified with kind permission from Springer Science + Business Media from Lugli (2010)

constraints in shallow water habitats. Theoretically, only sounds will propagate that have a wavelength shorter than approximately four times the water depth, a phenomenon known as the frequency cutoff phenomenon (Rogers and Cox 1988). For example, frequencies below 750 Hz will not propagate in water shallower than 50 cm, which means that almost all low-frequency sounds produced by fish such as drumming sounds will not propagate at all. Fine and Lenhardt (1983) studied sound propagation and transmission loss of the mating call of the oyster toadfish *Opsanus tau* in water 1 m in depth and found that the fundamental frequency (200 Hz) was 16 dB lower at 1 m and 29 dB lower at 3 m. They conclude that over a sandy bottom communication is restricted within a range of a several meters. Field measurements by Lugli and Fine (2003) on courtship sound transmission in *P. martensii* indicate an attenuation of 15–20 dB over 20 cm at a water depths of 50 cm. Due to the low amplitude of goby sounds (90–120 dB at 5–10 cm), call levels are below the noise level 50–60 cm from the source, even under quiet conditions.

In addition, acoustic windows are not typical in fish utilizing low frequencies for vocalizations. Coers et al. (2008) reported that the ambient noise in a tidal zone was most pronounced for frequencies below 250 Hz, thus overlapping most of the frequency range used by rock-pool blennies *Parablennius parvicornis* (family Blenniidae) for communication. Similarly to gobies, communication distances are quite short, reaching 25 cm under calm (low tide conditions) and no doubt less under high tide conditions.

In reefs, communication distances are obviously larger than in tidal zones or the very shallow creeks mentioned above. Mann and Lobel (1997) and Mann (2006) estimated that male damselfishes such as *Dascyllus albisella* (family Pomacentridae), which produce pulsed courtship sounds (chirps) to advertise their territories, will be detectable at or beyond 11–12 m from the source. At larger distances, reflection and refraction will affect the temporal, amplitude, and spectral patterns of fish sounds (Fig. 4.7). Studies on the short-range propagation of damselfish sounds showed that amplitude, pulse duration, and pulse frequency varied by as much as 50 % over 10 m (Mann and Lobel 1997). The pulse period of the sound varied the least (by 4 %) of the sound characteristics measured.

Detection distances were also calculated for the silver perch *Bairdiella chrysoura* (family Sciaenidae, drums or croakers) in North Carolina waters by Sprague and Luczkovich (2004). Source levels of individual fish in a chorus ranged from 128 to 135 dB. The maximum distance at which an individual silver perch could be detected by the hydrophone depends on the background noise level

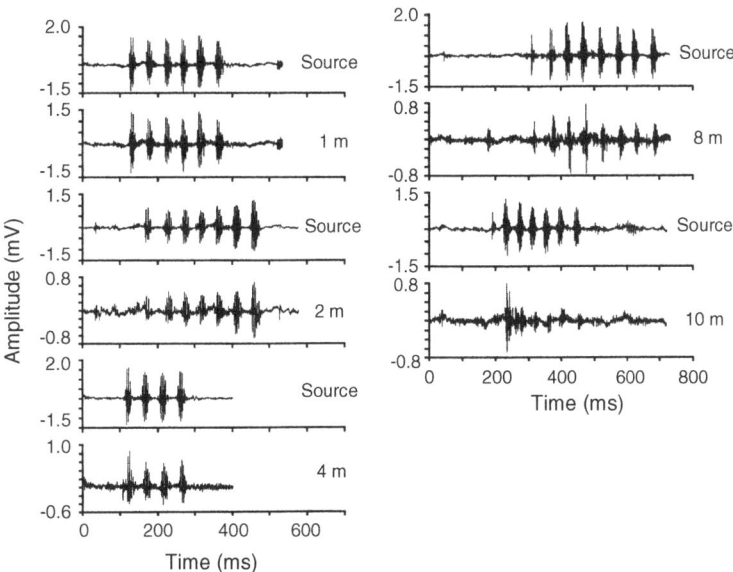

Fig. 4.7 Waveforms illustrating the propagation of damselfish sound over a range of distances. The source waveform shows the damselfish sound at 1 m from the hydrophone, the lower waveform the recording at distances from 1 up to 10 m. The lower waveforms show that amplitudes and pulse duration decrease with distance. Modified with permission from Mann and Lobel (1997). Copyright 1997, Acoustical Society of America

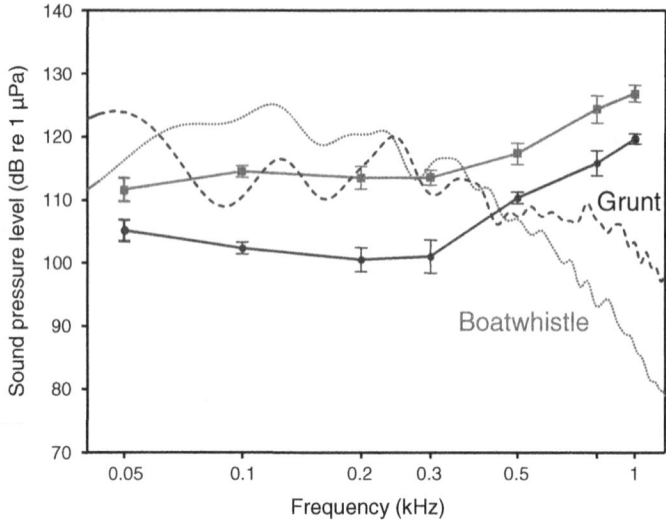

Fig. 4.8 Mean (±S.E.) hearing thresholds of the Lusitanian toadfish in the presence of ambient (*blue line/circles*) and ship noise (*red line/squares*) as well as power spectra of sounds (boatwhistle—*dotted green line*; grunt—*dashed brown line*). The boatwhistle spectrum was calculated from a distance of 20 cm, the grunt train 10 cm away from the calling animals. Modified after Vasconcelos et al. (2007)

and may vary considerably. For the loudest background level recorded a silver perch 1 m from the hydrophone would be undetectable. On the other side on a quiet morning an individual could be heard at more than 100 m. Under conditions recorded in the study, fish were detectable by the hydrophone at 1–7 m.

Investigations on fish communication distances are limited because direct observations are sparse and playback experiments were successful in only a few cases. Proving that fish detect sounds at a certain distance requires observing a phonotactic response (approach to the speaker): ideally, females should approach a speaker so that females are not attracted by visual signals. Field playback experiments in the damselfish *Stegastes partitus* showed that females approach conches, where male sounds emanate over a distance of approximately 10 m (Myrberg et al. 1986). Communication distances in fish beyond this distance have not been proven unambiguously so far. Some fishes, under certain conditions, might be able to communicate acoustically over much larger distances, but communication distances are typically much shorter, on average less than one meter, in many cases merely a few centimeters.

It is interesting to ask why acoustic communication distances in fish are much smaller than those of terrestrial animals such as frogs (Chaps. 5 and 6) or birds (Chaps. 7 and 8) and aquatic mammals such as whales (Chaps. 9 and 10). The reasons for this difference might be the lower levels of fish sounds (e.g., approximately 120 dB at 10 cm in croaking gouramis, family Osphronemidae—Ladich

2007; 126 dB at 1 m in toadfish, family Batrachoididae—Barimo and Fine 1998; 90–120 dB at 5–10 cm in gobiids—Lugli and Fine 2003; 130 dB in drums at 1 m distance—Sprague and Luczkovich 2004) as compared to whales, which reach up to 180 dB and more at 1 m distance (Chap. 9). The frequency cutoff phenomenon in shallow waters may also play a role, making it difficult to propagate low-frequency sound. Despite this frequency cutoff phenomenon, most fish concentrate their sound energies between 100 and 300 Hz (Amorim 2006; Ladich and Myrberg 2006). In contrast, baleen whales utilize low-frequency sounds to communicate over very long distances—hundreds of meters or even several kilometers (Chaps. 9 and 10)—quite the opposite of what is found in fishes. This discrepancy can be explained by differences in the biology of fishes and whales. Fish vocalize regularly close to substrates such as crevices, bottoms of their habitats, coral reefs, floating plants, etc., mostly in shallow waters (cm to m), whereas whales vocalize in open waters at much greater depths, where low frequencies propagate readily.

4.5 Communication

The previous section showed that the auditory sensitivities of fishes are adapted to the ambient noise (at calm conditions) and that fish with improved hearing are masked in noisier regions of their habitats or during noisier time periods (tides, wind, etc.). Communication is not only limited by masking, which decreases the hearing sensitivities, but also by restrictions in sound production and transmission. Most sound-production mechanisms emit low-amplitude low-frequency sounds, which limits the communication distances because of several physical factors in the environments. Factors include the high levels of ambient noise at low frequencies (Figs. 4.1, 4.2, 4.8) and the frequency cutoff phenomenon. Quiet windows at low frequencies may improve communication distances in some habitats, but communication distances remains quite short (<0.5 m). Any increase in the noise level will lower the communication distances even further.

4.5.1 Animal Acoustic Adaptations to Anthropogenic Noise

Animals exhibit strategies to cope with anthropogenic noise in their environment. Several animal taxa such as frogs, birds, and mammals, including whales, are able to adapt their vocalizations to increasing noise levels. Frogs can decrease their calling rate and time calling in the presence of anthropogenic noise (Chaps 5. and 6). A number of songbird species such as great tits, nightingales, blackbirds, and robins change their singing behavior in cities as compared to forests and other habitats that are minimally affected by traffic or industrial noise. The main strategies include increasing the SPL or dominant frequencies of songs or shifting singing to quiet periods of the day (Chaps. 7 and 8). Increasing the sound level is a well-known phenomenon termed Lombard effect (Brumm and Zollinger 2011).

Mechanisms to compensate for increased noise have also been observed in aquatic habitats, where human-made noise has increased significantly over the past century. Aquatic mammals such as whales and manatees change frequencies, sound levels, or call duration in the presence of noise (Chaps. 9 and 10).

4.5.2 Anthropogenic Noise and Communication in Fish

Do certain fish species react similarly to birds and whales in the presence of noise? So far none of these behavioral responses has been described in fishes. We do not know if fish are able to adapt their vocal output to increasing noise levels by calling louder, longer, or at higher frequencies. Our lack of information could reflect the inability of fish to adapt to different conditions in ways similar to birds and mammals or perhaps the inability of researchers to collect long-term data or conduct appropriate experiments in the field or in the lab. Physiological experiments indicate that it is unlikely that there is a Lombard effect in fishes which utilize swim bladder muscles for sound production. Fine et al. (2001) found a small dynamic range in electrically stimulated toadfish sounds. Therefore, toadfish will not be able to increase the amplitude of their sounds. As long as a Lombard effect has not been shown in fishes, we have to assume that increasing noise levels will reduce communication distances. Two recent papers point into this direction. Vasconcelos et al. (2007) investigated the hearing abilities and the ability to detect conspecific sound in the Lusitanian toadfish *Halobatrachus didactylus* (family Batrachoididae) in the Tagus River estuary in the presence of ambient noise and ferry-boat noise. This species has best hearing sensitivities at low frequencies between 50 and 200 Hz, and the main energies of the ferry-boat noise were within the most sensitive hearing range, considerably masking their hearing abilities (Fig. 4.7). Comparisons between masked hearing thresholds and sound spectra of the toadfish's mating and agonistic vocalizations revealed that ship noise decreased the ability to detect conspecific acoustic signals and thus reduced communication distance. Accordingly, we must assume that acoustic communication, which is essential in nest advertisement, during nest defence and mate attraction, is restricted in coastal environments in the presence of human-made noise.

Codarin et al. (2009) examined the effects of hearing and the detection of conspecific sounds in the presence of boat noise in vocal representatives of different families in the Adriatic Sea near Trieste. They investigated the auditory sensitivities, in the presence of boat noise, of the brown meagre *S. umbra*, the Mediterranean damselfish *C. chromis*, and the red-mouthed goby *G. cruentatus*. The thresholds to conspecific sounds were 98 dB for *S. umbra* and 101 dB for *C. chromis* under both quiet lab noise and ambient noise conditions (calm sea), but increased in the presence of boat noise by approximately 20 dB (Fig. 4.9). The

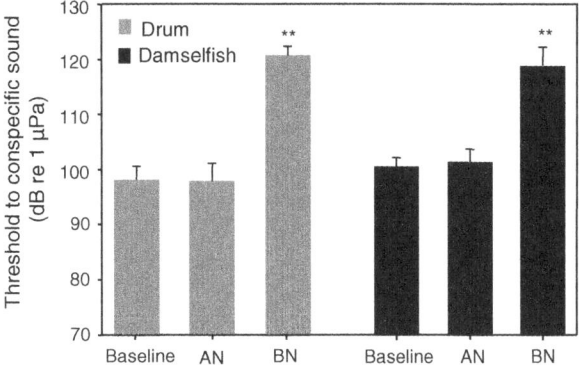

Fig. 4.9 Mean hearing thresholds (±S.E.) to conspecific sounds under lab (*baseline*), ambient noise (AN) and boat noise (BN) conditions in the drum *S. umbra* and the damselfish *C. chromis*. Asterisks indicate significant differences between BN and other conditions. Modified from Codarin et al. (2009). Copyright 2009, with permission from Elsevier

authors estimated that the detection distance for the drum's sounds will decrease from more than 100 m down to less than 1 m under boat noise conditions. In the damselfish, which has lower hearing sensitivities, they calculate that, under ambient noise conditions, sounds will be detectable up to 10 m; boat noise, however, would completely mask the signal even at a distance to the vocalizing fish of less than 1 m.

We know little about the responses of fish to increasing ambient noise levels. We do know that fish can modify their vocalizations in response to con- or heterospecifics. Fish such as male haddock *Melanogrammus aeglifinus* (family Gadidae) can modify their fundamental frequency during courtship to a certain degree (Hawkins 1993; Ladich 2004). Note that these frequency modulations were found during courtship or agonistic encounters and reflect different motivational levels. It is unknown if fish can increase their sound frequencies to avoid interference with low-frequency ambient noise. Other potential behavioral responses such as shifting the calling activity to more quiet periods of the day, postponing calling until the noise fades away or alternatively lengthening call durations have also not been described so far. Decreasing the calling activity in the presence of another sound source has been observed in the silver perch *Bardiella chrysoura* (family Sciaenidae) and the gulf toadfish *Opsanus beta* (family Batrachoididae). Luczkovich et al. (2000) found that bottlenose dolphin whistles suppress mating choruses of silver perch and Remage-Healey et al. (2006) reported that the call rate of the gulf toadfish declines when pop sounds of the bottlenose dolphin were played back.

While acoustic responses to noise have not been observed, we know that fish may avoid loud sound sources. Some flee from rapidly approaching loud underwater noise sources. Underwater video recordings of roaches *Rutilus rutilus* and rudds *Scardinius erythrophthalmus* (both family Cyrinidae) in the Meuse River in

Belgium showed that the fishes actively avoided high-speed boats (Boussard 1981). The flight reactions started at distances of approximately 5 m. Similar responses have been reported regarding fishing vessels: cods *Gadus morhua* significantly altered their behavior during and after the passage of a bottom trawling vessel. According to Handegard et al. (2003), cods initially reacted by diving, then with horizontal movements away from the ship. Besides triggering avoidance behavior, noise can affect the foraging behavior and cause stress in fishes. Purser and Radford (2011) found strong evidence that adding noise affects the attention of fish and increases food handling errors. Furthermore, Wysocki et al. (2006) observed that the common carp *C. carpio*, the gudgeon *Gobio gobio* (both family Cyprinidae), and the European perch *Perca fluviatilis* (family Percidae) responded with increased cortisol secretion when exposed to ship noise (Wysocki et al. 2006). The data indicate that ship noise, characterized by amplitude and frequency fluctuations, constitutes a potential stressor for all three species independently of their different hearing sensitivities (for the implications of anthropogenic noise for the conservation of fish and other animals see Chap. 14).

4.6 Summary and Conclusion

Bony fishes evolved a large number of sonic organs, indicating the importance of sound communication in these animals. Moreover, at least one-third of fishes possess structures enabling them to extend their hearing range to several kilohertz and low sound levels. Numerous studies showed that hearing sensitivities, in particular of taxa with hearing enhancements, decreased when exposed for longer periods to high noise levels or in the presence of moderate noise levels due to masking. Aquatic habitats are characterized by large differences in noise levels and spectral profiles due to numerous abiotic and biotic factors such as running water, wind, tides, vocalizing animals, etc. Currently, we do not know if and how ambient noise and physical constraints such as the frequency cutoff phenomenon limit acoustic communication in fishes. Some light has been shed on these questions; recent studies showed that fish are adapted to ambient noise under calm conditions and that their hearing is masked under more noisy conditions (severe weather conditions, running water).

Most vocal fishes communicate over short distances (<0.5 m), probably because of low sound levels and low sound frequencies produced and because of the ambient noise conditions. Some species, e.g., gobies, partly overcome these limitations by exploiting quiet windows in the ambient noise (frequencies of low spectral noise levels) for acoustic communication. However, it remains unclear whether these are adaptations or coincidences, and many more groups need to be investigated to answer this question.

Human-made noise such as ship noise masks hearing and potentially hinders acoustic communication in several marine fish families. We do not yet know if fish are able to cope with noise pollution similar to songbirds and whales by modifying sound characteristics or calling behavior. It will be an important goal to close these gaps in our knowledge on acoustic communication in fishes in near future. This will help us to assess the impacts of aquatic noise pollution on the fitness of fishes and on fish populations.

Acknowledgments Thanks to Michael L. Fine and an anonymous reviewer for critically reading the manuscript. Support from the Austrian Science Fund (FWF grant 22319).

References

Amorim MCP (2006) Diversity in sound production in fish. In: Ladich F, Collin SP, Moller P, Kapoor BG (eds) Communication in fishes, vol 1. Science Publisher, Enfield, pp 71–105

Amoser S, Ladich F (2003) Diversity in noise-induced temporary hearing loss in otophysines fishes. J Acoust Soc Am 113:2170–2179

Amoser S, Ladich F (2005) Are hearing sensitivities of freshwater fish adapted to the ambient noise in their habitats? J Exp Biol 208:3533–3542

Amoser S, Wysocki LE, Ladich F (2004) Noise emission during the first powerboat race in an Alpine lake and potential impact on fish communities. J Acoust Soc Am 116:3789–3797

Amoser S, Ladich F (2010) Year-round variability of ambient noise in temperate freshwater habitats and its implications for fishes. Aquat Sci 72:371–378

Andrew RK, Howe BM, Mercer JA, Dzieciuch MA (2002) Ocean ambient sounds: comparing the 1960s with the 1990s for a receiver off the Californian coast. Acoust Res Lett Online 3:65–70

Andrew RK, Howe BM, Mercer JA (2011) Long-time trends in ship traffic noise for four sites off the North American West Coast. J Acoust Soc Am 129:642–651

Barimo JF, Fine ML (1998) Relationship of swim-bladder shape to the directionality pattern of underwater sound in the oyster toadfish. Can J Zool 76:134–143

Belanger AJ, Bobeica I, Higgs DM (2010) The effect of stimulus type and background noise on hearing abilities of the round goby *Neogobius melanostomus*. J Fish Biol 77:1488–1504

Boussard A (1981) The reactions of roach (*Rutilus rutilus*) and rudd (*Scardinus erythrophthalmus*) to noises produced by high speed boating. In: Proceeding of 2nd British freshwater fisheries conference, pp 188–200

Braun CB, Grande T (2008) Evolution of peripheral mechanisms for the enhancement of sound reception. In: Webb JF, Popper AN, Fay RR (eds) Fish bioacoustics. Springer, New York, pp 99–144

Brumm H, Zollinger SA (2011) The evolution of the Lombard effect: 100 years of psychoacoustic research. Behaviour 148:1173–1198

Chapman CJ, Hawkins AD (1973) A field study of hearing in the cod, *Gadus morhua* L. J Comp Physiol A 85:147–167

Chapman CJ (1973) Field studies of hearing in teleost fish. Helgol wiss Meeresunters 24:371–390

Codarin A, Wysocki LE, Ladich F, Picciulin M (2009) Effects of ambient and boat noise on hearing and communication in three fish species living in a marine protected area (Miramare, Italy). Mar Poll Bull 58:1880–1887

Coers A, Bouton N, Vincourt D, Slabbekoorn H (2008) Fluctuating noise conditions may limit acoustic communication distance in rock-pool blenny. Bioacoustics 17:63–64

Crawford JD, Jacob P, Benech V (1997) Sound production and reproductive ecology of strongly acoustic fish in Africa: *Pollimyrus isidori,* Mormyridae. Behaviour 134:677–725

Enger PS (1981) Frequency discrimination in teleosts—central or peripheral. In: Tavolga WN, Popper AN, Fay RR (eds) Hearing and sound communication in fishes. Springer, New York, pp 243–253

Fay RR (1974) Masking of tones by noise for the goldfish (*Carassius auratus*). J Comp Physiol Psychol 87:708–716

Fay RR (2009) Sound scapes and the sense of hearing in fishes. Integr Zool 4:26–32

Fine ML, Lenhardt ML (1983) Shallow-water propagation of the toadfish mating call. Comp Biochem Physiol 76A:225–231

Fine ML, Malloy KL, King CB, Mitchell SL, Cameron TM (2001) Movement and sound generation by the toadfish swimbladder. J Comp Physiol 187A:371–379

Handegard NO, Michalsen K, Tjřstheim D (2003) Avoidance behaviour in cod (*Gadus morhua*) to a bottom-trawling vessel. Aquat Living Resour 16:265–270

Hastings MC, Popper AN, Finneran JJ, Lanford PJ (1996) Effects of low-frequency underwater sound on hair cells of the inner ear and lateral line of the teleost fish *Astronotus ocellatus*. J Acoust Soc Am 99:1759–1766

Hawkins AD (1993) Underwater sound and fish behaviour. In: Pitcher TJ (ed) Behaviour of teleost fishes. Chapman and Hall, London, pp 129–169

Kennedy EV, Holderied MW, Mair JM, Guzman HM, Simpson SD (2010) Spatial patterns in reef-generated noise relate to habitats and communities: evidence from a Panamanian case study. J Exp Mar Biol Ecol 395:85–92

Ladich F (1988) Sound production by the gudgeon, *Gobio gobio* L., a common European freshwater fish (Cyprinidae, Teleostei). J Fish Biol 32:707–715

Ladich F, Bischof C, Schleinzer G, Fuchs A (1992) Intra- and interspecific differences in agonistic vocalization in croaking gouramis (genus: *Trichopsis*, Anabantoidei, Teleostei). Bioacoustics 4:131–141

Ladich F (1997) Comparative analysis of swimbladder (drumming) and pectoral (stridulation) sounds in three families of catfishes. Bioacoustics 8:185–208

Ladich F (2004) Sound production and acoustic communication. In: van der Emde G, Mogdans J, Kapoor BG (eds) The senses of fishes. Narosa Publishing House, New Delhi, pp 210–230

Ladich F (2007) Females whisper briefly during sex: context- and sex-specific differences in sounds made by croaking gouramis. Anim Behav 73:379–387

Ladich F (2008) Sound communication in fishes and the influence of ambient and anthropogenic noise. Bioacoustics 17:35–37

Ladich F (2010) Hearing: Vertebrates. In: Breed MD, Moore J (eds) Encyclopedia of animal behaviour, vol 2. Academic Press, Oxford, pp 54–60

Ladich F, Bass AH (2011) Sound production mechanisms and physiology. In: Farrell AP (ed) Encyclopedia of fish physiology: from genome to environment. Academic Press, San Diego, pp 321–329

Ladich F, Fine ML (2006) Sound generating mechanisms in fishes: a unique diversity in vertebrates. In: Ladich F, Collin SP, Moller P, Kapoor BG (eds) Communication in fishes, vol 1. Science Publishers, Enfield, pp 3–43

Ladich F, Myrberg AA (2006) Agonistic behaviour and acoustic communication. In: Ladich F, Collin SP, Moller P, Kapoor BG (eds) Communication in fishes, vol 1. Science Publishers, Enfield, pp 122–148

Ladich F, Popper AN (2004) Parallel evolution in fish hearing organs. In: Manley G, Popper AN, Fay RR (eds) Evolution of the Vertebrate auditory system. Springer, New York, pp 95–127

Lugli M (2010) Sounds of shallow water fishes pitch within the quiet window of the habitat ambient noise. J Comp Physiol A 196:439–451

Lugli M, Fine ML (2003) Acoustic communication in two freshwater gobies: ambient noise and short-range propagation in shallow streams. J Acoust Soc Am 114:512–521

Lugli M, Fine ML (2007) Stream ambient noise, spectrum and propagation of sounds in the goby *Padogobius martensii*: sound pressure and particle velocity. J Acoust Soc Amer 122:2881–2892

Lugli M, Yan HY, Fine ML (2003) Acoustic communication in two freshwater gobies: the relationship between ambient noise, hearing thresholds and sound spectrum. J Comp Physiol A 189:309–320

Luczkovich JJ, Dahle HJ, Hutchinson M, Jenkins T, Johnson SE et al (2000) Sounds of sex and death in the sea: bottlenose dolphins whistles suppress mating choruses of silver perch. Bioacoustics 10:323–334

Mann DA (2006) Propagation of fish sounds. In: Ladich F, Collin SP, Moller P, Kapoor BG (eds) Communication in fishes, vol 1. Science Publishers, Enfield, pp 107–120

Mann DA, Lobel PS (1997) Propagation of damselfish (Pomacentridae) courtship sounds. J Acoust Soc Am 101:3783–3791

Myrberg AA, Lugli M (2006) Reproductive behaviour and acoustical interactions. In: Ladich F, Collin SP, Moller P, Kapoor BG (eds) Communication in fishes, vol 1. Science Publishers, Enfield, pp 149–176

Myrberg AA, Mohler M, Catala JD (1986) Sound production by males of a coral reef fish (*Pomacentrus partitus*): its significance to females. Anim Behav 34:913–923

Myrberg AA, Spanier E, Ha SJ (1978) Temporal patterning in acoustical communication. In: Reese ES, Lighter FJ (eds) Contrasts in behaviour. Wiley, New York, pp 137–179

Popper AN (2003) Effects of anthropogenic sounds on fishes. Fish Res 28:24–31

Popper AN, Clarke NL (1976) The auditory system of the goldfish (*Carassius auratus*): effects of intense acoustic stimulation. Comp Biochem Physiol 53:11–18

Popper AN, Fay RR (2011) Rethinking sound detection by fishes. Hear Res 273:25–36

Popper AN, Hastings MC (2009a) The effects of human-generated sound on fish. Integr Zool 4:43–52

Popper AN, Hastings MC (2009b) The effects of anthropogenic sources of sound on fishes. J Fish Biol 75:455–489

Popper AN, Smith ME, Cott PA, Hanna BW, MacGillivray AO, Austin ME, Mann DA (2005) Effects of exposure to seismic airgun use on hearing of three fish species. J Acoust Soc Am 117:3958–3971

Purser J, Radford AN (2011) Acoustic noise induces attention shifts and reduces foraging performance in three-spined sticklebacks (*Gasterosteus aculeatus*). PLOS ONE 6:e17478

Radford CA, Stanley JA, Simpson SD, Jeffs AG (2011) Juvenile coral reef fish use sound to locate habitats. Coral Reefs 30:295–305

Ramcharitar J, Popper AN (2004) Masked auditory thresholds in sciaenid fishes: a comparative study. J Acoust Soc Am 116:1687–1691

Remage-Healey L, Nowacek DP, Bass AH (2006) Dolphins foraging sounds suppress calling and elevate stress hormone levels in prey species, the Gulf toadfish. J Exp Biol 209:4444–4451

Rogers PH, Cox H (1988) Underwater sound as a biological stimulus. In: Atema J, Fay RR, Popper AN, Tavolga WN (eds) Sensory biology of aquatic animals. Springer, New York, pp 131–149

Samuel Y, Morreale SJ, Clark CW, Greene CH, Richmond ME (2005) Underwater, low-frequency noise in a coastal sea turtle habitat. J Acoust Soc Am 117:1465–1472

Scholik AR, Yan HY (2001) Effects of underwater noise on auditory sensitivity of a cyprinid fish. Hear Res 152:17–24

Scholik AR, Yan HY (2002a) The effects of noise on the auditory sensitivity of the bluegill sunfish, *Lepomis macrochirus*. Comp Biochem Physiol A 133:43–52

Scholik AR, Yan HY (2002b) Effects of boat engine noise on the auditory sensitivity of the fathead minnow, *Pimephales promelas*. Environ Biol Fishes 63:203–209

Scholz K, Ladich F (2006) Sound production, hearing and possible interception under ambient noise conditions in the topmouth minnow *Pseudorasbora parva*. J Fish Biol 69:892–906

Slabbekoorn H, Bouton N, van Opzeeland I, Coers A, ten Cate C, Popper AN (2010) A noisy spring: the impact of globally rising underwater sound levels on fish. TREE 25:419–427

Smith ME, Kane AS, Popper AN (2003) Noise-induced stress response and hearing loss in goldfish. J Exp Biol 207:427–435

Smith M, Kane A, Popper A (2004) Acoustical stress and hearing sensitivity in fishes: does the linear threshold shift hypothesis hold water? J Exp Biol 207:3591–3602

Smith ME, Schuck JB, Gilley RR, Rogers BD (2011) Structural and functional effects of acoustic exposure in goldfish: evidence of tonotopy in the teleost saccule. BMC Neurosci 12:19

Speares P, Holt D, Johnston C (2011) The relationship between ambient noise and dominant frequency of vocalizations in two species of darters (Percidae: *Etheostoma*). Environ Biol Fish 90:103–110

Sprague MW, Luczkovich JJ (2004) Measurements of an individual silver perch *Bairdiella chrysoura* sound pressure level in a field recording. J Acoust Soc Am 116:3186–3191

Tonolla D, Lorang MS, Heutschi K, Tockner K (2009) A flume experiment to examine underwater sound generation by flowing water. Aquat Sci 71:449–462

Tolimieri N, Haine O, Jeffs A, McCauley R, Montgomery J (2004) Directional orientation of pomacentrid larvae to ambient reef sound. Coral Reefs 23:184–191

Urick RJ (1983) Principles of underwater sound. Chapter 7: the noise background of the sea: ambient noise. Peninsula Publishing, Los Altos, pp 202–236

Vasconcelos RO, Amorim MCP, Ladich F (2007) Effects of ship noise on the detectability of communication signals in the Lusitanian toadfish. J Exp Biol 210:2104–2112

Wenz GM (1962) Acoustic ambient noise in the ocean: spectra and sources. J Acoust Soc Am 34:1936–1956

Wysocki LE, Dittami JP, Ladich F (2006) Ship noise and cortisol secretion in European freshwater fishes. Biol Conserv 128:501–508

Wysocki LE, Ladich F (2005a) Hearing in fishes under noise conditions. JARO 6:28–36

Wysocki LE, Ladich F (2005b) Effects of noise exposure on click detection and temporal resolution ability of the goldfish auditory system. Hear Res 201:27–36

Wysocki LE, Amoser S, Ladich F (2007) Diversity in ambient noise in European freshwater habitats: noise levels, spectral profiles, and impact on fishes. J Acoust Soc Am 121:2559–2566

Chapter 5
Anuran Acoustic Signal Production in Noisy Environments

Joshua J. Schwartz and Mark A. Bee

Abstract Where they co-occur, male anurans of different species may signal from diverse locations and use vocalizations that differ spectrally. However, the relevance of such differences to the problem of signal masking, as well as their ubiquity and efficacy, may have been over-emphasized, especially given data from recent studies. Of greater significance are adjustments in signal timing, operating both within and among species, which can result in alternation of calling bouts or even rapid-fire alternation of notes among neighboring males. The possibility that frogs elevate call amplitude in response to noise deserves further study. Also requiring more research are the contributions to communication of seismic and ultrasonic signaling, employed in the presence of interfering biotic and abiotic noise, respectively, as well as the role played by signal redundancy to improved information transfer in loud choruses. Whether anthropogenic noise constitutes a significant threat to anurans remains an unresolved question.

5.1 Introduction

On August 15, 1965, the Beatles performed in Shea Stadium in New York to a crowd of more than 55,000 young fans. Many members of the audience screamed at the top of their lungs and together with the output of the Beatles' sound system created such a deafening acoustic background that each of the "Fab Four" could not hear themselves sing, let alone detect the voices of other members of the band.

J. J. Schwartz (✉)
Department of Biology, Pace University, Pleasantville, NY, USA
e-mail: jschwartz2@pace.edu

M. A. Bee
Department of Ecology, Evolution and Behavior, University of Minnesota, Saint Paul, MN 55108, USA

H. Brumm (ed.), *Animal Communication and Noise*,
Animal Signals and Communication 2, DOI: 10.1007/978-3-642-41494-7_5,
© Springer-Verlag Berlin Heidelberg 2013

John, Paul, George, and Ringo completed their set, albeit with some unorthodox behavior, because they were such accomplished musicians, stuck to their prearranged list of songs and used visual cues (Lewisohn 1992; Miles 2009).

We are reminded of the racket during that Shea Stadium concert almost every time we enter a dense chorus of frogs. During breeding season, males advertise for the attention of gravid females (Fig. 5.1). Operational sex ratios on any given night often are heavily biased toward males, so competition among males for potential mates is intense. In tropical areas, choruses can consist of a dozen or more species, and their advertisement signals create a cacophony that is not only frequently deafening, but also rich in spectral and temporal characteristics. To achieve

Fig. 5.1 Calling males of three species discussed in this review that have been a focus of research by one or both of the coauthors.
a *Dendropsophus microcephalus.*
b *Dendropsophus ebraccatus.*
c *Hyla versicolor*
(Photographs by Joshua J. Schwartz)

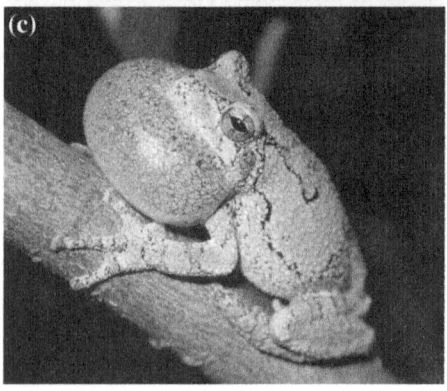

reproductive pairing, females especially, but often also males, must detect the calls of at least a subset of chorus members, localize call sources, possibly discriminate calls from those of nearby heterospecifics using their spectral and temporal features, and assess male calling performance. Background noise can render these tasks more difficult than would otherwise be the case (Wollerman and Wiley 2002; see Chap. 2) and so potentially result in wasted time, energy, gametes and increase the risk of predation (Gerhardt and Huber 2002; Wells 2007). How signalers and receivers communicate under such circumstances is a fascinating question, and one that we and many of our colleagues have attempted to answer over decades of research. Noise from abiotic sources can also potentially impair frog communication (Penna et al. 2005). Adaptations to the noise problem are anatomical, physiological, and behavioral, and involve the reception and neural processing of sound, as well as the production of signals (Feng and Ratnam 2000; Gerhardt and Huber 2002; Wells 2007).

In this chapter, we will focus on solutions to this problem related to acoustic signal production. There are important (and intimate) connections between production and reception that link, as well as constrain, evolutionary trajectories, and Chap. 6 by Vélez et al. addresses communication in noise from the perspective of signal perception. Although male anurans may use aggressive vocalizations to increase inter-male spacing, given the size of the topic, we will not address agonistic interactions here (for recent treatments see Gerhardt and Huber 2002; Wells 2007). We also will not review visual signaling (see reviews by Hödl and Amézquita 2001; Wells 2007; see Chap. 11), although we will discuss the use of sounds of very high frequencies in the vicinity of fast flowing water and seismic signaling. Calling location can influence active space in a variety of ways and, together with background noise, influences the signal-to-noise ratio for receivers (Brumm and Naguib 2009). Readers interested in venue-associated environmental effects on signal deterioration, as well as signal features facilitating transmission, are referred to Ryan and Kime (2003), Ey and Fisher (2009) and references therein. Lastly, because of the recent flux in the taxonomic nomenclature of anurans (Frost et al. 2006; Pyron and Wiens 2011) there is opportunity for confusion when the same species has multiple names in the literature. In this chapter, we adopt the nomenclature of Pyron and Wiens (2011).

5.1.1 Background and Overview

Male anurans are extremely sensitive to their acoustic environments, and observed changes in male calling may be linked to the imperative of maintaining relative attractiveness to females based on inherent characteristics of the advertisement display (e.g., call rate, call duration, and call complexity). Males also may eavesdrop on the calls of both conspecifics and heterospecifcs to more effectively gauge the risk of predation, and reduce or cease calling in response to reductions in calling by other males (the "predation rumors"; Phelps et al. 2007). Energetic constraints may limit the amount of time that males call each evening, and a modulated pattern of chorusing activity, sometimes referred to as "unison bout singing," can permit males to stretch their nightly chorusing over more hours than otherwise would be possible (Schwartz 1991). Finally, the dynamic aspect of male vocal activity may be a response to sounds that interfere with effective communication.

There are a number of ways that acoustic interference can be reduced in response to abiotic noise or noise created by an aggregation of calling males (Littlejohn 1977; Narins and Zelick 1988; Gerhardt and Huber 2002; Wells 2007). Calls of different individuals (or the constituent call notes) may be produced at different times, ranging from tens of milliseconds to days or months apart (Narins and Zelick 1988; Gerhardt and Huber 2002; Wells 2007). The interactive nature of rapid changes is often evident, and may stand a good possibility of being related to the threat of interference from conspecifics or heterospecifics. However, patterns of disjunction on a much greater time scale between heterospecifics may have a wider range of potential explanations (Gottsberger and Grubert 2004; Saenz et al. 2007).

In a multispecies assemblage, different species may use different frequencies for communication. The auditory systems of most chorus members probably respond to the signals of other species (Gerhardt and Schwartz 1995; but see Sect. 5.3.1.1 on ultrasonic signals), but the extent to which such spectral variation reduces the threat of masking for an individual depends on a suite of factors. These include features of the auditory system, number and location of callers, and signal intensities at the sources. Although individuals of some species can change the frequencies of their calls during male–male interactions (e.g., Wagner 1992; Bee et al. 2000), this is not known to be a response to the problem of acoustic interference among individuals. Thus, spectral separation among members of an assemblage is something that arises by chance, as a result of selection on call spectra (perhaps to reduce masking), or by virtue of which species join and remain in an anuran chorus. Call frequency structure also may be related to the risk of masking by natural abiotic or anthropogenic noise.

Under noisy circumstances, we humans typically speak more loudly, lengthen words, and repeat ourselves more frequently than in quiet conditions (Summers et al. 1988). Whether male anurans increase call intensity in response to noise remains uncertain, although the most recent research suggests that such behavior is rare (Love and Bee 2010; Brumm and Zollinger 2011). The calls of many species of frogs and toads consist of repeated elements (Gerhardt and Huber 2002). This feature renders them well adapted to communication under chorus conditions. By modulating signal redundancy in "real-time" (e.g., by changing signal duration, number of notes, or call rate), males may be able to improve the chances that a receiver will detect their vocalizations when noise levels are high (Wiley 2006).

Sensory systems of animals respond to a variety of physical phenomena, and, through the use of multicomponent or multimodal signaling, more than one system can be used either simultaneously or sequentially to facilitate communication (Brumm and Slabbekoorn 2005; see also Chap. 2). For example, an animal may augment communication using air-borne sound with communication using an entirely different modality (e.g., through use of substrate-borne vibrations or visual signals). Visual signaling as a possible solution to the problem of acoustic noise is best known for anurans that call near rushing water, although such signaling may also be helpful during storms and chorusing by heterospecifics (Amézquita and Hödl 2004). In some cases, vocal signals may alert potential receivers to a forthcoming visual signal (e.g., Grafe 2007; Preininger et al. 2009)!

All else being equal, increasing one's distance from a potential source of interference should be advantageous. Both within and among anuran species, aggressive calling and even physical combat can increase spacing among conspecific or heterospecific males (Gerhardt and Huber 2002; Wells 2007). Males of different species also could spatially segregate as a result of differences in microhabitat preferences. Such spatial partitioning could, when coupled with spectral differences, be exploited by directional characteristics of the auditory system. A reduction in the chance of mismatings would be another payoff if calling site differences among males reduces the proximity of females to heterospecific males as they move toward conspecific callers.

5.2 Potential Solutions to the Problem of Masking by Chorus Noise

5.2.1 Signal Separation in the Frequency Domain

In multispecies assemblages of frogs, differences among species in call spectrum could reduce the likelihood of masking by either the individual calls or call elements of a neighbor, or by the vocal activity of a group of males. The probability of masking depends not only on signal frequency, but also features of the auditory system (see Chap. 6). Spectral variation *potentially* could benefit anurans in other ways. For example, frequency differences can reduce the chance of mating with a

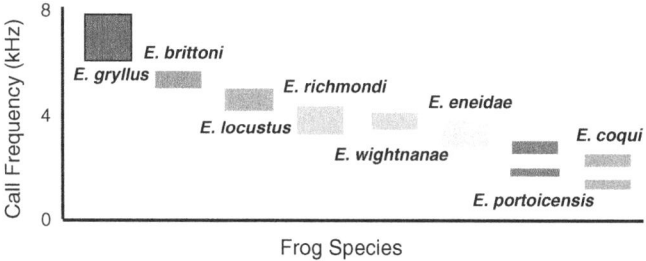

Fig. 5.2 Diagrammatic representation of call spectra of nine species of eleutherodactylid frogs that call in the forest near El Verde Puerto Rico. Note the stratification of the advertisement calls of these species along the frequency dimension (after Narins 1995)

member of another species (Gerhardt and Huber 2002; Wells 2007). Observations abound of communities of frogs in which species exhibit differences in call dominant frequency (Fig. 5.2; e.g., Hödl 1977; Duellman 1967, 1978; Schluter 1979, 1980; Narins 1995; Lüddecke et al. 2000; Garcia-Rutledge and Narins 2001; for reviews and additional references see Gerhardt and Schwartz 1995; Wells 2007) and so, on first blush, these appear consistent with the notion that species partition an "acoustic niche." Such partitioned communities could come about as a result of (1) divergence over time among species of sympatric/syntopic populations or (2) through selective assemblage whereby species with less frequency overlap are more likely to form communities or persist together than when there is more overlap in signal characteristics (Chek et al. 2003; Lemmon 2007). These processes and the reduced risk of mismating or acoustic interference also might account for changes in call frequencies of individual species across their geographic range (Lougheed et al. 2006).

Possible examples of acoustic niche partitioning are provided by Drewry and Rand (1983). In three forest sites and one meadow site in Puerto Rico, heterospecific males vocalizing in the same assemblage employed advertisement calls differing in dominant frequency. Thus, the limited overlap in the frequency domain (i.e., spectral stratification) might facilitate communication in an acoustically crowded environment. Nevertheless, this study, as well as others (e.g., Garcia-Rutledge and Narins 2001), indicates there can, in fact, be considerable spectral overlap among some members of the anuran community. Even so, partitioning of calling on a fine scale and spatial separation coupled with frequency differences are likely to yield at least some benefits to signalers (Narins and Zelick 1988; Narins 1995). Given the heterogeneity of species composition over time at sites and across even relatively small distances, such situations could be fortuitous rather than functionally-derived (Wells 2007). Furthermore, if males of different species differ in size for ecological or environmental reasons, then call frequency differences also would be present (Gerhardt and Huber 2002; Wells 2007).

There have been attempts to use sophisticated analytical and statistical approaches to test for acoustic niche partitioning. Duellman and Pyles (1983) stated, using cluster analysis, that call characteristics of closely related Neotropical hylid frogs ($n = 39$ species in three forest communities) differed more in sympatry than allopatry. However, because a suite of call variables was used in the analysis, it is unclear whether the result is meaningful from the perspective of frequency-based masking interference. Furthermore, results are not particularly convincing: among pairs of species with the most similar calls, more were sympatric ($n = 9$) than allopatric ($n = 6$).

In a study that also focused on a particular taxon, Lemmon (2007) compared a suite of features of the calls of 15 species of *Pseudacris* in North America. Controlling for phylogenetic relationship, she discovered that signals used in sympatry differed more from one-another than those used in allopatry. She also found that "physiology based" temporal variables such as call length, call rise time, call fall time, call duty cycle, and pulse number were less constrained by phylogeny than were "morphology based" call variables of a spectral nature (e.g., dominant frequency and relative spectral energy in different parts of the call). While the results are consistent with the notion that these anurans partition acoustic space with respect to temporal variables, they do not support the hypothesis that acoustic interference due to spectral overlap has driven signal divergence or facilitated coexistence of species.

A potential criticism of much of the work used as evidence of acoustic partitioning is that it lacks a rigorous test of whether observed call differences among species are any greater than one would expect by chance alone (assessed by, for example, assembling a community of frog species by randomly plucking animals from the available species pool). In fact, in an earlier review, Gerhardt and Schwartz (1995) argued that future tests for acoustic partitioning needed to evaluate data relative to those generated from null models. Chek et al. (2003) performed such a much-needed analysis using their own and others' previously published data on 11 anuran assemblages, mostly from the New World tropics. They found statistically significant evidence of acoustic partitioning based on frequency for only three of these assemblages. There was also some support for partitioning based on call pulse rate which could reduce the probability that females would pair with heterospecific males. For the remaining eight assemblages, however, acoustic partitioning appeared absent. Similarly, Bourne and York (2001) obtained a mixed result for two frog assemblages in Guyana when comparing observed separation of spectral call features among species to those expected from a null model.

Vulnerability of sound signals to acoustic interference is related to the spectra of the signals (relative to "noise" sources) and the tuning characteristics of the receiver. Amézquita et al. (2006) studied both features in the Amazonian dendrobatid frog *Allobates femoralis* (formerly *Epipedobates femoralis*). Data were gathered at eight field sites, across a wide geographic range, and were used to test whether the threat of interference by *Ameerega trivittata* (formerly *Epipedobates trivittatus*), as well as the presence of other species, influenced the spectral form of the call and behavioral responses of territorial males to playbacks of calls of different frequency. *Ameerega trivittata* is a potentially important source of interference because the upper frequency region of its advertisement call overlaps the lower frequency region of the call of *A. femoralis*. The timing of vocal activity of the two species is similar when they occur together. In fact, masking interference may affect territory size and mating success of *A. femoralis* males (Amézquita et al. 2010). The researchers found that call frequency structure was not significantly affected by the presence of other anuran species in the various communities, but was correlated with geographic variation in body size. However, "receiver-response curves" of *A. femoralis*, an indicator of male sensitivity to different frequencies obtained from subjects' responses during the playback tests, exhibited an upward shift at lower frequencies at locations with as compared to without *A. trivittata*. The results thus fail to lend support to the spectral partitioning hypothesis, but are consistent with the notion that a most fundamental attribute of a communication system, receiver responses to different sound frequencies, is responsive to the problem of acoustic interference (see Sect. 6.4.1 in Chap. 6).

In summary, the data on assemblage composition, while sometimes suggestive, do not support the hypothesis that spectral separation of signals is a common means by which anurans mitigate the risk of signal masking. Perhaps it occurs only when the available signal space becomes crowded with species (Chek et al. 2003). Male anurans also have other "tools" at their disposal, especially in the form of adjustments in call or note timing. It is also possible that selection for spectral divergence is weakened by potential advantages of interspecific communication among heterospecific males (Cody 1974; Schwartz and Wells 1984a) and the availability of other call features to facilitate species discrimination by females (Gerhardt and Huber 2002). Features of the auditory system, such as frequency tuning (see Chaps. 6, 8, 10, 12) also could mitigate the interference problem. These features might enhance the contributions of any spatial cues (Schwartz and Gerhardt 1989; Bee 2007, 2008; Richardson and Lengagne 2010), or call temporal and/or small frequency differences that are present (see Chap. 6).

5.2.2 Signal Timing

The data on call timing illustrate dramatically how the sender in a communication system may respond to reduce the threat of acoustic interference. Indeed, timing of vocalizations is the most significant means by which male anurans reduce the threat of jamming (Klump and Gerhardt 1992). Such behavior occurs commonly among members of other taxa (e.g., see Chaps. 3, 7). In a dense and noisy chorus, although the region of male–male vocal interactions can change (Greenfield and Rand 2000), signaling adjustments typically are linked to calling by neighboring males, rather than distant males or the chorus as a whole. This is because the acoustic output of neighbors is relatively louder (and thus individual calls more easily detected; Gerhardt and Klump 1988), providing greater opportunities to exploit amplitude fluctuations than the din of the entire chorus (see Fig. 1 in Christie et al. 2010). The calls of neighbors also represent more potentially deleterious sources of signal interference (Schwartz 1993). Shifts in timing, by reducing signal overlap between males, can increase the likelihood that signals will be detected by receivers and, if so, that important aspects of signal structure will be discerned. This should increase the probability that nearby females will initiate phonotaxis and localize a signaler. The mating prospects of a male that adjusts call-timing also may be improved because he may more easily hear the calls of competitors if they do not overlap with his own calls (Schwartz 1987; Schwartz and Rand 1991; Narins 1992). This allows him to adjust his calling effort or aggressive calling accordingly. Thus, although mechanisms supporting call timing behavior (e.g., the neural circuitry) may incur some costs, signaling will be more effective (Ryan and Cummings 2005) and energy conserved.

5.2.2.1 Long-Term Timing Adjustments

In a natural chorus, rapid signal-timing shifts by neighboring conspecific or heterospecific males may require careful data analysis to detect. However, adjustments on a gross-time scale among different species can appear obvious to a human listener. This was Schwartz's experience during his first field season studying interspecific acoustic interactions of three hylid frogs in Panama. One of the species, the yellow cricket treefrog, *Dendropsophus microcephalus* (Fig. 5.1a; formerly *Hyla microcephala*), produced relatively loud multinote calls (mode = 106 dB Peak SPL at 50 cm; Schwartz and Wells 1984a) and was quite abundant at the study site. Acoustic output from choruses of this species tended to fluctuate dramatically. Males would call in bouts averaging about 17 s (range = 1.5–78.0, *n* = 259; Schwartz and Wells 1983a) and then quiet down for about 10 s (range = 1.5–58.5, *n* = 259). During these interbout intervals, males of the hourglass treefrog, *D. ebraccatus* (Fig. 5.1b; formerly *H. ebraccata*; a less loud and less abundant species at the study site) would call more vigorously than during bouts. Thus, for 4–5 h each night, there was a perceived alternation of calling by these two species. The vocalizations of *D. ebraccatus* and *D. microcephalus* overlap spectrally in the neighborhood of 3 kHz, and Schwartz and Wells hypothesized that the pattern of calling by *D. ebraccatus* had functional significance because it would reduce masking by calls of *D. microcephalus*. Analysis of recordings coupled with field playbacks (of chorus noise and filtered noise) to *D. ebraccatus* males demonstrated that the alternating activity pattern was due to an interaction of the treefrogs involving significant suppression of *D. ebraccatus* by *D. microcephalus* (Schwartz and Wells 1983a). Perhaps to compensate for their reduced calling when noise was broadcast within tests, males increased call rate (relative to pre-stimulus levels) during silent periods within tests. *Dendropsophus phlebodes* (formerly *Hyla phlebodes*), a species that often produces extremely long multinote calls, can inhibit calling by *D. ebraccatus* as well (Schwartz and Wells 1983b). Consistent with the masking hypothesis, *D. ebraccatus* reduced their proportion of multinote and aggressive vocalizations during periods of chorusing by *D. microcephalus* (Fig. 5.3) and filtered noise bursts of appropriate center frequency. These call types are given by males in response to the calls of nearby conspecifics. Furthermore, in phonotaxis tests, females of *D. ebraccatus* discriminated against conspecific advertisement calls overlapped by *D. microcephalus* chorus sounds relative to calls that were not overlapped (Schwartz and Wells 1983b).

Other researchers have described cases of interspecific inhibition of calling that resembles what Schwartz and Wells studied in Panama. Often there is an asymmetric relationship, with the species producing longer calls or vocalizing in bouts of greater duration inhibiting calling by the less loquacious species. For instance, in the now classic study by Littlejohn and Martin (1969), *Geocrinia victoriana* calls, which contain up to 100 notes, inhibited production of single-note calls given by *Pseudophryne semimarmorata* during playback tests. Males of *Crinia* (= *Ranidella*) *signifera*, also from Australia, could be quieted or induced to abandon

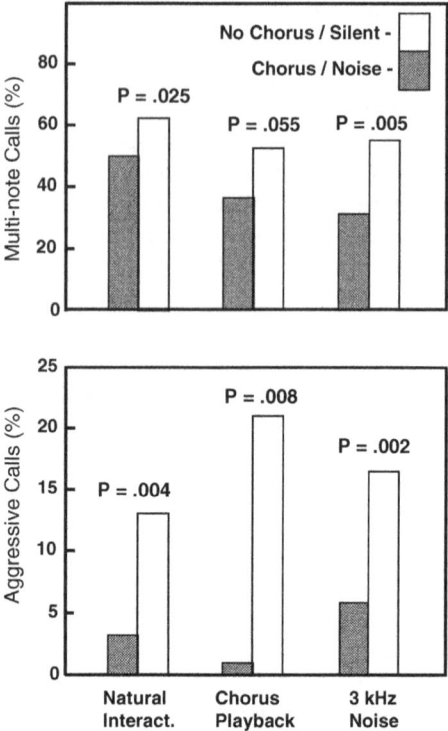

Fig. 5.3 Multinote and aggressive calling (given as % of total calls) by males of *Dendropsophus ebraccatus*. Data come from periods of relative quiet (e.g., interbout intervals of natural or recorded heterospecific chorusing; white bins) and exposure to chorusing by *D. microcephalus*, broadcasts of recorded choruses of *D. microcephalus* and broadcasts of one third-octave noise centered at 3 kHz (*shaded bins*). This is the dominant frequency of the calls of *D. ebraccatus* and the lower frequency band in the calls of *D. microcephalus*. Data for all playback amplitudes have been pooled (chorus broadcasts 85, 96, and 102 dB; noise broadcasts 80, 85, 90, 95, 100, and 104 dB). P-values are for a one-sided sign test (Data from Schwartz and Wells (1983a); figure modified from Fig. 7.20 in Wells (2007) with permission of the author)

calling sites in response to the longer calls of *C. parinsignifera* (Mac Nally 1982; Littlejohn et al. 1985). *Crinia signifera* produces long strings of much louder and more rapidly delivered calls than another member of this genus, *C. (= Ranidella) riparia*, and it has been hypothesized that the challenges posed by interference may constrain the latter's geographic distribution and occupation of preferred habitat (Odendall et al. 1986). The bullfrog, *Rana catesbeiana* (= *Lithobates catesbeianus*), and the green frog *R. clamitans* are common frogs in ponds in the northeastern USA with the males of the former species producing advertisement signals that acoustically dominate those of the latter species (bullfrog calls are typically louder and longer—often being given as a series of croaks = notes). Herrick (2013) found that although bullfrogs and green frogs in Connecticut have approximately the same summer breeding seasons (May–August) and call during

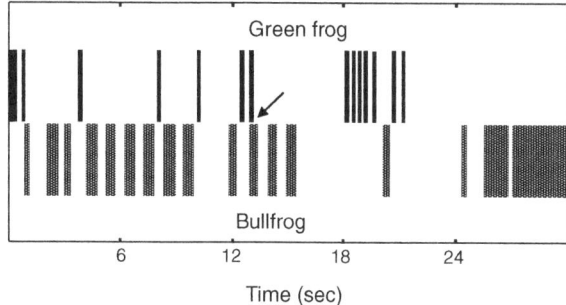

Fig. 5.4 Avoidance of call overlap by male green frogs of bullfrog calls in a Connecticut pond during a 30 s time interval. Just 1 of the 15 green frogs calls (indicated by arrow) exhibits overlap. Blocks and clear areas indicate the presence or absence of calls, respectively, in successive 100 ms periods and so resolution of individual calls was not always possible (Redrawn with permission from Herrick 2013)

the same hours of the night, *R. clamitans* avoided calling during bouts of *R. catesbeiana* croaking (confirmed by comparing overlap in randomized datasets with those observed). Male green frogs frequently inserted single calls in small gaps between the calls of nearby bullfrogs (Fig. 5.4). Different male bullfrogs also alternate bouts. During these interactions, reduced call overlap occurs among more widely separated individuals relative to that exhibited by clusters of males in closer proximity (Simmons et al. 2008; Bates et al. 2010). The adaptive significance of such spatial–temporal substructuring of the choruses is not certain. However, because overlap elevates signal amplitude and yields accentuated patterns of amplitude modulation that could more effectively stimulate the auditory system of females, it may be cooperative (Bates et al. 2010).

Acoustically signaling insects often create appreciable levels of background noise at times and places where frogs call (Narins 1995; Römer 1998; Gerhardt and Huber 2002). If the dominant frequencies of insect signals overlap those of syntopic anurans, one might expect the calling of insects to inhibit calling by frogs (Narins 1982; Zelick and Narins 1982; Schwartz and Wells 1983b; Penna and

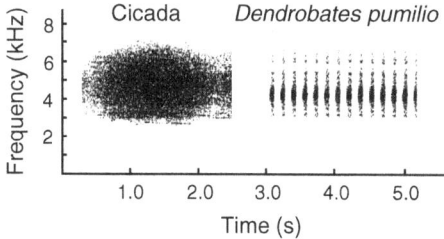

Fig. 5.5 Sonograms of a cicada chorus and the call of a male strawberry poison-dart frog (*Dendrobates pumilio*) recorded in the Arboretum of the La Selva Biological Station, Costa Rica indicating the frequency overlap of insects and the frog (Redrawn from Paez et al. 1993, with permission of the senior author and John Wiley and Sons Ltd)

Hamilton-West 2007). In fact, only two studies provide examples of such an interaction, both on the same frog species. At la Selva, Costa Rica, male *Dendrobates* (= *Oophaga*) *pumilio* call year round from defended territories on the forest floor. Cicadas also call throughout the year, and the broad spectrum of their choruses overlap the call frequencies of the frogs (Fig. 5.5). Paez et al. (1993) presented playbacks of calls to territorial males over a series of amplitudes during periods of both loud cicada chorusing and reduced cicada activity. Orientation changes, approach, and vocal responses clearly demonstrated the masking potential of the insect-generated noise. Wong et al. (2009) followed up on the earlier study with research on *D. pumilio* in Nicaragua. Consistent with Paez et al., calling by males was suppressed profoundly during playbacks of cicada sounds and also declined during those of other insects (a ground cricket and tree cricket). In addition to a reduction in calling rate, the pattern of calling (e.g., bout duration, percent time calling) changed selectively, depending on the type of stimulus.

Some syntopic species may exhibit little or no signal overlap if they concentrate their calling during different hours of the day or evening (Drewry 1970; Crump 1974; Bowker and Bowker 1979; Kuramoto 1980; Telford 1982; Drewry and Rand 1983; Given 1987; Shimoyama 1989; Bevier 1997). Temporal partitioning on this scale could have a number of explanations (e.g., phylogenetic, ecological, climatological, physiological, reproductive isolation, or chance; Bridges and Dorcas 2000; Gerhardt and Huber 2002; Oseen and Wassersug 2002; Wells 2007). Assessing whether such patterns represent an adaptation to reduce acoustic interference by some species is extremely difficult. In addition, *clearcut* and consistent patterns of such temporal disjunction seem relatively rare and typically involve just a subset of the community of anuran species (e.g., Fig. 24 in Telford 1982). For example, at the Thai site studied by Garcia-Rutledge and Narins (2001), hourly profiles of calling activity by males of eight species reveal a fairly crude degree of temporal segregation. Some species concentrate the bulk of their calling in the evening hours before midnight to about 2:00 AM and others either throughout the night, later or with multiple peaks of activity. Examination of the call spectra and calling locations (provided for one night) suggest that, absent fine-scale timing adjustments, call interference among many species is unavoidable. Species removal experiments or inspection of other areas missing one or more species could be particularly informative in studies of this type (see Hsu et al. 2006 for an example). If those species that, based on call spectra, would be expected to benefit by such absences, shift or broaden the time periods during which they call, this would be consistent with the idea that the threat of masking helps shape hourly patterns of vocal activity.

5.2.2.2 Fine-Scale Timing Adjustments

Pairs of males isolated from a chorus, or males in the midst of a chorus, may modify the timing of their calls or call elements relative to those of one or more individuals on a time scale of under 1 s (Zelick and Narins 1985; Schwartz and

Wells 1985; Grafe 2003; Martínez-Rivera and Gerhardt 2008). Although rapidly timed responses often result in call, note, or even pulse alternation, as well as leader–follower timing without interference, this behavior can sometimes produce signal overlap (Gerhardt and Huber 2002). At the *proximate* level, which form the interaction among males takes may be determined by features of the neural circuits associated with the rate of call generation, the extent to which call production is inhibited by auditory input, and the rate of recovery (for details of potential mechanisms see Klump and Gerhardt 1992; Greenfield, 1994; Gerhardt and Huber 2002; Klump and Gerhardt 1992 also review diagnostic experimental and analytical techniques). However, the details of such interactions *ultimately* are largely shaped by the preferences of females (Höbel and Gerhardt 2007)—although what it is about a particular timing pattern that influences females can vary, even possibly within the same species (Höbel and Gerhardt 2007). For example, although females of some species may show no preference for leading or following callers (Forester and Harrison 1987; Klump and Gerhardt 1992; Ibanez 1993), females of other species may *inherently* prefer calls that lead (Dyson and Passmore 1988a, b; Howard and Palmer 1995; Grafe 1996; Greenfield et al. 1997; Greenfield and Rand 2000; Bosch and Marquez 2002; Tárano and Herrera 2003) or follow those of a neighbor (Wells and Schwartz 1984a; Bosch and Marquez 2002). In some cases a leader preference, typically manifest when signals partially overlap, may reflect a true precedence effect, whereby the source of a following call or call element is localized at the source of the leading signal (Marshall and Gerhardt 2010). Female preference also may be associated with a perceived disruption of critical temporal information in the call (Schwartz 1987; Ibanez 1993), masking of the calls or notes of one male by another (Wells and Schwartz 1984; Grafe 1999), or possibly male quality (Richardson et al. 2008). In most cases, we lack data indicating which explanations account for particular forms of fine-scale call timing in a species. Accordingly, we will discuss what we find to be particularly interesting examples below with the caveat that the consequences of masking may or may not have been involved in shaping male behavior.

An impressive display of fine-scale call timing during playback tests was documented in males of the Coqui frog, *Eleutherodactylus coqui*, by Zelick and Narins (1985) in Puerto Rico. A stimulus was broadcast of variable-duration tones that were interspersed with silent gaps that were each just large enough for a male to place a "Co" and "Qui" note. The gaps were pseudorandomly distributed so that subjects would not be able to predict their occurrence in time. Nevertheless, males succeeded in placing their calls in the quiet intervals. Opportunistic behavior was also demonstrated in another experiment when test tone intensities were varied within 2.5 s test and control tone periods (Zelick and Narins 1983). Some animals evidently were able to discern differences in relative tone level as small as 4 dB and concentrated their calling in these less intense 1.5-s sound "windows." Transitory declines in background noise can trigger Zimbabwean males of *Hyperolius marmoratus broadleyi* to call, as demonstrated with playback tests using gap-containing tones (Fig. 5.6; Grafe 1996). Although males of not all species employ a signal-timing mechanism that allows them to exploit temporal gaps

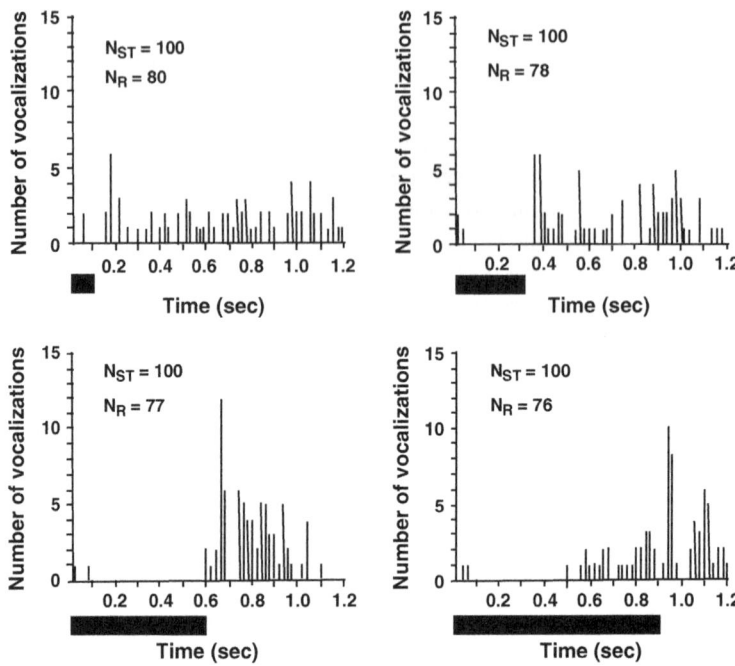

Fig. 5.6 Vocal responses of a male *Hyperolius marmoratus broadleyi* to tone bursts of different duration (indicated by black bars below each histogram; tone period = 1.2 s). Histogram bars (bin width = 20 ms) are for call onset times. N_{ST} = number of tone bursts presented, N_R = number of vocalizations by the subject during broadcast of the stimulus. A significantly ($P < 0.001$) greater percentage of calls were given during the quiet gap between tone bursts than expected (based on random calling) for all tone durations but 0.1 s (data for eight males) (Redrawn with permission from Grafe 1996)

(Klump and Gerhardt 1992), such behaviors are of obvious advantage if an objective is to reduce call overlap with the most potent threats of acoustic interference in a multimale chorus or communicate in an environment in which noise amplitude is modulated (Vélez and Bee 2010; see Chaps. 6, 7). Similar selectivity has been demonstrated in *D. microcephalus* (Schwartz 1993).

As described above, *D. microcephalus* frequently chorus in dense aggregations and cluster their multinote calls in bouts with those of other males. This on–off pattern of calling (often referred to as 'unison bout singing') appears to be a consequence of mutual stimulation of calling by males coupled with a need to stretch out energy reserves during the course of an evening's chorusing. Although experiments revealed that termination of bouts was likely not tied to acoustic interference (Schwartz 1991), this problem has shaped fine-scale aspects of male signal-timing. In fact, inspection of the call and note timing during pairwise interactions revealed that although calls frequently overlap, acoustic interference is largely absent! Rather, because male note timing is so precise, the notes of the interacting individuals interleave. Inhibition of note production by interrupting

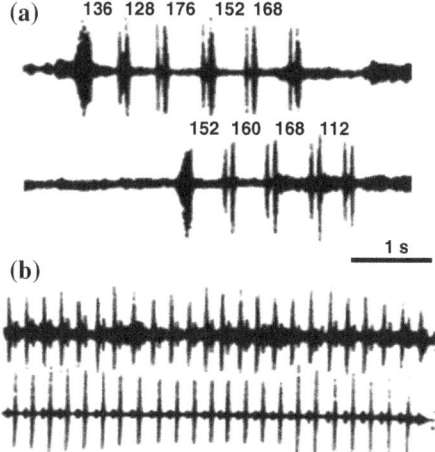

Fig. 5.7 Call overlap without acoustic interference in males of *Dendropsophus microcephalus* (**a**) and *D. phlebodes* (**b**) during pairwise interactions. *D. microcephalus* rapidly increase internote spacing when an interval is interrupted by the note of another male (time given in ms). *Dendropsophus phlebodes* mutually stimulate one-another to produce overlapping calls containing very long trains of notes (**a** Reprinted from Schwartz and Wells (1985) with permission of the American Society of Ichthyologists and Herpetologists. **b** Reprinted from Schwartz and Wells (1984b) with permission of the Herpetologists' League)

sound contributes to this achievement, with the individual notes of each male eliciting an increase in the duration of the internote intervals of his neighbor (Fig. 5.7a; Schwartz and Wells 1985; for a species in which males increase interpulse intervals during overlap see Martínez-Rivera and Gerhardt 2008). Schwartz (1991) found, using note-triggered interrupting stimuli, that inhibition does not begin to wane until interruptions reach durations of over 200 ms, more than twice the duration of an advertisement call note. For males to rapidly alternate notes they also must be sensitive to the drop in signal amplitude that accompanies the end of a neighbor's notes. That they are sensitive in this way was demonstrated using broadcasts of interrupting 200 ms notes incorporating a central gap. In such situations, test males began their next note before the interruption had ended (Schwartz 1993). The probability of call overlap should increase with the number of callers (Schneider et al. 1988; Schwartz 1993; Grafe 1996; Martínez-Rivera and Gerhardt 2008; Fig. 5.8) and a perspicacious reader will, no doubt, be wondering how interference is reduced in an aggregation of many males. Data gathered with an 8-channel call-monitoring system and from interrupting stimulus broadcasts revealed that males of *D. microcephalus* usually selectively adjust their note timing with respect to their one or two loudest (and closest) neighboring males but permit their call notes to overlap with the other, less potent, sources of interference (Schwartz 1993). What is especially impressive is that disparities in perceived intensities of neighbors can be very small (estimated at <2 dB peak SPL at the location of the receiver), yet differentially affect the timing response of a male.

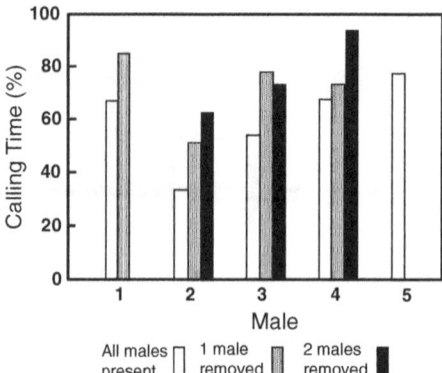

Fig. 5.8 The percent of the total calling time of each of five male *Dendropsophus microcephalus* that was not obstructed by calls of any of the other frogs in his group during a sequential male-removal experiment. Male 5 was first removed and subsequently male 1 was removed (Redrawn from Schwartz (1993))

Brush and Narins (1989) and Greenfield and Rand (2000) also observed selective attention to a subset of callers by males of *E. coqui* and the túngara frog, *Engystomops pustulosus* (= *Physalaemus pustulosus*), respectively. Work with the latter species indicates that the "rules" that govern calling behavior of males are responsive to chorus conditions, and the domain of attention depends on the number of neighboring males and the relative amplitude of their vocalizations at the receiver. Indications of dynamism (based on location within a group) were also observed with *D. microcephalus* in which more centrally located males attended to more neighbors than did males at the edge of a group. Curiously, at least when arranged around the periphery of an octagonal artificial pond, males of the eastern gray treefrog, *H. versicolor*, (Fig. 5.1c) exhibited greater call overlap with their immediate neighbors as compared to those further away. Males also showed statistically significant reductions in call overlap in two-male but not larger groups (Schwartz et al. 2002). Clearly, more work on the extent of this phenomenon among species of anurans and its mechanisms is warranted.

Although intentional overlap of calls would seem to be a poor calling strategy (Alexander 1975), there are situations in which it may be advantageous. For example, call overlap by males of the canyon treefrog, *H. arenicolor* could increase signal amplitude relative to nonoverlapped calls, and increase the likelihood that males calling near streams will be detected by gravid females (Marshall personal communication). This explanation is speculative and calls of males could become clustered in time because short-latency response facilitates matching calling effort of neighbors. Males of the neotropical *Smilisca sila*, studied by Tuttle and Ryan (1982), also prefer to advertise in particularly noisy locations at streams and near waterfalls. Calls of neighbors are answered with short latencies such that call overlap frequently occurs among two or more individuals. In this example, synchronous calling, as well as calling by flowing water, may be primarily an adaptation to reduce the hunting success of frog-eating bats. However, as is the case for the stream-breeding *Cochranella granulosa* (Ibanez 1991), the price tag of call synchrony could be lowered allure of males to females (Wells 2007).

In contrast to the cases of possible cooperation, call overlap among pairs of interacting *D. ebraccatus* may be a selfish attempt of individuals to mask important signal elements of a competitor. Wells and Schwartz (1984) found that males frequently answer a neighbor rapidly enough so that the primary note of their advertisement call overlaps a secondary note of the neighbor's calls. Choice experiments with females not only demonstrated that calls with secondary notes are more attractive than calls without these notes, but in such cases of overlap, the following male has the advantage. Evidently, disruption of the leading male's secondary notes by acoustic interference is more detrimental to his chances of attracting a mate than is disruption of primary notes for a following male. Whether the overlapping behavior evolved for this reason, or is simply a consequence of a call timing response that has another explanation, is not known. There is some recent evidence that by switching to aggressive calls, a leading male of *D. ebraccatus* can extend the end point of his calls beyond that of an answering male and eliminate his attractiveness deficit (Reichert 2011). Limited overlap (e.g., 10–25 %) of calls also can provide following males of African *Kassina fusca* with an advantage relative to leaders, due perhaps to forward fringe masking of the end of the leading call by the beginning of the following call. Although leaders can be favored (possibly due to backward fringe masking or a precedence effect) if followers answer so rapidly as to cause considerably more overlap (e.g., 75–90 %), following males rarely allow this to happen (Grafe 1999). Self-serving timing behavior by males has also been reported in *H. cinerea* by Höbel and Gerhardt (2007). Grafe argued that such behavior in *K. fusca* indicates that the form of call timing need not be an epiphenomenon rigidly prescribed by the proximate call timing mechanisms described by Greenfield (1994).

Grafe's (1999) work with *K. fusca* also demonstrates that the form of male timing relative to a neighbor can depend on structural characteristics of the neighbor's signal. Male responses are probably triggered by the onset of conspecific calls (facilitating overlap); however, rapid responses triggered by signal offset facilitated overlap avoidance to heterospecific calls and noise bursts. Playbacks of conspecific and heterospecific calls to males of *D. ebraccatus* indicated that rise time was a particularly influential signal feature, with males more effectively stimulated to give short-latency responses when amplitude onset was fast (Schwartz and Wells 1984a).

5.2.3 Signal Redundancy

A dense aggregation of male frogs is not only a venue in which background noise levels may be high, but also one in which competition for mates can be especially intense. Accordingly, if acoustic criteria are used during mate choice, the imperative of attracting a female will select for male vocal displays that are favored by females. These are often vocalizations or patterns of calling incorporating high levels of sound energy (Gerhardt and Huber 2002; Wells 2007) that may more effectively stimulate the auditory system (Ryan and Keddy-Hector 1992), be more memorable (Akre and Ryan 2010), or otherwise advertise desirable male attributes (Andersson 1994). Changes males make in response to greater competition that increase acoustic energy include elevation in rate of calling, increasing complexity of calls (e.g., by appending notes of one or more types; Wells and Schwartz 1984), and lengthening of calls or notes (e.g., Wells and Taigen 1986; Martínez-Rivera and Gerhardt 2008; for a discussion of shifts in call amplitude see Sect. 2.4). These kinds of changes can increase the serial redundancy (Brumm and Slabbekoorn, 2005) of male signals. Therefore, they can improve the odds that a potential mate can be detected using sound (Wells and Schwartz 1984; Pallett and Passmore 1988; Halliday and Tejedo 1995; Ronacher et al. 2000) and, if so, facilitate localization and even recognition or discrimination of callers using important signal features (Wiley 1983; Kime 2001; see Chap. 6). However, the changes in serial redundancy of the calls of conspecific and heterospecific males could themselves increase background noise levels and acoustic clutter and so, via positive-feedback among males in a chorus, increase the challenges for acoustic communication.

In this section, we describe some specific examples of frogs increasing the serial redundancy of their signals. There may be multiple advantages for males that increase redundancy in a chorus environment (Wiley 2006), and we usually lack data that would allow us to determine the adaptive significance of these responses. Other structural attributes also may increase the ability of receivers to detect or localize anuran calls (Ryan and Kime 2003). For example, frequency modulation may enhance detectability by concentrating energy at any particular moment in a relatively narrow bandwidth while traversing a wider range of frequencies over the duration of the call or note (Ryan 1985; Bosch and De la Riva 2004). This may not only be the case under circumstances where other species contribute to different frequency regions of a time-varying background noise spectrum, but also during out-of-phase signal overlap with conspecifics (Schwartz and Rand 1991). Anuran acoustic signals can incorporate multiple components (e.g., spectral, temporal) that facilitate species discrimination by females. Such simultaneous redundancy could be especially helpful for species in noisy multispecies environments (Wollerman and Wiley 2002). Unfortunately, noisy-environment related advantages of structural features are largely putative and speculation needs to be followed by focused investigations.

Perhaps the most common phonoresponse of male anurans to the calling of other individuals (sometimes even that of heterospecifics) is to increase the rate of calling or note production (Gerhardt and Huber 2002; Wells 2007), although such behavior is not universal (Harrison and Littlejohn 1985; Gerhardt and Huber 2002; Humfeld 2003; Tobias et al. 2004; Penna and Velásquez 2011). For example, in *D. microcephalus*, males increased calling rates relative to pre-stimulus levels during playbacks of conspecific and heterospecific stimuli (although with low note-number stimuli the result was significant for only some heterospecific call types;

Schwartz and Wells 1985). Male frogs tend to be more "permissive" than are females (Bernal et al. 2007), and responding to heterospecific calls would not be even a trivial mistake if it increases the probability of detection (c.f. Chap. 2). The signaling increases were less pronounced or absent for *D. microcephalus* males that had relatively high call rates before broadcasts (Schwartz and Wells 1985; Schwartz 1986). This suggests that males are little inclined to do more than is necessary to remain competitive (Gerhardt et al. 2000) or that constraints on further elevation of calling were present (Wells and Taigen 1989; Schwartz et al. 1995). Call rate increases could be accompanied by an increase in the proportion of multinote calls, especially if responses were given shortly after stimulus calls. During dyadic interactions, males of this species are predisposed to roughly match note numbers (Schwartz 1986) as they answer one another in overlapping calls with interleaving notes (see above). *Dendropsophus phlebodes* (Schwartz and Wells 1984b) and *D. microcephalus* (Wells and Schwartz 1984) also append notes to their calls following acoustic stimulation. The former species is especially impressive in that pairs of neighboring males can alternate over 20 notes in overlapping calls (Fig. 5.7b). The note matching observed in *D. phlebodes*, *D. microcephalus*, as well as other species (e.g., Arak 1983; Pallett and Passmore 1988; Jehle and Arak 1998; Gerhardt et al. 2000), suggests that the increase in note rate is primarily driven by competition for mates based on inherent signal attractiveness rather than because it improves detection of signals in a noisy environment. Changes in redundancy by males in response to stimulation by a *single* real or simulated competitor (whose calls would typically do little to impair detection of a signaler) also suggest that vocal changes are at least in part competition-driven. Close coupling in time of better matched calls also is consistent with this explanation, as the time period over which females assess males (e.g., cognitively evaluate relative to one another) may be relatively brief (Schwartz et al. 2004; Akre and Ryan 2010; Baugh and Ryan 2010). Interestingly, when two speakers where used in playback tests with male Australian quacking frogs, *C. georgiana*, the note-numbers of the responses of males suggested they added the notes from consecutive 4-note calls from different sources. Males also reduced note production when the second 4-note stimulus call was reduced in intensity. In this case, both behaviors jive with a noise-based explanation for calling adjustments, although the former summing response may be a product of a failure to segregate spatially discrete sound sources (Farris et al. 2002; Wells and Schwartz 2007; see Chap. 6)—perhaps due to experimental idiosyncracies (e.g., identical stimulus call frequencies; Gerhardt et al. 2000). Interestingly, in natural choruses, calls with more notes do not yield dependable mating advantages to male quacking frogs (Smith and Roberts 2003).

Hyla versicolor and *H. chrysoscelis* increase the duration of their pulsed advertisement calls while reducing call rates in response to the calls of other individuals in the chorus and during playback tests (Wells and Taigen 1986; Gerhardt 2001; Schwartz et al. 2002; Love and Bee 2010). The change in rate is typically compensatory such that "pulse effort" (number of pulses per call x call rate) varies little. In arena-based tests of discrimination, females of *H. versicolor* exhibit a robust preference for longer relative to shorter calls and the preference is often, *but not always*, maintained when pulse efforts are equal (Schwartz et al. 2001, 2008) or nearly so (Klump and Gerhardt 1987; Gerhardt et al. 1996). An advantage for long callers was reduced considerably when females were allowed to choose among up to eight males in an artificial pond or in 8-speaker choice test (with un-handled females) conducted at the edge of a natural pond (Schwartz et al. 2001). In fact, females discriminated little against all but the briefest vocalizations, and in the artificial pond, less than 10 % of the variance in male pairing success was explained by call duration. Schwartz et al. (2002) also discovered, through manipulation of chorus size, that, while adjusting call duration, males maintained their relative rank in the chorus for this call feature. Thus, they willingly produced calls of lower duration than they were capable of producing and so potentially failed to exploit an opportunity to move to a higher rank in the call-duration hierarchy. This behavior, together with some of the aforementioned observations (including those on call interference among neighbors mentioned in the section on call timing, an inability of males to perform temporal induction (Schwartz et al. 2010), and the 1997 results of Grafe on energetic substrate utilization) led Schwartz to propose two other explanations for the pattern of calling dynamism in gray treefrogs (Schwartz et al. 2008). Both hypotheses are relevant to problem of communication in a sound-cluttered environment (also see Chap. 6).

The first idea (the Interference Risk Hypothesis, IRH) proposes that even at lower call rates, producing longer calls in a dense chorus increases the chances that females will perceive a sufficient number of unobscured pulses and interpulse intervals to elicit phonotaxis (see Fig. 8 in Schwartz et al. 2001). This is a reasonable expectation because females discriminate strongly against both calls in which the inherent pulse structure is degraded by call interference (Schwartz 1987; Schwartz and Marshall 2006) and also very short calls (Gerhardt et al. 2000). The second idea (the Call Detection Hypothesis, CDH) proposes that longer calls, even at lower call rates, are more easily detected in background noise than are shorter calls. Why this might work and the relationship to the integration characteristics of the auditory system are discussed in Vélez et al. (see Sect. 6.3.4 in Chap. 6) and Schwartz et al. (2013).

Some data from the artificial pond choruses were consistent with the IRH. These were obtained from pair-wise comparisons of males in the same choruses who had nearly equivalent (within about 10 %; $\bar{x} = 4.6$ %) pulse efforts but very different call durations (>25 %; $\bar{x} = 57.1$ %). In these cases, the male producing longer calls had a greater total number of nonoverlapped pulses in his calls (beyond a six pulse putative attractiveness threshold) than the male producing shorter calls significantly more often (19:7; $P = 0.029$, two-tailed binomial test). However, a set of female discrimination experiments designed to specifically test whether males would realize an advantage by shifting to longer calls at lower call rates under chorus-like acoustic conditions failed to support predictions of the IRH (Schwartz et al. 2008). With respect to the CDH, lone males of *H. versicolor* calling in an artificial pond modified their calling behavior in the expected way when white noise, filtered to mimic the spectrum of a natural conspecific chorus, was broadcast. Increasing noise levels resulted in an increase in call duration and a decline in call rate and decreasing noise levels the opposite response (Schwartz et al. 2013). Love and Bee (2010) have obtained similar results with Cope's gray treefrog, *H. chrysoscelis*. However, longer calls were no more easily detected (assessed with single-speaker tests of phonotaxis) under noisy conditions than were shorter calls (both at equal and unequal call rates; Schwartz et al. 2013). It may therefore be that males use noise levels in the chorus as a proxy for the overall degree of vocal competition for mates. Adjustments in call rate and duration may be made to increase the *inherent* attractiveness of male's vocalizations, rather than to improve detection, even though the benefits may often be small.

Kime (2001) tested and rejected the hypotheses that call complexity increases the female response likelihood and reduces response latency for calls imbedded in noise in Northern cricket frogs, *Acris crepitans*, and the túngara frog, *E. pustulosus*. Both of these species give vocalizations with repeating elements, and their number varies with the social environment. In the cricket frog, pulsatile advertisement calls are given in "call groups" and individuals add pulses to calls and calls to call groups in response to the calls of other males. Pulses also may be aggregated into two rather than one "pulse group." These complexity changes may be associated with male–male aggression (Wagner 1989), and they also modulate relative attractiveness to females (Kime et al. 2004). The latter is also the case in túngara frogs (Ryan 1985), in which males append secondary chuck notes to introductory whines during interactions. Kime (2001) employed two signal-to-noise ratios with túngara frogs and one with cricket frogs. In addition, the calls compared for cricket frogs (one versus two pulse groups) had identical numbers of pulses. It would be especially interesting to expand the study by estimating masked thresholds of different calls in both species and to test calls with different numbers of pulses in cricket frogs.

5.2.4 Changes in Call Amplitude

When people converse, an almost reflexive response to loud background noise is to increase the amplitude of their voices. This behavior is known as the 'Lombard effect,' and such upward adjustments in signal amplitude have been reported for a range of other vertebrate species (Brumm and Zollinger 2011 and references therein, see Chaps. 7, 9). To date, the phenomenon has not been reported in fishes, reptiles, or insects (Brumm and Zollinger 2011; see Chap. 3). Given that a high level of chorus noise is the status quo for many species of anurans, it seems that shifts in call intensity should be one of the many possible adaptive responses to this sonic milieu. Indeed, males of three species of leptodactylid frogs, *Leptodactylus albilabris* of Puerto Rico (Lopez et al. 1988) and *Eupsophus calcaratus* (Penna et al. 2005) and *E. emiliopugini* (Penna and Hamilton-West 2007) of Chile, are capable of adjusting the amplitude of their calls under experimental conditions. Their responses therefore appear consistent with such adaptive behavior. However,

Fig. 5.9 Mean ± SE call
amplitude of male *Hyla
chrysoscelis* as a function of
the playback level of noise
filtered to resemble the long-
term spectrum of conspecific
choruses. Values did not
differ significantly (Redrawn
with permission from Love
and Bee (2010) and Elsevier)

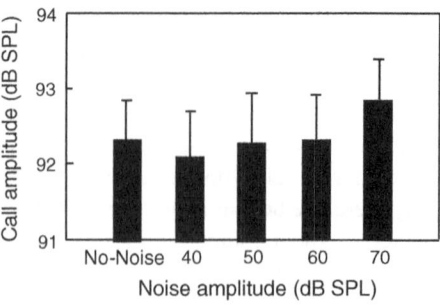

methodological details of these studies cast doubt on whether a true Lombard effect was witnessed (Love and Bee 2010). For this reason, and to help test the hypothesis that "voice amplitude adjustment" constitutes a "generic vertebrate response to coping with environmental noise", Love and Bee (2010) performed playback experiments using males of Cope's gray treefrog, *H. chrysoscelis*. Computer-generated noise digitally filtered such that its spectrum closely resembled that of a natural chorus was broadcast to subjects at 40–70 dB SPL (10 dB steps in random order). Males clearly responded to changes in noise amplitude by adding pulses to their calls and reducing call rate. This is a characteristic response of gray treefrogs to an increase in chorus density (Gerhardt 2001; Wells and Taigen 1986; Schwartz et al. 2002). However, the results also demonstrated, unambiguously, behavior inconsistent with the Lombard effect (Fig. 5.9). Mechanistic constraints may be one explanation for the findings. Love and Bee also suggested that species of anurans, such as gray treefrogs, in which males are under intense selection to maximize call energy content, may have little freedom to change call amplitude. Accordingly, they suggest that research on the Lombard effect continue with a variety of species including those where males advertise in more rarified choruses and use lower amplitude vocalizations.

5.2.5 Seismic Signaling

Another way that anurans could potentially communicate when airborne signals are subject to acoustic intereference is to use a quieter channel via substrate-borne vibrations. Peter Narins, Edwin Lewis, and their students and colleagues have investigated this possibility in the white-lipped frog, *L. albilabris*, in Puerto Rico. Males of this species can be found over much of the island, including areas populated by many other species of frogs. Thus, the environment in which males advertise may render information transfer via airborne sound difficult. The frogs often vocalize from burrows or depressions on moist ground such that the accompanying expansion of the vocal sac causes it to strike the substrate (Lewis et al. 2001). Furthermore, males are extraordinarily sensitive to an approaching observer. Narins (1990) reported that even tapping the ground with a finger can cause a male 5 m away to cease calling! These observations provided the impetus for an extraordinarily challenging and technically sophisticated effort to identify and quantify the necessary components of a seismic communication system in the white-lipped frog. The researchers employed a geophone array to record the substrate-borne vibrations produced when the frogs vocalized (Fig. 5.10a; also see Fig. 1 in Lewis and Narins 1985), constructed a "thumper" to present seismic signals to males in the field (see Fig. 17 in Lewis et al. 2001), and made electrophysiological recordings from neurons innervating the sacculus (an organ of the inner ear) of subjects within a custom room isolated from all but infinitesimal environmental vibration. Consistent with the seismic communication scenario, the spectrum of male thumps produced during calling is well matched to the vibratory frequencies to which low-frequency neurons are most sensitive. Amazingly, the saccular afferents were roughly 10 times as sensitive as similar fibers in the bullfrog, *R. catesbeiana* (Narins and Lewis 1984). No data have been published on female responses, but during thumper playbacks, some males modified their call timing. However, such responses were most pronounced only when airborne thumper sounds were masked by broadcast of noise (Fig. 5.10b; Lewis et al. 2001). This suggests that seismic communication becomes biologically meaningful when it most needed; that is, when background noise levels preclude the

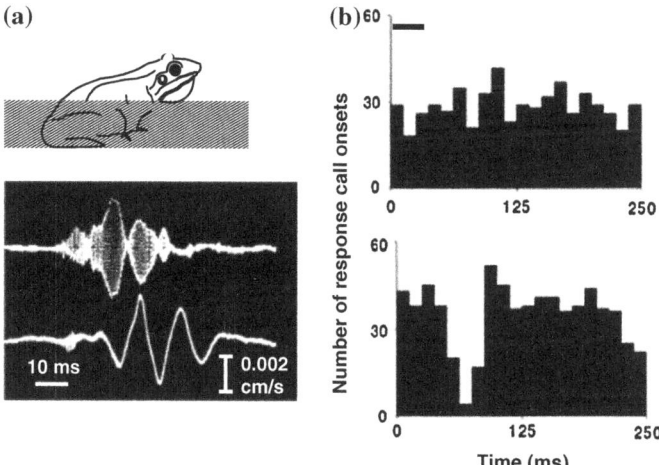

Fig. 5.10 a Illustration of a white-lipped frog (*Leptodactylus albilabis*) calling in a depression in the ground. The upper trace below the male shows the sound pressure waveform of a chirp call recorded with a microphone. The lower trace shows the vibrational velocity waveform (*vertical component*) of the corresponding seismic thump recorded with a geophone. Recordings were made about 1 m from the male. **b** Histograms compiled over the stimulus period showing the distribution of chirp timing of a male in response to playback of thumps produced by an artificial "thumper" (*upper panel*) or thumps played back together with airborne masking noise (*lower panel*). Horizontal bar indicates timing and approximate duration of the thumps. Note the gap in the lower panel reflecting the propensity of males to postpone chirping for approximately 30 or more ms starting about 30–40 ms after the onset of the stimulus. This indicates a change in chirp timing with masking noise (Reprinted with permission from Lewis et al. (2001) and by Oxford University Press. Frog (redrawn with the author's permission) based on Narins (1990))

more conventional form of information transfer. Because substrate-borne Rayleigh waves produced during thumping would travel more slowly than the simultaneously generated airborne signals, when noise levels are lower, theoretically *L. albilabris* could use time of arrival differences to localize callers (Narins 1990). Whether this is the case, is not yet known. There are reports on other species suggesting vibratory communication or at least generation of vibratory signals (see Caldwell et al. 2010 and references therein). However, we are not aware of the extent to which these frogs might use substrate-borne vibrations specifically to facilitate communication in noisy choruses.

5.3 Responses to Abiotic Environmental and Anthropogenic Noise

Although their function is not understood (Wells 2007), some species have been reported to produce "rain calls" in response to such precipitation or in anticipation of a storm (Bogart 1960). However, as compared to noise generated by other organisms (and in particular anurans), the effect of noise produced by nonliving entities on communication by frogs has received relatively little attention. Very recent work has begun to address this deficiency with a growing focus on noise created by machines. Below we discuss findings related to potential natural abiotic sources of masking (e.g., wind, rain, and flowing water) as well those associated with human activities.

5.3.1 Noise of Wind, Rain, and Flowing Water

Although observation-derived data based on activity patterns indicate that abiotic environmental noise can *correlate* with reduced vocal activity (e.g., Weir et al. 2005), research employing noise playbacks are necessary to convincingly demonstrate causation. Penna et al. (2005) and Penna and Hamilton-West (2007) performed such studies with *Eupsophus calcaratus* and *E. emiliopugini*, respectively using wind, rain, and sea surf noise. Males of these leptodactylids vocalize from within burrows that they create within bog vegetation of temperate zone forests of southern Chile. Noise generated by other species of frogs is absent or low, but abiotic noise may be a significant source of acoustic interference. Penna et al. (2005) found that broadcasts of creek and rain noises of intermediate amplitude (66 dB RMS SPL, fast weighting) elicited an elevation in call rate in *E. calcaratus*, which could be further elevated by addition of call playbacks. Creek noise also caused males to increase call duration. What is particularly intriguing is that similar playbacks to *E. emiliopugini* failed to elicit such responses to abiotic noise (and call rate could even decline with increasing noise intensity during call playback). Penna and Hamilton-West speculated that the difference among species is related to natural differences in abiotic noise levels or differences in call intensities. When *E. emiliopugini* breeds, abiotic noise levels are, on average, 8–9 dB SPL lower although its calls are about 15 dB SPL more intense as compared to the other species. Thus, the behavior of *E. calcaratus* could reflect adaptation to counter greater vulnerability of its communication to background noise (Penna and Hamilton-West 2007).

Rapidly flowing water, in the vicinity of streams and waterfalls, typically creates an acoustic milieu that has the potential to mask the calls of nearby animals. In contrast with some other kinds of abiotic noise (e.g., wind, rustling of plant parts, and thunder), this kind of noise is relatively stable in amplitude. Thus, animals attempting to communicate do not have the option of taking advantage of short-term fluctuations in noise level via rapid shifts in the timing of their signals. Rather, the form of acoustic signals may have evolved in ways that make detection by receivers more likely. Use of special visual signals also may be utilized to transmit information. Here we will mainly discuss the putative contributions of call spectrum, form, and delivery pattern to acoustic communication near flowing water and only briefly address visual signaling.

5.3.1.1 Frogs That Employ Ultrasonic Signaling

An especially fascinating discovery in the past decade is that frogs of some species produce purely ultrasonic calls, or those with ultrasonic harmonics, and communicate using these very high signal frequencies (Feng et al. 2006; Arch et al. 2009). The principal selective agent appears to be broadband sound produced by rapidly flowing water, such as found in waterfalls (Fig. 5.11; Narins et al. 2004; Arch and Narins 2008). This kind of environmental noise has a disproportionate amount of energy at low frequencies, so masking of calls potentially can be reduced, and the signal-to-noise ratio increased, by using frequency channels that are shifted up relative to background noise. This requires that the auditory system of the animals be sufficiently sensitive to high frequencies within the range audible to humans and ultrasound. Auditory system sensitivity to ultrasound has been demonstrated using behavior, neurophysiology, anatomy, and laser Doppler vibrometry (Feng et al. 2006; Feng and Narins 2008; Gridi-Papp et al. 2008; Arch et al. 2009) for three ranid species in Southeast Asia, the concave-eared torrent frog, *Odorrana tormota* (formerly *Amolops tormotus*), the Hole-in-the-Head Frog, *Huia cavitympanum*, and the large odorous frog, *R. livida* (= *O. graminea, O. livida*). The most work has been done on the first two of these species, and we discuss them below.

Production of the spectrally complex "bird-like melodic" calls of *O. tormota* depends on an idiosyncratic vocal cord anatomy that can exhibit nonlinear oscillatory dynamics (including chaos) (Suthers et al. 2006; Feng and Narins 2008). This sonic complexity makes possible patterns of call variation (Feng et al. 2009a) that facilitate individual recognition (Feng et al. 2009b). *Odorrana tormota* calls are composed of a mix of frequencies and are not purely ultrasonic (Fig. 5.11).

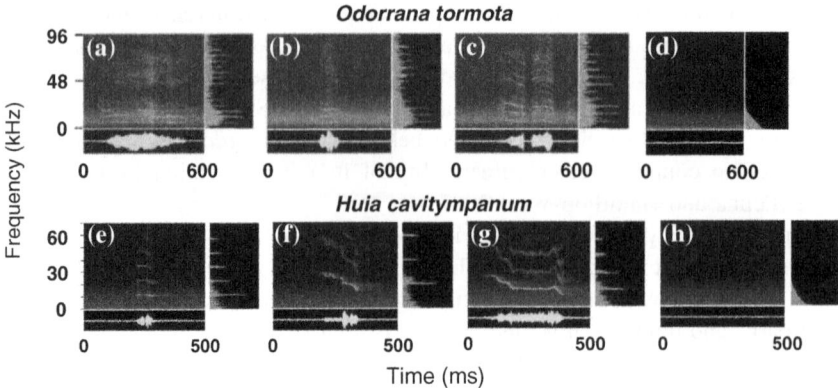

Fig. 5.11 a–c Waveforms (*bottom panel*), spectrograms (*upper panel*) and average amplitude spectra (*right panel*; relative amplitude scale −90–0 dB) of calls of three male *Odorrana tormota*. **d** Same for a recording of the background noise (absent calling frogs) in the vicinity of a creek where this species is found in China. The noise actually has a flat spectrum below 10 kHz, but this is not visible because the creek was recorded using an ultrasonic microphone (high-pass cutoff frequency of 15 kHz and 10 dB per octave roll-off). Modified from Feng and Narins (2008) (frog calls) with permission of the senior author and Feng personal communication (creek panels). **e–g** Displays for calls of *Huia cavitympanum* and **h** ambient background near river adjacent to calling sites of males in Borneo. Recordings made with a wideband microphone (Modified from Arch et al. (2008) (Figure 1a, 1b, 1c, 1f) with permission of the senior author and Royal Society Publishing)

Rather, the bulk of spectral energy is in the audible range of humans (dominant frequencies: 5–7 kHz) and is augmented by energy at high frequency harmonics. Because high frequencies suffer from greater excess attenuation than low frequencies (Ryan and Kime 2003), these dominant frequencies may enable the frogs to communicate acoustically over longer distances than otherwise would be possible. Alternatively, a response to selection favoring the use of calls with more energy in higher frequencies than presently occur may be underway but not yet complete. Developmental or fitness-related constraints on the evolution of morphological attributes that would facilitate production of calls with more high frequency energy are other nonmutually exclusive possibilities. The use of high frequencies may pose additional problems for males. The higher frequency call types less susceptible to masking are also those that are less individually distinctive (their fundamental frequency is above the frequency expected to benefit from vocal tract resonances, the source of much of the individual variation). Therefore, there may be a tradeoff between recognition and detection by receivers in the vicinity of flowing water (Feng et al. 2009a).

Although frogs appear to be most sensitive to frequencies relatively close to the dominant frequency of advertisement calls (Gerhardt and Schwartz 2001), recordings of both auditory evoked potentials and single units in the auditory midbrain (torus semicircularis) of *O. tormota* revealed responses extending well into the ultrasonic range (Feng et al. 2006). Short-latency vocal responses were elicited during playback tests with calls high-pass filtered to restrict sound energy to high frequency harmonics. A curious part of the story is that *O. tormota* exhibits sexual dimorphism in ear anatomy, with females lacking the recessed tympanum of males. Thus, females may not be sensitive to ultrasound (Shen et al. 2011). If this is so, then why is ultrasonic communication important in male–male inter-actions but not in male–female communication (Feng and Narins 2008)? Possibly females assess potential mates at short distances, where masking by water-gen-erated sounds is not a major problem. Females also might use visual cues to select mates. When gravid, females produce a high frequency courtship call with a fundamental frequency about 2–3 kHz above that of the males' calls (Shen et al. 2008). There also are ultrasonic harmonics present in this signal. Males answer the female call with calls with fundamental frequencies (6.2–7.8 kHz) that are posi-tively correlated with the level of ambient noise. This behavior suggests a dynamic response of males to reduce masking of their answers to females (Shen et al. 2008). During phonotaxis experiments using female calls, males displayed a remarkable ability to localize the sound source and exhibited mean jump angle errors of less than 1 degree. This is an order of magnitude better than the performance of other species of anurans and may be connected to their sensitivity to high frequencies.

Unlike males of *O. tormota*, which use ultrasound as well as lower frequencies, the dominant frequencies of the vocalizations of *H. cavitympanum* occupy just the high frequency portion of the spectrum (Fig. 5.11). Some calls (14 % of recorded samples) in the repertoire are purely ultrasonic (Arch et al. 2008; 2009). Never-theless, the variation in dominant frequency among calls could insure that the active space for communication by males is less limited than might otherwise be the case (Arch et al. 2008). In playback tests using ultrasonic calls, males responded similarly to playbacks of audible, lower frequency calls, by elevating their call rates. Thus, based on current knowledge, this anuran is unique among nonmammalian vertebrates in its use of purely ultrasonic signals in intraspecific communication (Arch et al. 2009). Like *O. tormota*, *H. cavitympanum*, as its name indicates, has a recessed tympanum. In contrast to *O. tormota*, however, sensitivity to very high frequencies is not dependent on closure of the Eustachean tubes (Arch et al. 2009). In spite of their differences, these two species provide a fascinating likely example of how background noise can be a potent selective agent and drive convergent evolution. Species using ultrasound may also benefit through an improved ability to localize sound sources and also through energetic savings resulting from the better match of call frequency wavelengths to the diameter of the vocal sac (Ryan 1988; Prestwich et al. 1989; Arch et al. 2009).

5.3.1.2 Other Frogs that Call near Rapidly Flowing Water

The evolution of spectral characteristics of a less extreme nature than those of ultrasonic calls could also be a result of selection to reduce masking by the noise produced by flowing water (but see Vargas-Salinas and Amézquita, 2013). For example, the advertisement vocalizations of the Brazilian torrent frog, *Hylodes asper*, are of narrow bandwidth and higher frequency ($\sim$5–6.5 kHz dominant frequency) than the background abiotic noise (Haddad and Giaretta 1999). The Indian stream frog, *Micrixalus saxicolus* has a high frequency advertisement call that can be detected by humans near loud fast-flowing water (Krishna and Krishna 2006). Species might also evolve lower call frequencies in response to stream-generated noise, as was recently suggested following analysis of data available for 116 east Australian frogs. Stream-breeding hylids use call dominant frequencies (mean = 1.7 kHz, n = 16) that are lower than those of pool-breeding hylids (mean = 2.7 kHz, n = 32) when body size is accounted for (Hoskin et al. 2009). The reported average dominant frequency of stream noise was 2.5 kHz (obtained from Goosem et al. 2007).

Ranid frogs in the genus *Staurois* from Borneo to the Philippines often are found near fast-flowing water. Visual signaling, that may include use of noncalling deployment of the vocal pouch, foot-flagging, and arm-waving, constitutes an important part of their behavioral repertoire. Boeckle et al. (2009) have recently made the case that the high frequency calls of some of these, and other, species are tied to the distinct acoustic properties of their habitat. *Staurois latopalmatus*, for example, has a call dominant frequency greater than 5 kHz. Sound pressure levels of its calls (at 2 m from males) were higher than those of the noise generated by waterfalls when determined at the dominant frequency of the calls. Similar results have been recently obtained with *S. parvus* by Grafe et al. (2012).

Signal form and pattern of delivery also may enhance detection and localization of calling males near streams and waterfalls. Calls or call notes tend to be brief, repetitive, and often narrow-band (Wells 2007), and similarities in signal structure may cross vertebrate classes. Dubois and Martens (1984) described resemblances between the calls of some Nepalese species of *Rana* and a warbler (*Phylloscopus magnirostris*) that conform to these characteristics and thus render them easy to hear in the vicinity of Himalayan torrents. Therefore, the use of high frequency signals, although advantageous because it reduces spectral overlap with the background noise, is, as mentioned above, not required for acoustic communication near flowing water, and other signal features may be adequate or helpful. For example, the calls of the aforementioned Nepalese *Rana* spp. have most of their energy near 1–2 kHz, and the calls of *Smilisca sila*, discussed in a previous section, have peaks in call energy below 3 kHz (Tuttle and Ryan 1982). Whether call synchrony among males of *S. sila* enhances the ability of females to detect males near flowing water is not known.

5.3.2 Anthropogenic Noise

During the twentieth century, not only did the size of the human population more than triple but the internal combustion engine invaded the land, sea, and air. The developing world has not been immune to the onslaught. Recently, there has been growing attention focused on how animals are affected by noise generated by humanity's raucous contraptions (Rabin and Greene 2002; Warren et al. 2006; Slabbekoorn and Ripmeester 2008; Barber et al. 2009; Van der Ree et al. 2009–2010; Slabbekoorn et al. 2010; see Chap. 14). In particular, researchers are interested in whether anthropogenic noise can lead to changes in species' distributions, affect reproductive output, alter competitive relationships, increase risk of predation, and have implications for conservation. From the perspective of communication, we need to know the extent to which such noise impairs the effectiveness of information transfer using sound, and, if so, whether acoustically signaling animals are altering their behavior in ways that might mitigate the deleterious effects of anthropogenic noise. Anurans are a particularly interesting group in which to address this issue because loud noise is a natural reality for many species and, for which, as described herein and in the chapter by Vélez et al. (Chap. 6), they possess a number of communication-related adaptations.

Only a handful of studies in peer-reviewed journals have investigated the potential impact of man-made noise on anuran behavior (for an unpublished study see Barrass (1985), and for a report published by James Cook University on behalf of the Australian Government's Marine and Tropical Sciences Research Facility, see Goosem et al. (2007). Five of these address signaler responses and will be discussed briefly here. Bee and Swanson (2007) presented data on how traffic noise could influence phonotaxis to advertisement calls by female receivers and their work is discussed by Vélez et al. (Chap. 6).

The study by Sun and Narins (2005) was the first research published on the effect of anthropogenic noise on vocalizing frogs. This study is particularly interesting because it demonstrates species differences in behavior that reflect responses to the temporally fluctuating noise background of natural choruses. The species of main focus was *Rana taipehensis*, a frog already rendered vulnerable or threatened over parts of its range in Taiwan and Hong Kong. The study site in Thailand was a small pond, and frogs were exposed to noise produced by automobiles and motorcycles from a road about 20 m away. The site also was subjected to noise from aircraft flying overhead. During exposure to flight noise (mean duration 91.4 s), males of *R. taipehensis* increased their rate of calling. This was in contrast to males of two other species common at the pond (*Microhyla butleri* and *R. nigrovittata*), which lowered their call rates—often quite dramatically (see their Fig. 2). Playback tests, which investigated the effects of relatively brief (29.7 s) episodes of motorcycle noise, revealed call rate responses similar to those observed with aircraft noise for *M. butleri* and *R. nigrovitatta*. However, in this situation, *R. taipehensis* males postponed elevation of their call rates until the stimulus ended. Males of *R. taipehensis* also altered the types of calls they produced as a function of the ambient acoustics.

Sun and Narins interpreted their data as follows. Low frequency noise from airplanes inhibits calling by most species at the pond. The elevation of call rate in *R. taipehensis* observed under such circumstances was a response to the accompanying drop in the biotic noise level (which would be a decline of more than 10 dB SPL) in the males' vicinity, rather than the aircraft noise per se. The difference in behavior among this and other species, in part, also may be due to the use of calls of higher frequency by *R. taipehensis*; such calls have less spectral overlap with the engine noise of airplanes than calls of the other species. Motorcycle noise also inhibited calling by heterospecifics, to which *R. taipehensis* reacted. However, because the duration of the playback stimulus was much briefer than the average duration of aircraft noise, the elevation in calling by *R. taipehensis* in response to the drop in chorus noise typically occurred after the motorcycle stimulus ended. Thus, the differences in behavior observed among species may be tied to differing susceptibilities to acoustic interference (related to call frequencies) and a shared ability to take advantage of temporal fluctuations in noise level produced in natural choruses.

Males of the hylid *Dendropsophus triangulum* advertise for mates in open and forest edge habitats in Amazonia. In the study sites of Kaiser and Hammers (2009) near Iquitos, Loreto, Perú, the treefrogs also occasionally are exposed to noise from motorcycles and aircraft flying overhead. These researchers quantified the effects of motorcycle noise (both continuous and intermittent), human song, and conspecific chorus noise using playback tests. Playback intensities were higher (75 dB SPL) than the anthropogenic noise levels to which the treefrogs are normally exposed (<60 dB SPL) at the sites. Therefore, the findings demonstrate responses that are possible (and could likely occur at other venues) as opposed to those that occur more commonly at the study sites. In response to all experimental stimuli, there was a dramatic increase ($\sim 2 \times$) in rate of advertisement calling with responses similar to those observed in response to chorus noise (continuous motorcycle versus chorus). During exposure to the intermittent engine sound, males concentrated more of their calling in the 15 s noise bursts than in the 30 s quiet intervals between bursts. Thus, responses to punctuated noise were opposite to what one might expect based on the behavior of some other tested species such as *D. ebraccatus* calling close to groups of chorusing *D. microcephalus* (Schwartz and Wells 1983) or *P. semimarmorata* near *G. victoriana* (Littlejohn and Martin 1969). The authors also report that observations of syntopic species that were not tested indicate heterogeneity in response during human-generated noise, with some species increasing and others decreasing calling.

It is not clear whether the behavior of these frog species to anthropogenic noise has biologically significant consequences. In the case of *D. triangulum*, males rapidly returned to baseline levels of calling following stimulus presentation. Hence, there were no prolonged modifications of calling rate to noise exposure. Neither the impact of playbacks on subjects' call complexity (e.g., note number) nor data on fine-scale timing of calls relative to calls other males were presented, nor were effects on female behavior tested. These could potentially influence mating success of some males. Kaiser and Hammers proposed that species in which males naturally elevate call rate in response to chorus noise may be more likely to tolerate noise in areas subject to human disturbance.

Lengagne (2008) investigated the responses of male European treefrogs (*Hyla arborea*) to playbacks of traffic noise in an experimental tank. Presentations of anthropogenic noise were made at relatively low (72 dB SPL) and high amplitude (88 dB SPL) and with and without the simultaneous broadcast of a recording of a natural chorus (at ∼ 74 dB SPL). Males showed no tendency to move toward or away from the sources of traffic noise. Nor was there a tendency to change call frequency or other aspects of signal structure. However, Lengagne found that the acoustic conditions were quite important in determining whether males altered their calling effort and calling bout duration. When the chorus was silent, males reduced their effort and bout duration from pre-stimulus (72 dB traffic noise) to post-stimulus periods. With louder traffic noise, there was an additional dramatic decline in calling activity during as well as after stimulus presentation. In contrast, when Lengagne broadcast the chorus background together with traffic noise, there were no significant changes in calling behavior unless the traffic background was very loud. The results suggest that we cannot assume that anthropogenic noise will elicit significant modifications in male vocal advertisement or affect individual male fitness. However, where noise levels are high and relatively steady (rather than strongly modulated), and choruses relatively small (and thus of relatively low amplitude), human-generated noise has a potential to alter male vocal behavior in important ways.

Green frogs (*R. clamitans*), northern leopard frogs (*R. pipiens*), eastern gray treefrogs (*H. versicolor*), and American toads (*Bufo americanus*) are extremely common North American anurans, and it is not unusual to find them in well-populated areas of the United States and Canada. Cunnington and Fahrig (2010) studied vocal behavior of these species in venues in southern Ontario with low ($\bar{x} =$ 43.8 dBA, SD = 2.66) and high levels ($\bar{x} =$ 73.2 dBA, SD = 4.91) of traffic noise and in response to playbacks of traffic noise of at least 76 dBA at the location of test subjects. Males of the two ranid species, whose call spectra render calls more susceptible to masking by the low-frequency sound energy in traffic noise than calls of the toads or treefrogs, exhibited higher dominant frequencies at sites with the higher anthropogenic noise levels and when exposed to broadcasts of such noise. Curiously, call amplitude was reported to be sometimes lower under noisier conditions and so could increase the vulnerability of calls to masking. No significant changes in dominant frequencies or call amplitude were observed in the remaining two species. All species but *B. americanus* vocalized at lower rates when exposed to traffic noise. Although the researchers did not measure call durations, it seems likely that the gray treefrogs increased duration because call rate and pulses per call typically change in opposite directions in this species (e.g., Wells and Taigen 1986). If so, the male gray treefrogs would have responded qualitatively to traffic noise as they do to the calls and chorus noise generated by conspecifics.

Recent work has demonstrated that the frequency of the song of some species of birds differs between urban environments with relatively high and low levels of ambient noise (Nemeth and Brumm 2010 and citations therein). Parris et al. (2009) tested whether this is true for two species of Australian frogs, *Litoria ewingii* and *Crinia signifera*. Traffic noise has a frequency profile with the bulk of sound energy below 2 kHz, and Parris et al. predicted that, should it occur, males would show elevated, rather than lowered, frequencies of their advertisement call to reduce acoustic interference. The data, obtained from a large number of sites that differed in the level of traffic noise, indicated that for *L. ewingii* the dominant frequency of calls increased by 123 Hz over the range of site noise levels (equivalent to 4.1 Hz per dB of noise), but results for the other species were inconclusive. Some of the observed difference in frequency was due to a reduction in the size of males between sites and additional information will be required to pin down the association between noise and SVL. Moreover, there was no evidence that frequency differences resulted from short-term behavioral responses, as has recently been demonstrated in birds (Gross et al. 2010). Putative advantages to males of using calls of higher frequency include more effective communication to both females and males as a result of reduced masking and thus an increase in active space (which was calculated). This increase paled, however, when contrasted with the reduction in active space imposed by traffic noise itself. However, it should be remembered that noise generated by conspecific and perhaps heterospecific males may usually impose more significant constraints on communication between individuals than that generated by vehicles. A noisy chorus may also attenuate any additional effects of anthropogenic noise on vocal behavior (Lengagne 2008) because of the masking effect of the natural background and because males are already exhibiting their response to noise.

Because so many species of anurans rely heavily on acoustic communication—especially to bring the sexes together for reproduction, human-generated noise is clearly a potentially harmful agent worthy of investigation. However, because the dataset is so sparse, we need to be cautious drawing sweeping conclusions from the recent studies. What is obvious is that there can be variation among species in their responses to anthropogenic noise. How changes in male behavior translate into changes in fitness is unclear. What would be especially helpful in the future are results from experiments using similar test stimuli and protocols. These studies should look for changes in call rate, types of calls, frequency, and amplitude (but see Love and Bee 2010). Researchers need also to use experimental designs and analyses that consider modulation in the amplitude of anthropogenic noise as would be expected near roads and airports. When possible the work with males should be followed by research that can assess the consequences of calling changes (and background noise) for female choice if such data are not already available. Although phonotaxis experiments have demonstrated that certain changes (e.g., in call rate and complexity) can be significant in the context of mate attraction (reviewed in Schwartz 2001; Gerhardt and Huber 2002; Wells and Schwartz 2007; Wells 2007), the advantages of frequency shifts may be minimal (Parris et al. 2009; Nemeth and Brumm 2010). While possibly important, the negative consequences for anurans of anthropogenic noise may be considerably less than other environmental changes that often go hand-in-hand with roads (e.g., habitat destruction, road salt, rolling tires; Van der Ree 2009–2010; Collins and Russell 2009; Langen et al. 2009). Given enough time, near a very busy road, these other factors could often doom a population of amphibians irrespective of noise levels. Therefore, it would be very interesting to conduct long-term studies to test for long-term negative effects, if any, of noise exposure in sites less vulnerable to nonacoustic damage. Relatively pristine areas away from well-traveled roads but exposed to considerable aircraft noise would be ideal. Finally, it would be especially interesting to take a comparative approach when looking at the consequences of exposure to chronic as well as impulsive noise of very high amplitude (such as that which can occur on military bases). There might be differences in susceptibility to auditory system damage among chorusing and non-chorusing species that differ naturally in their exposure to loud noise.

5.4 Conclusions

Anuran amphibians are often found in dense and noisy assemblages in which males of one or more species advertise vocally for mates while also communicating with each other. Selection pressures to reduce signal masking are doubtless powerful because signal overlap can seriously impair the abilities of males to attract a mate. Nevertheless, it is important to remember that selection to reduce detection by illegitimate receivers, such as predators, parasites, and competitors, or to communicate with heterospecifics as well as nonselective forces can influence the path of signal evolution.

From the perspective of production, increases in the signal-to-noise ratio can potentially be achieved in a variety ways. These include frequency and temporal separation of sound sources, increases in signal redundancy, elevation of call amplitude, and use of an advantageous calling venue. In spite of decades of research, it is clear that solutions to the noise problem require more study. Future efforts would benefit from the use of relatively new (Schwartz 2001) and developing technologies (both hardware and software; e.g., Jones and Ratnam 2009; Mizumoto et al. 2011) supplemented with experiments using more traditional approaches (e.g., playback tests). This research could help us define the domains of timing interactions in choruses, determine how timing patterns are shaped by the male spatial distribution (and or perceived amplitude) and assess the stability of patterns of interaction. This kind of information is now quite limited and available on just a few species. Males of different species can respond in different ways to sources of acoustic interference (especially with respect to changes in call rate and timing). Why? Are there conditions that favor one kind of response over others? In some cases, explanations offered are little better than hand waving. Phylogeny, although its signal may be obscured by a range of factors, influences calling behavior and call attributes (Erdtmann and Amézquita 2006; Lemmon 2007). For example, leptodactylid and rhacophorid frogs may utilize different sized pools of note types and note-sequencing "schemes" for assembling calls leading to greater repertoire diversity in members of the latter family (Narins et al. 2000). To what extent, however, are aspects of signal production *putatively linked to the masking problem* correlated with or constrained by phylogeny? Given its implied importance, we need more work addressing the reality and significance of spectral stratification within anuran communities. More research is also needed on the Lombard effect. Some studies are suggestive, but due to methodological details they really need to be repeated. If male anurans do not increase call amplitude with background noise level, why not? Information on a range of species could help us tease apart contributions of proximate and ultimate level factors to limitations in this ability (Love and Bee 2010). Experiments with males need to be coupled with those using females in order to assess the role of various factors in shaping calling behavior. For example, when males increase signal redundancy during acoustic stimulation is it because such calls are inherently more attractive to females or because such calls are more easily detected under noisy conditions? If both explanations apply, what is their relative influence on the male behavior? Finally, we need to know the extent and importance of the threat posed by anthropogenic noise to anurans. These animals as well as other amphibians are vulnerable to a suite of abiotic and biotic agents. The danger posed by some perturbations may be relatively small while that posed by others may be profound at a local or even a global scale. Although man-made noise acting on its own may not be sufficient to precipitate extirpation, it might prove to be the pro-verbial "straw that breaks the camel's back" under some stressful circumstances. We just do not know whether this can be the case. Given the paucity of studies, work in this area needs to be expanded - especially with research that uses standardized approaches and considers interacting factors.

Acknowledgments We are especially grateful to Kentwood Wells and Henrik Brumm for comments on the manuscript. Some of the material described in this review was based upon work supported by the National Science Foundation (under Grant Nos. 0342183 and 9727623), Pace University Scholarly Research Awards and Smithsonian Institution Short-Term Visitor Awards to JJS. Work by MAB was supported by the National Science Foundation (under Grant No. 0842759), the National Institute on Deafness and Other Communication Disorders (under Grant Nos. R03DC009582 and 5R01DC009582), and a Grant-in-Aid from the University of Minnesota Graduate School.

References

Akre KL, Ryan MJ (2010) Complexity increases working memory for mating signals. Curr Biol 20:502–505

Alexander RD (1975) Natural selection and specialized chorusing behavior in acoustical insects. In: Pimental D (ed) Insects, science and society. Academic Press, New York, pp 35–77

Amézquita A, Hödl W (2004) How, when, and where to perform visual displays? The case of the Amazonian frog *Hyla parviceps*. Herpetologica 60:20–29

Amézquita A, Hödl W, Castellanos L, Lima A, Erdtmann L, De Araújo MC (2006) Masking interference and the evolution of the acoustic communication system of the Amazonian poison frog *Epipedobates femoralis*. Evolution 60:1874–1887

Amézquita A, Lima AP, Hödl W (2010) Noisy neighbors: a role for masking interference in the evolution of communication systems in neotropical dendrobatid frogs. Abstract 4aAB4. J Acoust Soc Am 128(4, Pt. 2):2413

Andersson M (1994) Sexual selection. Princeton University Press, Princeton, NJ

Arak A(1983) Vocal interactions, call matching and territoriality in a Sri Lankan treefrog, *Philautus leucorhinus* (Rhacophoridae). Anim Behav 31:292–302

Arch VS, Narins PM (2008) Silent signals: selective forces acting on ultrasonic communication systems in terrestrial vertebrates. Anim Behav 76:1423–1428

Arch VS, Grafe TU, Narins PM (2008) Ultrasonic signalling by a Bornean frog. Biol Lett 4:19–22

Arch VS, Grafe TU, Gridi-Papp M, Narins PM (2009) Pure ultrasonic communication in an endemic bornean frog. PLoS ONE 4:e5413. doi:10.1371/journal.pone.0005413

Barrass AN (1985) The effects of highway traffic noise on the phonotactic and associated reproductive behavior of selected anurans. Unpublished Ph.D. Dissertation, Vanderbuilt University

Barber JR, Crooks KR, Fristrup KM (2009) The costs of chronic noise exposure for terrestrial organisms. Trends Ecol Evol 25:180–189

Bates ME, Cropp B, Gonchar M, Knowles J, Simmons JA, Simmons AM (2010) Spatial location influences acoustic interactions in chorusing bullfrogs. J Acoust Soc Am 127:2664–2677

Baugh AT, Ryan MJ (2010) Mate choice in response to dynamic presentation of male advertisement signals in túngara frogs. Anim Behav 79:145–152

Bee MA (2007) Sound source segregation in the grey treefrog: spatial release from masking by the sound of a chorus. Anim Behav 74:549–558

Bee MA (2008) Finding a mate at a cocktail party: spatial release from masking improves acoustic mate recognition in grey treefrogs. Anim Behav 75:1781–1791

Bee MA, Swanson EM (2007) Auditory masking of anuran advertisement calls by road traffic noise. Anim Behav 74:1765–1776

Bee MA, Perrill SA, Owen PC (2000) Male green frogs lower the pitch of acoustic signals in defense of territories: a possible dishonest signal of size? Behav Ecol 11:169–177

Bernal X, Rand AS, Ryan MJ (2007) Sex differences in response to non-conspecific advertisement calls: receiver permissiveness in male and female túngara frogs. Anim Behav 73:955–964

Bevier CR (1997) Breeding activity and chorus tenure of two Neotropical hylid frogs. Herpetologica 53:297–311

Boeckle M, Preininger D, Hödl W (2009) Communication in noisy environment I. Acoustic signals of Staurois latopalmatus Boulenger 1879. Herpetologica 65:154–165

Bogert CM (1960) The influence of sound on the behavior of amphibians and reptiles. In: Lanyon WE, Tavolga WN (eds) Animal sounds and communication, Publ 7. AIBS, Washington, pp 137–320

Bosch J, Márquez R (2002) Female preference function related to precedence effect in an amphibian anuran (Alytes cisternasii): tests with non-overlapping calls. Behav Ecol 13:149–153

Bosch J, De la Riva I (2004) Are frog calls modulated by the environment? An analysis with anuran species from Bolivia. Can J Zool 82:880–888

Bowker RG, Bowker MH (1979) Abundance and distribution of anurans in a Kenyan pond. Copeia 1979:278–285

Bourne GR, York H (2001) Vocal behaviors are related to nonrandom structure of anuran breeding assemblages in Guyana. Ethol Ecol Evol 13:313–329

Bridges AS, Dorcas ME (2000) Temporal variation in anuran calling behavior: implications for surveys and monitoring programs. Copeia 2000:587–592

Brumm H, Naguib M (2009) Environmental acoustics and the evolution of bird song. Adv Study Behav 40:1–33

Brumm H, Slabbekoorn H (2005) Acoustic communication in noise. Adv Stud Behav 35:151–209

Brumm H, Zollinger SA (2011) The evolution of the Lombard effect: 100 years of psychoacoustic research. Behaviour 148:1173–1198

Brush JS, Narins PM (1989) Chorus dynamics of a Neotropical amphibian assemblage: comparison of computer simulation and natural behaviour. Anim Behav 37:33–44

Caldwell MS, Johnston GR, McDaniel JG, Warkentin KM (2010) Vibrational signaling in the agonistic interactions of red-eyed treefrogs. Curr Biol 20:1–6

Chek AA, Bogart JP, Lougheed SC (2003) Mating signal partitioning in multi-species assemblages: a null model test using frogs. Ecol Lett 6:235–247

Christie K, Schul J, Feng AS (2010) Phonotaxis to male's calls embedded within a chorus by female gray treefrogs, Hyla versicolor. J Comp Physiol A 196:569–579

Cody ML (1974) Competition and the structure of bird communities. Princeton University Press, Princeton

Collins SJ, Russel RW (2009) Toxicity of road salt to Nova Scotia amphibians. Environ Poll 157:320–324

Crump ML (1974) Reproductive strategies in a tropical anuran community. Misc Pub Mus Nat Hist Univ Kansas 61:1–68

Cunningham GM, Fahrig L (2010) Plasticity in the vocalizations of anurans in response to traffic noise. Acta Oecologica 36:463–470

Drewry G (1970) Factors affecting activity of rain forest frog populations as measured by electrical recording of sound pressure levels. In: Odum HT (ed) A tropical rain forest. A study of Irradiation and Ecology at El Verde, Puerto Rico. U.S. Atomic Energy Commission. Oak Ridge, Tennessee, Chapter E-4, pp E-55–E-68

Drewry GE, Rand AS (1983) Characteristics of an acoustic community: Puerto Rican frogs of the genus *Eleutherodactylus*. Copeia 1983:639–649

Dubois A, Martens JM (1984) A case of possible vocal convergence between frogs and a bird in Himalayan torrents. J für Ornithologie 125:455–463

Duellman WE (1967) Courtship isolating mechanisms in Costa Rican hylid frogs. Herpetologica 23:169–183

DuellmanWE (1978) The biology of equatorial herpetofauna in Amazonian Ecuador. Misc Pub Mus Nat Hist Univ Kansas 65:1–352

Duellman WE, Pyles RA (1983) Acoustic resource partitioning in anuran communities. Copeia 1983:639–649

Dyson ML, Passmore NI (1988a) Two-choice phonotaxis in *Hyperolius marmoratus*: the effect of temporal variation in presented stimuli. Anim Behav 36:648–652

Dyson ML, Passmore NI (1988b) The combined effect of intensity and the temporal relationship of stimuli on the phonotactic responses of female painted reed frogs (*Hyperolius marmoratus*). Anim Behav 36:1555–1556

Erdtmann L, Amézquita A (2006) Differential evolution of advertisement call traits in dart-poison frogs (Anura: Dendrobatidae). Ethology 115:801–811

Ey E, Fischer J (2009) The acoustic adaptation hypothesis—a review of the evidence from birds, anurans and mammals. Bioacoustics 19:21–48

Farris EH, Rand AS, Ryan MJ (2002) The effects of spatially separated call components on phonotaxis in Túngara frogs: evidence for auditory grouping. Brain Behav Evol 60:181–188

Feng AS, Narins PM (2008) Ultrasonic communication in concave-eared torrent frogs (*Amolops tormotus*). J Comp Physiol 194:159–167

Feng AS, Ratnam R (2000) Neural basis of hearing in real-world situations. Annu Rev Psychol 51:699–725

Feng AS, Narins PM, Xu CH, Lin WY, Yu ZL, Qiu Q, Xu ZM, Shen JX (2006) Ultrasonic communication in frogs. Nature 440:333–336

Feng AS, Riede T, Arch VS, Yu ZL, Xu ZM, Yu XJ, Shen JX (2009a) Diversity of the vocal signals of concave-eared torrent frogs (*Odorrana tormota*): evidence for individual signatures. Ethology 115:1015–1028

Feng AS, Arch VS, Yu ZL, Yu XJ, Xu ZM, Shen JX (2009b) Neighbor-stranger discrimination in concave-eared torrent frogs, *Odorrana tormota*. Ethology 115:851–856

Forester DC, Harrison WK (1987) The significance of antiphonal vocalisation by the spring peeper, *Pseudacris crucifer* (Amphibia, Anura). Behaviour 103:1–15

Frost DR, Grant T, Faivovich J, Bain RH, Haas A, Haddad CFB, de Sá RO, Channing A, Wilkinson M, Donnellan SC, Raxworthy CJ, Campbell JA, Blotto BL, Moler P, Drewes RC, Nussbaum RA, Lynch JD, Green DM, Wheeler WC (2006) The amphibian tree of life. Bull Am Mus Nat Hist 297:1–370

Garcia-Rutledge E, Narins PM (2001) Shared acoustic resources in an old world frog community. Herpetologica 57:103–116

Gerhardt HC (2001) Acoustic communication in two groups of closely related treefrogs. Adv Study Behav 30:99–167

Gerhardt HC, Huber F (2002) Acoustic communication in insects and frogs: common problems and diverse solutions. University of Chicago Press, Chicago

Gerhardt HC, Klump GM (1988) Masking of acoustic signals by the chorus background noise in the green treefrog: a limitation on mate choice. Anim Behav 36:1247–1249

Gerhardt HC, Schwartz JJ (1995) Interspecific interactions and species recognition. In: Heatwole H, Sullivan BK (eds) Amphibian Biology, vol 2. Social Behavior. Surrey Beatty, Chipping Norton, UK, pp 603–632

Gerhardt HC, Schwartz JJ (2001) Auditory tuning, frequency preferences and mate choice in anurans. In: Ryan MJ (ed) Anuran Communication. Smithsonian Institution Press, Washington, DC, pp 73–85

Gerhardt HC, Dyson ML, Tanner SD (1996) Dynamic acoustic properties of the advertisement calls of gray tree frogs: patterns of variability and female choice. Behav Ecol 7:7–18

Gerhardt HC, Roberts JD, Bee MA, Schwartz JJ (2000) Call matching in the quacking frog (Crinia georgiana). Behav Ecol Sociobiol 48:243–251

Given MF (1987) Vocalizations and acoustic interactions of the carpenter frog, Rana virgatipes. Herpetologica 43:467–481

Goosem M, Hoskin C, Dawe G (2007). Nocturnal noise Levels and edge impacts on amphibian habitats adjacent to kuranda range road. Report to the Marine and Tropical Sciences Research Facility, Reef and Rainforest Research Centre Limited, James Cook University, Cairns

Grafe TU (1996) The function of call alternation in the African reed frog Hyperolius marmoratus: precise call timing prevents auditory masking. Behav Ecol Sociobiol 38:149–158

Grafe U (1997) Use of metabolic substrates in the gray treefrog Hyla versicolor: implications for calling behavior. Copeia 1997:356–362

Grafe TU (1999) A function of synchronous chorusing and a novel female preference shift in an anuran. Proc R Soc London B 266:2331–2336

Grafe TU (2003) Synchronised interdigitated calling in the Kuvangu running frog (Kassina kuvangensis). Anim Behav 66:127–136

Grafe TU, Wanger TC (2007) Multimodal signaling in male and female foot-flagging frogs Staurois guttatus (Ranidae): an alerting dunction of calling. Ethology 113:772–781

Grafe TU, Preininger D, Sztatecsny M, Kasah R, Dehling JM, Proksch S, Hödl W (2012) Multimodal communication in a noisy environment: a case study of the Bornean Rock frog Staurois parvus. PLoS ONE 7(5):e37965. doi:10.1371/journal.pone.0037965

Greenfield MD (1994) Cooperation and conflict in the evolution of signal interactions. Ann Rev Ecol Syst 25:97–126

Greenfield MD, Rand AS (2000) Frogs have rules: selective attention algorithms regulate chorusing in Physalaemus pustulosus (Leptodactylidae). Ethology 106:331–347

Greenfield MD, Tourtellot MK, Snedden WA (1997) Precedence effects and the evolution of chorusing. Proc Roy Soc London B 264:1355–1361

Gottsberger B, Grubert E (2004) Temporal partitioning of reproductive activity in a neotropical anuran community. J Trop Ecol 20(2):71–280

Gridi-Papp M, Feng AS, Shen JX, Yu ZL, Rosowski JJ, Narins PM (2008) Active control of ultrasonic hearing in frogs. Proc Natl Acad Sci U S A 105:11014–11019

Gross K, Pasinelli G, Kunc HP (2010) Behavioral plasticity allows short-term adjustment to a novel environment. Am Nat 176:456–464

Haddad CFB, Giaretta AA (1999) Visual and acoustic communication in the Brazilian torrent frog, Hylodes asper (Anura, Leptodactylidae). Herpetologica 55:324–333

Halliday TR, Tejedo M (1995) Intrasexual selection and alternative mating behavior. In: Heatwole H, Sullivan BK (eds) Amphibian Biology, vol 2., Social BehaviourSurrey Beatty, Chipping Norton, UK, pp 419–468

Harrison PA, Littlejohn MJ (1985) Diphasy in the advertisement calls of Geocrinia laevis (Anura: Leptodactylidae): vocal responses of males during field playback experiments. Behav Ecol Sociobiol 18:67–73

Herrick SZ (2013) Ecological and behavioral interactions between two closely related north American frogs (Rana catesbeiana and R. clamitans). Unpublished PhD Dissertation, University of Connecticut, Storrs

Höbel G, Gerhardt HC (2007) Sources of selection on signal timing in a treefrog. Ethology 113:973–982

Hödl W (1977) Call differences and calling site segregation in anuran species from central Amazonian floating meadows. Oecologia 28:351–363

Hödl W, Amezquita A (2001) Visual signaling in anuran amphibians. In: Ryan MJ (ed) Anuran communication. Smithsonian Institution Press, Washington, DC, pp 121–141

Hoskin CJ, James S, Grigg CG (2009) Ecology and taxonomy-driven deviations in the frog call-body size relationship across the diverse Australian frog fauna. J Zool 278:36–41

Howard RD, Palmer JG (1995) Female choice in *Bufo americanus*: effects of dominant frequency and call order. Copeia 1995:212–217

Hsu MY, KamYC Fellers GM (2006) Temporal organization of an anuran acoustic community in a Taiwanese subtropical. J Zool 269:331–339

Humfeld SC (2003) Signaling, intersexual dynamics and the adoption of alternative male mating behaviors in green treefrogs, *Hyla cinerea*. Unpublished PhD Dissertation, University of Missouri, Columbia

Ibañez R (1991) Synchronized calling in *Centrolenella granulosa* and *Smilisca sila* (Amphibia, Anura). Unpublished PhD Dissertation, University of Connecticut, Storrs

Ibañez R (1993) Female phonotaxis and call overlap in the Neotropical glassfrog, *Centrolenella granulosa*. Copeia 1993:846–850

Jehle R, Arak A (1998) Graded call variation in the Asian cricket frog *Rana nicobariensis*. Bioacoustics 9:35–48

Jones DL, Ratnam R (2009) Blind location and separation of callers in a natural chorus using a microphone array. J Acoust Soc Am 126:895–910

Kaiser K, Hammers JL (2009) The effect of anthropogenic noise on male advertisement call rate in the neotropical treefrog, *Dendropsophus triangulum*. Behaviour 146:1053–1069

Kime MN (2001) Female mate choice for socially variable advertisement calls in the cricket frog, *Acris crepitans*. Unpublished PhD Dissertation, University of Texas, Austin

Kime NM, Burmeister SS, Ryan MJ (2004) Female preferences for socially variable call characters in the cricket frog, *Acris crepitans*. Anim Behav 68:1391–1399

Klump GM, Gerhardt HC (1987) Use of non-arbitray acoustic criteria in mate choice by female gray treefrogs. Nat 326:286–288

Klump GM, Gerhardt HC (1992) Mechanisms and function of call-timing in male–male interactions in frogs. In: McGregor PK (ed) Playback and studies of animal communication. Plenum, New York, pp 153–174

Krishna SN, Krishna SB (2006) Visual and acoustic communication in an endemic stream frog, *Micrixalus saxicolus* in the western ghats, India. Amphibia-Reptilia 27:143–147

Kuramoto M (1980) Mating calls of treefrogs of genus *Hyla* in the Far East, with description of a new species from Korea. Copeia 1980:100–108

Langen TA, Ogden K, Schwarting L (2009) Predicting hotspots of herpetofauna road mortality along highway networks. J Wildlife Manage 73:104–114

Lemmon EM (2007) Patterns and processes of speciation in north American Chorus frogs. Unpublished PhD Dissertation, University of Texas, Austin

Lengagne T (2008) Traffic noise affects communication behaviour in a breeding anuran, *Hyla arborea*. Biol Conserv 141:2023–2031

Lewis ER, Narins PM (1985) Do frogs communicate with seismic signals? Science 227:187–189

Lewis ER, Narins PM, Cortopassi KA, Yamada WM, Poinar EH, Moore SW, Yu XL (2001) Do male white-lipped frogs use seismic signals for intraspecific communication? Am Zool 41:1185–1199

Lewisohn M (1992) The complete beatles chronicles. Harmony Books, New York

Miles B (2009) The beatles: a diary: an intimate day by day history. Omnibus Press, London

Littlejohn MJ (1977) Long-range acoustic communication in anurans: an integrated and evolutionary approach. In: Taylor DH, Guttman SI (eds) The reproductive biology of amphibians. Plenum Press, New York, pp 263–294

Littlejohn MJ, Martin AA (1969) Acoustic interaction between two species of leptodactylid frogs. Anim Behav 17:785–791

Littlejohn MJ, Harrison PA, Mac Nally RC (1985) Interspecific acoustic interactions in sympatric populations of *Ranidella signifera* and *R. parinsignifera* (Anura: Leptodactylidae). In: Grigg

G, Shine R, Ehrman H (eds) The biology of Australian frogs and reptiles. Royal Zoological Society of New South Wales, Sydney, Australia, pp 287–296

Lopez PT, Narins PM, Lewis ED, Moore SW (1988) Acoustically induced call modification in the white-lipped frog, *Leptodactylus albilabris*. Anim Behav 36:1295–1308

Lougheed SC, Austin JD, Bogart JP, Boag PT, Chek AA (2006) Multi-character perspectives on the evolution of intraspecific differentiation in a neotropical hylid frog. BMC Evol Biol 6:23. doi:10.1186/1471-2148-6-23

Love EK, Bee MA (2010) An experimental test of noise-dependent voice amplitude regulation in Cope's grey treefrog (*Hyla chrysoscelis*). Anim Behav 80:509–515

Lüddecke H, Amézquita A, Bernal X, Guzmán F (2000) Partitioning of vocal activity in a Neotropical highland-frog community. Stud Neotrop Fauna Environ 35:185–194

Mac Nally, RC (1982) Ecological, behavioural, and energy dynamics of two sympatric species of *Ranidella* (Anura). Unpublished Ph.D. Dissertation, University of Melbourne, Parkville, Australia

Marshall VT, Gerhardt HC (2010) A precedence effect underlies preferences for calls with leading pulses in the grey treefrog, *Hyla versicolor*. Anim Behav 80:139–145

Martínez-Rivera CC, Gerhardt HC (2008) Advertisement-call modification and female preference in the bird-voiced treefrog, *Hyla avivoca*. Behav Ecol Sociobiol 63:195–208

Mizumoto T, Aihara I, Otsuka T, Takeda R, Aihara K, Okuno HG (2011) Sound imaging of nocturnal animal calls in their natural habitat. J Comp Physiol A 197:915–921

Narins PM (1982) Effects of masking noise on evoked calling in the Puerto Rican Coqui (Anura: Leptodactylidae). J Comp Physiol 147:438–446

Narins PM (1990) Seismic communication in anuran amphibians. Bioscience 40:268–274

Narins PM (1992) Reduction of tympanic membrane displacement during vocalization of the arboreal frog, *Eleutherodactylus coqui*. J Acoust Soc Am 91:3551–3557

Narins PM (1995) Frog communication. Sci Am 273:78–83

Narins PM, Lewis ER (1984) The vertebrate ear as an exquisite seismic sensor. J Acoust Soc Am 76:1384–1387

Narins PM, Zelick R (1988) The effects of noise on auditory processing and behavior in amphibians. In: Fritszch B, Wilczynski W, Ryan MJ, Hetherington T, Walkowiak W (eds) The evolution of the amphibian auditory system. Wiley, New York, pp 511–536

Narins PM, Lewis ER, McClelland BE (2000) Hyperextended call repertoire of the endemic Madagascar treefrog, *Boophis madagascariensis* (Rhacophoridae). J Zool Lond 250:283–298

Narins PM, Feng AS, Lin WY, Schnitzler HU, Denzinger A, Suthers RA, Xu C (2004) Old world frog and bird vocalizations contain prominent ultrasonic harmonics. J Acoust Soc Am 115:910–913

Nemeth E, Brumm H (2010) Birds and anthropogenic noise. Am Nat 176:465–475

Odendaal FJ, Bull CM, Telford SR (1986) Influence of the acoustic environment on the distribution of the frog *Ranidella riparia*. Anim Behav 34:1836–1843

Oseen KL, Wassersug RJ (2002) Environmental factors influencing calling in sympatric anurans. Oecologia 133:616–625

Paez VP, Bock BC, Rand AS (1993) Inhibition of evoked calling of *Dendrobates pumilio* due to acoustic interference from cicada calling. Biotropica 25:242–245

Pallett JR, Passmore NI (1988) The significance of multi-note advertisement calls in a reed frog, *Hyperolius tuberilinguis*. Bioacoustics 1:13–23

Parris KM, Velik-Lord M, North JMA (2009) Frogs call at a higher pitch in traffic noise. Ecology and Society 14:25

Penna M, Velásquez N (2011) Heterospecific vocal interactions in a frog from the southern temperate forest, *Batrachyla taeniata*. Ethology 117:63–71

Penna M, Hamilton-West C (2007) Susceptibility of evoked vocal responses to noise exposure in a frog of the temperate austral forest. Anim Behav 74:45–56

Penna M, Pottstock H, Velásquez N (2005) Effect of natural and synthetic noise on evoked vocal responses in a frog of the temperate austral forest. Anim Behav 70:639–651

Phelps S, Rand AS, Ryan MJ (2007) The mixed-species chorus as public information: Túngara frogs eavesdrop on a heterospecific. Behav Ecol 18:108–114

Preininger D, Boeckle M, Hödl W (2009) Communication in noisy environment II. Visual signaling behavior of male foot-flagging frogs *Staurois latopalmatus*. Herpetologica 65:166–173

Prestwich KN, Brugger KE, Topping M (1989) Energy and communication in three species of Hylid frogs: power input, power output and efficiency. J Exp Biol 144:53–80

Pyron RA, Wiens JJ (2011) A large-scale phylogeny of Amphibia including over 2800 species, and a revised classification of extant frogs, salamanders, and caecilians. Mol Phylogenet Evol 61:543–583

Rabin LA, Greene CM (2002) Changes to acoustic communication systems in human-altered environments. J Comp Psychol 116:137–141

Reichert M (2011) Aggressive calls improve leading callers' attractiveness in the treefrog *Dendropsophus ebraccatus*. Behav Ecol 22:951–959

Richardson C, Léna J-P, Joly P, Lengagne T (2008) Are leaders good mates? A study of call timing and male quality. Anim Behav 76:1487–1495

Richardson C, Lengagne T (2010) Multiple signals and male spacing affect female preference at cocktail parties in treefrogs. Proc R Soc B 277:1247–1252

Römer H (1998) The sensory ecology of acoustic communication in insects. In: Hoy RR, Popper AN, Fay RR (eds) Comparatve hearing: insects. Springer-Verlag, New York

Ronacher B, Krahe B, Hennig RM (2000) Effects of signal duration on the recognition of masked communication signals by the grasshopper *Chorthippus biguttulus*. J Comp Physiol A 186:1065–1072

Ryan MJ (1985) The Tungara frog. The University of Chicago Press, Chicago

Ryan MJ (1988) Constraints and patterns in the evolution of anuran acoustic communication. In: Fritszch B, Hetherington T, Ryan MJ, Wilczynski W, Walkowiak W (eds) The evolution of the amphibian auditory system. Wiley, New York, pp 637–677

Ryan MJ, Cummings ME (2005) Animal signals and the overlooked costs of efficacy. Evolution 59:1160–1161

Ryan MJ, Keddy-Hector A (1992) Directional patterns of female mate choice and the role of sensory biases. Amer Nat 139:S4–S35

Ryan MJ, Kime NM (2003) Selection on long-distance acoustic signals. In: Simmons AM, Popper AN, Fay RR (eds) Acoustic communication. Springer-Verlag, Berlin, pp 225–273

Saenz D, Fitzgerald LA, Baum KA, Conner RN (2007) Abiotic correlates of anuran calling phenology: the importance of rain, temperature, and season. Herpetological Monogr 20:64–82

Schlüter A (1979) Bio-akustische Untersuchungen an Hyliden in einem begrenzten Gebiet des tropischen Regenwaldes von Peru (Amphibia: Salientia: Hlidae). Salamandra 15:211–236

Schlüter A (1980) Bio-akustische Untersuchungen an Leptodactyliden in einem begrenzten Gebiet des tropischen Regenwaldes von Peru (Amphibia: Salientia: Leptodactylidae). Salamandra 16:227–247

Schneider H, Joermann G, Hödl W (1988) Calling and antiphonal calling in four Neotropical anuran species of the family Leptodactylidae. Zool Jb Physiol 92:77–103

Schwartz JJ (1986) Male calling behavior and female choice in the Neotropical treefrog *Hyla microcephala*. Ethology 73:116–127

Schwartz JJ (1987) The function of call alternation in anuran amphibians: a test of three hypotheses. Evolution 41:461–471

Schwartz JJ (1991) Why stop calling? A study of unison bout singing in a Neotropical treefrog. Anim Behav 42:565–577

Schwartz JJ (1993) Male calling behavior, female discrimination and acoustic interference in the Neotropical treefrog *Hyla microcephala* under realistic acoustic conditions. Behav Ecol Sociobiol 32:401–414

Schwartz JJ (2001) Call monitoring and interactive playback systems in the study of acoustic interactions among male anurans. In: Ryan MJ (ed) Anuran communication. Smithsonian Institution Press, Washington, DC, pp 183–204

Schwartz JJ, Gerhardt HC (1989) Spatially-mediated release from masking in an anuran amphibian. J Comp Physiol A 166:37–41

Schwartz JJ, Marshall VT (2006) Forms of call overlap and their impact on advertisement call attractiveness to females of the gray treefrog, *Hyla versicolor*. Bioacoustics 16:39–56

Schwartz JJ, Rand AS (1991) The consequences for communication of call overlap in the tungara frog, a Neotropical anuran with a frequency-modulated call. Ethology 89:73–83

Schwartz JJ, Wells KD (1983a) An experimental study of acoustic interference between two species of Neotropical treefrogs. Anim Behav 31:181–190

Schwartz JJ, Wells KD (1983b) The influence of background noise on the behavior of a Neotropical treefrog, *Hyla ebraccata*. Herpetologica 39:121–129

Schwartz JJ, Wells KD (1984a) Interspecific acoustic interactions of the Neotropical treefrog *Hyla ebraccata*. Behav Ecol Sociobiol 14:211–224

Schwartz JJ, Wells KD (1984b) Vocal behavior of the Neotropical treefrog *Hyla phlebodes*. Herpetologica 40:452–463

Schwartz JJ, Wells KD (1985) Intra- and interspecific vocal behavior of the Neotropical treefrog *Hyla microcephala*. Copeia 1985:27–38

Schwartz JJ, Ressel S, Bevier CR (1995) Carbohydate and calling: depletion of muscle glycogen and the chorusing dynamics of the Neotropical treefrog *Hyla microcephala*. Behav Ecol Sociobiol 37:125–135

Schwartz JJ, Buchanan B, Gerhardt HC (2001) Female mate choice in the gray treefrog (*Hyla versicolor*) in three experimental environments. Behav Ecol Sociobiol 49:443–455

Schwartz JJ, Buchanan B, Gerhardt HC (2002) Acoustic interactions among male gray treefrogs (*Hyla versicolor*) in a chorus setting. Behav Ecol Sociobiol 53:9–19

Schwartz JJ, Brown R, Turner S, Dushaj K, Castano M (2008) Interference risk and the function of dynamic shifts in calling in the gray treefrog. J Comp Psych 122:283–288

Schwartz JJ, Huth K, Hutchin T (2004) How long do females really listen? Assessment time for female mate choice in the gray treefrog, *Hyla versicolor*. Anim Behav 68:533–540

Schwartz JJ, Huth K, Jones SH, Brown R, Marks J, Yang X (2010) Tests for call restoration during signal overlap in the gray treefrog, *Hyla versicolor*. Bioacoustics 20:59–86

Schwartz JJ, Crimarco NC, Bregman Y, Umeoji K (2013) An investigation of the functional significance of responses of the gray treefrog (*Hyla versicolor*) to chorus noise. J Herp 47:354–350

Shen JX, Feng AS, Xu ZM, Yu ZL, Arch VS, Yu XJ, Narins PM (2008) Ultrasonic frogs show hyperacute phonotaxis to the female's courtship calls. Nature 453:914–917

Shen JX, Xu ZM, Yu ZL, Wang S, Zheng DZ, Fan SC (2011) Ultrasonic frogs show extraordinary sex differences in auditory frequency sensitivity. Nat Commun 2:342. doi:10.1038/ncomms1339

Shimoyama R (1989) Breeding ecology of a Japanese pond frog, *Rana porosa*. In: Matsui M, Hikada T, Goris RC (eds) Current herpetology in east Asia. Herpetological Society of Japan, Kyoto, pp 323–331

Simmons AM, Simmons JA, Bates ME (2008) Analyzing acoustic interactions in natural bullfrog choruses. J Comp Psych 122:274–282

Slabbekoorn H, Ripmeester EA (2008) Birdsong and anthropogenic noise: implications and applications for conservation. Mol Ecol 17:72–83

Slabbekoorn H, Bouton N, van Opzeeland I, Coers A, ten Cate C, Popper AN (2010) A noisy spring: the impact of globally rising underwater sound levels on fish. Tree 25:419–427

Smith MJ, Roberts JD (2003) Call structure may affect male mating success in the quacking frog *Crinia georgiana* (Anura: Myobatrachidae). Behav Ecol Sociobiol 53:221–226

Summers WV, Pisoni DB, Bernacki RH, Pedlow RI, Stokes MA (1988) Effects of noise on speech production: acoustic and perceptual analyses. J Acoust Soc Am 84:917–928

Sun JWC, Narins PM (2005) Anthropogenic sounds differentially affect amphibian call rate. Biol Conserv 121:419–427

Suthers RA, Narins PM, Lin WY, Schnitzler HU, Denzinger A, Xu CH, Feng AS (2006) Voices of the dead: complex nonlinear vocal signals from the larynx of an ultrasonic frog. J Exp Biol 209:4984–4993

Tárano Z, Herrera EA (2003) Female preferences for call traits and mating success in the Neotropical frog, *Physalaemus enesefae*. Ethology 109:121–134

Telford SR (1982) Aspects of mate recognition and social behavior in a sub-tropical frog-community. Unpublished Ph.D. Dissertation, University of the Witwatersrand, Johannesburg, South Africa

Tobias ML, Barnard C, O'Hagan R, Horng SH, Rand M, Kelley DB (2004) Vocal communication between male *Xenopus laevis*. Anim Behav 67:353–365

Tuttle MD, Ryan MJ (1982) The role of synchronized calling, ambient light, and ambient noise, in anti-bat predator behavior of a treefrog. Behav Ecol Sociobiol 11:125–131

Van der Ree R, Jaeger JAG, Van der Grift E, Clevenger AP (guest editors) (2009–2010) Effects of roads and traffic on wildlife populations and landscape function. Special issue of ecology and society 14–15 http://www.ecologyandsociety.org/viewissue.php?sf=41

Vargas-Salinas F, Amézquita A (2013) Abiotic noise, call frequency and stream-breeding anuran assemblages. Evol Ecol. doi:10.1007/s10682-013-9675-6

Vélez A, Bee MA (2010) Signal recognition by frogs in the presence of temporally fluctuating chorus-shaped noise. Behav Ecol Sociobiol 64:1695–1709

Wagner WE Jr (1989) Graded aggressive signals in Blanchard's cricket frog: vocal responses to opponent proximity and size. Anim Behav 38:1025–1038

Wagner WE Jr (1992) Deceptive or honest signalling of fighting ability? A test of alternative hypotheses for the function of changes in call dominant frequency by male cricket frogs. Anim Behav 44:449–462

Warren PS, Katti M, Ermann M, Brazel A (2006) Urban bioacoustics: it's not just noise. Anim Behav 71:491–502

Weir LA, Royle AJ, Nanjappa P, Jung R (2005) Modeling anuran detection and site occupancy on North American Amphibian Monitoring (NAAMP) routes in Maryland. J Herpetol 39:627–63

Wells KD (2007) The ecology and behavior of amphibians. University of Chicago Press, Chicago, IL

Wells KD, Schwartz JJ (1984) Vocal communication in a Neotropical treefrog, *Hyla ebraccata*: advertisement calls. Anim Behav 32:405–420

Wells KD, Schwartz JJ (2007) The behavioral ecology of anuran communication. In: Narins PM, Feng AS, Fay RR, Popper AN (eds) Hearing and sound communication in amphibians. Springer, New York, pp 44–86

Wells KD, Taigen TL (1986) The effect of social interactions on calling energetics in the gray treefrog (*Hyla versicolor*). Behav Ecol Sociobiol 19:9–18

Wells KD, Taigen TL (1989) Calling energetics of a Neotropical treefrog, *Hyla microcephala*. Behav Ecol Sociobiol 25:13–22

Wiley RH (1983) The evolution of communication: information and manipulation. In: Halliday TR, Slater PJB (eds) Animal behaviour, vol 2. Communication. W. H. Freeman and Company, pp 156–189

Wiley RH (2006) Signal detection and animal communication. Adv Study Behav 36:217–247

Wollerman L, Wiley RH (2002) Possibilities for error during communication by Neotropical frogs in a complex acoustic environment. Behav Ecol Sociobiol 52:465–473

Wong S, Parada H, Narins PM (2009) Heterospecific Acoustic Interference: effects on calling in the frog *Oophaga pumilio* in Nicaragua. Biotropica 41:74–80

Zelick RD, Narins PM (1982) Analysis of acoustically evoked call suppression behaviour in a Neotropical treefrog. Anim Behav 30:728–733

Zelick RD, Narins PM (1983) Intensity discrimination and the precision of call timing in two species of Neotropical treefrogs. J Comp Physiol A 153:403–412

Zelick RD, Narins PM (1985) Characterization of the advertisement call oscillator in the frog *Eleutherodactylus coqui*. J Comp Physiol A 156:223–229

Chapter 6
Anuran Acoustic Signal Perception in Noisy Environments

Alejandro Vélez, Joshua J. Schwartz and Mark A. Bee

Abstract Choruses of acoustically signaling frogs and toads are among the most impressive acoustic spectacles known from the natural world. They are loud, raucous social environments that form for one purpose and one purpose only: sex. The loud sexual advertisement signals that males produce are often necessary and sufficient to elicit responses from reproductive females, and they also function in communicating with other males during interactions over calling sites and territories. Frogs listening in a chorus must detect, recognize, localize, and discriminate among competing signals amid high levels of biotic, and often abiotic, background noise. In essence, frogs must solve a biological analog of the human cocktail party problem. In this chapter, we describe the frog's cocktail party problem in functional terms relevant to frog reproduction and communication. We then describe results from experimental studies, mostly of behavior, that elucidate how the frog auditory system goes about solving problems related to auditory masking and auditory scene analysis.

6.1 Introduction

Most people reading this chapter will have had first-hand experience trying to converse in noisy social gatherings, such as in a popular bar or at a professional sporting event. Difficulty understanding speech in these sorts of social environments is aptly named the "cocktail party problem" (Cherry 1953) in the literature on

A. Vélez · M. A. Bee (✉)
Department of Ecology, Evolution and Behavior, University of Minnesota,
Saint Paul, MN 55108, USA
e-mail: mbee@umn.edu

J. J. Schwartz
Department of Biology, Pace University, Pleasantville, NY 10570, USA

H. Brumm (ed.), *Animal Communication and Noise*,
Animal Signals and Communication 2, DOI: 10.1007/978-3-642-41494-7_6,
© Springer-Verlag Berlin Heidelberg 2013

human hearing and speech communication (reviewed in Bronkhorst 2000; McDermott 2009). Acoustic communication in noisy environments like cocktail parties places heavy demands on receivers' abilities to detect and discriminate among signals (Chap. 2). However, communicating in noisy environments requires that receivers do more than merely detect a signal's presence and determine whether two or more signals differ. The problem we and other animals have hearing in such environments—and indeed in any environment where there are multiple concurrent sound sources—stems ultimately from the physics of sound. Unlike objects in a visual scene, which can occlude light emitted or reflected by background objects, the sound sources in an "acoustic scene" generate sound pressure waves that add linearly to form a single, composite waveform that impinges on the ears. A primary function of the auditory system is to organize this sensory input into coherent perceptual representations of the various sound sources present in the environment (Bregman 1990; Yost et al. 2008). Perceptually organizing complex acoustic scenes requires that listeners decompose the composite waveform and assign its constituent parts to their correct sources. These latter perceptual tasks—variously referred to as "sound source segregation" (Brown and Cooke 1994), "sound source perception" (Yost et al. 2008), "auditory grouping" (Darwin and Carlyon 1995; Darwin 1997), "auditory streaming" (Shamma and Micheyl 2010), or "auditory object formation" (Griffiths and Warren 2004)—represent key elements of what has been more broadly termed "auditory scene analysis" (Bregman 1990). While issues of auditory perception related to signal detection (Chap. 2), sound pattern recognition (Gerhardt and Huber 2002), source localization (Christensen-Dalsgaard 2005), and noise (Brumm and Slabbekoorn 2005) have featured prominently in studies of animal acoustic communication, this is less so for auditory scene analysis (Hulse 2002; Bee and Micheyl 2008). Consequently, we are just beginning to uncover the mechanisms by which nonhuman animals perceptually organize complex acoustic scenes.

The perceptual challenges faced by breeding anurans parallel the difficulty we have following one person speaking in a noisy social gathering. Both their auditory systems and the diversity present in their vocal communication systems make frogs particularly interesting animal models for studying perceptual mechanisms for hearing and sound communication in noisy and acoustically complex environments (Feng and Ratnam 2000; Feng and Schul 2007). Vocal communication in frogs functions primarily in the context of reproduction and it commonly takes place in social and physical environments characterized by multiple concurrent sound sources and high levels of biotic and abiotic background noise (Chap. 5; see also Narins and Zelick 1988; Ryan 2001; Gerhardt and Huber 2002; Narins et al. 2007; Wells 2007). Reproduction usually depends on a female frog's ability to respond correctly to the advertisement signals of a conspecific male. Vocal signals also mediate agonistic interactions, enabling male frogs to estimate their opponent's proximity, size and fighting ability, and even recognize him as a familiar individual (Gerhardt and Bee 2007). All of these behaviors require that receivers accomplish a number of key perceptual tasks that include detecting signals, recognizing them as conspecific calls, localizing their source, and discriminating among different signal variants. But how do frogs successfully complete these tasks in the cacophonous acoustic scene of a breeding chorus? How do they determine distinct sound sources? How do frogs perceptually organize complex acoustic scenes and solve their own cocktail-party-like problem? As we hope to make clear in this chapter, answering these questions remains both a fundamental challenge in understanding hearing and sound communication in frogs, as well a primary goal of modern research on these remarkable little animals.

6.2 Frogs as Model Systems for Studies of Acoustic Communication in Noise

Anuran systematics is an area of very active research and taxonomic nomenclature in this group has undergone some revisions and re-revisions in recent years (Frost et al. 2006; Pyron and Wiens 2011). In review chapters like this one, there is considerable potential for confusion when one species has multiple scientific names in the literature. In this chapter, we adopt the nomenclature of (Pyron and Wiens 2011) and point out where we are using new names.

6.2.1 Experimental Methods

Frogs have long served as model organisms for investigating the mechanisms, function, and evolution of animal acoustic communication. Excellent reviews of anuran hearing and communication can be found in previous volumes edited by Fritzsch et al. (1988), Ryan (2001), and Narins et al. (2007), and in books by Gerhardt and Huber (2002) and Wells (2007). Three primary experimental approaches based on behavioral responses have been used successfully to study anuran hearing and communication. The two most commonly used methods involve assessing the animal's natural behavior in response to playbacks of either natural or synthetic calls (Gerhardt 1992a). Female frogs in reproductive condition, as well as males defending calling sites or territories, exhibit positive *phonotaxis* toward speakers broadcasting conspecific calls (Gerhardt 1995). While there are studies of phonotaxis behavior in the field (e.g., Schwartz et al. 2001; Narins et al. 2003, 2005; Amézquita et al. 2005, 2006; Ursprung et al. 2009), most phonotaxis studies have been conducted with female subjects under controlled conditions in laboratory sound chambers that provide high levels of control over acoustic test environments. Generally, two types of phonotaxis test designs are used, and both involve presenting repeated sound stimuli in a systematic way that simulates one or more naturally calling individuals (Gerhardt 1995). In *multiple-stimulus tests,* subjects are presented with two or more alternating or overlapping stimuli, and experimenters assess the proportions of subjects choosing each competing alternative. If a proportion of females greater than that expected by chance responds to a particular stimulus, the interpretation is that they can discriminate among stimuli, localize the source of at least one of them, and have a preference for that kind of signal. Multiple-stimulus tests are sometimes referred to as *choice tests* (e.g., Bush et al. 2002) or *discrimination tests* (e.g., Ryan and Rand 2001). By far, the most common type of multiple-stimulus test in studies of anuran communication has been the two-alternative choice test, which pairs two alternating stimuli against each other. In the second main type of phonotaxis test, *single-stimulus tests*, the experimenter measures behavior (e.g., latency of approach) in response to a single stimulus. If females respond in a single-stimulus test by approaching the speaker, the interpretation is that they can detect the sound, recognize it as the call of an appropriate (or at least acceptable) mate, and localize it. Single-stimulus tests have also been referred to as *no-choice tests* (e.g., Bush et al. 2002) and *recognition tests* (e.g., Ryan and Rand 2001) in the literature. Another natural behavior important in experimental studies of anuran hearing and communication is the *evoked vocal response* (EVR) (Capranica 1965). When stimulated with playbacks of conspecific calls, male frogs commonly call back in response. While more common in studies of territorial aggression (e.g., Bee and Gerhardt 2002) and call site defence (e.g., Wagner 1989), several studies of hearing in noise (e.g., Narins 1982; Penna et al. 2005; Penna and Hamilton-West 2007) and perceptual organization (e.g., Simmons and Bean 2000) have measured the EVR as well.

A third but less common approach used to study frog hearing involves measuring *prepulse inhibition*, or *reflex modification* (Yerkes 1904; Hoffman and Ruppen 1996). In this psychophysical technique, the amplitude of a reflex (e.g., leg flexion) elicited by one stimulus (e.g., mild shock) is modified by prior presentation of a brief, neutral stimulus (e.g., a tone) that does not elicit the reflex by itself (reviewed in Simmons and Moss 1995). By varying the neutral stimulus (e.g., tone amplitude) and measuring the resulting changes in reflex inhibition, it is possible to assess the animal's sensitivity to the neutral stimulus. The method is not currently used widely in research on anurans, but it has proven very useful as a way to probe perceptual organization in this group. Unfortunately, more traditional psychophysical methods involving classical or operant conditioning, as well as measures of other unconditioned responses (e.g., the galvanic skin response), have not been very successful tools in the study of frog hearing (reviewed in Simmons and Moss 1995; but see Elepfandt et al. 2000). The establishment of rigorous methods to study frog hearing based on conditioned responses would be a welcomed development in the field.

Two cautionary points about studies of hearing and sound communication in frogs are worth bearing in mind. First, a major disadvantage of natural responses to acoustic signals as behavioral assays (i.e., phonotaxis and the EVR) is the difficulty (perhaps impossibility) of distinguishing between signal *detection* and signal *recognition*, though it might be interesting or desirable to do so. Motivated subjects that fail to respond may do so because they did not detect the sounds *or* because the sounds were not recognized as conspecific signals. Likewise, animals may not behaviorally discriminate among signals even though they can perceive acoustic differences among them. Therefore, frog studies employing phonotaxis or the EVR provide useful information about "just meaningful differences," but not about "just noticeable differences" (Nelson and Marler 1990). This general point, of course, potentially applies to all animal playback studies and is an important consideration when interpreting results.

Second, some frogs perceive the sounds of a chorus not as "noise" (to be ignored) but as a "signal" of interest. Both male and female frogs can be attracted by the sounds of breeding choruses (Gerhardt and Klump 1988a; Bee 2007a; Swanson et al. 2007; Christie et al. 2010). One advantage of this type of behavior is that it makes frogs ideal systems for studies of "soundscape orientation" (Slabbekoorn and Bouton 2008). The major disadvantage, however, is that the behavioral significance of chorus sounds as a signal can potentially confound their salience as a masker in experimental studies. In some cases, the lack of a behavioral response to a target signal in the presence of chorus noise could mean that the "masker" actually acted as a relatively more attractive signal. Therefore, it is important to assess the attractiveness of chorus noises or other sounds used as maskers or distractors in control experiments (Vélez and Bee 2010, 2011, 2013; Nityananda and Bee 2011; Vélez et al. 2012, 2013).

6.2.2 Anuran Breeding Choruses as Cocktail Parties

Three aspects of anuran communication are particularly relevant when thinking of frog breeding choruses as cocktail-party-like acoustic scenes. First, frog calls can be exceedingly loud. In a detailed study of sound pressure levels and sound pattern radiation in 21 species of frogs from North America, Gerhardt (1975) reported peak sound pressure levels (peak SPL) at a distance of 1 m ranging between about 90–110 dB (see also Loftus-Hills and Littlejohn 1971; Passmore 1981; Penna and Solís 1998). As careful practitioners of frog bioacoustics research will attest, it is not always an insignificant technological challenge to reproduce frog calls with high fidelity (i.e., low noise, no distortion) at natural call amplitudes.

Second, breeding choruses are noisy, multisource acoustic environments (Fig. 6.1). Choruses can easily include hundreds of loudly calling males (Murphy 2003) and can be heard from distances of up to 1–2 km (Griffin 1976; Arak 1983). The noise generated by the aggregation of signaling males can reach maximum levels of up to 90 dB SPL (re 20 μPa, RMS) in frequency regions corresponding to spectral components of advertisement calls (Narins 1982). Moreover, anuran choruses usually include multiple species, perhaps numbering a dozen or more in the tropics. Sometimes, though certainly not always, the advertisement calls of the different species composing mixed-species choruses occupy different regions of the frequency spectrum in a way suggesting "acoustic niche" partitioning (see Sect. 5.2.1 in Chap. 5).

Third, frog calls typically are made up of temporal sequences of distinct sound elements (e.g., notes or pulses) with spectral energy (e.g., harmonics) occurring simultaneously across the frequency spectrum (Fig. 6.1). Information related to

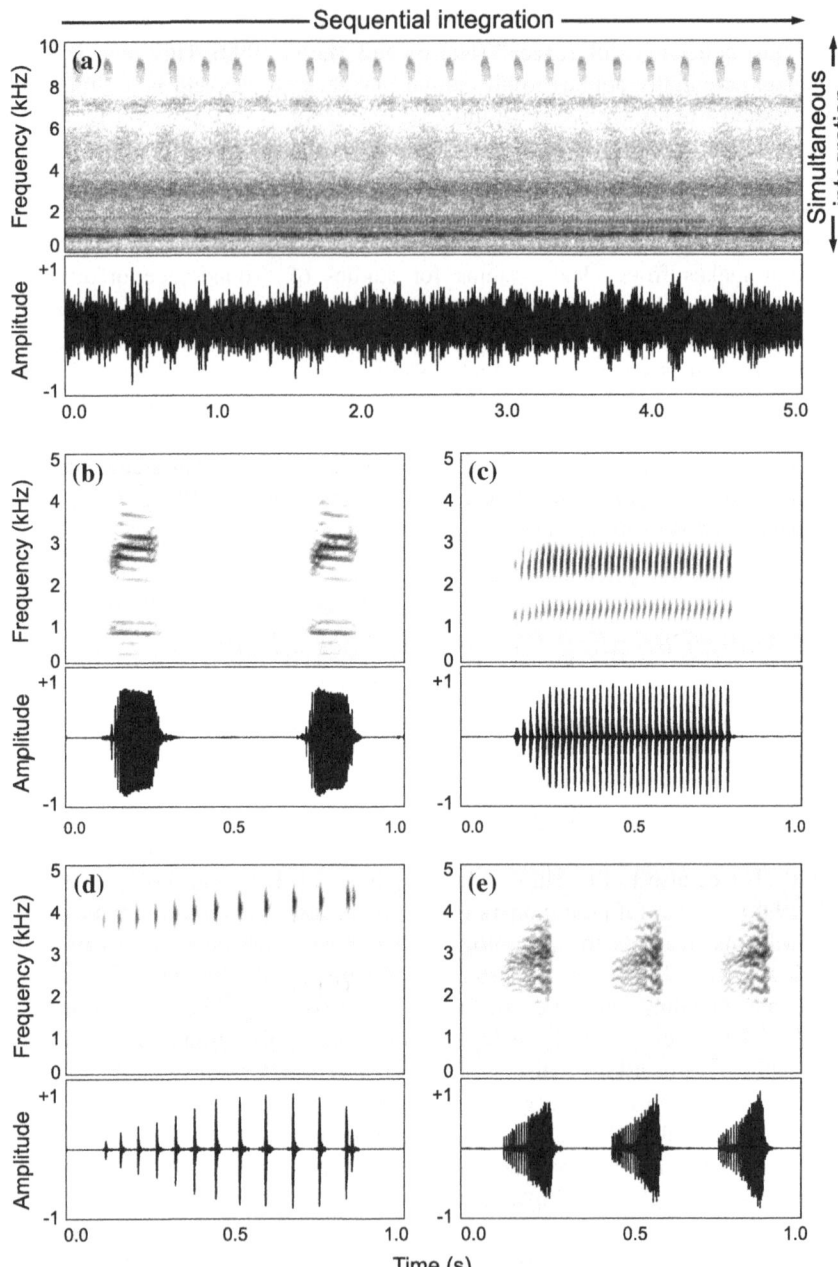

Fig. 6.1 The multisource acoustic scenes of frog breeding choruses. Spectrograms (*top panels*) and waveforms (*bottom panels*) of (**a**) the sound of a frog breeding chorus and the vocalizations of (**b**) *Hyla cinerea*, (**c**) *H. chrysoscelis*, (**d**) *Pseudacris maculata*, and (**e**) *Dendrobates histrionicus*. The acoustic scene of the breeding chorus in (**a**) includes calls of green treefrogs (*H. cinerea*) with spectral energy around 0.85 kHz and between 2.5 and 3.0 kHz, Gulf Coast toads (*Bufo valliceps*) consisting of a series of pulses with spectral energy around 1.5 kHz, and an orthopteran insect with spectral energy between 8.5 and 9 kHz. *Arrows* highlight temporal and spectral acoustic elements that the auditory system must integrate sequentially and simultaneously

species identity, body size and fighting ability, aggressive motivation, physiological condition, genetic quality, and individual identity can be conveyed by the distinctive temporal and spectral properties of frog calls (Gerhardt 1992b; Gerhardt and Bee 2007). Of course, not all of the frog calls in a chorus are generated by potential mates or sexual competitors. Sometimes the call of most immediate concern may be that of a predatory frog species (Schwartz et al. 2000; Bernal et al. 2007). Decoding the acoustic information in calls and generating adaptive behavioral responses requires that receivers assign temporal and spectral elements of vocalizations to their correct source.

Importantly, conspecific and heterospecific frogs are rarely the only sources of noise in the local environment. In addition to other signaling animals (e.g., insects, Paez et al. 1993; Fig 6.1a), anuran choruses often form in areas with high levels of abiotic noise from sound sources such as rivers and streams (Feng et al. 2002; Narins et al. 2004), waterfalls (Boeckle et al. 2009), rain (Penna et al. 2005) and wind (Penna et al. 2005). Sometimes abiotic "noise" may serve as a potent sound stimulus. West African reed frogs (*Hyperolius nitidulus*), for instance, exhibit negative phonotaxis in response to the sound of fire by fleeing and searching for cover (Grafe et al. 2002).With the increasing human population and the concomitant expansion of urban areas, anuran breeding habitats are also being invaded by anthropogenic noise (Chap. 5 and 14).

6.2.3 Some Relevant Features of the Anuran Ear

6.2.3.1 Frequency Tuning in the Peripheral Auditory System

All auditory systems exhibit frequency selectivity or "tuning" that ultimately arises from spectral filtering by the periphery (e.g., insects: Chap. 3; birds: Chap. 8 ; mammals: Chap. 10). In frogs, airborne sounds with frequencies characteristic of acoustic signals are encoded by two anatomically distinct sensory papillae in the inner ears (reviewed in Capranica 1976; Zakon and Wilczynski 1988; Lewis and Narins 1999; Simmons et al. 2007). The amphibian papilla exhibits tonotopic organization and the auditory nerve fibers innervating this papilla have best excitatory frequencies ranging from as low as 80 Hz up to 600–1,600 Hz, depending on the species. In contrast, the basilar papilla lacks tonotopic organization and is sensitive to higher frequencies than the amphibian papilla. Although frogs tend to have somewhat wider filter bandwidths compared to other vertebrates, behavioral studies have also reported enhanced selectivity in frequency regions that match the spectral components of conspecific advertisement calls (Moss and Simmons 1986; Fay and Simmons 1999).

Neurophysiological studies have shown the tuning of the basilar papilla, and in some cases that of the amphibian papilla, to match closely one or more frequency components in conspecific advertisement calls in many anurans (reviewed in Zakon and Wilczynski 1988; Lewis and Narins 1999; Gerhardt and Schwartz 2001).

The discovery of enhanced sensitivity to frequencies emphasized in conspecific signals gave rise to the view of the anuran peripheral auditory system as a "matched filter" (Capranica and Moffat 1983; Simmons 2013). Given the dispersion in frequency sometimes seen in multispecies frog choruses (Chap. 5), one primary function of a matched filter ear would be to reduce auditory masking by calls of heterospecific signalers and perhaps also abiotic noise. Unfortunately, behavioral audiograms exist for only a handful of species, and they offer limited support for the idea that frogs have their most sensitive hearing in the range of frequencies found in their own calls (Fig. 6.2). In green treefrogs (*Hyla cinerea*; Megela-Simmons et al. 1985) and African clawed frogs (*Xenopus laevis*; Elepfandt et al. 2000), audiograms measured using reflex modification and conditioning, respectively, exhibited increased sensitivity to frequencies emphasized in conspecific calls (Fig. 6.2a). This is not the case, however, for North American bullfrogs (*Rana catesbeiana*), which have harmonically rich calls with a bimodal spectrum characterized by two predominant peaks in acoustic energy, one centred between 200 and 300 Hz and the other between 1,200 and 1,600 Hz; there is very little acoustic energy in the range of 500–1,000 Hz (Capranica 1965; Bee and Gerhardt 2001). Using reflex modification, Megela-Simmons et al. (1985) measured broad, U-shaped behavioral audiograms with maximal sensitivity (thresholds of 10–20 dB SPL) in the range of 400–1,600 Hz (Fig. 6.2b). Sensitivity declined at rates of 26 dB/octave on the low end and 32 dB/octave on the high end. Interestingly, the lowest thresholds of the two animals for which audiogram data are available were at 600 and 800 Hz, precisely in the range of frequencies between the two peaks of the advertisement call's spectrum. Subsequent measures of the bullfrog's masked threshold and critical ratio function also were not closely tied to the spectrum of the advertisement call (Simmons 1988b). These findings with bullfrogs do not strongly support the matched filter hypothesis. Additional studies measuring behavioral audiograms in a greater diversity of frogs in quiet and noisy conditions are still badly needed.

6.2.3.2 Directionality of the Peripheral Auditory System

In order to reproduce, female frogs must often locate males in structurally complex habitats characterized by dense vegetation and other obstacles. Males have to localize intruders into their territory or calling site in these same habitats. Moreover, it may often be necessary to localize other individuals under very low light levels (e.g., in a rainforest on a cloudy night). Therefore, sound localization has obvious fitness consequences for anurans and, not surprisingly, it has been investigated in several behavioral and physiological studies over the years (reviewed in Eggermont 1988; Rheinlaender and Klump 1988; Klump 1995; Gerhardt and Huber 2002; Christensen-Dalsgaard 2005; Christensen-Dalsgaard 2011). In most species, directional acuity in the horizontal plane is on the order of 10°–20°, though acuity may be considerably better in some species (Shen et al. 2008). Three-dimensional directional acuity is typically lower than that in azimuth (e.g., Passmore et al. 1984).

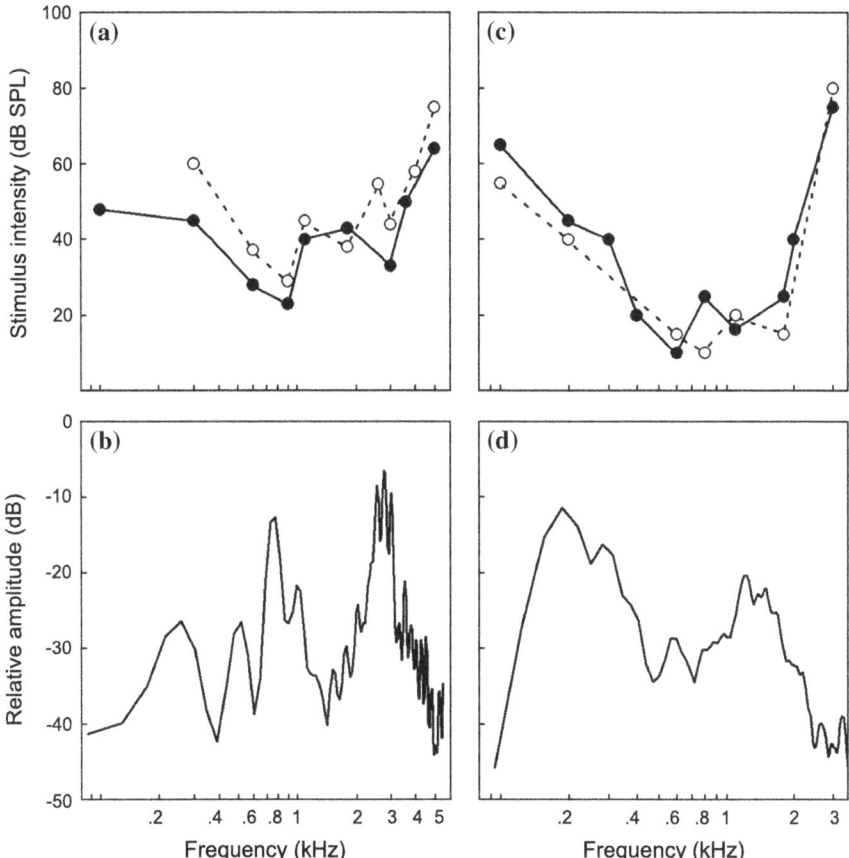

Fig. 6.2 Auditory sensitivity in relation to the spectral content of advertisement calls. Behavioral audiograms determined using reflex modification techniques (**a, c**) and the frequency spectra of advertisement calls (**b, d**) for (**a, b**) two green treefrogs (*Hyla cinerea*) and (**c, d**) two North American bullfrogs (*Rana catesbeiana*). As these two species illustrate, some frogs show increased hearing sensitivity to frequencies emphasized in conspecific calls (**a, b**) but others do not (**c, d**). Audiograms redrawn from Megela-Simmons et al. (1985) and reprinted with permission from the Acoustical Society of America

Two aspects of spatial hearing in frogs are worth bearing in mind. First, unlike mammalian ears, which are decoupled pressure receivers, anuran ears are pressure-gradient receivers. The two middle ears on opposite sides of the head are coupled through the mouth cavity and Eustachian tubes (Narins et al. 1988; Christensen-Dalsgaard 2005, 2011). This has important implications for directional hearing. Frogs are small animals and the dominant sound frequencies present in most frog calls (e.g., in the range of 0.2–7 kHz) typically have wavelengths (e.g., 172–5 cm, respectively) much longer than the width of most frogs' heads. Consequently, interaural time and level differences are quite small at the external surfaces of the tympanic membranes, which sit flush with the side of the head in most species. However, the internal coupling of the two middle ears gives rise to inherent directionality due to the interaction of sounds arriving at both surfaces of the tympanic membranes. Second, there are extratympanic inputs into the anuran auditory system (reviewed in Mason 2007). Sounds that enter via extratympanic pathways like the floor of the mouth and the lungs account for most of the sensitivity and directionality to low frequency sounds (reviewed in Christensen-Dalsgaard 2005). As discussed below, spatial hearing may be important in some aspects of the perceptual organization of acoustic scenes in frogs, rendering the anuran auditory system a particularly interesting model for studies of hearing and communication in noisy environments.

6.3 Functional Consequences of Communicating in a Chorus

6.3.1 Decreased Signal Active Space

Active space is a fundamental concept in the study of communication referring to the spatial extent over which signals effectively elicit appropriate behavioral responses from receivers (Bradbury and Vehrencamp 2011). High noise levels and interfering sounds are among the many factors that influence the active space of acoustic signals. For frogs, this means receivers may only be able to hear and assess the calls of a few males in their immediate vicinity. One of the first studies to specifically address the problems imposed by chorus noise was conducted with green treefrogs by Gerhardt and Klump (1988b). In single-stimulus tests, females responded to male advertisement calls when the signal-to-noise ratio (SNR) was at least 0 dB. Based on this result and on previous field observations of the distances moved by females in the chorus (Gerhardt et al. 1987), they suggested that, in natural settings, female green treefrogs might sample just a few nearby males whose calls reach or exceed the background noise generated by the aggregation. Wollerman (1999) used two-alternative choice tests to estimate the SNR at which females of the hourglass treefrog, *Dendropsophus ebraccatus* (formerly *Hyla ebraccata*), responded to advertisement calls in the presence of chorus noise (Fig. 6.3a). Females were allowed to choose between two speakers separated by 180°; one speaker broadcast only chorus noise and the other speaker broadcast chorus noise and advertisement calls at SNRs of 0 dB, +1.5 dB, +3 dB, or +6 dB. The minimum SNR necessary to elicit differential approaches to the speaker broadcasting the call was between +1.5 and +3 dB (Fig. 6.3a). By considering the

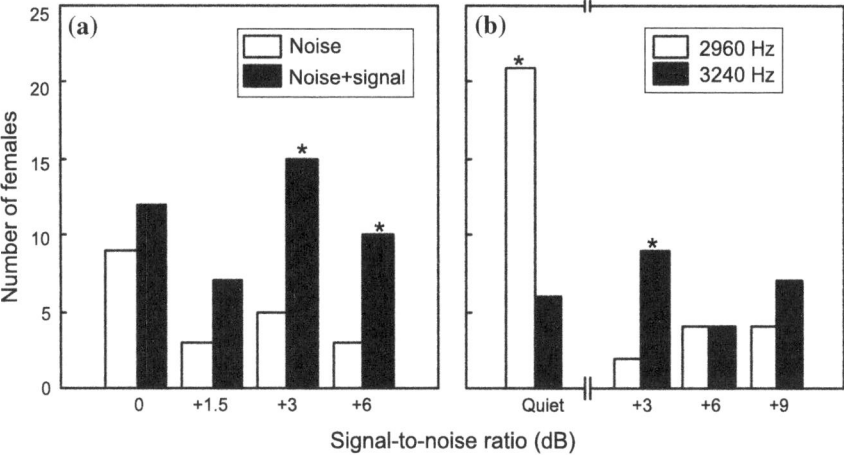

Fig. 6.3 Signal recognition and discrimination in noise by females of the hourglass treefrog, *Dendropsophus ebraccatus*. Results of two-alternative choice experiments showing the numbers of females choosing (**a**) speakers broadcasting chorus noise alone versus chorus noise + signal at the indicated signal-to-noise ratios and (**b**) speakers broadcasting a call with a dominant frequency of either 2,960 or 3,240 Hz at the indicated signal-to-noise ratios. Asterisks indicate significant differences. Graphs redrawn from (**a**) Wollerman (1999) and (**b**) Wollerman and Wiley (2002) and reprinted with permission from Elsevier

distribution and density of calling males in the chorus, Wollerman (1999) esti-mated that, at any given location in the chorus, females could recognize the calls of only the nearest male. If *D. ebraccatus* females were to exert mate choice under these conditions, they would have to move relatively large distances to sample several males. Field observations suggest that female *D. ebraccatus* move and sample up to seven males in a chorus before mating (Morris 1991).

More recently, Bee and Schwartz (2009) investigated the effects of a "chorus-shaped" masker (i.e., an artificial noise with the long-term spectrum of a chorus) on the ability of females of Cope's gray treefrogs (*Hyla chrysoscelis*) to recognize advertisement calls. They compared three different methods of estimating "signal recognition thresholds," which they operationally defined as the minimum signal level required to elicit reliable positive phonotaxis. Defined in this manner, the signal recognition threshold is conceptually analogous to the "speech reception threshold" common in studies of the human cocktail party problem (Plomp and Mimpen 1979a, b). In a single-stimulus experiment, the target signal (a conspecific advertisement call) was broadcast across different tests at one of nine different fixed levels ranging between 37 and 85 dB SPL in 6 dB steps. In a parallel two-alternative choice experiment, females were given a choice between the same target signal and a decoy signal (calls of the closely related eastern gray treefrog, *Hyla versicolor*). Across different tests, the two signals were broadcast at the same fixed levels used in the single-stimulus experiment. Estimates of signal recognition thresholds were calculated based on the proportion of females approaching the target signal, the latency to reach the source of the target signal, and each subject's angular orientation relative to the target signal 20 cm from the subject's initial position. In a third experiment, Bee and Schwartz (2009) used an adaptive tracking procedure to estimate signal recognition thresholds; in this procedure, thresholds are determined by adjusting the SNR up or down between consecutive tests depending on the subject's response in a previous test (see Bee and Schwartz 2009 for a description). In the absence of noise, most estimates of signal recognition thresholds in all three experiments ranged between 35 and 42 dB SPL, which was similar to that determined for the eastern gray treefrog by Beckers and Schul (2004). Signal recognition thresholds in noise typically increased to between 65 and 71 dB SPL, which corresponded to threshold SNRs ranging between about -5 to $+1$ dB, values similar to, or slightly lower than, those obtained for green treefrogs (Gerhardt and Klump 1988a) and hourglass treefrogs (Wollerman 1999). These studies of gray treefrogs indicate that females can detect and recognize calls at drastically lower sound levels in the absence of noise compared with in the presence of chorus-like noise. Consequently, the background noise in a chorus has substantial impact on reducing the *potential* active space of a male's signal.

While a few studies have investigated the impacts of anthropogenic noise on the signaling behaviors of male frogs (reviewed in Chap. 5), only one published study has investigated its impacts on signal active space (Bee and Swanson 2007; see Barrass 1985 for an unpublished account). Bee and Swanson (2007) recorded road traffic noise near highways adjacent to two wetlands where frogs breed. In single-stimulus phonotaxis experiments with Cope's gray treefrog, they estimated signal recognition thresholds in three conditions: a no masker condition, a chorus-shaped noise masking condition, and a traffic-shaped noise masking condition. The traffic-shaped noise was a broadband noise with most of its energy below 1 kHz. Masked recognition thresholds were generally similar in the two masking conditions and about 25 dB higher than in the no-masker control condition. These results suggest that traffic noise could impose limitations on communication. An interesting study yet to be conducted is to compare female performance under the combined influences of chorus noise and anthropogenic noise. In addition, traffic noise could have even greater effects in species with advertisement calls having lower frequencies. However, given the numbers of dead frogs seen on roads during seasonal migrations into suitable breeding habitats, car tires may already do more damage to anuran populations than traffic noise ever will. While investigations into the effects of anthropogenic noise have become increasingly important in conservation biology in recent years (Barber et al. 2010; Brumm 2010), anthropogenic noise may be relatively far down on the list of threats to anurans living in a world with chytrid fungus, ranavirus, rampant habitat loss, invasive species, chemical toxins, and water pollution (reviewed in Semlitsch 2003).

6.3.2 Impaired Proximity Assessment

Male frogs maintain non-random spacing in choruses (Martof 1953; Gerhardt et al. 1989). One of the primary acoustic cues for doing so is the perceived amplitude of nearby neighbors' calls (Wilczynski and Brenowitz 1988). When the amplitude of the calls of a neighbor or an intruder exceeds some threshold, males commonly switch from producing advertisement calls to aggressive calls (Rose and Brenowitz 1991). Lemon (1971) and Passmore and Telford (1981) hypothesized that neighboring male frogs avoid call overlap through precise call timing interactions to preserve their abilities to judge neighbors' proximities. Schwartz (1987) found support for this prediction in field playback experiments with the eastern gray treefrog (*H. versicolor*), spring peepers (*Pseudacris crucifer*), and the yellow cricket treefrog (*Dendropsophus microcephalus*, formerly *Hyla microcephala*). He used a portable frog-call synthesizer that could be triggered to produce a call after various delays in response to a subject's own calls. On different trials, and over a range of stimulus amplitudes, synthetic calls were triggered so that they either overlapped the subject's calls or alternated with them in time. Males gave significantly more aggressive calls in the alternating condition compared to when stimulus calls overlapped the subjects' calls. This result indicates that males cannot as accurately estimate a neighbor's proximity when their calls overlap in time. Therefore, call alternation might function in allowing males to maintain optimal inter-male distances in the chorus. To our knowledge, no study has investigated the influences of overall background noise levels on aggressive behavior and the maintenance of inter-male spacing.

6.3.3 Impaired Source Localization

As described above, frogs tested in the laboratory are generally quite accurate at localizing sound sources. Therefore, one *potential* problem that frogs might be expected to encounter in the noisy, multisource environment of a breeding chorus involves increased difficulty localizing sources. Two studies of phonotaxis behavior in female frogs have investigated the effects of call overlap on source localization. Passmore and Telford (1981) showed that females of the painted reed frog (*Hyperolius marmoratus*) were equally good at localizing sources that broadcast either alternating or synchronous calls. Schwartz (1987) reported similar findings in his studies of eastern gray treefrogs, yellow cricket treefrogs, and spring peepers. Interestingly, there appear to have been no previous studies of frogs' abilities to localize sounds in the presence of high levels of background noise (Feng and Schul 2007). Additional work is needed to assess the influences of chorus noise on the acuity of sound localization in anurans in order to determine the extent to which impaired localization constitutes part of the frog's cocktail party problem.

6.3.4 Impaired Sound Pattern Recognition

High levels of background noise and overlapping signals can impact the ability of receivers to recognize spectral and temporal properties that identify conspecific calls. Consider, for example, the calls of the closely related eastern gray treefrog (*H. versicolor*) and Cope's gray treefrog (*H. chrysoscelis*), which differ primarily in pulse rate. Under quiet conditions, females of both species are highly selective for calls with pulses produced at normal conspecific rates, and in two choice tests they practically never choose calls of the wrong species (Littlejohn et al. 1960; Gerhardt and Doherty 1988; Bush et al. 2002; Schul and Bush 2002; Bee 2008a; Nityananda and Bee 2011). In the presence of chorus-shaped noise, however, responsive females surprisingly choose the correct and incorrect calls in similar proportions at low SNRs (e.g., -9 dB), indicating that high noise levels potentially impair species recognition in choruses (Bee 2008a). High noise levels might also contribute to the mating mistakes females occasionally (though rarely) make in real chorus environments (Gerhardt et al. 1994). Call overlap from neighboring males is also a serious problem impairing recognition of conspecific calls in these frogs (Schwartz 1987; Marshall et al. 2006; Schwartz and Marshall 2006).

Schwartz (1987) used multiple-stimulus experiments to test the hypothesis that acoustic interference in the form of call overlap disrupts a female's perception of temporal information critical for species discrimination. He tested females of the eastern gray treefrog, yellow cricket treefrog, and spring peeper. Males of the first two species produce pulsatile advertisement calls and pulse timing information is important for sound pattern recognition. In contrast, male spring peepers produce advertisement calls lacking a pulsed structure and consisting instead of a single frequency modulated tone ("peep"). In a four-alternative choice test, Schwartz (1987) gave females a choice between four stimuli presented from a four-speaker array, with each speaker separated by 90° and oriented toward the centre of a circular arena. Two stimuli were presented as alternating calls broadcast from speakers on opposite sides of the arena (separated by 180°); the other two stimuli were broadcast as overlapping calls from the other pair of speakers (also separated by 180°). In different tests, the two overlapping calls were either presented "in-phase" (i.e., precisely synchronized) or "out-of-phase." When overlapping calls were in-phase, females of all three species chose alternating and overlapping calls in similar proportions. In the out-of-phase conditions, however, the temporal structure of pulsed calls was no longer preserved. Females of eastern gray treefrogs and yellow cricket treefrogs, the two species with pulsed calls, strongly preferred alternating to overlapping calls in the out-of-phase condition. In contrast, spring peepers, which do not produce pulsed advertisement calls, showed no preferences for alternating calls versus overlapping calls presented out of phase. These results supported the hypothesis that call overlap obscures temporal features of calls necessary for sound pattern recognition.

Given the findings of Schwartz (1987), it would seem advantageous for male eastern gray treefrogs (*H. versicolor*) to avoid call overlap with their neighbors. Surprisingly, however, mesocosm-scale experiments creating choruses of different group sizes in an artificial pond showed that groups larger than two males do not avoid call overlap (Schwartz et al. 2002). Instead, males of *H. versicolor* (and also its close relative, *H. chrysoscelis*) increase the number of pulses in their calls while reducing call rates in response to the calls of other individuals in the chorus and during playbacks of both calls and chorus noise (Wells and Taigen 1986; Klump and Gerhardt 1987; Schwartz et al. 2002; Love and Bee 2010; Ward et al. 2013b). Males usually maintain a constant "pulse effort" (number of pulses per call × call rate). Schwartz et al. (2008; 2013) have tested two hypotheses that might account for this interesting signaling behavior (Chap. 5). According to the "interference risk hypothesis" (Schwartz et al. 2001), the observed correlated changes in call duration and rate function to increase the chances that females will perceive enough unmasked pulses and interpulse intervals per call necessary to allow call recognition. This hypothesis predicts that, whenever there is risk of call overlap, producing longer calls at lower call rates should attract more females than producing shorter calls at higher call rates because, on average, the number of unobscured pulses and interpulse intervals should be higher in the former combination. To test this prediction, Schwartz et al. (2008) gave *H. versicolor* females a choice between short and long calls with equal pulse efforts, which had, on average, either 50 or 67 % of the call overlapped by another call or by a burst of chorus-shaped noise. Their results strongly refuted the interference risk hypothesis. Females showed no preferences for the longer alternatives under any of the circumstances in which overlapping calls or bursts of noise interfered with the temporal structure of the advertisement calls.

More recently, Schwartz et al. (2013) have tested an alternative hypothesis (the "call detection hypothesis") to explain the correlated changes in call duration and rate in male signaling behavior in the presence of noise and competing calls. The key prediction of this hypothesis is that longer calls are easier to detect in a noisy environment. This prediction is based on a fundamental feature of auditory processing known as temporal integration, which refers to the ability to integrate acoustic features over time (reviewed in Brumm and Slabbekoorn 2005; Recanzone and Sutter 2008). It is well established that signal detection thresholds decrease (over a limited range) as a function of increasing signal duration (Heil and Neubauer 2003), and that other animals besides gray treefrogs also lengthen their signals as a function of ambient noise levels (e.g., Brumm et al. 2004). In quiet conditions, a minimum number of consecutive pulses (e.g., 3–6 in *H. versicolor* and 6–9 in *H. chrysoscelis*) are required to elicit positive phonotaxis from females of both gray treefrog species (Bush et al. 2002; Vélez and Bee 2011). These findings are generally consistent with temporal integration by neurons in the frog midbrain that require a minimum number of correct interpulse intervals before firing (Alder and Rose 1998; Edwards et al. 2002; Schwartz et al. 2010a). Importantly, different pulse-integrator neurons fire in response to different threshold numbers of pulses (Alder and Rose 1998; Edwards et al. 2002).

Therefore, signals with more pulses are more likely to activate larger populations of pulse-integrator neurons. Might temporal integration by these or similar neurons provide an advantage to producing longer calls in noise? In a series of single-stimulus phonotaxis tests, Schwartz et al. (2013) used an adaptive tracking procedure (Bee and Schwartz 2009) to measure response thresholds of female *H. versicolor* during broadcasts of calls of different durations (10, 20, 30, or 40 pulses) in the presence of chorus shaped noise. In contrast to predictions of the call detection hypothesis, response thresholds did not vary as a function of call duration. These results are inconsistent with the hypothesis that changes in the signaling behaviors of male eastern gray treefrogs in acoustically cluttered environments function to take advantage of temporal integration in the receiver's auditory system. At present, an entirely satisfactory functional explanation of this interesting signaling behavior remains elusive.

6.3.5 Constraints on Mate Choice Preferences

Anurans have featured prominently in studies of sexual selection (Ryan 1991; Ryan and Rand 1993; Gerhardt and Huber 2002). Consistent with findings from numerous other taxa (reviewed in Andersson 1994), female frogs often exhibit directional preferences for various signal features, such as longer or louder calls, faster rates of calling, more complex calls, and lower sound frequencies (reviewed in Ryan and Keddy-Hector 1992; Gerhardt and Huber 2002). Much of what we currently know about the intraspecific mate choice preferences of female frogs comes from phonotaxis tests conducted under optimal listening conditions. However, it is also a well-established fact that chorus noise and call overlap can render a male's calls less attractive than they would be in quiet conditions (Schwartz and Wells 1983; Wells and Schwartz 1984). Current opinion is that mate choice preferences identified in quiet laboratory sound chambers are usually constrained by the overlapping calls and high levels of background noise present in the natural environment.

The abilities of call overlap and high noise levels to constrain mate choice preferences based on temporal call features is well illustrated by studies of gray treefrogs. In two-alternative choice tests conducted in quiet conditions, female eastern gray treefrogs (*H. versicolor*) and Cope's gray treefrogs (*H. chrysoscelis*) exhibit strong, directional preferences for longer calls that contain more pulses (Gerhardt et al. 1996; Schwartz et al. 2001; Bee 2008b; Vélez et al. 2013; Ward et al. 2013b). Evidence from quantitative breeding studies of the former species indicate that preferences for longer calls provide females with indirect benefits associated with increased offspring fitness (Welch et al. 1998). The expression of preferences for longer calls, however, is constrained under acoustic conditions designed to simulate the "real world" acoustic environments of breeding choruses. In laboratory two-alternative choice tests conducted in the presence of chorus noise, females of both species exhibit reduced preferences for longer calls

(Schwartz et al. 2001; Bee 2008b). Under some conditions, noise was able to completely abolish or even reverse preferences for longer over shorter calls in *H. chrysoscelis* (Bee 2008b). Schwartz et al. (2001) extended their investigations of *H. versicolor* by testing female preferences for call duration under more realistic conditions. In a mesocosm-scale experiment, they created real choruses consisting of between four and eight males calling in an artificial pond constructed in a greenhouse. Male calling behavior was monitored while individual females were released from a holding cage at the centre of the pond and allowed to choose a mate. Overlapping calls were common. Logistic regression analyses revealed only weak discrimination by females based on differences in the average number of pulses in male calls. Results from a field experiment conducted in a natural chorus were similar. In that experiment, Schwartz et al. (2001) placed eight speakers housed in cylindrical screen cages around the edge of a pond in which *H. versicolor* males called and formed a chorus. Typical calls contain between 8 and 28 pulses, with an average pulse number of about 16 pulses per call. The eight speakers at the pond broadcast synthetic calls that had 6, 9, 12, 15, 18, 21, 24, or 27 pulses. The results were clear; females discriminated strongly against the shortest, six pulse call, but there was little evidence that females discriminated among calls with nine or more pulses. In summary, experiments with gray treefrogs conducted in the laboratory, in mesocosm choruses in artificial ponds, and in natural choruses indicate that high levels of background noise and call overlap impair the ability of females to exercise adaptive mate choice preferences for males that produce longer calls with more pulses.

Chorus noise can also impact the choices females make between calls with different spectral properties (Fig. 6.3b). In two-alternative choice tests conducted in the absence of noise, female hourglass treefrogs preferred calls with a dominant frequency of 2,960 Hz to those with a dominant frequency near the population mean of 3,240 Hz (Wollerman 1998; Wollerman and Wiley 2002). Because dominant frequency is inversely related to body size, preferences for lower frequency calls might allow females to choose larger males (Wollerman 1998). In the presence of chorus noise, however, the preference for the low frequency call was abolished at SNRs of +6 and +9 dB (Wollerman and Wiley 2002). Interestingly, the preference was actually reversed in favor of the higher-frequency call at a SNR of +3 dB (Fig. 6.3b). This result was interpreted as a possible change in strategy used by females in very noisy conditions. At an ultimate level, shifting from a discrimination task to one of detecting the most common call produced by conspecific males (e.g., calls with near-average values) might allow females to avoid mating with heterospecific males when high levels of background noise introduce uncertainty about species identity (Wollerman and Wiley 2002). A possible proximate-level hypothesis for these data is that signal recognition thresholds in noise vary as a function of dominant frequency. Recall that (Wollerman 1999) estimated a SNR for signal recognition between +1.5 and +3 dB for an average call with a dominant frequency of 3,240 Hz (Fig. 6.3a). If the masked recognition threshold for a 2,960 Hz call is higher, the observed change in female preference in favor of the 3,240 Hz call at a low SNR might be related to differences in the ability of females to detect or recognize both calls in the presence of chorus noise.

As a counter example to studies of gray treefrogs and hourglass treefrogs, Schwartz and Gerhardt (1998) have shown using two-alternative choice tests with another treefrog that the addition of chorus-like noise can also reveal significant preferences for calls differing in frequency that were not exhibited in quiet conditions. In that study, female spring peepers failed to discriminate behaviorally between two calls with different dominant frequencies in the absence of noise, but preferred a higher-frequency alternative when tested in the presence of artificial chorus noise. Multiunit recordings from the torus semicircularis of females demonstrated that this result was associated with a de-sensitization of the auditory system in response to loud noise (i.e., a threshold shift). At a high stimulus amplitude (85 dB SPL) without noise, calls of different frequency elicited similarly strong neural responses. This was likely due to broadening of eighth nerve tuning curves and rate saturation. However, isointensity neural response profiles became more peaked as stimulus amplitudes were reduced to 55 dB SPL. Interestingly, when noise simulating a loud chorus of males accompanied stimulus calls presented at 85 dB SPL, neural responses were more similar to those obtained at 55–65 dB SPL in quiet conditions. Moreover, such noisy conditions were the only ones during which there was a significant association between both the neural response strength to, and behavioral discrimination of, calls of different frequency for individual females.

6.3.6 Summary and Future Directions

The challenges associated with communicating in the acoustic scene of a breeding chorus impose functional consequences on receivers. To date, most of this work has examined the consequences for female frogs in the contexts of sound pattern recognition and choosing preferred mates. By comparison, we know less about how chorus noise and interfering signals impact source localization, not only in the horizontal and vertical planes, but also in terms of source proximity. An important direction for future studies will be to assess how frogs determine source location and proximity in noisy, multisource conditions. Another goal for future research will be to understand better the effects that noise has on the neural processing of communication sounds in relation to its impacts on perception (e.g., Schwartz and Gerhardt 1998). For example, auditory nerve fibers exhibit shifts in tone-evoked rate-level functions in noise that appear to function as a gain control mechanism that allows the auditory system to encode intensity information in the presence of noise (e.g., Narins 1987). Likewise, there is neurophysiological evidence from recordings in the frog midbrain to suggest noise can enhance the encoding of important information (e.g., amplitude modulation) through stochastic resonance (e.g., Bibikov 2002). Precisely how these sorts of neural phenomenon correspond to a receiver's perceptual experience in noise remains poorly understood. While the studies reviewed in this section illustrate the *problems* for communication posed by listening in breeding choruses, an important next step for future research will be to understand the potential *solutions* by which receivers overcome or ameliorate them. We turn to these issues in the next three sections.

6.4 Release from Auditory Masking

6.4.1 Shifts in Frequency Tuning Associated with Heterospecific Signalers

As discussed above (Sect. 6.2.3.1), there is evidence that in some frog species, the peripheral auditory system acts as a "filter" tuned to the spectral content of conspecific calls. As a result, the periphery can reduce the potential for auditory masking by filtering out frequencies in heterospecific calls. We return to this idea here to highlight how the labile nature of behavioral sensitivity to different sound frequencies in frogs may be related to hearing and communication in a chorus environment.

Males of *Allobates femoralis* (formerly *Epipedobates femoralis*), a neotropical frog common throughout Amazonia, defend territories on the forest floor against intrusion by other conspecific males; an important component of a defensive response is phonotaxis toward a calling intruder (Narins et al. 2003, 2005; Amézquita et al. 2005, 2006). In some geographic locations, but not others, *A. femoralis* occurs syntopically with another frog species, *Ameerega trivittata* (formerly *E. trivittatus*). Both species produce calls composed of a series of pulses, and *A. trivittata* calls have lower, but overlapping, frequencies that could potentially mask the frequency content of *A. femoralis* calls (Amézquita et al. 2005). There is no evidence to suggest the spectral content of the two species' calls have diverged in sympatry (Amézquita et al. 2006). In a study of territorial behavior in *A. femoralis*, Amézquita et al. (2006) used a phonotaxis assay to generate behavioral "frequency-response curves" that measured the magnitude of response as a function of the carrier frequency of a synthetic call (Fig. 6.4). By comparing responses from male *A. femoralis* from populations that were sympatric and allopatric with *A. trivittata*, they tested the hypothesis that male responsiveness to calls with different frequencies is shaped in ways consistent with evolutionary shifts in frequency tuning in the auditory system. That is, they were interested in testing whether the frequency sensitivity of the auditory system is evolutionarily labile and can change in ways that reduce auditory masking by heterospecifics. The results unequivocally showed that the low-frequency tail of the behavioral frequency-response curve of *A. femoralis* was shifted toward higher frequencies in areas of sympatry, as expected if the tuning of the auditory system had shifted over evolutionary time to filter out the calls of *A. trivittata* (Fig. 6.4). More recently, Amézquita et al. (2011) showed that the signal recognition space, both in the spectral and temporal domains, is shaped in ways that reduces acoustic interference by heterospecific calls in a complex acoustic environment of ten vocally active species.

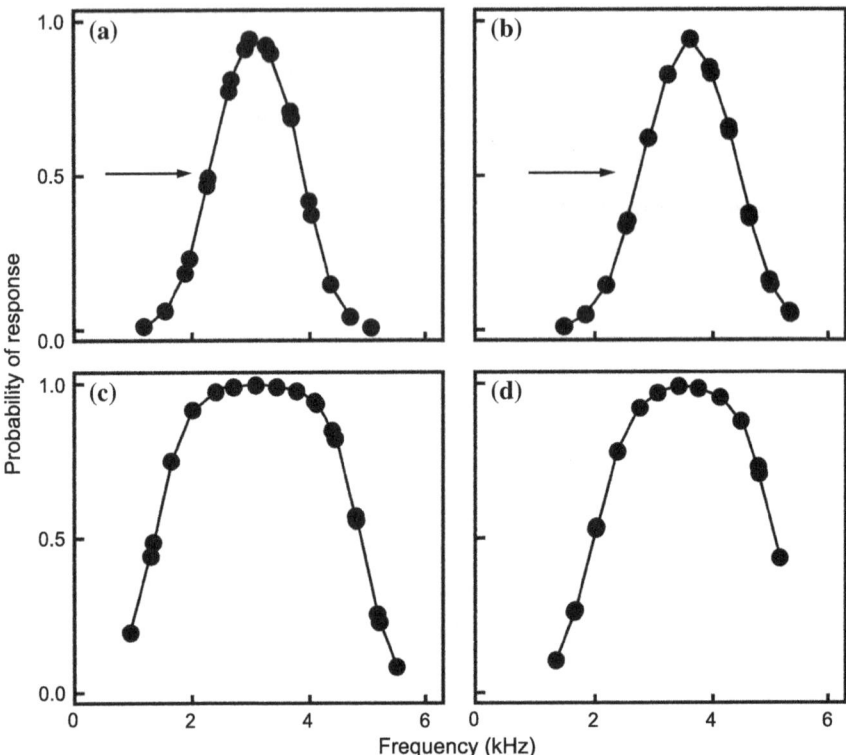

Fig. 6.4 Putative shifts in frequency tuning in the Amzonian dendrobatid frog, *Allobates femoralis*. Males of this species produce a call composed of a series of frequency modulated pulses sweeping upward between about 2.6 and 3.4 kHz. This species has a geographic distribution partially overlapping that of another frog, *Ameerega trivittata*. In *A. trivittata*, male calls have slightly lower frequency (e.g., 2.3–3.0 kHz) that partially overlaps the lowest frequencies in *A. femoralis* calls and thus represent a potential source of masking and interference in sympatric populations. The data depicted here represent behavioral frequency response curves showing the probability that males of *A. femoralis* responded to playbacks at the given frequency interpolated from logistic regression analysis of binary responses. As illustrated here, the low-frequency end of behavioral frequency response curves were shifted to higher frequencies in (**a, b**) sympatric populations compared with (**c, d**) allopatric populations. Whether these shifts represent shifts in auditory tuning or shifts in behavioral decision rules remains an interesting and important question for future study. Redrawn from Amézquita et al. (2006) with permission from John Wiley and Sons

If the behavioral results of (Amézquita et al. 2006) can be confirmed with *physiological* measures of frequency sensitivity, it would represent a very exciting confirmation that the "matched filter" tuning of frog auditory systems is evolutionarily labile in ways that can bring about a release from auditory masking in mixed-species choruses without concomitant (co-evolutionary) shifts in the spectral content of conspecific signals. At present, however, it is not possible to draw this conclusion without physiological measures of auditory tuning because the results appear equally consistent with shifts in either frequency tuning or behavioral decision rules (Chap. 2). That is, it is possible that the behavioral differences between sympatric and allopatric populations illustrated in Fig. 6.4 reflect some form of stimulus-specific behavioral plasticity, such as decreases in aggressive responsiveness (e.g., Bee 2003) or increases in aggressive response thresholds (e.g., Humfeld et al. 2009).

6.4.2 Shifts in Frequency Tuning Associated with Environmental Noise

Among the predicted evolutionary responses to high levels of abiotic background noise in a habitat would be to use a frequency channel free from the noise or a different sensory modality all together. Frogs exhibit both types of solutions in response to noise generated by fast-flowing water. The use of visual signals by frogs breeding in such habitats is well known (reviewed in Hödl and Amézquita 2001). The discovery that some frogs breeding near sources of loud water noise also communicate with ultrasonic signals is arguably the most important recent finding in studies of anuran acoustic communication (Feng et al. 2006; Arch et al. 2008, 2009). In some habitats, the frequency spectrum of sound generated by fast-flowing water is characterized by high levels of acoustic energy below 30 kHz, with most energy present at the very low end (e.g., 100 Hz) of this frequency range. An anuran acoustic communication system in this habitat might benefit from using ultrasonic frequencies to reduce auditory masking (see Sect. 5.3.1.1 in Chap. 5).

Sensitivity to ultrasound has been demonstrated in a few species of frogs, such as the concave-eared torrent frog, *Odorrana tormota* (formerly *Amolops tormotus*), the hole-in-the-head frog, *Huia cavitympanum*, and the large odorous frog, *Rana livida*. Ultrasonic signaling has been studied most extensively in the first two of these species. Recordings of evoked potentials and single units from the auditory midbrain of these two species reveal sensitivity to frequencies above 20 kHz (Feng et al. 2006; Arch et al. 2009). Doppler vibrometry measurements of the tympanic membrane of *H. cavitympanum* revealed a broad peak in sensitivity that extended into the ultrasonic frequency range (Arch et al. 2009). At a behavioral level, sensitivity to ultrasonic signals has been documented through playback experiments with males of both species. In field playback experiments, Arch et al. (2009) found that *H. cavitympanum* males respond antiphonally to playbacks of both audible (<20 kHz) and ultrasonic (>20 kHz) components of conspecific calls. Similarly, audible and ultrasonic components of calls also evoke vocal responses from *O. tormota* (Feng et al. 2006). In the latter species, ultrasonic hearing involves an apparently derived ability to close the Eustachian tubes, which are thought to be permanently open in most other frogs. The accompanying increase in the impedance of the middle ear both boosts tympanic vibrations for higher sound frequencies and lowers them for lower frequencies (Gridi-Papp et al. 2010). Also likely helpful for ultrasonic hearing are recessed tympana and unusually thin tympanic membranes (Feng et al. 2006; Feng and Narins 2008).

6.4.3 Spatial Release from Masking

In natural settings, including multisource social aggregations, sound sources are often spatially separated. Human listeners experience significant improvements in speech intelligibility when the sources of target speech and competing sources of noise are spatially separated (reviewed in Bronkhorst 2000). Compared to a "co-localized" condition, in which target speech and masking noise originate from the same location, a release from masking of about 6–10 dB is observed when signals and maskers are displaced in azimuth. Under binaural listening conditions, the major cues for this *spatial release from masking* in humans are an improvement in the signal-to-noise ratio at one ear and disparities in the interaural time and level differences of signals and maskers (Bronkhorst 2000). Spatial release from masking is not unique to humans and has been demonstrated in several birds (Chap. 8) and mammals (Chap. 10). Several studies have also investigated the effect of spatial separation between signals and noise in anurans.

Using two-alternative choice tests, Schwartz and Gerhardt (1989) measured the ability of female green treefrogs to detect and discriminate between advertisement calls and aggressive calls in the presence of broadband maskers that were either co-localized with the signals or separated by either 45° or 90° in azimuth. Spatial separation led to observable improvements in the ability of females to detect calls, but not to discriminate between the two call types. Consistent with signal detection theory, this result suggests signal discrimination is a more difficult task than signal detection (Chap. 2). The magnitude of spatial unmasking was estimated to be about 3 dB. More recent studies of Cope's gray treefrog (*H. chrysoscelis*) have investigated spatial release from masking in the presence of chorus-shaped noise. Using single-stimulus phonotaxis experiments, Bee (2007b) presented females with advertisement calls at SNRs between −12 and +12 dB (in 6 dB steps) in the presence of chorus-shaped noise. The target signal and masker were broadcast from two speakers that were either adjacent (angular separation of 7.5°) or spatially separated by 90°. Based on measures of normalized response latencies, Bee (2007b) estimated the magnitude of spatial unmasking to be on the order of 6–12 dB (Fig. 6.5). A more recent study using an adaptive tracking procedure to measure signal recognition thresholds in co-localized and separated (90°) conditions revealed about 4 dB of masking release in the separated condition (Fig. 6.5; Nityananda and Bee 2012). These estimated magnitudes of masking release are biologically relevant, as females are known to discriminate differences in SNR as small as 2 dB (Bee et al. 2012). Bee (2008a) found that spatial separation between signals and chorus-shaped noises also influenced the ability of female Cope's gray treefrogs to discriminate between conspecific advertisement calls and those of the

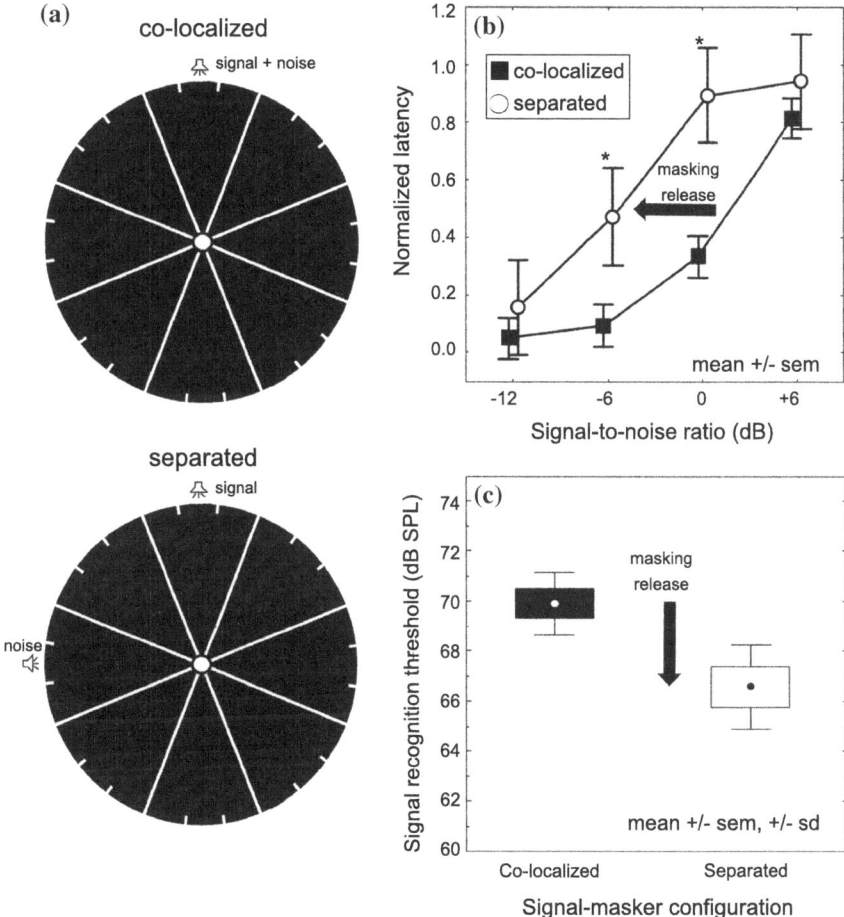

Fig. 6.5 Spatial release from masking in Cope's gray treefrog, *Hyla chrysoscelis*. **a** Schematic top view of a phonotaxis test arena (2 m diameter) showing the positions of speakers, signals, and noises in a co-localized condition (signal and noise broadcast from the same or immediately adjacent speakers) and a separated condition (signal and noise broadcast from two speakers separated by 90°). **b** Normalized latency to reach the target speaker as a function of signal-to-noise ratio (SNR) and angular separation between the target signal and the chorus-shaped masker. Normalized latencies were calculated relative to the latency to reach the target signal in reference conditions in the absence of masking noise; values equal to 1.0 represent latencies similar to those in reference conditions, whereas values close to 0 represent very slow behavioral responses or no response at all. Compared to the co-localized condition, normalized latencies were significantly higher in the separated (90°) condition at SNRs of −6 and 0 dB (asterisks). **c** Signal recognition thresholds, estimated using an adaptive tracking procedure, are significantly lower in the separated (90°) condition compared with the co-localized condition. Redrawn from (**b**) Bee (2007b) and (**c**) Nityananda and Bee (2012) and reprinted with permission from Elsevier

eastern gray treefrog (*H. versicolor*), a closely related species that often breeds synchronously and syntopically with Cope's gray treefrogs. It was recently shown that improved discrimination in the separated conditions resulted from better recognition of temporal sound patterns (Ward et al. 2013a). Spatial separation between sound sources may also influence intraspecific mate choice. Richardson and Lengange (2010) recently showed that increased spatial separation between signals enhanced the ability of female European treefrogs, *Hyla arborea*, to discriminate between calls in the presence of background noise.

Spatial separation between sound sources also confers benefits when calls are masked by other temporally overlapping calls. As discussed earlier (Sect. 6.3.4), call overlap can degrade the temporal structure of pulsed calls, thereby hindering sound pattern recognition. When given a choice between overlapping calls broadcast from adjacent or spatially separated speakers, female eastern gray treefrogs discriminated in favor of the separated calls when the angle of separation was 120°, but not when it was 45° (Schwartz and Gerhardt 1995). Two other studies, however, have shown that spatial separation of sound sources may offer limited benefits. A spatial separation of 120° was insufficient for females of the yellow cricket treefrog to discriminate in favor of separated calls over adjacent calls (5°) that were presented such that temporal overlap degraded the pulse structure of the call (Schwartz 1993). In the two closely related gray treefrog species, *H. versicolor* and *H. chrysoscelis*, females showed strong preferences for conspecific calls when they alternated with heterospecific calls separated by 90° (Marshall et al. 2006). Interestingly, however, when the two calls overlapped, *H. chrysoscelis* females still showed a strong preference for the conspecific call, while *H. versicolor* females actually approached the speaker broadcasting the *heterospecific* call (Marshall et al. 2006). This study highlights the fact that sound source segregation mechanisms might operate differently even among closely related species (see also Vélez et al. 2012; Vélez and Bee 2013).

Spatial unmasking in anurans has also been investigated at a neurophysiological level in northern leopard frogs (*Rana pipiens*). Ratnam and Feng (1998) found that increasing the angular separation between the sources of a masking noise and an amplitude-modulated signal resulted in lower signal detection thresholds in neurons in the torus semicircularis. Similarly, Lin and Feng (2001) found evidence for spatial release from masking in the responses of both single auditory nerve fibers and torus semicircularis neurons. In the auditory nerve, the magnitude of spatial unmasking was about 3 dB, while that in midbrain neurons was on the order of about 9 dB (Lin and Feng 2001). On the one hand, these results indicate that central neural processing contributes to enhancing the effect of spatial separation measured at the periphery (Lin and Feng 2001, 2003). On the other hand, however, the magnitude of spatial unmasking at the periphery is similar to that seen in behavior, although this may represent differences between the species tested (treefrogs versus leopard frogs). Additional work integrating behavioral and physiological measures of spatial unmasking in the same species will be required to resolve this issue.

These findings serve to emphasize that spatial release from masking may be one important mechanism by which frogs solve problems associated with breeding in choruses. Given heterogeneity in the spatial distributions of calling males in the habitat, it seems likely that receivers in a chorus often encounter situations in which signals of interest are well separated from other concurrent signals and dominant sources of background noise in the environment. Under such conditions, spatial release from masking may contribute to a listener's ability to hear individual calls. As the studies above illustrate, there may sometimes be considerable variation in auditory processing strategies among species, emphasizing the need for rigorous comparative studies to understand the evolution of mechanisms for hearing in noisy natural settings.

6.4.4 Dip Listening and Comodulation Masking Release

A well-known feature of natural soundscapes, including those generated in frog breeding choruses, is that sound levels fluctuate through time, that is, they are *amplitude modulated* (Richards and Wiley 1980; Nelken et al. 1999; Vélez and Bee 2010). Furthermore, these amplitude modulations are often correlated through time across different regions of the frequency spectrum; that is, natural sounds are often *comodulated* (Klump 1996; Nelken et al. 1999). Psychophysical studies of a phenomenon called *dip listening* indicate that human listeners are often much better at detecting and recognizing target signals, including speech, when maskers fluctuate in amplitude compared to steady-state maskers with stationary envelopes. The magnitude of masking release due to dip listening commonly ranges between 5 dB and 20 dB, depending on the temporal properties of the maskers and the target signals, and has been attributed to the listener's ability to catch brief "glimpses" of target signals at moments when the amplitude of the masker dips to low levels (Buus 1985; Gustafsson and Arlinger 1994; Bacon et al. 1998; Füllgrabe et al. 2006). Studies of a related phenomenon called *comodulation masking release* (CMR, reviewed in Verhey et al. 2003) indicate that listeners are also sensitive to spectro-temporal correlations in the fluctuating envelopes of masking noise. For example, when a tone signal is masked by a narrow band noise with a fluctuating envelope, the addition of a second narrowband noise at a remote frequency can produce several dB of masking release when its envelope is comodulated with that of the on-signal band compared to when it fluctuates independently. Dip listening has also been demonstrated in insects (Chap. 3) and CMR has been demonstrated in songbirds (Chap. 8), gerbils (Klump et al. 2001), and dolphins (Chap. 10). Might these processes also function in hearing and sound communication in frogs?

The sounds generated by frog choruses exhibit species-specific patterns of amplitude fluctuation (Vélez and Bee 2010). A recent study using single-stimulus tests has shown that female Cope's gray treefrogs experience a release from masking by listening in the dips of sinusoidally amplitude modulated (SAM) chorus-shaped noise (Fig. 6.6, Vélez and Bee 2011). Interestingly, evidence for dip listening was found only in SAM maskers that fluctuated at slow rates [e.g., 0.625 (Fig. 6.6b), 1.25, and 2.5 Hz], for which signal recognition thresholds were about 2-4 dB lower than those obtained in the presence of a steady-state masker (Fig. 6.6d). At intermediate rates [e.g., 5 (Fig. 6.6c) to 20 Hz], signal recognition thresholds were not different from those measured in the presence of the steady-state masker (Fig. 6.6d). And at faster rates of fluctuation (e.g., 40–80 Hz), SAM maskers caused increases in signal recognition thresholds of about 4–6 dB compared with steady-state maskers (Fig. 6.6d). Given that advertisement calls in Cope's gray treefrogs have pulse rates of about 35–50 pulses/s, impaired recognition at faster rates of modulation is consistent with the idea that the temporal structure of the masker interfered with the subjects' perception of the temporal pulse structure of the signal. The masking release seen at slow rates of masker modulation could be attributed to dip listening. Analyses of the target signals and maskers revealed that the maximum number of consecutive pulses fitting between the 6-dB down points of the sinusoidal modulation was 32 pulses in the most slowly fluctuating masker, decreasing exponentially to just one pulse in the most rapidly fluctuating maskers (Fig. 6.6e). Significant masking release was observed in the masking conditions for which the number of consecutive pulses occurring in a dip was nine pulses or more. In other words, females benefited from dip listening when they could catch acoustic glimpses of about nine pulses. This result was consistent with parallel tests conducted in quiet showing the threshold number of consecutive pulses required to elicit positive phonotaxis was between six and nine pulses (Fig. 6.6e; Vélez and Bee 2011). Schwartz et al. (2013) have also demonstrated dip listening in the eastern gray treefrog. Together, these results suggest that the ability of female gray treefrogs to listen in the dips of amplitude modulated noise may be constrained by sensory mechanisms responsible for encoding temporal properties critical for species recognition, such as neurons in the midbrain that "count" interpulse intervals (Alder and Rose 1998; Edwards et al. 2002). Interestingly, however, not all frogs may benefit from dip listening. Parallel tests of call recognition in fluctuating noise with green treefrogs have so far failed to uncover evidence for dip listening (Vélez et al. 2012; Vélez and Bee 2013).

The anuran auditory system, with its two anatomically distinct sensory papillae for encoding airborne sounds, offers a superb model for studying CMR. The two sensory papillae can be considered separate "channels" for sensory input, providing a unique perspective on questions of within-channel versus across-channel mechanisms (Verhey et al. 2003). At a behavioral level, there is evidence to suggest that frogs exploit common envelope fluctuations across frequency channels in the recognition of advertisement calls. A release from masking of approximately 3 dB to 5 dB in the presence of comodulated maskers has been reported for females of *H. chrysoscelis* (Bee and Vélez 2008). At a physiological level, neural correlates of CMR have been documented for neurons in the auditory

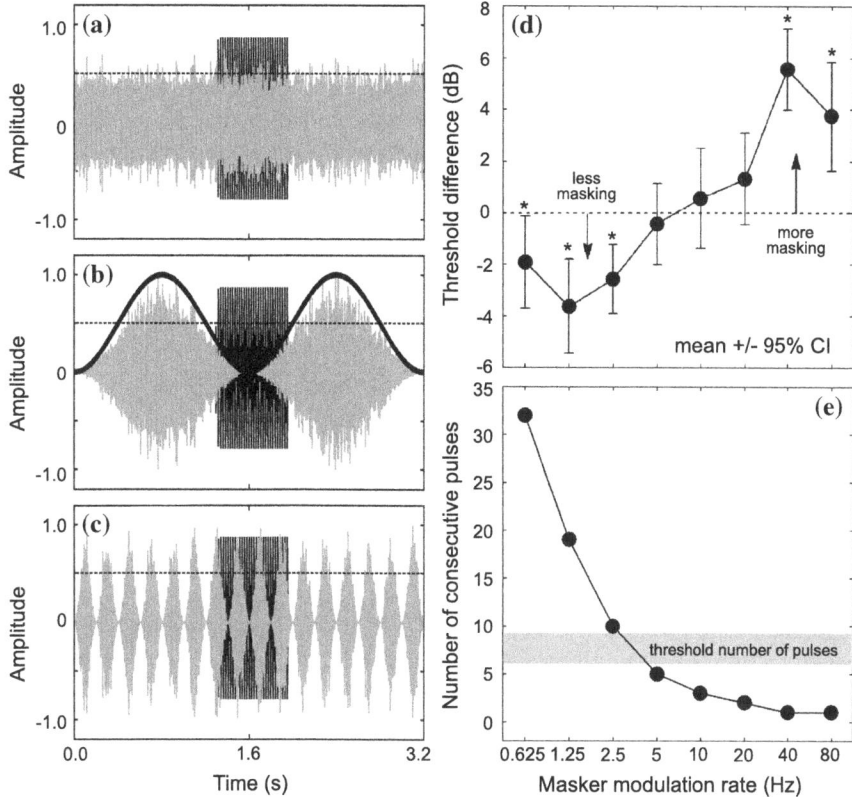

Fig. 6.6 Dip listening in Cope's gray treefrog, *Hyla chrysoscelis*. Waveforms of 3.2 s segments of the target signal (*black*) in the presence of chorus-shaped maskers (*gray*) representing **a** the steady-state control condition, and conditions in which maskers fluctuated sinusoidally (SAM) at rates of **b** 0.625 Hz and **c** 5 Hz. The target signal was a synthetic advertisement call composed of 32 pulses delivered at a rate of 45.5 pulses/s. The *solid black line* in **b** depicts the sine wave used to modulate the masker. The *dashed lines* in **b** and **c** illustrate the values at which the amplitude of the fluctuating maskers reaches 50 % of the maximum amplitude and mark the 6 dB down-points used to measure the number of pulses falling within dips of fluctuating maskers (**e**). **d** Threshold differences as a function of masker fluctuation rate; these differences are relative to the threshold measured in the control condition with a non-fluctuating, steady-state masker. The dashed line represents no-difference (i.e., 0 dB) from the control condition. **e** Maximum number of consecutive pulses falling in the dips of SAM maskers as a function of masker fluctuation rate. A pulse was considered to fall in a dip when its maximum amplitude fell between the 6 dB down points of the masker. Redrawn from Vélez and Bee (2011) and reprinted with permission from Elsevier

midbrain of the northern leopard frog (Goense and Feng 2012). Additional studies on the relative role of within and across channel contributions to CMR, at both the behavioral and physiological levels, would make valuable contributions to our understanding of the mechanisms for hearing in noisy environments in anurans.

6.4.5 Summary and Future Directions

As illustrated in this section, evolution has equipped anurans with a number of tricks and tools to reduce the impact of masking noise in their environment. Interesting and important questions remain about the ability of evolution to fine-tune auditory tuning in ways that filter out the calls of heterospecifics. A goal for future studies should be to determine at a physiological level whether there are differences in frequency sensitivity between sympatric and allopatric populations of species having calls that spectrally overlap. Likewise, there are tantalizing data from cricket frogs (*Acris crepitans*) to suggest the hypothesis that population differences in auditory tuning may sometimes reflect the operation of natural selection arising from differences in habitat acoustics (Witte et al. 2005). Hypotheses about population differences in auditory tuning might be easily tested using portable instruments to generate audiograms based on the auditory brainstem response (e.g., Schrode et al. in press), or by using distortion product otoacoustic emissions (e.g., Meenderink et al. 2010). Another important question for future studies will be to assess the evolution of ultrasonic hearing in frogs. Have we already discovered all species that communicate using ultrasound, or do many more fascinating discoveries await the herpetologist equipped with the right equipment for recording and reproducing ultrasonic frequencies (Arch and Narins 2008)? Clearly, it will be necessary to investigate ultrasonic communication in a phylogenetic framework. Studies of spatial release from masking, dip listening, and CMR indicate that frogs and humans may exploit some of the same perceptual cues for listening in noisy environments. Several features of the anuran auditory system – e.g., ears that function as pressure-gradient receivers, inner ears with multiple sensory papillae, and midbrain neurons that count interpulse intervals – make studies of the physiological mechanisms involved in achieving masking release particularly important (Bee 2012). While often smaller than the masking release observed in other vertebrates, frogs clearly benefit from exploiting spatial separation between signals and noise and temporal fluctuations in noise. Whether differences in magnitude between frogs and other vertebrates reflect real species differences or stem from differences in methodology remains an unanswered question. The use of more traditional psychophysical measures with frogs, or the development of entirely new techniques (e.g., Márquez et al. 2008) that measure phonotaxis behavior in stationary subjects (e.g., similar to trackballs and walking compensators used in many insect studies; Gerhardt and Huber 2002) might shed much needed light on this issue.

6.5 Auditory Scene Analysis

Acoustic scenes are often quite complex. They may comprise multiple sources concurrently producing sequences of spectrally rich sounds. Somehow, the auditory system has to make sense of this type of sensory input, and the processes by which it does are commonly studied under the rubrics of *auditory scene analysis*

(Bregman 1990) or *auditory grouping* (Darwin and Carlyon 1995; Darwin 1997). The major question of interest concerns how the auditory system integrates the sounds generated by one source into a coherent representation (often termed an auditory *stream* or *object*) that is distinct from the sounds produced by other sources in the environment. In other words, how do auditory systems put together sounds that belong together and keep apart sounds that do not?

As discussed previously, frog calls often consist of discrete sound elements (e.g., notes or pulses) produced in sequence, and each element often has simultaneous sound energy across the frequency spectrum (Fig. 6.1). In this section, we take up the question of how receivers perceptually bind or integrate sounds arising from the same source. Two forms of perceptual integration can be distinguished (Bregman 1990): (1) the *sequential integration* of temporally separated sound elements produced by the same source (e.g., pulses, notes, and calls) and (2) the *simultaneous integration* of different components of the frequency spectrum of a sound originating from the same source (e.g., harmonics, formants). The human auditory system accomplishes sequential and simultaneous integration by exploiting a relatively small number of commonalities in the acoustic properties of sounds arising from a single source. Sounds produced by a given source are more likely to be grouped into one auditory stream when they share common acoustic properties (Bregman 1990; Darwin and Carlyon 1995; Darwin 1997; Carlyon 2004). In contrast, sounds with acoustic properties that differ substantially are more likely to be assigned to different auditory streams. In this section, we review the current state of knowledge concerning sequential and simultaneous integration in the context of hearing and sound communication in frogs. In addition, we review recent studies of frogs' abilities to perceptually reconstruct auditory objects when signals are partially masked and to exploit schema-based cues in auditory grouping.

6.5.1 Sequential Integration

6.5.1.1 Auditory Streaming Based on Frequency Differences

The term *auditory streaming* is commonly used to refer to the ability to integrate sequences of sounds from one source into a coherent auditory stream that can be attended and followed through time (Bregman 1990; Carlyon 2004; Shamma and Micheyl 2010). Examples of auditory streaming at work are when we follow a melody line in polyphonic music or one person speaking in a noisy restaurant. Studies of auditory streaming in humans have made extensive use of simple sound sequences of two repeated, interleaved tones (A and B) differing in frequency or some other salient acoustic property (e.g., ABABAB...; Fig. 6.7a, b) (reviewed in Moore and Gockel 2002; Carlyon 2004). Frequency is an important cue for organizing such sequences. When the frequency separation (ΔF) between A and B tones is small, we hear a "trilled" sound jumping up and down in frequency (Fig. 6.7a). When ΔF is larger, something very different is perceived. The two

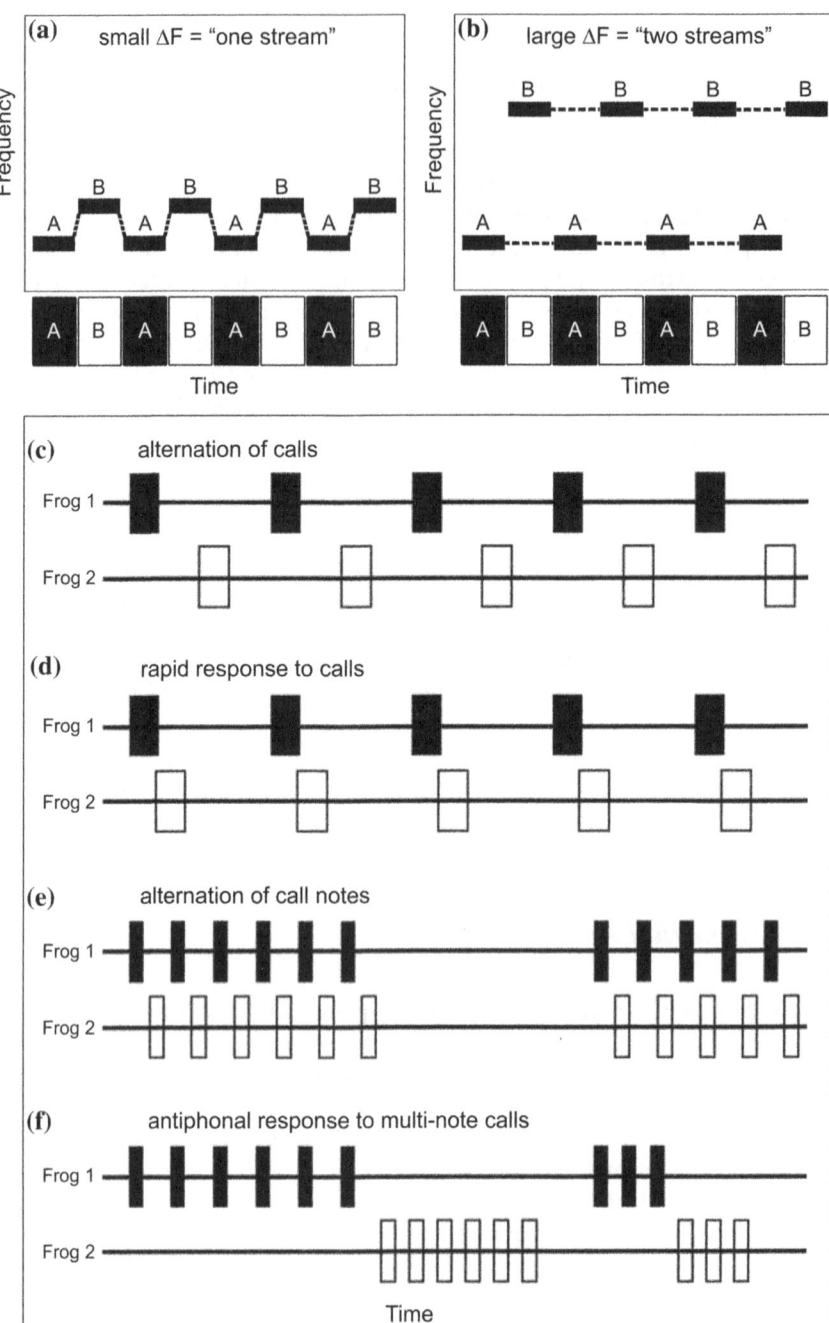

◄ **Fig. 6.7** Call timing and the utility of auditory streaming in frogs. **a, b** Schematic illustration (*top* spectrograms, *bottom* waveforms) of auditory streaming based on differences in frequency (ΔF) in the ABAB stimulus paradigm common in human psychoacoustic studies. **a** At small frequency separations between tones A and B, the tones are more likely to be perceptually "grouped" together to form one stream and listeners tend to hear a "trilled" sound jumping up and down in frequency (e.g., ABABABAB…). **b** In contrast, at large frequency separations, the A and B tones become perceptually separated into two different auditory streams and listeners tend to hear two tone sequences broadcast at half the rate, one comprising all A tones (e.g., A–A–A–A–…) and the other all B tones (e.g., –B–B–B–B…). **c–f** Schematic waveforms illustrating several types of call timing interactions that occur between neighboring male frogs in breeding choruses (see Chap. 5). Differences in frequency or other acoustic properties might allow receivers to perceptually separate interleaved calls (**c, d, f**) or notes (**e**) into different auditory streams. **c–f** Reprinted from Wells and Schwartz (2007) with permission from Springer

tones perceptually "split" into different streams, one comprising all A tones (e.g., A–A–A–…) and one comprising all B tones (–B–B–B…), each perceived at half the rate of the original sequence (Fig. 6.7b). Tones similar in frequency are grouped together into a coherent stream, while tones different in frequency are assigned to different streams. Psychophysical studies using this or similar stimulus paradigms with fish, songbirds, and monkeys have revealed auditory streaming to be common among vertebrates (reviewed in Fay 2008). In humans, some of the same mechanisms responsible for ΔF-based streaming of tone sequences may underlie our abilities to exploit differences in F_0 to assign concurrent voices to separate auditory streams (e.g., Brokx and Nooteboom 1982; Bird and Darwin 1998).

Auditory streaming based on frequency differences might be important for anuran communication, but this question has so far received little attention. Across species, male frogs calling in close proximity in a chorus exhibit a diversity of call timing behaviors to avoid obscuring the temporal structure of their calls (reviewed in Chap. 5). The end result is a sequence of temporally interleaved sounds arising from different sources (Fig. 6.7c–f). To what extent might receivers exploit individual differences or species differences in frequency (or other acoustic properties) to perceptually organize such sequences into coherent auditory streams that correspond to different sources? Nityananda and Bee (2011) took up this question in a study of Cope's gray treefrog. In single-stimulus tests, they measured female responses to a target signal that simulated a conspecific call and consisted of a short pulse train with a pulse rate of 45.5 pulses/s (Fig. 6.8a). In the absence of other sounds, this signal elicited robust phonotaxis. The key to the experiment was that in some tests, the pulses of the target signal were temporally interleaved with a behaviorally neutral "distractor" consisting of a continuous pulse train (also 45.5 pulses/s). Each time the signal was presented with the distractor, the instantaneous pulse rate was effectively doubled to 91 pulses/s (Fig. 6.8a). Importantly, control tests had demonstrated that females were selective for the conspecific pulse rate of 45.5 pulses/s and discriminated strongly against signals with the faster pulse rate (see also Bush et al. 2002; Schul and Bush 2002). Therefore, Nityananda and Bee (2011) reasoned as follows. If females assigned target and distractor pulses to different auditory streams, then they should exhibit positive phonotaxis toward the

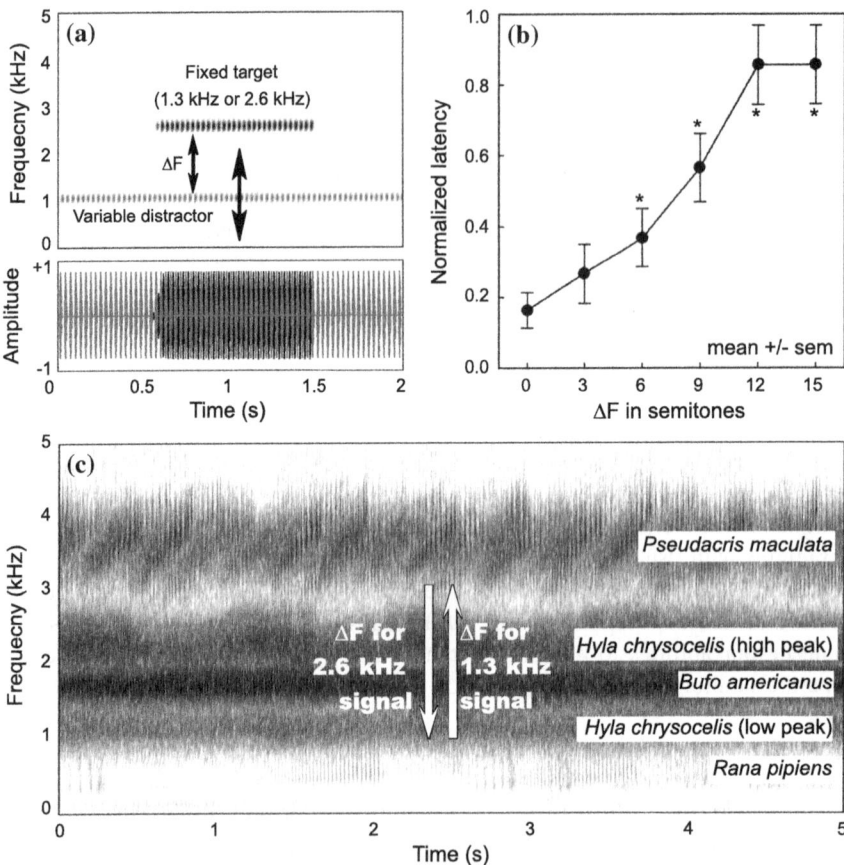

Fig. 6.8 Auditory streaming in Cope's gray treefrog, *Hyla chrysoscelis*. **a** Spectrogram (*top*) and waveform (*bottom*) of the interleaved target signal and distractor pulse trains. When the two pulse trains are perceptually integrated into a single auditory stream, the resulting pulse rate is 91 pulses/s and unattractive to females. When the target signal and the distractor are segregated into two streams, the target signal has an attractive pulse rate of 45.5 pulses/s. **b** Normalized latency to reach the target speaker as a function of frequency separation (Δ*F*). Values equal to 1.0 represent latencies similar to those in reference conditions lacking a distractor, whereas values close to 0 represent very slow behavioral responses or no response at all. Compared to the condition in which both pulse trains had the same frequency (Δ*F* = 0 semitones), responses were significantly faster when the frequency separation was equal to or greater than six semitones (asterisks). **c** Spectrogram of the acoustic environment of a mixed-species chorus that included Cope's gray treefrogs and several other frog species. Arrows depict the frequency range of the distractors when the target signal was the low spectral peak of the call (1.3 kHz, *right arrow*) and the high spectral peak of the call (2.6 kHz, *left arrow*). Redrawn from Nityananda and Bee (2011) and reprinted with permission from the authors

attractive percept of the target signal. If, however, they perceptually "fused" or integrated pulses from the target and the distractor into the same stream, this should result in an unattractive percept based on pulse rate. When the target and distractor had the same carrier frequency ($\Delta F = 0$ semitones), females exhibited little interest in the target (Fig. 6.8b). But on trials when the carrier frequency of the distractor was sufficiently far removed (e.g., $\Delta F \geq 6$ semitones) from that of the target, but still within the empirically determined hearing range, females exhibited phonotaxis toward the target (Fig. 6.8b). These data are consistent with the hypothesis that auditory streaming was possible based on differences in frequency. That improvements in performance were observed at frequency separations of 6 semitones was important, as this approximates the difference in frequency between the two spectral peaks present in conspecific advertisement calls and the dominant spectral peak in the calls of the synchronously and syntopically breeding American toad (*Bufo americanus*) (Fig. 6.8c).

6.5.1.2 Common Spatial Origin

In humans, spatially related cues can be effective in sequential integration when there is only one sound source present in the environment (reviewed in Darwin 2008). Thus far, only a few studies have investigated the role of spatial cues in sequential integration in frogs. Currently available evidence from several studies suggests that frogs may be willing to group temporally separated call elements over fairly large spatial separations. For example, one study of the EVR in males of the Australian quacking frog (*Crinia georgiana*) indicated that receivers perceptually group sounds coming from opposite directions. In this species, males produce a multinote call that sounds very much like a quacking duck. During episodes of vocal competition with neighbors, males attempt to match the number of "quacks" in their neighbors' calls. In a field playback test, Gerhardt et al. (2000) presented males with two sequential four-note calls from speakers separated by 180°. The timing of the two calls was such that they had the same overall temporal pattern as an eight-note call. Somewhat surprisingly, males responded to the playbacks as if they had heard a single eight-note call coming from one location, instead of two consecutive four-note calls coming from opposite directions. Males continued to show evidence of grouping the two four-note calls together even when the second call in the sequence was attenuated by 6 dB. In terms of auditory grouping, male quacking frogs are fairly permissive of spatial separation between sounds comprising behaviorally meaningful temporal sequences. As studies of túngara frogs (*Engystomops pustulosus*, formerly *Physalaemus pustulosus*) and Cope's gray treefrogs (*H. chrysoscelis*) indicate, female frogs are also willing to group widely separated sounds.

Male túngara frogs produce a simple call consisting of a whine only, and a complex call consisting of a whine followed by one or more chucks (Fig. 6.9a). Female túngara frogs exhibit positive phonotaxis toward speakers broadcasting whines alone but not chucks alone (Ryan 1985; Ryan and Rand 1990). Based on these findings, Farris et al. (2002, 2005) tested the hypothesis that common spatial origin promotes the sequential integration of whines and chucks into coherent representations of complex calls. In a circular arena, they broadcast whines and

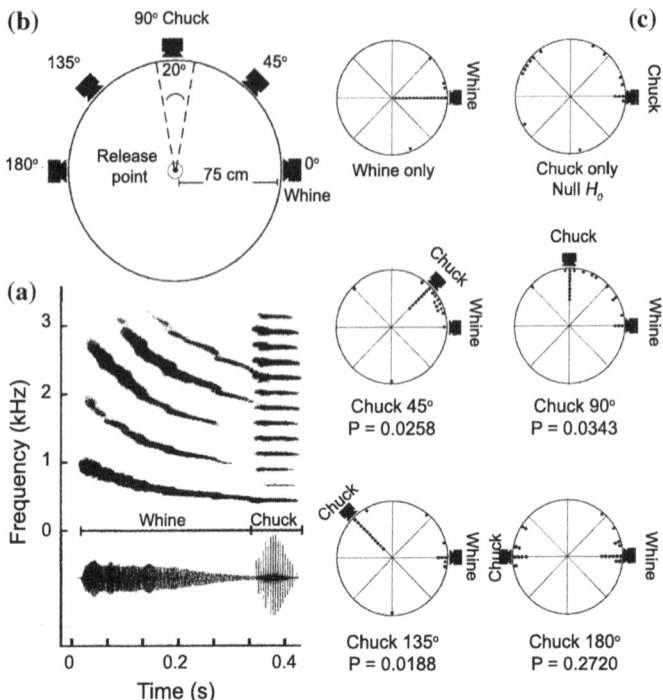

Fig. 6.9 Sequential integration in túngara frogs, *Engystomops pustulosus*. **a** Spectrogram and waveform of a túngara frog complex call composed of a whine and a chuck. **b** Schematic illustration of the test arena and the positions of the speakers that broadcast the whine and chucks from a common spatial origin (0°) or from different spatial origins (45°, 90°, 135°, and 180°). **c** Angles at which females exited the test arena in response to each condition. Females showed significant orientation toward the whine alone but not toward the chuck alone. When the whine and the chuck were broadcast from different locations, orientation toward the chuck was significant at angular separations of 45°, 90°, and 135° but not 180°. Reprinted from Farris et al. (2002) with costly permission from Karger AG

chucks in the natural temporal sequence from either the same speaker (angular separation of 0°) or from two speakers separated by 45°, 90°, 135°, or 180° (Fig. 6.9b) Females significantly oriented toward the chuck in conditions in which the two components of the call were separated by 45°, 90°, and 135°, but not by 180° (Fig. 6.9c) These results provide strong evidence for auditory grouping by frogs, but they refute the hypothesis that common spatial origin is necessary for grouping signal components separated in time. Based on their results, Farris et al. (2002) suggested whines and chucks are weighed differently in making "what" and "where" decisions; information about species identity ("what") is primarily encoded in the whine, while information about location ("where") is primarily encoded in the chuck. More recently, Farris and Ryan (2011) demonstrated females make relative comparisons and group whines and chucks in relatively closer proximity when multiple chucks are separated from a single whine.

Bee and Riemersma (2008) showed that common spatial origin is also not necessary for sequential integration in Cope's gray treefrogs. Females of this species are highly selective for calls with conspecific pulse rates (approximately 35–50 pulses/s), which are about twice as fast as the pulse rate of calls produced by males of the eastern gray treefrog (*H. veriscolor*). Bee and Riemersma (2008) presented females of *H. chrysoscelis* with two interleaved pulse sequences in which the pulses from each sequence were interdigitated. Each sequence had a pulse rate of 25 pulses/s (similar to *H. versicolor* calls), but the composite of both sequences combined had the preferred pulse rate of 50 pulses/s (as in conspecific calls). Hence, if the two sequences were perceptually integrated, the percept should have been one of an attractive conspecific call, whereas perceptual segregation should have resulted in the percept of two unattractive calls. On separate trials, the two interleaved sequences were separated by 0°, 45°, 90°, or 180°. The results showed that females were very permissive of spatial separation, and were able to integrate the two pulse trains even at a spatial separation of 180°. Together with results from quacking frogs and túngara frogs, these results suggest that common spatial origin may be a relatively weak acoustic cue for sequential integration in frogs. The permissiveness of sequential integration based on spatial cues may explain the failure of spatial separation to improve call recognition in some species (Schwartz 1993; Schwartz and Gerhardt 1995).

6.5.2 Simultaneous Integration

6.5.2.1 Harmonicity

Acoustic signals often contain harmonics that are multiple integers of the fundamental frequency (F_0). Instead of hearing simultaneous pure tones at each harmonic frequency, we tend to hear harmonic complexes as single, fused sounds with one pitch corresponding to F_0. Thus, our auditory systems group simultaneous sounds sharing a common F_0 (i.e., "harmonicity") into a single coherent percept. The role of harmonicity as an auditory grouping cue has been studied quite extensively in humans using vowel sounds and complex tones (reviewed in Darwin and Carlyon 1995; Darwin 1997; Carlyon and Gockel 2008). A slight "mistuning" of one harmonic can cause listeners to "hear out" that spectral component, or perceptually segregate it from the rest of the harmonic complex. Under these conditions, listeners hear two simultaneous sounds: one complex tone and one separate pure tone corresponding to the mistuned harmonic. Therefore, inharmonic relationships between the spectral components of concurrent sounds favor their segregation into different auditory streams. Not surprisingly, then, differences in F_0 between simultaneously spoken sentences facilitate identification and recognition of target speech (e.g., Brokx and Nooteboom 1982; Bird and Darwin 1998).

Frogs are also sensitive to harmonicity. Using reflex modification, Simmons (1988a) demonstrated that thresholds for detecting a two-tone complex in noise were about 10 dB higher in green treefrogs when the two tones were inharmonically related compared with detection of harmonic tone complexes. These results confirmed that frogs process harmonic sounds differently than inharmonic ones. In a subsequent study of this species, Gerhardt et al. (1990) failed to find evidence that harmonicity influenced female preferences in phonotaxis experiments. When given a choice between harmonic and inharmonic synthetic calls, females chose randomly between the two alternatives. In the presence of background noise, there was no evidence for higher detectability of the harmonic alternative, nor preferences for the harmonic stimulus. In contrast to green treefrogs, two-alternative choice tests with the closely related barking treefrog (*H. gratiosa*) revealed that females of this species discriminate between harmonic and inharmonic synthetic calls (Bodnar 1996). Interestingly, females actually preferred inharmonic alternatives to harmonic ones when no frequency modulation (FM) was present in the signals. When FM was added, females preferred the harmonic alternative. Bodnar's (1996) study also demonstrated that females were very sensitive to harmonicity, with a mistuning of one spectral component by 1.1 % sufficient for call discrimination. Results from studies of male frogs also reveal conflicting patterns of results. Simmons et al. (1993) used the EVR of male green treefrogs to investigate harmonicity as a call recognition cue in the laboratory. There were no significant differences in male vocal behavior (number of evoked calls and latency to first vocal response) in response to harmonic and inharmonic synthetic calls. In contrast, Simmons and Bean (2000) conducted field experiments testing the EVR in North American bullfrogs and found that they could discriminate between harmonic calls and calls with one spectral component mistuned by 2.8 %.

6.5.2.2 Common Onsets and Offsets

Another cue that contributes to simultaneous integration in humans is onset and offset synchrony (reviewed in Darwin and Carlyon 1995; Darwin 1997). Spectral components that start and end at the same times are more commonly grouped together into one auditory object. In contrast, frequency components that start, or end, at sufficiently different times from the other components are usually assigned to different auditory objects. To our knowledge, no study of anurans has investigated the effects of onset/offset synchrony. Indeed, we are aware of only one study that has investigated the role of common onsets/offsets in the communication system of a nonhuman animal. Geissler and Ehret (2002) showed that synthetic pup wriggling calls with a harmonic having an asynchronous onset or offset reduced the probability that female mice (*Mus domesticus*) would respond appropriately. Similar studies should be conducted with male and female frogs to investigate the role of common onsets/offset as an auditory grouping cue in anuran communication.

6.5.2.3 Common Spatial Origin

There is a general consensus that spatial cues play relatively minor roles in simultaneous integration in humans (reviewed in Darwin 2008). To date only one study has investigated the role of common spatial origin in allowing frogs to integrate the simultaneous spectral components common in multiharmonic acoustic signals. Bee (2010) took advantage of the spectral preferences of female Cope's gray treefrogs to test the hypothesis that common spatial origin promotes simultaneous integration. The advertisement calls of this species consist of pulses with two harmonically related spectral peaks centred around 1.1–1.4 and 2.2–2.8 kHz (Fig. 6.10a). Females will approach calls with only one or the other spectral peak, but they generally prefer calls having both spectral peaks (Gerhardt 2005; Gerhardt et al. 2007; Bee 2010). Using two-alternative choice tests, Bee (2010) offered females a choice between two calls that were either spatially coherent or incoherent and that alternated in time with each other from opposite sides of a test arena (Fig. 6.10a). In the spatially coherent alternative, both harmonics were broadcast simultaneously from the same speaker on one side of the arena. In the spatially incoherent alternative, each harmonic was broadcast from one speaker in a pair of speakers located on the opposite side of the arena from the spatially coherent call. Across different trials, the separation between the sources of the harmonics in the spatially incoherent alternative was 7.5°, 15°, 30°, or 60° (Fig. 6.10a). At all angular separations tested, females significantly (or nearly so) preferred the spatially coherent alternative. In fact, females preferred the spatially coherent alternative to the incoherent one in proportions not different from their preferences for calls with two spectral peaks over those with just one spectral peak (Fig. 6.10b). These results support the hypothesis that common spatial origin promotes simultaneous integration in gray treefrogs.

6.5.3 Auditory Induction

In noisy environments, receivers may occasionally have to deal with signals that are partially masked by short, intermittent, loud sounds in the environment. While at a cocktail party, for instance, part of one person's sentence may be momentarily masked by the loud cough or sneeze of a nearby guest. As studies of *phonemic restoration* (Warren 1970) have shown, our auditory system is quite good at perceptually reconstructing speech elements that are partially masked. Studies of a phenomenon known as the *continuity illusion* have generalized these results by showing, for example, that inserting a silent gap into a tone and then filling the gap with noise induces the perceptual illusion of an uninterrupted tone that continues through the noise (reviewed in King 2007). In both instances, it is as if our auditory system is able to "fill in" missing sound elements. Together, phonemic restoration and the continuity illusion represent examples of something called *auditory induction*, which refers to the auditory system's ability to reconstruct or restore

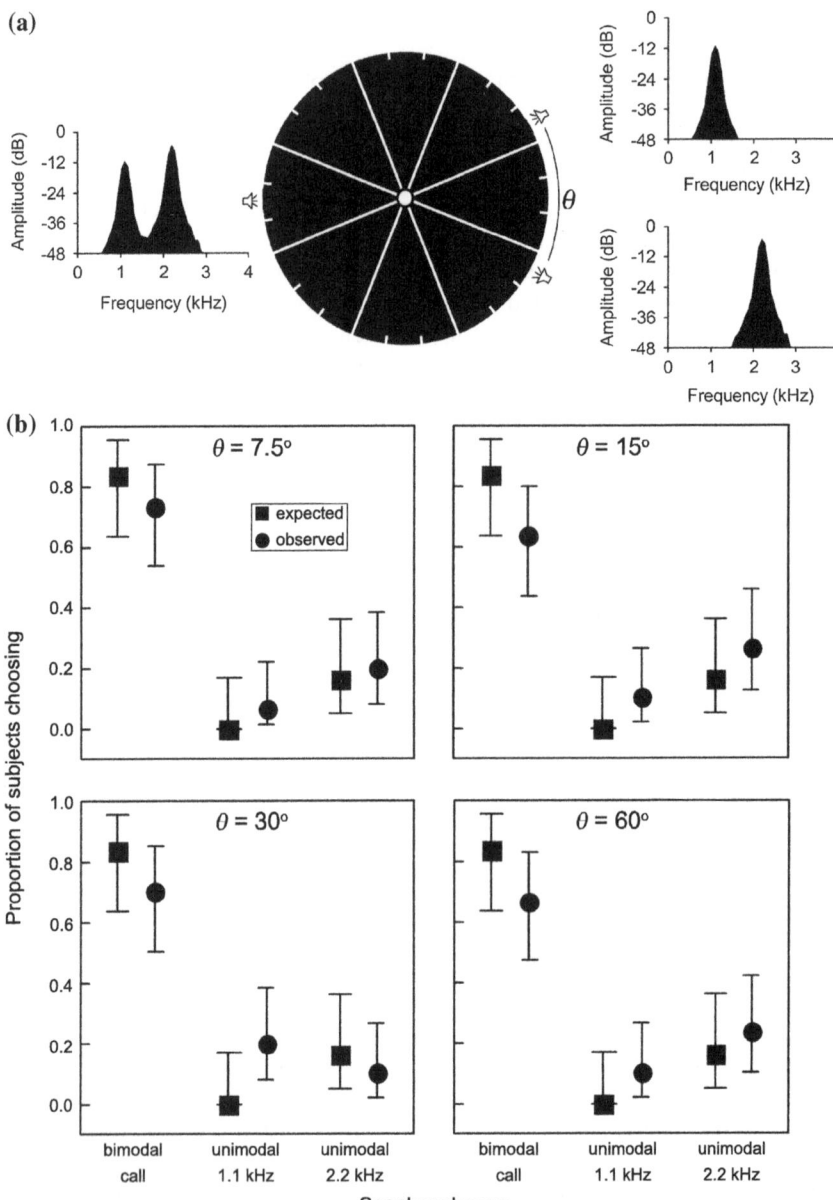

Fig. 6.10 Simultaneous integration in Cope's gray treefrog, *Hyla chrysoscelis*. **a** Schematic representation of the circular test arena, the locations of the speakers, and the power spectra of the spatially coherent bimodal call (*left*) and the unimodal 1.1 kHz (*top right*) and 2.2 kHz (*bottom right*) unimodal calls. **b** Expected (*squares*) and observed (*circles*) proportions of subjects choosing each speaker when the harmonics of the spatially incoherent stimuli were separated by angles (θ) equal to 7.5°, 15°, 30°, or 60°. Expected proportions were based on results from two-alternative choice tests pairing a spatially coherent bimodal call against a unimodal call with the specified frequency. Redrawn from Bee (2010) and reprinted with permission from the American Psychological Association

masked or missing elements of sound (King 2007). Importantly, auditory induction is not uniquely human and has been demonstrated using vocalizations in songbirds (Braaten and Leary 1999; Seeba and Klump 2009) and monkeys (Miller et al. 2001; Petkov et al. 2003). In contrast, two recent studies failed to find strong evidence that frogs experience auditory induction (Schwartz et al. 2010b; Seeba et al. 2010).

Using two-alternative choice tests, Seeba et al. (2010) took advantage of female preferences for longer calls (Fig. 6.11a) to test the hypothesis that female Cope's gray treefrogs (*H. chrysoscelis*) perceptually reconstruct discrete pulses of male advertisement calls. When females were given a choice between a call with a normal pulse structure and a call with silent gaps produced by removing groups of pulses, females unanimously chose the complete call over the "gap call" (Fig. 6.11b). Importantly, a "gap-filled call" created by filling the gaps with bursts of band-limited noise was chosen unanimously over a gap call (Fig. 6.11c). On the surface, this result seemed to support the auditory induction hypothesis. However, the question remained as to whether subjects actually perceived illusory pulses during the noise bursts. Seeba et al. (2010) reasoned that if females perceptually restored the missing pulses in the gap-filled call, then they should preferentially choose it (20 real pulses +15 illusory pulses) over a shorter 20 pulse call, and it should be equally attractive as an equivalent duration 35 pulse call. These predictions were not supported. Females chose the longer gap-filled call and the shorter 20 pulse call in proportions not significantly different from chance expectations (Fig. 6.11d), but exhibited a significant preference for a 35 pulse call over the gap-filled call of equivalent duration (Fig. 6.11e). Schwartz et al. (2010b) similarly exploited female preferences for longer duration calls to investigate auditory induction in eastern gray treefrogs (*H. versicolor*), but they also found no evidence that females heard illusory pulses when real pulses were replaced with noise.

These results with gray treefrogs suggest the provisional conclusion that frogs, unlike birds and mammals, may be incapable of perceptually reconstructing signals momentarily interrupted by noise. Seeba et al. (2010) and Schwartz et al. (2010b) discuss various hypotheses for these negative results. One such hypothesis is that frogs cannot restore temporally discrete elements in pulsatile calls. It was suggested that frogs might be able to restore missing portions of continuous sounds, such as a long call note. Preliminary data from studies of auditory induction in túngara frogs, however, have so far provided little evidence that females are able to perceptually restore short segments that are deleted from the normally continuous whine portion of the call and filled with noise (AT Baugh, MA Bee, and MJ Ryan, unpublished data).

6.5.4 Schema-Based Auditory Grouping

Thus far, we have considered only forms of auditory scene analysis based on commonalities in grouping cues present in the spectral and temporal properties of sound elements composing vocal signals. But the formation of auditory groups can also

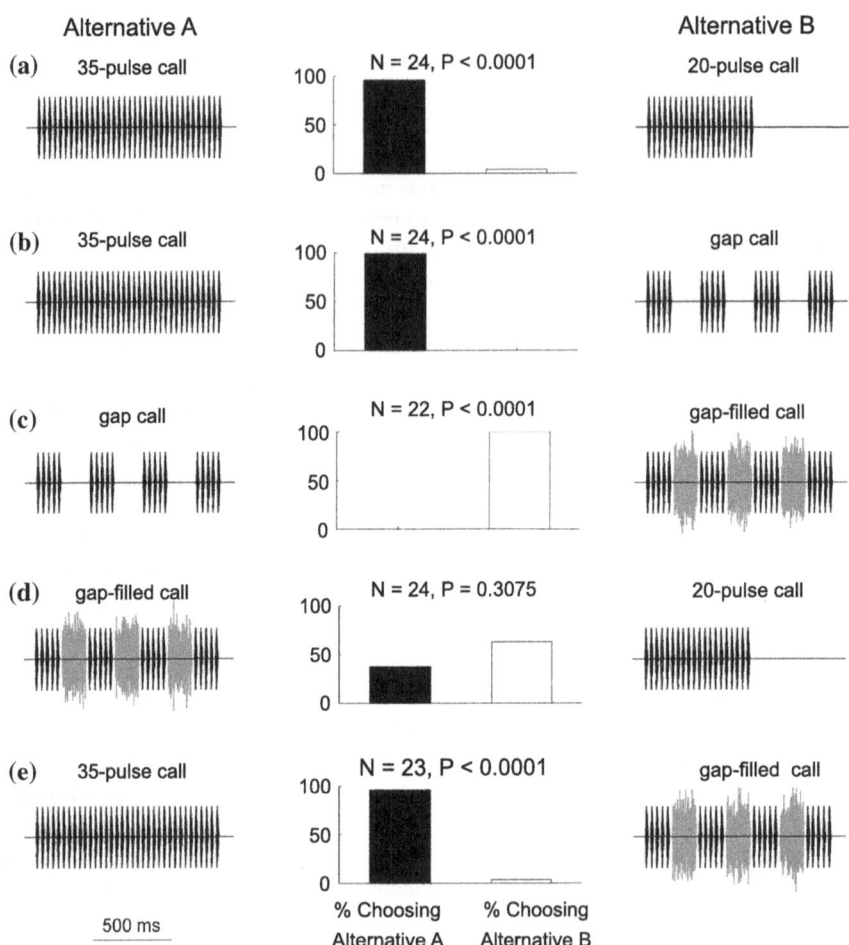

Fig. 6.11 A test of auditory induction in Cope's gray treefrogs, *Hyla chrysoscelis*. Schematic waveforms and percentage of females choosing each alternative stimulus (A or B) in two-alternative choice tests. **a** Females preferred longer calls with more pulses (35 pulses vs. 20 pulses). **b** Females also preferred a 35 pulse call over an equivalent-duration 20 pulse call in which 15 pulses (3 groups of 5) were removed (gap call). **c** Consistent with the auditory induction hypothesis, females preferred calls in which the gaps were filled with noise bursts (gap-filled call) over the gap call. As illustrated in (**d, e**), however, females did not perceive the gap-filled call as a call with 35 pulses. This conclusion follows because (**d**) females chose randomly between the gap-filled call and a shorter 20 pulse call, but strongly preferred a 35 pulse call to a gap-filled call of equivalent overall duration. Redrawn from Seeba et al. (2010) and reprinted with permission from Elsevier

occur based on a listener's prior experiences or expectations in a process Bregman (1990) referred to as *schema-based* scene analysis. There is little a priori reason to limit schema-based analyses to experiential influences. Evolution might also equip listeners with hard-wired schema for analyzing acoustic scenes. Thus far, this general question has received little attention in frogs. Some frogs produce complex vocalizations comprising distinctly different sound elements that follow statistical rules of ordering (e.g., Larson 2004; Gridi-Papp et al. 2006). Schema-based auditory grouping might occur if receivers have evolved templates for call recognition that incorporate the same rules governing signal production. For example, the complex call of the túngara frog consists of a whine followed by one or more chucks due to the morphological constraints of call production (Gridi-Papp et al. 2006). Farris and Ryan (2011) have shown that female túngara frogs are sensitive to this temporal order when it comes to grouping whines and chucks together, indicating for the first time that schema-based auditory grouping might be important in frogs.

6.5.5 Summary and Future Directions

The rigorous study of auditory scene analysis is still in its infancy in frogs, yet scene analysis represents an important aspect of perceiving acoustic signals in complex and noisy acoustic environments (Bee 2012). Many important questions remain for future work. It will be particularly important to determine the spectral, temporal, and spatial cues that frogs use to perceptually organize acoustic scenes. For example, do other cues, besides frequency separation, promote the formation of multiple auditory streams? To what extent do multiple cues interact? Is auditory streaming necessary for receivers to make sense of the call timing interactions of nearby males? Much work also remains before we understand the importance of harmonicity as an auditory grouping cue in frogs. Previous results from studies of harmonicity in treefrogs and bullfrogs are important because they highlight two fundamental principles in the study of hearing and sound communication in frogs. First, sometimes receivers may be able to detect acoustic differences (in harmonic structure in this case) that may or may not be used in call discrimination tasks. Studies of communication behaviors may not always tell the whole story about anuran hearing. Second, similar studies in multiple species, sometimes even closely related ones, can yield contrasting outcomes. These findings highlight the need for comparative studies of hearing in these animals. Questions about the importance of common onsets/offsets as an auditory grouping cue have yet to be addressed in frogs. Studies of the influence of onset/offset synchrony on call recognition, and its potentially correlated effects on combination-sensitive neurons in the auditory system (e.g., Fuzessery and Feng 1982), are needed to understand this potentially important auditory grouping cue.

The few presently available studies suggest some potentially interesting differences between humans and frogs in terms of the role spatial cues play in sequential and simultaneous integration. While spatial cues appear to play a limited role in simultaneous integration in humans, they may be important in frogs. In contrast, spatial cues appear to play a role during sequential integration in humans, but perhaps only a minor one in frogs. Efforts to understand how the frog's pressure-gradient receiver contributes to exploiting spatial cues in the perceptual organization of acoustic scenes represent an exciting frontier in research on anuran hearing. An important direction for future studies should be to rigorously quantify the spatial heterogeneity present in frog choruses. Recent developments in microphone array technology provide the technological basis for doing so (Jones and Ratnam 2009; Bates et al. 2010). Similarly, it is presently unclear why frogs appear not to exhibit auditory induction. Until additional data from more frog species become available, however, conclusions about differences in auditory induction between humans, frogs, and other animals must remain provisional. Additional studies of schema-based auditory scene analysis in frogs are also badly needed.

6.6 Multimodal Cues

In dense social aggregations, or any time background noise levels are high, receivers may benefit from using multiple sensory modalities for communication. In human listeners, for instance, speech detection and intelligibility improve when speech sounds are accompanied by corresponding lip gestures (Grant and Seitz 2000; Schwartz et al. 2004). Indeed, environmental noise may generally select for the use of multimodal signals in animals (Hebets and Papaj 2005). Early work established that some frogs, such as the Coqui frog (*Eleutherodactylus coqui*) and the white-lipped frog (*Leptodactylus albilabris*) of Puerto Rico, are sensitive to acoustic signals and coincident seismic signals produced when the vocal sac "thumps" the substrate (Lewis and Narins 1985; Lewis et al. 2001; see also Sect. 5.2.5 in Chap. 5). A few recent studies indicate that frogs may make more use of multimodal information than previously thought by showing that receivers of both sexes also attend to visual cues associated with the vocal sac (Narins et al. 2003, 2005; Rosenthal et al. 2004; Taylor et al. 2008, 2011; Taylor and Ryan 2013; Gomez et al. 2009; Richardson et al. 2010). Beyond establishing that frogs attend to visual cues associated with vocalizations, these studies firmly establish the use of robotics and video playbacks as important new tools in the study of hearing and sound communication in anurans. To the best of our knowledge, no published study has investigated whether reliance on visual cues associated with an inflating vocal sac improves the ability of frog receivers to solve cocktail-party-like problems. Determining the extent to which male and female frogs may be lip-reading (or vocal-sac-reading, to be more accurate) will be an important next step in understanding how frogs perceptually organize complex audio-visual scenes.

6.7 Conclusions

In this chapter, we have adopted what might be called a "cocktail party perspective" on anuran acoustic signal perception in noisy environments. Given that essentially modern frogs already hopped the earth while dinosaurs still roamed it

(Wells 2007), the evolutionary success of this vertebrate lineage is a testament to their ability to overcome cocktail-party-like problems associated with breeding in noisy social environments. Experimental studies conducted over the last three decades, many of them in the last few years, have begun to uncover the mechanisms by which frogs cope with high levels of masking noise and competing sound sources in complex acoustic scenes. Current evidence suggests that frogs exploit some of the same spectral, temporal, and spatial cues that humans also use to achieve a release from auditory masking and to form perceptually coherent auditory groups. There is, however, evidence to suggest that at least some of these cues may differ in relative importance or be processed differently in frogs and humans. This is not surprising given some of the evolved differences between the peripheral and central auditory systems of amphibians and mammals. And it is precisely the evolutionary history of the vertebrate auditory system that makes the study of anuran acoustic signal perception in noisy environments so important. Many of the basic features involved in hearing in noisy, multisource environments probably arose early during the evolution of vertebrate hearing (Popper and Fay 1997; Fay and Popper 2000; Lewis and Fay 2004). But it is important to also bear in mind that some key features of the vertebrate auditory system have had multiple evolutionary origins (Manley et al. 2004). Tympanic hearing, for example, appears to have arisen independently in each major lineage of tetrapod vertebrates (Christensen-Dalsgaard and Carr 2008). Evolution by natural selection is well known for finding a diversity of solutions to common problems. Therefore, it is certainly not unreasonable to expect (and in fact, it would be unreasonable *not* to expect) that different vertebrate lineages may possess different suites of hearing mechanisms comprising some ancient ones inherited from our last common ancestor, as well as some novel ones that have been derived or elaborated in a particular lineage since their divergence from other lineages. Sometimes solutions to cocktail-party-like problems may be evolutionarily homologous across taxa; other times, evolution may have created analogous solutions to the problem. This has profound implications for how we study animal acoustic communication (Bee and Micheyl 2008). The only way to examine both ancient, shared mechanisms and more recently derived novel mechanisms is to take a broad comparative approach to understand how animals in different lineages solve similar problems. Given the number of questions that remain concerning how frogs perceive sounds and acoustic signals in noisy environments, anuran amphibians will continue to be an important taxon for this line of comparative research on hearing and sound communication over the coming decades.

Acknowledgments We are especially grateful to Peter Narins and Henrik Brumm for comments on the manuscript. Some of the material described in this review was based upon work supported by the National Science Foundation (under Grant No. 0842759), the National Institute on Deafness and Other Communication Disorders (under Grant Nos. R03DC009582 and R01DC009582), and a Grant-in-Aid from the University of Minnesota Graduate School to MAB. Work by JJS was supported by the National Science Foundation (under Grant Nos. 0342183 and 9727623), Pace University Scholarly Research Awards and Smithsonian Institution Short Term Visitor Awards.

References

Alder TB, Rose GJ (1998) Long-term temporal integration in the anuran auditory system. Nat Neurosci 1:519–523

Amézquita A, Castellanos L, Hödl W (2005) Auditory masking of male *Epipedobates femoralis* (Anura: Dendrobatidae) under field conditions. Anim Behav 70:1377–1386

Amézquita A, Hödl W, Lima AP, Castellanos L, Erdtmann L, De Araújo MC (2006) Masking interference and the evolution of the acoustic communication system in the Amazonian dendrobatid frog Allobates femoralis. Evolution 60:1874–1887

Amézquita A, Flechas SV, Lima AP, Gasser H, Hödl W (2011) Acoustic interference and recognition space within a complex assemblage of dendrobatid frogs. Proc Natl Acad Sci USA 108:17058–17063

Andersson M (1994) Sexual selection. Princeton University Press, Princeton

Arak A (1983) Sexual selection by male–male competition in natterjack toad choruses. Nature 306:261–262

Arch VS, Narins PM (2008) 'Silent' signals: selective forces acting on ultrasonic communication systems in terrestrial vertebrates. Anim Behav 76:1423–1428

Arch VS, Grafe TU, Narins PM (2008) Ultrasonic signalling by a Bornean frog. Biol Lett 4:19–22

Arch VS, Grafe TU, Gridi-Papp M, Narins PM (2009) Pure ultrasonic communication in an endemic Bornean frog. PLoS ONE 4:e5413

Bacon SP, Opie JM, Montoya DY (1998) The effects of hearing loss and noise masking on the masking release for speech in temporally complex backgrounds. J Speech Lang Hear R 41:549–563

Barber JR, Crooks KR, Fristrup KM (2010) The costs of chronic noise exposure for terrestrial organisms. Trends Ecol Evol 25:180–189

Barrass AN (1985) The effects of highway traffic noise on the phonotactic and associated reproductive behavior of selected anurans. In: Environmental and water resources engineering. Vanderbilt University, Nashville, p 108

Bates ME, Cropp BF, Gonchar M, Knowles J, Simmons JA, Simmons AM (2010) Spatial location influences vocal interactions in bullfrog choruses. J Acoust Soc Am 127:2664–2677

Beckers OM, Schul J (2004) Phonotaxis in *Hyla versicolor* (Anura, Hylidae): the effect of absolute call amplitude. J Comp Physiol A 190:869–876

Bee MA (2003) Experience-based plasticity of acoustically evoked aggression in a territorial frog. J Comp Physiol A 189:485–496

Bee MA (2007a) Selective phonotaxis by male wood frogs (*Rana sylvatica*) to the sound of a chorus. Behav Ecol Sociobiol 61:955–966

Bee MA (2007b) Sound source segregation in grey treefrogs: spatial release from masking by the sound of a chorus. Anim Behav 74:549–558

Bee MA (2008a) Finding a mate at a cocktail party: spatial release from masking improves acoustic mate recognition in grey treefrogs. Anim Behav 75:1781–1791

Bee MA (2008b) Parallel female preferences for call duration in a diploid ancestor of an allotetraploid treefrog. Anim Behav 76:845–853

Bee MA (2010) Spectral preferences and the role of spatial coherence in simultaneous integration in gray treefrogs (*Hyla chrysoscelis*). J Comp Psychol 124:412–424

Bee MA (2012) Sound source perception in anuran amphibians. Curr Opin Neurobiol 22:301–310

Bee MA, Gerhardt HC (2001) Neighbour-stranger discrimination by territorial male bullfrogs (*Rana catesbeiana*): I. Acoustic basis. Anim Behav 62:1129–1140

Bee MA, Gerhardt HC (2002) Individual voice recognition in a territorial frog (*Rana catesbeiana*). P Roy Soc B Biol Sci 269:1443–1448

Bee MA, Swanson EM (2007) Auditory masking of anuran advertisement calls by road traffic noise. Anim Behav 74:1765–1776

Bee MA, Micheyl C (2008) The cocktail party problem: what is it? How can it be solved? And why should animal behaviorists study it? J Comp Psychol 122:235–251

Bee MA, Riemersma KK (2008) Does common spatial origin promote the auditory grouping of temporally separated signal elements in grey treefrogs? Anim Behav 76:831–843

Bee MA, Vélez A (2008) Comodulation masking release in the perception of vocalizations by gray treefrogs. Abstr Assoc Res Otolaryngol 31:#812

Bee MA, Schwartz JJ (2009) Behavioral measures of signal recognition thresholds in frogs in the presence and absence of chorus-shaped noise. J Acoust Soc Am 126:2788–2801

Bee MA, Vélez A, Forester JD (2012) Sound level discrimination by gray treefrogs in the presence and absence of chorus-shaped noise. J Acoust Soc Am 131:4188–4195

Bernal XE, Rand AS, Ryan MJ (2007) Sexual differences in the behavioral response of túngara frogs, *Physalaemus pustulosus*, to cues associated with increased predation risk. Ethology 113:755–763

Bibikov NG (2002) Addition of noise enhances neural synchrony to amplitude-modulated sounds in the frog's midbrain. Hear Res 173:21–28

Bird J, Darwin CJ (1998) Effects of a difference in fundamental frequency in separating two sentences. In: Palmer AR, Rees A, Summerfield AQ, Meddis R (eds) Psychophysical and physiological advances in hearing. Whurr, London, pp 263–269

Bodnar DA (1996) The separate and combined effects of harmonic structure, phase, and FM on female preferences in the barking treefrog (*Hyla gratiosa*). J Comp Physiol A 178:173–182

Boeckle M, Preininger D, Hödl W (2009) Communication in noisy environments I: acoustic signals of *Staurois latopalmatus* Boulenger 1887. Herpetologica 65:154–165

Braaten RE, Leary JC (1999) Temporal induction of missing birdsong segments in European starlings. Psychol Sci 10:162–166

Bradbury JW, Vehrencamp SL (2011) Principles of animal communication, 2nd edn. Sinauer Associates, Sunderland

Bregman AS (1990) Auditory scene analysis: the perceptual organization of sound. MIT Press, Cambridge

Brokx JPL, Nooteboom SG (1982) Intonation and the perceptual separation of simultaneous voices. J Phonetics 10:23–36

Bronkhorst AW (2000) The cocktail party phenomenon: a review of research on speech intelligibility in multiple-talker conditions. Acustica 86:117–128

Brown GJ, Cooke M (1994) Computational auditory scene analysis. Comput Speech Lang 8:297–336

Brumm H (2010) Anthropogenic noise: implications for conservation. In: Breed MD, Moore J (eds) Encyclopedia of animal behavior. Academic Press, Oxford

Brumm H, Slabbekoorn H (2005) Acoustic communication in noise. Adv Stud Behav 35:151–209

Brumm H, Voss K, Kollmer I, Todt D (2004) Acoustic communication in noise: regulation of call characteristics in a New World monkey. J Exp Biol 207:443–448

Bush SL, Gerhardt HC, Schul J (2002) Pattern recognition and call preferences in treefrogs (Anura: Hylidae): a quantitative analysis using a no-choice paradigm. Anim Behav 63:7–14

Buus S (1985) Release from masking caused by envelope fluctuations. J Acoust Soc Am 78:1958–1965

Capranica RR (1965) The evoked vocal response of the bullfrog: a study of communication by sound. M.I.T Press, Cambridge

Capranica RR (1976) Morphology and physiology of the auditory system. In: Llinas R, Precht W (eds) Frog neurobiology. Springer, New York, pp 551–575

Capranica RR, Moffat JM (1983) Neurobehavioral correlates of sound communication in anurans. In: Ewert JP, Capranica RR, Ingle DJ (eds) Advances in vertebrate neuroethology. Plenum Press, New York, pp 701–730

Carlyon RP (2004) How the brain separates sounds. Trends Cog Sci 8:465–471

Carlyon RP, Gockel H (2008) Effects of harmonicity and regularity on the perception of sound sources. In: Yost WA, Popper AN, Fay RR (eds) Auditory perception of sound sources. Springer, New York, pp 191–213

Cherry EC (1953) Some experiments on the recognition of speech, with one and with two ears. J Acoust Soc Am 25:975–979

Christensen-Dalsgaard J (2005) Directional hearing in nonmammalian tetrapods. In: Popper AN, Fay RR (eds) Sound source localization. Springer, New York, pp 67–123

Christensen-Dalsgaard J (2011) Vertebrate pressure-gradient receivers. Hear Res 273:37–45

Christensen-Dalsgaard J, Carr CE (2008) Evolution of a sensory novelty: tympanic ears and the associated neural processing. Brain Res Bull 75:365–370

Christie K, Schul J, Feng AS (2010) Phonotaxis to male's calls embedded within a chorus by female gray treefrogs, *Hyla versicolor*. J Comp Physiol A 196:569–579

Darwin CJ (1997) Auditory grouping. Trends Cog Sci 1:327–333

Darwin CJ (2008) Spatial hearing and perceiving sources. In: Yost WA, Popper AN, Fay RR (eds) Auditory perception of sound sources. Springer, New York, pp 215–232

Darwin CJ, Carlyon RP (1995) Auditory grouping. In: Moore BCJ (ed) Hearing. Academic Press, New York, pp 387–424

Edwards CJ, Alder TB, Rose GJ (2002) Auditory midbrain neurons that count. Nat Neurosci 5:934–936

Eggermont JJ (1988) Mechanisms of sound localization in anurans. In: Fritzsch B, Ryan MJ, Wilczynski W, Hetherington T, Walkowiak W (eds) The evolution of the amphibian auditory system. Wiley, New York, pp 307–336

Elepfandt A, Eistetter I, Fleig A, Gunther E, Hainich M, Hepperle S, Traub B (2000) Hearing threshold and frequency discrimination in the purely aquatic frog *Xenopus laevis* (pipidae): measurement by means of conditioning. J Exp Biol 203:3621–3629

Farris HE, Ryan MJ (2011) Relative comparisons of call parameters enable auditory grouping in frogs. Nat Commun 2:410

Farris HE, Rand AS, Ryan MJ (2002) The effects of spatially separated call components on phonotaxis in túngara frogs: evidence for auditory grouping. Brain Behav Evol 60:181–188

Farris HE, Rand AS, Ryan MJ (2005) The effects of time, space and spectrum on auditory grouping in túngara frogs. J Comp Physiol A 191:1173–1183

Fay RR (2008) Sound source perception and stream segregation in nonhuman vertebrate animals. In: Yost WA, Popper AN, Fay RR (eds) Auditory perception of sound sources. Springer, New York, pp 307–323

Fay RR, Simmons AM (1999) The sense of hearing in fishes and amphibians. In: Fay RR, Popper AN (eds) Comparative hearing: fish and amphibians. Springer, New York, pp 269–318

Fay RR, Popper AN (2000) Evolution of hearing in vertebrates: the inner ears and processing. Hear Res 149:1–10

Feng AS, Ratnam R (2000) Neural basis of hearing in real-world situations. Annu Rev Psychol 51:699–725

Feng AS, Schul J (2007) Sound processing in real-world environments. In: Narins PA, Feng AS, Fay RR, Popper AN (eds) Hearing and sound communication in amphibians. Springer, New York, pp 323–350

Feng AS, Narins PM (2008) Ultrasonic communication in concave-eared torrent frogs (*Amolops tormotus*). J Comp Physiol A 194:159–167

Feng AS, Narins PM, Xu CH (2002) Vocal acrobatics in a Chinese frog, *Amolops tormotus*. Naturwissenschaften 89:352–356

Feng AS, Narins PM, Xu CH, Lin WY, Yu ZL, Qiu Q, Xu ZM, Shen JX (2006) Ultrasonic communication in frogs. Nature 440:333–336

Fritzsch B, Wolkowiak W, Ryan MJ, Wilczynski W, Hetherington T (1988) The evolution of the amphibian auditory system. Wiley, New York

Frost DR, Grant T, Faivovich J, Bain RH, Haas A, Haddad CFB, De Sa RO, Channing A, Wilkinson M, Donnellan SC, Raxworthy CJ, Campbell JA, Blotto BL, Moler P, Drewes RC, Nussbaum RA, Lynch JD, Green DM, Wheeler WC (2006) The amphibian tree of life. B Am Mus Nat Hist 297:8–370

Füllgrabe C, Berthommier F, Lorenzi C (2006) Masking release for consonant features in temporally fluctuating background noise. Hear Res 211:74–84

Fuzessery ZM, Feng AS (1982) Frequency selectivity in the anuran auditory midbrain: single unit responses to single and multiple tone stimulation. J Comp Physiol 146:471–484

Geissler DB, Ehret G (2002) Time-critical integration of formants for perception of communication calls in mice. Proc Natl Acad Sci USA 99:9021–9025

Gerhardt HC (1975) Sound pressure levels and radiation patterns of vocalizations of some North American frogs and toads. J Comp Physiol 102:1–12

Gerhardt HC (1992a) Conducting playback experiments and interpreting their results. In: McGregor PK (ed) Playback and studies of animal communication: problems and prospects. NATO advanced research workshop. Plenum Press, New York, pp 59–77

Gerhardt HC (1992b) Multiple messages in acoustic signals. Sem Neurosci 4:391–400

Gerhardt HC (1995) Phonotaxis in female frogs and toads: execution and design of experiments. In: Klump GM, Dooling RJ, Fay RR, Stebbins WC (eds) Methods in comparative psychoacoustics. Birkhäuser Verlag, Basel, pp 209–220

Gerhardt HC (2005) Acoustic spectral preferences in two cryptic species of grey treefrogs: implications for mate choice and sensory mechanisms. Anim Behav 70:39–48

Gerhardt HC, Doherty JA (1988) Acoustic communication in the gray treefrog, *Hyla versicolor:* evolutionary and neurobiological implications. J Comp Physiol A 162:261–278

Gerhardt HC, Klump GM (1988a) Phonotactic responses and selectivity of barking treefrogs (*Hyla gratiosa*) to chorus sounds. J Comp Physiol A 163:795–802

Gerhardt HC, Klump GM (1988b) Masking of acoustic signals by the chorus background noise in the green treefrog: a limitation on mate choice. Anim Behav 36:1247–1249

Gerhardt HC, Schwartz JJ (2001) Auditory tuning, frequency preferences and mate choice in anurans. In: Ryan MJ (ed) Anuran communication. Smithsonian Institution Press, Washington DC, pp 73–85

Gerhardt HC, Huber F (2002) Acoustic communication in insects and anurans: common problems and diverse solutions. Chicago University Press, Chicago

Gerhardt HC, Bee MA (2007) Recognition and localization of acoustic signals. In: Narins PM, Feng AS, Fay RR, Popper AN (eds) Hearing and sound communication in amphibians. Springer, New York, pp 113–146

Gerhardt HC, Diekamp B, Ptacek M (1989) Inter-male spacing in choruses of the spring peeper, *Pseudacris* (*Hyla*) *crucifer*. Anim Behav 38:1012–1024

Gerhardt HC, Allan S, Schwartz JJ (1990) Female green treefrogs (*Hyla cinerea*) do not selectively respond to signals with a harmonic structure in noise. J Comp Physiol A 166:791–794

Gerhardt HC, Dyson ML, Tanner SD (1996) Dynamic properties of the advertisement calls of gray tree frogs: patterns of variability and female choice. Behav Ecol 7:7–18

Gerhardt HC, Daniel RE, Perrill SA, Schramm S (1987) Mating behavior and male mating success in the green treefrog. Anim Behav 35:1490–1503

Gerhardt HC, Ptacek MB, Barnett L, Torke KG (1994) Hybridization in the diploid-tetraploid treefrogs *Hyla chrysoscelis* and *Hyla versicolor*. Copeia 1994:51–59

Gerhardt HC, Roberts JD, Bee MA, Schwartz JJ (2000) Call matching in the quacking frog (*Crinia georgiana*). Behav Ecol Sociobiol 48:243–251

Gerhardt HC, Martinez-Rivera CC, Schwartz JJ, Marshall VT, Murphy CG (2007) Preferences based on spectral differences in acoustic signals in four species of treefrogs (Anura : Hylidae). J Exp Biol 210:2990–2998

Goense JBM, Feng AS (2012) Effects of noise bandwidth and amplitude modulation on masking in frog auditory midbrain neurons. PLoS ONE 7:e31589

Gomez D, Richardson C, Lengagne T, Plenet S, Joly P, Lena JP, Thery M (2009) The role of nocturnal vision in mate choice: females prefer conspicuous males in the European tree frog (Hyla arborea). P Roy Soc B Biol Sci 276:2351–2358

Grafe TU, Dobler S, Linsenmair KE (2002) Frogs flee from the sound of fire. P Roy Soc B Biol Sci 269:999–1003

Grant KW, Seitz PF (2000) The use of visible speech cues for improving auditory detection of spoken sentences. J Acoust Soc Am 108:1197–1208

Gridi-Papp M, Rand AS, Ryan MJ (2006) Animal communication: complex call production in the túngara frog. Nature 441:38

Gridi-Papp M, Arch VS, Narins PM (2010) Ultrasound transmission and behavioral tuning in the middle ears of Asian frogs. Hear Res 263:244–245

Griffin DR (1976) The audibility of frog choruses to migrating birds. Anim Behav 24:421–427

Griffiths TD, Warren JD (2004) What is an auditory object? Nat Rev Neurosci 5:887–892

Gustafsson HA, Arlinger SD (1994) Masking of speech by amplitude-modulated noise. J Acoust Soc Am 95:518–529

Hebets EA, Papaj DR (2005) Complex signal function: developing a framework of testable hypotheses. Behav Ecol Sociobiol 57:197–214

Heil P, Neubauer H (2003) A unifying basis of auditory thresholds based on temporal summation. Proc Natl Acad Sci USA 100:6151–6156

Hödl W, Amézquita A (2001) Visual signaling in anuran amphibians. In: Ryan MJ (ed) Anuran communication. Smitshsonian Institution Press, Washington DC, pp 121–141

Hoffman HS, Ruppen F (1996) An apparatus for the assessment of prepulse inhibition in the frog. Behav Res Meth Ins C 28:357–359

Hulse SH (2002) Auditory scene analysis in animal communication. Adv Stud Behav 31:163–200

Humfeld SC, Marshall VT, Bee MA (2009) Context-dependent plasticity of aggressive signalling in a dynamic social environment. Anim Behav 78:915–924

Jones DL, Ratnam R (2009) Blind location and separation of callers in a natural chorus using a microphone array. J Acoust Soc Am 126:895–910

King AJ (2007) Auditory neuroscience: filling in the gaps. Curr Biol 17:R799–R801

Klump GM (1995) Studying sound localization in frogs with behavioral methods. In: Klump GM, Dooling RJ, Fay RR, Stebbins WC (eds) Methods in comparative psychoacoustics. Birkhäuser Verlag, Basel, pp 221–233

Klump GM (1996) Bird communication in the noisy world. In: Kroodsma DE, Miller EH (eds) Ecology and evolution of acoustic communication in birds. Cornell University Press, Ithaca, pp 321–338

Klump GM, Gerhardt HC (1987) Use of non-arbitrary acoustic criteria in mate choice by female gray tree frogs. Nature 326:286–288

Klump GM, Kittel M, Wagner E (2001) Comodulation masking release in the Mongolian gerbil. Abstracts of the Association for Research on Otolaryngology 25:#84

Larson KA (2004) Advertisement call complexity in northern leopard frogs, Rana pipiens. Copeia 2004:676–682

Lemon RE (1971) Vocal communication by the frog Eleutherodactylus martinicensis. Can J Zool 49:211–217

Lewis ER, Narins PM (1985) Do frogs communicate with seismic signals? Science 227:187–189

Lewis ER, Narins PM (1999) The acoustic periphery of amphibians: anatomy and physiology. In: Fay RR, Popper AN (eds) Comparative hearing: fish and amphibians. Springer, New York, pp 101–154

Lewis ER, Fay RR (2004) Environmental variables and the fundamental nature of hearing. In: Manley GA, Popper AN, Fay RR (eds) Evolution of the vertebrate auditory system. Springer, New York, pp 27–54

Lewis ER, Narins PM, Cortopassi KA, Yamada WM, Poinar EH, Moore SW, Yu XL (2001) Do male white-lipped frogs use seismic signals for intraspecific communication? Am Zool 41:1185–1199

Lin WY, Feng AS (2001) Free-field unmasking response characteristics of frog auditory nerve fibers: comparison with the responses of midbrain auditory neurons. J Comp Physiol A 187:699–712

Lin WY, Feng AS (2003) GABA is involved in spatial unmasking in the frog auditory midbrain. J Neurosci 23:8143–8151

Littlejohn MJ, Fouquette MJ, Johnson C (1960) Call discrimination by female frogs of the Hyla versicolor complex. Copeia 1960:47–49

Loftus-Hills JJ, Littlejohn MJ (1971) Mating-call sound intensities of anuran amphibians. J Acoust Soc Am 49:1327–1329

Love EK, Bee MA (2010) An experimental test of noise-dependent voice amplitude regulation in Cope's grey treefrog, *Hyla chrysoscelis*. Anim Behav 80:509–515

Manley GA, Popper AN, Fay RR (2004) Evolution of the vertebrate auditory system, vol 22. Springer, New York

Márquez R, Bosch J, Eekhout X (2008) Intensity of female preference quantified through playback setpoints: call frequency versus call rate in midwife toads. Anim Behav 75:159–166

Marshall VT, Schwartz JJ, Gerhardt HC (2006) Effects of heterospecific call overlap on the phonotactic behaviour of grey treefrogs. Anim Behav 72:449–459

Martof BS (1953) Territoriality in the green frog, *Rana clamitans*. Ecology 34:165–174

Mason MJ (2007) Pathways for sound transmission to the inner ear in amphibians. In: Narins PM, Feng AS, Fay RR, Popper AN (eds) Hearing and sound communication in amphibians. Springer, New York, pp 147–183

McDermott JH (2009) The cocktail party problem. Curr Biol 19:R1024–R1027

Meenderink SWF, Kits M, Narins PM (2010) Frequency matching of vocalizations to inner-ear sensitivity along an altitudinal gradient in the coqui frog. Biol Lett 6:278–281

Megela-Simmons A, Moss CF, Daniel KM (1985) Behavioral audiograms of the bullfrog (*Rana catesbeiana*) and the green tree frog (*Hyla cinera*). J Acoust Soc Am 78:1236–1244

Miller CT, Dibble E, Hauser MD (2001) Amodal completion of acoustic signals by a nonhuman primate. Nat Neurosci 4:783–784

Moore BCJ, Gockel H (2002) Factors influencing sequential stream segregation. Acta Acustica United Acustica 88:320–333

Morris MR (1991) Female choice of large males in the treefrog *Hyla ebraccata*. J Zool 223:371–378

Moss CF, Simmons AM (1986) Frequency selectivity of hearing in the green treefrog, *Hyla cinerea*. J Comp Physiol A 159:257–266

Murphy CG (2003) The cause of correlations between nightly numbers of male and female barking treefrogs (*Hyla gratiosa*) attending choruses. Behav Ecol 14:274–281

Narins PM (1982) Effects of masking noise on evoked calling in the Puerto Rican coqui (Anura, Leptodactylidae). J Comp Physiol 147:439–446

Narins PM (1987) Coding of signals in noise by amphibian auditory nerve fibers. Hear Res 26:145–154

Narins PM, Zelick R (1988) The effects of noise on auditory processing and behavior in amphibians. In: Fritzsch B, Ryan MJ, Wilczynski W, Hetherington TE, Walkowiak W (eds) The evolution of the amphibian auditory system. Wiley, New York, pp 511–536

Narins PM, Ehret G, Tautz J (1988) Accessory pathway for sound transfer in a neotropical frog. Proc Natl Acad Sci USA 85:1508–1512

Narins PM, Hödl W, Grabul DS (2003) Bimodal signal requisite for agonistic behavior in a dart-poison frog, *Epipedobates femoralis*. Proc Natl Acad Sci USA 100:577–580

Narins PM, Feng AS, Fay RR, Popper AN (2007) Hearing and sound communication in amphibians, vol 28. Springer, New York

Narins PM, Grabul DS, Soma KK, Gaucher P, Hödl W (2005) Cross-modal integration in a dart-poison frog. Proc Natl Acad Sci USA 102:2425–2429

Narins PM, Feng AS, Lin WY, Schnitzler HU, Denzinger A, Suthers RA, Xu CH (2004) Old World frog and bird, vocalizations contain prominent ultrasonic harmonics. J Acoust Soc Amer 115:910–913

Nelken I, Rotman Y, Bar Yosef O (1999) Responses of auditory-cortex neurons to structural features of natural sounds. Nature 397:154–157

Nelson DA, Marler P (1990) The perception of birdsong and an ecological concept of signal space. In: Berkley MA, Stebbins WC (eds) Comparative perception, vol II. Wiley, New York, pp 443–478

Nityananda V, Bee MA (2011) Finding your mate at a cocktail party: frequency separation promotes auditory stream segregation of concurrent voices in multi-species frog choruses. PLoS ONE 6:e21191

Nityananda V, Bee MA (2012) Spatial release from masking in a free-field source identification task by gray treefrogs. Hear Res 285:86–97

Paez VP, Bock BC, Rand AS (1993) Inhibition of evoked calling of *Dendrobates pumilio* due to acoustic interference from cicada calling. Biotropica 25:242–245

Passmore NI (1981) Sound levels of mating calls of some African frogs. Herpetologica 37:166–171

Passmore NI, Telford SR (1981) The effect of chorus organization on mate localization in the painted reed frog (*Hyperolius marmoratus*). Behav Ecol Sociobiol 9:291–293

Passmore NI, Capranica RR, Telford SR, Bishop PJ (1984) Phonotaxis in the painted reed frog (*Hyperolius marmoratus*): the localization of elevated sound sources. J Comp Physiol 154:189–197

Penna M, Solís R (1998) Frog call intensities and sound propagation in the South American temperate forest region. Behav Ecol Sociobiol 42:371–381

Penna M, Hamilton-West C (2007) Susceptibility of evoked vocal responses to noise exposure in a frog of the temperate austral forest. Anim Behav 74:45–56

Penna M, Pottstock H, Velasquez N (2005) Effect of natural and synthetic noise on evoked vocal responses in a frog of the temperate austral forest. Anim Behav 70:639–651

Petkov CI, O'Connor KN, Sutter ML (2003) Illusory sound perception in macaque monkeys. J Neurosci 23:9155–9161

Plomp R, Mimpen AM (1979a) Improving the reliability of testing the speech reception threshold for sentences. Audiology 18:43–52

Plomp R, Mimpen AM (1979b) Speech reception threshold for sentences as a function of age and noise level. J Acoust Soc Am 66:1333–1342

Popper AN, Fay RR (1997) Evolution of the ear and hearing: issues and questions. Brain Behav Evol 50:213–221

Pyron RA, Wiens JJ (2011) A large-scale phylogeny of amphibia including over 2800 species, and a revised classification of extant frogs, salamanders, and caecilians. Mol Phylogenet Evol 61:543–583

Ratnam R, Feng AS (1998) Detection of auditory signals by frog inferior collicular neurons in the presence of spatially separated noise. J Neurophysiol 80:2848–2859

Recanzone GH, Sutter ML (2008) The biological basis of audition. Annu Rev Psychol 59:119–142

Rheinlaender J, Klump GM (1988) Behavioral aspects of sound localization. In: Fritzsch B, Ryan MJ, Wilczynski W, Hetherington T (eds) The evolution of the amphibian auditory system. Wiley, New York, pp 297–305

Richards DG, Wiley RH (1980) Reverberations and amplitude fluctuations in the propagation of sound in a forest: implications for animal communication. Am Nat 115:381–399

Richardson C, Lengagne T (2010) Multiple signals and male spacing affect female preference at cocktail parties in treefrogs. P Roy Soc B Biol Sci 277:1247–1252

Richardson C, Gomez D, Durieux R, Thery M, Joly P, Lena JP, Plenet S, Lengagne T (2010) Hearing is not necessarily believing in nocturnal anurans. Biol Lett 6:633–635

Rose GJ, Brenowitz EA (1991) Aggressive thresholds of male Pacific treefrogs for advertisement calls vary with amplitude of neighbors' calls. Ethology 89:244–252

Rosenthal GG, Rand AS, Ryan MJ (2004) The vocal sac as a visual cue in anuran communication: an experimental analysis using video playback. Anim Behav 68:55–58

Ryan MJ (1985) *The túngara frog:* a study in sexual selection and communication. Chicago University Press, Chicago

Ryan MJ (1991) Sexual selection and communication in frogs. Trends Ecol Evol 6:351–355

Ryan MJ (2001) Anuran communication. Smithsonian Institution Press, Washington DC

Ryan MJ, Rand AS (1990) The sensory basis of sexual selection for complex calls in the túngara frog, *Physalaemus pustulosus* (sexual selection for sensory exploitation). Evolution 44:305–314

Ryan MJ, Keddy-Hector A (1992) Directional patterns of female mate choice and the role of sensory biases. Am Nat 139:S4–S35

Ryan MJ, Rand AS (1993) Species recognition and sexual selection as a unitary problem in animal communication. Evolution 47:647–657

Ryan MJ, Rand AS (2001) Feature weighting in signal recognition and discrimination by túngara frogs. In: Ryan MJ (ed) Anuran communication. Smithsonian Institution Press, Washington DC, pp 86–101

Schrode KM, Buerkle NP, Brittan-Powell EF, Bee MA Auditory brainstem responses in Cope's gray treefrog (Hyla chrysoscelis): effects of frequency, level, sex and size. J Comp Physiol A in press

Schul J, Bush SL (2002) Non-parallel coevolution of sender and receiver in the acoustic communication system of treefrogs. P Roy Soc Lond B Biol Sci 269:1847–1852

Schwartz JJ (1987) The function of call alternation in anuran amphibians: a test of three hypotheses. Evolution 41:461–471

Schwartz JJ (1993) Male calling behavior, female discrimination and acoustic interference in the Neotropical treefrog Hyla microcephala under realistic acoustic conditions. Behav Ecol Sociobiol 32:401–414

Schwartz JJ, Wells KD (1983) The influence of background noise on the behavior of a neotropical treefrog, Hyla ebraccata. Herpetologica 39:121–129

Schwartz JJ, Gerhardt HC (1989) Spatially mediated release from auditory masking in an anuran amphibian. J Comp Physiol A 166:37–41

Schwartz JJ, Gerhardt HC (1995) Directionality of the auditory system and call pattern recognition during acoustic interference in the gray treefrog, Hyla versicolor. Aud Neurosci 1:195–206

Schwartz JJ, Gerhardt HC (1998) The neuroethology of frequency preferences in the spring peeper. Anim Behav 56:55–69

Schwartz JJ, Marshall VT (2006) Forms of call overlap and their impact on advertisement call attractiveness to females of the gray treefrog, Hyla versicolor. Bioacoustics 16:39–56

Schwartz JJ, Bee MA, Tanner SD (2000) A behavioral and neurobiological study of the responses of gray treefrogs, Hyla versicolor, to the calls of a predator, Rana catesbeiana. Herpetologica 56:27–37

Schwartz JJ, Buchanan BW, Gerhardt HC (2001) Female mate choice in the gray treefrog (Hyla versicolor) in three experimental environments. Behav Ecol Sociobiol 49:443–455

Schwartz JJ, Buchanan BW, Gerhardt HC (2002) Acoustic interactions among male gray treefrogs, Hyla versicolor, in a chorus setting. Behav Ecol Sociobiol 53:9–19

Schwartz JJ, Huth K, Hunce R, Lentine B (2010a) Effect of anomalous pulse timing on call discrimination by females of the gray treefrog (Hyla versicolor): behavioral correlates of neurobiology. J Exp Biol 213:2066–2072

Schwartz JJ, Crimarco NC, Bregman Y, Umeoji K (2013) Responses of the gray treefrog (Hyla versicolor) to chorus noise and an investigation of their functional significance. J Herp 47:354–360

Schwartz JJ, Brown R, Turner S, Dushaj K, Castano M (2008) Interference risk and the function of dynamic shifts in calling in the gray treefrog (Hyla versicolor). J Comp Psychol 122:283–288

Schwartz JJ, Huth K, Jones SH, Brown R, Marks J (2010b) Tests for call restoration in the gray treefrog, Hyla versicolor. Bioacoustics 20:59–86

Schwartz JL, Berthommier F, Savariaux C (2004) Seeing to hear better: evidence for early audio-visual interactions in speech identification. Cognition 93:B69–B78

Seeba F, Klump GM (2009) Stimulus familiarity affects perceptual restoration in the European starling (Sturnus vulgaris). PLoS ONE 4:e5974

Seeba F, Schwartz JJ, Bee MA (2010) Testing an auditory illusion in frogs: perceptual restoration or sensory bias? Anim Behav 79:1317–1328

Semlitsch RD (2003) Amphibian conservation. Smithsonian, Washington DC

Shamma SA, Micheyl C (2010) Behind the scenes of auditory perception. Curr Opin Neurobiol 20:361–366

Shen JX, Feng AS, Xu ZM, Yu ZL, Arch VS, Yu XJ, Narins PM (2008) Ultrasonic frogs show hyperacute phonotaxis to female courtship calls. Nature 453:914–916

Simmons AM (1988a) Selectivity for harmonic structure in complex sounds by the green treefrog (*Hyla cinerea*). J Comp Physiol A 162:397–403

Simmons AM (1988b) Masking patterns in the bullfrog (*Rana catesbeiana*). I. Behavioral effects. J Acoust Soc Am 83:1087–1092

Simmons AM (2013) "To Ear is Human, to Frogive is Divine": Bob Capranica's legacy to auditory neuroethology. J Comp Physiol A 199:169–182

Simmons AM, Moss CF (1995) Reflex modification: a tool for assessing basic auditory function in anuran amphibians. In: Klump GM, Dooling RJ, Fay RR, Stebbins WC (eds) Methods in comparative psychoacoustics. Birkhäuser Verlag, Basel, pp 197–208

Simmons AM, Bean ME (2000) Perception of mistuned harmonics in complex sounds by the bullfrog (*Rana catesbeiana*). J Comp Psychol 114:167–173

Simmons AM, Buxbaum RC, Mirin MP (1993) Perception of complex sounds by the green treefrog, *Hyla cinerea*: envelope and fine-structure cues. J Comp Physiol A 173:321–327

Simmons DD, Meenderink SWF, Vassilakis PN (2007) Anatomy, physiology, and function of the auditory end-organs in the frog inner ear. In: Narins PA, Feng AS, Fay RR, Popper AN (eds) Hearing and sound communication in amphibians. Springer, New York, pp 184–220

Slabbekoorn H, Bouton N (2008) Soundscape orientation: a new field in need of sound investigation. Anim Behav 76:E5–E8

Swanson EM, Tekmen SM, Bee MA (2007) Do female anurans exploit inadvertent social information to locate breeding aggregations? Can J Zool 85:921–932

Taylor RC, Klein BA, Stein J, Ryan MJ (2008) Faux frogs: multimodal signalling and the value of robotics in animal behaviour. Anim Behav 76:1089–1097

Taylor RC, Ryan MJ (2013) Interactions of multisensory components perceptually rescue túngara frog mating signals. Science 341:273–274

Taylor RC, Klein BA, Stein J, Ryan MJ (2011) Multimodal signal variation in space and time: how important is matching a signal with its signaler? J Exp Biol 214:815–820

Ursprung E, Ringler M, Hödl W (2009) Phonotactic approach pattern in the neotropical frog *Allobates femoralis*: a spatial and temporal analysis. Behaviour 146:153–170

Vélez A, Bee MA (2010) Signal recognition by frogs in the presence of temporally fluctuating chorus-shaped noise. Behav Ecol Sociobiol 64:1695–1709

Vélez A, Bee MA (2011) Dip listening and the cocktail party problem in grey treefrogs: signal recognition in temporally fluctuating noise. Anim Behav 82:1319–1327

Vélez A, Höbel G, Gordon NM, Bee MA (2012) Dip listening or modulation masking? Call recognition by green treefrogs (*Hyla cinerea*) in temporally fluctuating noise. J Comp Physiol A 198:891–904

Vélez A, Bee MA (2013) Signal recognition by Cope's gray treefrogs (*Hyla chrysoscelis*) and green treefrogs (*H. cinerea*) in naturally fluctuating noise. J Comp Psychol 127:166–178

Vélez A, Gu Y, Sun Y, Bee MA (2013) Pulse-number discrimination by females of Cope's gray treefrog (*Hyla chrysoscelis*) in modulated and unmodulated noise. J Acoust Soc Am 134:3079–3089

Verhey JL, Pressnitzer D, Winter IM (2003) The psychophysics and physiology of comodulation masking release. Exp Brain Res 153:405–417

Wagner WE (1989) Graded aggressive signals in Blanchard's cricket frog: vocal responses to opponent proximity and size. Anim Behav 38:1025–1038

Ward JL, Buerkle NP, Bee MA (2013a) Spatial release from masking improves sound pattern discrimination along a biologically relevant pulse-rate continuum in gray treefrogs. Hear Res in press.

Ward JL, Love EK, Vélez A, Buerkle NP, O'Bryan LR, Bee MA (2013b) Multitasking males and multiplicative females: dynamic signalling and receiver preferences in Cope's grey treefrog (*Hyla chrysoscelis*). Anim Behav 86:231–243

Warren RM (1970) Perceptual restoration of missing speech sounds. Science 167:392–393

Welch AM, Semlitsch RD, Gerhardt HC (1998) Call duration as an indicator of genetic quality in male gray treefrogs. Science 280:1928–1930

Wells KD (2007) The ecology and behavior of amphibians. University of Chicago Press, Chicago

Wells KD, Schwartz JJ (1984) Vocal communication in a neotropical treefrog, *Hyla ebraccata*: advertisement calls. Anim Behav 32:405–420

Wells KD, Schwartz JJ (2007) The behavioral ecology of anuran communication. In: Narins PM, Feng AS, Fay RR, Popper AN (eds) Hearing and sound communication in amphibians. Springer, New York, pp 44–86

Wells KD, Taigen TL (1986) The effect of social interactions on calling energetics in the gray treefrog (*Hyla versicolor*). Behav Ecol Sociobiol 19:9–18

Wilczynski W, Brenowitz EA (1988) Acoustic cues mediate inter-male spacing in a neotropical frog. Anim Behav 36:1054–1063

Witte K, Farris HE, Ryan MJ, Wilczynski W (2005) How cricket frog females deal with a noisy world: habitat-related differences in auditory tuning. Behav Ecol 16:571–579

Wollerman L (1998) Stabilizing and directional preferences of female *Hyla ebraccata* for calls differing in static properties. Anim Behav 55:1619–1630

Wollerman L (1999) Acoustic interference limits call detection in a Neotropical frog Hyla ebraccata. Anim Behav 57:529–536

Wollerman L, Wiley RH (2002) Background noise from a natural chorus alters female discrimination of male calls in a Neotropical frog. Anim Behav 63:15–22

Yerkes RM (1904) Inhibition and reinforcement of reaction in the frog, *Rana clamitans*. J Comp Neurol Psychol 13:124–137

Yost WA, Popper AN, Fay RR (2008) Auditory perception of sound sources. Springer, New York

Zakon HH, Wilczynski W (1988) The physiology of the anuran eighth nerve. In: Fritzsch B, Wolkowiak W, Ryan MJ, Wilczynski W, Hetherington T (eds) The evolution of the amphibian auditory system. Wiley, New York, pp 125–155

Chapter 7
Avian Vocal Production in Noise

Henrik Brumm and Sue Anne Zollinger

Abstract Birds use acoustic signals to mediate a number of crucial social inter-actions such as territorial defence, mate attraction and predator avoidance. Thus, differences in signalling efficiency are likely to have major fitness consequences. Acoustic signal transmission is considerably constrained by noise, e.g. sounds in the environment that interfere with the detection, discrimination or recognition of a signal. In this chapter, we discuss noise sources encountered by birds, and the diverse ways birds use to make their signals heard in this noisy world. One concept of signal evolution suggests that bird vocalisations undergo microevolutionary adaptations over time that tailor their sounds to the specific noise profiles of their species-typical habitats. On the individual level, birds across many different taxa also possess the vocal plasticity to make short-term adjustments to their signals to reduce masking in response to changing environmental noise conditions. Such adjustments can take different forms in different species. However, the widespread problem of acoustic communication in noise has also led to the evolution of one shared solution in birds: the Lombard effect, i.e. a noise-dependent regulation of vocal amplitude. In addition, birds may also change the frequency, the duration, the timing, and/or the redundancy of their vocal signals in noise, although in many cases it is not yet clear whether these additional changes are achieved through ontogenetic plasticity or through short-term regulation. In recent years, there has been a flurry of new studies reporting correlations between increased levels of anthropogenic noise and a variety of changes in the vocal behaviour of birds. While many of these studies have focused on increases in song or call frequency in birds exposed to high levels of traffic noise, it is not yet known whether these differences in vocal pitch are actually adaptive. We encourage future research studies to take a more rigorous and inte-grative approach to the study of vocal signalling in noise. Finally, we note the need for more research on the impact of noise on the evolution and usage of multi-component signals that combine vocal and visual signals.

H. Brumm (✉) · S. A. Zollinger
Communication and Social Behaviour Group, Max Planck Institute for Ornithology,
Seewiesen, Germany
e-mail: brumm@orn.mpg.de

H. Brumm (ed.), *Animal Communication and Noise*,
Animal Signals and Communication 2, DOI: 10.1007/978-3-642-41494-7_7,
© Springer-Verlag Berlin Heidelberg 2013

7.1 Introduction

The songs of birds are probably the most complex and arguably some of the most beautiful sounds in nature. Bird song has inspired artists and scientists alike, and during the last half century bird song has become an important model in the study of animal behaviour (Slater 2003). In addition to songs, there is an astonishing variety of functionally different bird calls, and both songs and calls can be used in a host of sophisticated ways which are probably only surpassed by the faculty of human speech.

A substantial body of research suggests that the two main functions of bird song are territory defence and mate attraction, and that these evolutionary pressures are what led to the stunning diversity that can be observed in the songs of different bird species throughout the world (Catchpole and Slater 2008). In the temperate zones, it is usually the males that sing. Females are attracted by male song, whereas rival males are repelled. This pattern shows striking parallels with the male advertisement signals of anurans and insects, in which males produce acoustic mating signals that are used in female choice and male–male competition (Chaps. 3, 5 and 6). In the tropics, however, female birds sing much more commonly than in the more temperate latitudes, and they may use their songs in both territory defence and attraction of male mates (Slater and Mann 2004).

In addition to songs, birds also use a variety of calls to communicate with each other. Calls are usually shorter and simpler than songs, and often comprise only one syllable type. Because the function of song is linked to reproduction, songs are often seasonal in areas where reproduction occurs only at certain times of the year. Quite the opposite, calls are usually produced throughout the year and by both sexes. Bird calls are typically used in particular contexts that can be related to specific functions such as the announcement and exchange of food, maintenance of social proximity and group cohesion, predator alarm and so on (Marler 2004). Interestingly, birds not only understand the calls of conspecifics but they can also glean information from the alarm calls of other bird species (e.g. Krams and Krama 2002; Magrath et al. 2007; Goodale and Kotagama 2008; Haff and Magrath 2013; Wheatcroft and Price 2013) and even of primates (Rainey et al. 2004).

In oscine birds, songs are individually acquired through vocal production learning (Hultsch and Todt 2008). This enables songbirds to adjust their signals more quickly to the acoustic properties of their habitats, because song structure is not only shaped by environmental and sexual selection but also cultural evolution and ontogenetic changes. Thus, the songs of birds that are learnt may be more flexible in evolutionary and individual terms compared to those which are not (Lachlan and Servedio 2004; Tumer and Brainard 2007; Ríos-Chelén et al. 2012a). Learning, however, is not necessary to develop complex communication systems with many different signal types. Even phylogenetically very basal bird species that do not learn their vocal signals can have complex call repertoires (Schuster et al. 2012). The particularly rich vocal repertoires of birds may have something to do with the lack of sophisticated chemical communication in this group. Whereas many arthropods use pheromones for long-distance signalling (Chap. 13), sound is the primary modality in birds.

Because so many important aspects in the life of birds are mediated by acoustic signals, interference within the auditory communication channel can have particularly severe consequences. However, before we investigate the potential costs of signal masking in more detail, we will briefly review some of the many sources of noise that birds encounter in their habitats.

7.1.1 Noise Sources in Bird Habitats

As discussed in Chap. 2, noise in communication can be regarded as anything in the communication channel, or in the nervous system of the receiver that leads to errors by receivers. In this chapter we will consider a particular form of noise that is equivalent to the common definition of the word, i.e. interfering sound in the acoustic communication channel. Birds are remarkably adaptable and they inhabit a wide range of biotopes on all continents, from the vast tropical rainforests to the tundra and from the open seas to high-altitude mountains. Some species are even found in cities. However, all of these strikingly different habitats have one thing in common—they are noisy places. Even when one discounts the ever-increasing levels of man-made noise that pollute more and more of the landscape, nature itself produces a lot of sound.

First of all, there are those sounds that already filled the air long before any birds inhabited the earth. These sounds are generated by physical processes, such as the weather. Wind and rain can produce considerable noise levels, and in many bird habitats they are indeed the major abiotic noise sources.

Even a gentle breeze with wind speeds of 4 m/s has been found to generate noise levels between 30 and 40 dB(A) in coniferous forests and between 45 and 55 dB(A) in deciduous woods (Fegeant 1999). In open grasslands, the same wind speed results in a L_{A95} (i.e. the A-weighted sound pressure level which is exceeded for 95 % of the measurement time) of about 35 dB (Boersma 1997). Wind noise is generated by the movement of air past the ground and vegetation. In addition, wind can also produce sound when passing an animal's head. This means that species may differ in the levels of wind noise they perceive because of differences in the size, surface characteristics and, most importantly, shape of their heads. (Differences in wind-induced head noise are probably not very large between different bird species, which typically have very aerodynamic heads, selected for flying. In mammals, however, this may be more important, especially when we consider the different shapes and sizes of pinnae in this group).

The spectrum of wind noise varies mainly with the wind speed and the type of vegetation over which the wind is passing. In both coniferous and deciduous forests, wind-induced vegetation noise can have substantial amounts of energy in the frequency band between 1 and 4 kHz (Fegeant 1999; Bolin 2009), which overlaps with the lower frequency range of most bird songs. In addition to masking, wind can also affect acoustic communication by scattering sound, as turbulence from the slightest winds can create considerable amplitude fluctuations in the received signal (Wiley and Richards 1978).

Wind has long been recognised as a constraint on acoustic communication in birds (e.g. Jilka and Leisler 1974; Morton 1975; Wiley and Richards 1978, 1982; Brenowitz 1982; Ryan and Brenowitz 1985; Wiley 1991; Lengagne et al. 1999) but sound generated by rain has received much less attention. Nevertheless, rain noise can easily reach 50 dB(A) in forests, and can cover a broad frequency band from 0.5 to 8 kHz (Miller 1978). In fact, in forests with high rainfall, rain noise is often much more disrupting than wind. The work by Keast (1994) on the vocal activity of an Australian forest bird community and the study by Lengagne and Slater (2002) on calling in tawny owls (*Strix aluco*) are two of the few studies that have looked at the effects of rain on vocal behaviour in birds.

In some bird habitats, both wind and rain noise can occur at certain predictable times of day, and it has been hypothesised that avoidance of noise masking may explain diurnal patterns of singing activity in some species. Along these lines, Henwood and Fabrick (1979) suggested that the reason that so many birds show a marked peak of singing activity around dawn is that wind and air turbulence are particularly low at this time of day. However, other studies could not find a notable sound transmission advantage at dawn and thus there may also be other benefits for singing at daybreak (reviewed in Catchpole and Slater 2008; Brumm and Naguib 2009).

In addition to rain, other forms of moving water, e.g. surf, waterfalls, torrents and rocky streams, can also generate substantial noise that may mask bird vocalisations. These abiotic noise sources are much less ubiquitous than wind and rain but in the habitats where they do occur, this particular noise is often present over very long periods, if not 24 hours a day. Because waterfall and torrent noise incessantly covers a broad frequency band with very high sound levels, they provide an excellent opportunity to investigate the long-term effects of noise on animal communication in a natural environment. Several studies on frogs (Chap. 5) and birds (see Sect. 7.2) have demonstrated the powerful effects of these massive noise sources on the design of acoustic signals.

Abiotic sounds are far from being the only noise sources that impede the exchange of acoustic signals between birds. Additional noises can come from other animals, and these biotic noises often lead to much more unfavourable signal-to-noise ratios than abiotic noises. Experiments with captive birds show that temporal and spectral overlap with background noise results in masking that reduces signal detection (Chap. 8). Likewise, experiments on vocal production have shown that noise within the frequency band of their own vocalisations is most effective in inducing birds to adjust their songs and calls (Manabe et al. 1998; Brumm and Todt 2002). In some biotopes, this crucial frequency band overlaps with those used by sound-producing insects and anurans. The sound pressure levels produced by aggregations of these animals can be deafening (Greenfield 2005) and it has been suggested that avoidance of insect chorus noises may be important for vocalising birds (Ryan and Brenowitz 1985; Kirschel et al. 2009; Weir et al. 2012). However, bird songs are often masked most potently by the songs of other birds because of their spectral and temporal similarities (Chap. 8).

In tropical rainforests, more than a hundred different bird species may be found in a single square kilometre, and especially during the dawn chorus, their various songs may mutually mask each other. One of the loudest is the very appropriately named screaming piha *Lipaugus vociferans* from South America, which blasts out its call at more than 110 dB(A) (Nemeth 2004). However, although it is often overlooked, biotic noise from birds can also be substantial in the temperate zones as well. Many temperate species have loud voices, that can reach 85–90 dB(A) at 1 m distance (e.g. Eurasian wren *Troglodytes troglodytes* (Holland 2000), great tits *Parus major* (Blumenrath and Dabelsteen 2004), common nightingales *Luscinia megarhynchos* (Brumm 2004), common chaffinches *Fringilla coelebs* (Brumm and Ritschard 2011)). In a series of measurements during the dawn chorus in a mixed forest in North-Eastern Germany we recorded an average sound level of 48 dB(A) with peak values around 56 dB(A) (Fig. 7.1). This level of biotic noise is similar

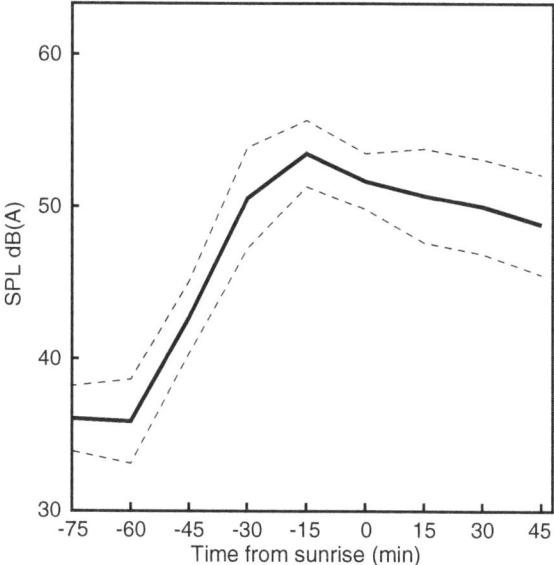

Fig. 7.1 Time course of the dawn chorus noise in early spring (last week of April) in a mixed forest in North-Eastern Germany. Average peak levels (*solid line*) and standard deviations (*dashed lines*) are shown for six different recording sites. The total number of bird species in the choruses varied between 12 and 15 between the sites (the most common species present at all sites were blue tit (*Cyanistes caeruleus*), chaffinch (*Fringilla coelebs*), common blackbird (*Turdus merula*), common wood pigeon (*Columba palumbus*), Eurasian blackcap (*Sylvia atricapilla*), Eurasian nuthatch (*Sitta europaea*), European robin (*Erithacus rubecula*), great tit (*Parus major*), and song thrush (*Turdus philomelos*). The maximum sound level within a 1-min interval was measured every 15 min. The onsets of the dawn choruses were between 60 and 45 min before sunrise. Unpublished data collected by Henrik Brumm, Davide Dominoni, Stefan Greif and Erwin Nemeth

to that of the dawn chorus in a Venezuelan rainforest (Ellinger and Hoedl 2003). Indeed, the songs from different temperate forest species mutually mask each other, and it has been predicted that they constrain the distance over which a given bird can communicate (Nemeth and Brumm 2010).

Yet, it is not only vocalisations from other species that can impair signal transmission but also those from conspecifics. This is the case in instances where signalling is targeted at a particular individual (or a few) but where there are many other individuals calling in the area, such as in parent-offspring communication through the noisy hubbub of seabird colonies or the competing calls of nestlings begging for food (e.g. Aubin and Jouventin 2002; Leonard and Horn 2005). In such cases it is unlikely that individual signal changes can solve the masking problem because the competing conspecifics will also change their call characteristics, and this will eventually lead to an escalation of the scenario. Therefore, strong selection acts on the receivers' side to evolve effective auditory mechanisms for call detection and discrimination in noise (Chap. 8).

An evolutionarily very recent, but at the same time very substantial, form of noise is the din produced by humans. Urbanisation and global traffic are projected to increase over the next decades (United Nations 2012), and as a result man-made noise levels will continue to rise and to affect ever more remote areas that as yet have been relatively undisturbed. Traffic noise produced by cars and lorries is usually low-pitched with its main energy typically below 1 kHz (Can et al. 2010; Bocharov et al. 2012). However, noise emission levels in urban areas or close to motorways can be very high, often reaching levels of 65 dB(A) or more (Barrigon et al. 2002; Zannin et al. 2002; Tsai et al. 2009). This means that the high-frequency components of traffic noise can still be loud enough to reduce the active space of bird songs (Nemeth and Brumm 2010). Of course, intense anthropogenic noise is not only a potential signal masker but might also pose a threat to the health of birds, similar to the serious effects it has on humans (WHO 2011). In Sect. 7.1.2, we will argue that acoustic signal masking alone, be it from man-made noise or any other noise source, can have potential fitness consequences for birds (Fig. 7.2).

7.1.2 Consequences of Signal Masking

Because acoustic signals are so important for the survival and reproduction of many bird species, interruption of signal exchange by noise can potentially have major fitness costs. According to Signal Detection Theory, receiver errors can be classified as missed detections or as false alarms (Chap. 2). The potential costs of missed detections seem rather self-evident given the functions of bird vocalisations (see Sect. 7.1), but it can be more difficult to evaluate the consequences of false alarms. Of course one could argue that by spending time responding to an irrelevant signal, a bird cannot perform other important behaviours, such as foraging. However, this argument could be made about any pair of behaviours and thus it is a rather unproductive behavioural ecology cliché.

Fig. 7.2 A great thrush *Turdus fuscater* looks out over the busy, and noisy, city of Quito, Ecuador. Like common blackbirds, great thrushes have songs with low-pitched motif elements below 2.5 kHz, which are particularly vulnerable to masking by traffic noise (Nemeth and Brumm 2010). As more and more natural habitats become polluted by anthropogenic noise, birds must find ways to adjust or adapt their vocal behaviour to maintain communication. We are just beginning to understand a few of the impacts that increased noise levels have on the vocalisations, behaviour, health and fitness of birds, but there are still many unanswered questions

The missed detection of a mating signal from an optimal mate can have fitness consequences for both the sender and the receiver, thus we can plausibly assume that there is strong selection for effective signal transmission in bird song (see Sect. 8.2, Chap. 8). As mentioned above, bird songs function not only in mate selection but also in territorial defence. Territorial defence becomes problematic when the songs of a rival cannot be detected or when the songs of a rival cannot be discriminated from those of a neighbour who poses only a minor threat (Catchpole and Slater 2008). It is unclear though whether such missed detections would eventually result in the loss of a territory, but it seems possible that they might at least lead to a loss of paternity when rivals are able to "sneak in" undetected and engage in extra-pair copulations. Another potential consequence of the masking of territorial songs is that the active space of the signal decreases and therefore territories become smaller and potentially contain fewer resources, leading to a reduction in fitness.

False alarms in mating signals may also bear potential costs, e.g. when an animal selects a suboptimal mate because it failed to recognise the quality of the mate due to signal masking. Such a scenario has been suggested in fishes in which females accepted matings from suboptimal males in turbid waters where visual signals are impaired (Järvenpää and Lindström 2004). However, it is unclear whether such false alarms occur in acoustic signals in birds, which, during courtship displays, usually exchange multiple signals over short distances. Indeed, the elaborateness of courtship displays in birds, which sometimes also include multimodal signals, can be interpreted as a strategy to reduce false alarms and hence ensure optimal mating decisions (Chap. 2). As to the costs of false alarms in territorial signalling, we can only speculate. Birds often reply to territorial challenges by increasing their song output, but singing in birds does not seem to entail significant energetic costs (Oberweger and Goller 2001; Ward et al. 2003, 2004; Zollinger et al. 2011). Instead of energetic constraints, it seems that song production is more likely to be constrained by social aggression (Brumm and Ritschard 2011). Accordingly, a false alarm in territorial signalling that results in increased song rates and/or amplitudes may evoke costs through increased aggression by conspecifics.

In contrast to songs that are used in reproduction, bird calls sometimes deal with immediate issues of life and death, e.g. when birds warn each other of danger. In this context, false alarms obviously have very different costs from missed detections. When, for example, predator detection is impaired by the masking of alarm calls, the error of the receiver is potentially lethal. Although hindrance of predator detection by noise is often mentioned in the literature (Quinn et al. 2006; Barber et al. 2009; Brumm 2010), studies are lacking that actually quantify the increase of bird predation rates due to noise-induced missed alarm call detection.

As illustrated in Chap. 2, it is advisable to differentiate between receiver responses involving signal detection and those involving signal discrimination. In birds, auditory thresholds for signal detection are generally lower than those for signal discrimination (Klump 1996; Brumm and Slabbekoorn 2005), and thus different levels of signal masking will result in different receiver errors. When the signal-to-noise ratio is below detection threshold, the receiver fails to detect the signal. At higher contrasts between signal and noise (i.e. lower noise levels or higher signal amplitudes) there is a range where the receiver detects the signal but is unable to recognise it. In this case, receivers might attempt to restore signal transmission, for example by moving to other locations where signal-to-noise ratios are higher and thus the signal can be recognised.

In areas with intense chronic noise, signal masking may not only reduce the fitness of individuals but might ultimately affect the dynamics and viability of entire populations (Brumm 2010; Kight and Swaddle 2011). With regard to anthropogenic noise, this raises conservation concerns, not only in birds but also in other taxa as well (Chap. 14). Indeed, recent studies suggest a link between an increase in anthropogenic noise and a decrease of reproductive success in birds (Halfwerk et al. 2011; Kight et al. 2012; Schroeder et al. 2012). However, it remains to be seen whether the reduced reproduction in noisy areas is causally linked to impaired communication or whether it is the outcome of other noise-related effects, for example, increased physiological stress in breeding birds and their young.

In cases in which bird vocalisations are individually acquired through vocal learning, noise may not only disrupt signal transmission but may also affect communication in an indirect way by impacting on song learning. The latter relates to possible changes in song structure due to impaired vocal learning. In this regard, birds may produce poor song imitations if the perception of tutor songs is impaired by noise. Moreover, birds need to hear themselves during song development when they gradually match their own vocal output to the memorised tutor song. This means that a disruption of auditory feedback due to noise could also lead to inaccurate copies of tutor songs (Brainard and Doupe 2000; Funabiki and Konishi 2003; Zevin et al. 2004; Funabiki and Funabiki 2009).

However, birds are not necessarily helpless when faced with the cold reality of signal masking. Birds have evolved sophisticated perceptual mechanisms that enhance signal detection and recognition in noise (Chap. 8). They may also change the structure and performance of their vocalisations to maintain signal transmission in noisy environments.

7.2 Signal Changes to Maintain Information Transfer in Noise

Changes in signalling behaviour to maintain information transfer in noise can be categorised on two different temporal scales. (1) Long-term changes are evolutionary adaptations during which signals are shaped to stand out against the masking noise. These evolutionary dynamics operate within populations and species, and lead to signal characteristics that are typical for a habitat or a species. Therefore, they should be more precisely termed microevolutionary changes, in contrast to macroevolution, which deals with evolution above the species level and with broad relationships between clades. In the case of oscine bird song, selection acts on signals that are individually acquired through vocal learning, which means that signal changes at the microevolutionary level are intertwined with signal development at the individual level. Thus, microevolutionary responses to noise are the outcome of interplay between cultural evolution and environmental selection in this group of birds. (2) Short-term changes relate to signal plasticity based on individual signal adjustments. These can be observed in single animals, which maintain communication in fluctuating noise by changing the structure or performance of their vocalisations. Ontogeny plays a role here too; as the ontogenetic trajectories set the default level around which individual signal plasticity can operate. Likewise, both major levels—microevolutionary changes and individual plasticity—are linked because microevolutionary adaptations to environmental acoustics may constrain the scope for individual plasticity. In the following sections we will review the evidence for the two different ways in which birds may change their vocalisations to keep up signal transmission in noise, and we will also highlight areas that we feel are particularly promising for future research.

7.2.1 Microevolutionary Changes

Bird songs are an exceptionally well-studied example of how animal signals are adapted to the environment, a phenomenon known as the acoustic adaptation hypothesis (Brumm and Naguib 2009; Ey and Fischer 2009). The acoustic properties of a given habitat can exert selection pressure on acoustic signals because they affect signal transmission and thus the probability of receivers to respond. This is particularly important in long-range signals, such as many types of bird song. Therefore, one would expect populations and species occurring in acoustically different habitats to have vocalisations adapted to transmit efficiently in each environment (Morton 1975). Overall, there is good evidence for a match between habitat and bird songs, although some studies did not find such an effect (Catchpole and Slater 2008).

Research on the acoustic adaptation hypothesis has very often addressed the effects of reverberation and frequency-dependent attenuation in a given habitat. Nevertheless, acoustic signal transmission can also be crucially constrained by background noise. Therefore, it is conceivable that consistent noise in a particular frequency range acts as a selection pressure driving the evolution of signals with reduced spectral overlap or other features that improve signal reception. Several studies have suggested that this kind of microevolutionary change is indeed what has shaped certain bird vocalisations. For example, species that occur along torrents in the Himalayas, such as the large-billed leaf warbler, *Phylloscopus magnirostris* (Fig. 7.3d), were found to produce comparably high-pitched, tonal songs that literally rise above the background noise (Dubois and Martens 1984; Martens and Geduldig 1990). Likewise, the relatively high-pitched calls of white-throated dippers *Cinclus cinclus* (Fig. 7.3b) are also suspected to be an adaptation to the noise of the fast running streams that form the bird's habitat (Brumm and Slabbekoorn 2005). Another bird that occurs along noisy streams, the rufous-faced warbler (*Abroscopus albogularis*, Fig. 7.3a), has taken the frequency shift a good deal further: Narins et al. (2004) discovered that the songs of this species contain prominent ultrasonic harmonics. Provided that rufous-faced warblers can actually perceive such high frequencies, the ultrasonic song components would be likely to increase the signal-to-noise ratio and thus help to maintain information transfer. Ultrasonic signalling has also been found in frogs in similar habitats (Chap. 5), which suggests that the strong environmental selection pressure of torrent noise might have led to a possible convergent evolution of ultrasonic signalling in birds and anurans.

As we have shown in Sect. 7.1.1, birds have not only to cope with abiotic noise, such as the sounds of running water, but also with the sounds produced by other animals. This biotic noise may exert a selective force on the evolution of bird songs, just as abiotic noise does. Ryan and Brenowitz (1985) analysed environmental noise levels in Neotropical forests and grasslands and concluded that the evolution of comparatively low-pitched bird songs in forests was probably partly due to the masking of higher frequency bands by the incredible din produced by insects. This notion has recently been corroborated by a similar study by Weir et al. (2012), who surveyed a large number of bird species from the New World and found that species breeding in tropical forests sing at lower frequencies than species breeding in open habitats. Like Ryan and Brenowitz (1985), they suggest that this restriction on bird song frequency may be due to the presence of high-frequency insect noise in tropical forests (together with greater degradation of high-frequency sounds in this habitat).

Fig. 7.3 Some species that are thought to have vocalisations adapted to the noise profiles of their unique habitats. **a** Rufous-faced warbler, *Abroscopus albogularis* (Narins et al. 2004), *photograph* Tang Jun © 2012. **b** White-throated dipper, *Cinclus cinclus* (Brumm and Slabbekoorn 2005), *photograph* Stefan Greif © 2012. **c** Little greenbul, *Andropadus virens* (Slabbekoorn and Smith 2002), *photograph* Volker Deecke © 2013. **d** Large-billed leaf warbler, *Phylloscopus magnirostris* (Martens and Geduldig 1990), *photograph* Michelle and Peter Wong © 2011

Assemblages of closely related species that occur in the same geographic area and thus regularly encounter one another are notable for their diversity in ecological niches (Chesson 2000; Losos 2010). The same may also be true for "acoustic niches", i.e. the occupied parameter space of acoustic signals. An impressive example of this idea comes from the long-term study of Darwin's finches by Grant and Grant (2010). They have found that the songs of two Darwin's finch species diverged over several decades after the arrival of a new species, whose song overlapped the frequency ranges of the two resident species. The divergence of the songs of the two resident species may have reduced acoustic interference, or, in other words, the masking noise introduced by the songs of the new species lead to a microevolutionary character shift.

In the Darwin's finch example, the diverging traits were temporal song features, especially trill rate and song duration. It is also conceivable that in order to reduce mutual masking the songs of different bird species may be shifted by selection to different frequency bands, so that they eventually partition the acoustic space. One recent study found that black-capped chickadees *Poecile atricapillus* can make real-time frequency shifts to avoid masking by overlapping tones (Goodwin and Podos 2013), and thus could presumably make similar shifts to avoid overlapping with neighbours. On the population level, several studies have found that within sympatric communities of birds, species can occupy discrete acoustic niches (Seddon 2005; Luther 2009). Similar phenomena are also observed in frogs (Chap. 5) and insects (Chap. 3), suggesting that biotic noise may be a strong selective force that affects the evolution of acoustic signals across taxa. However, the degree to which bird songs can be shaped (or modified) by selective forces is limited by morphological and phylogenetic constraints (Ryan and Brenowitz 1985). Thus, even strong competition between species may not always result in discernible separation of acoustic niches.

It is tempting to assume that song features that reduce potential masking, such as the use of certain frequency bands, are an adaptation to environmental noise. However, it pays to be cautious because it is often rather difficult to show that the correlations between bird song characteristics and increased signal-to-noise ratios that are observed in the field are actually the outcome of direct selection for certain signal traits. Moreover, one should bear in mind that some observations that seem to support the acoustic adaptation hypothesis at first glance could in fact be the outcome of individual signal adjustments. Therefore, it is necessary to disentangle possible evolutionary changes in populations or species from individual signal plasticity. In contrast to evolutionary changes, individual plasticity is easily accessible through experimental manipulations, and there is a growing body of evidence demonstrating how individual birds can adjust their vocalisations to reduce masking in noise.

7.2.2 Signal Plasticity

At the individual level, many birds across different taxa possess the vocal plasticity to adjust their signals in response to changes in environmental noise conditions in ways that can improve communication efficacy. A bird producing an acoustic signal in an environment filled with potentially masking noise has a variety of mechanisms at its disposal that can help to improve the chances of its signal being heard. It can be useful to consider how these different mechanisms work to improve the effectiveness of a signal by thinking about how they affect signal-to-noise ratios. In the broadest sense, we can lump the different mechanisms birds use to boost the signal-to-noise ratio into two categories: (1) reducing the noise level, and (2) increasing the signal level. Within the existing body of research on vocal signalling in birds, we find many examples of changes in signalling behaviour that fall into each of these categories. In addition to spectral modifications of their signals, individuals can also improve the signal-to-noise ratio of their vocalisations by changing the timing or location of their acoustic behaviours.

7.2.2.1 Reducing the Noise

While it is unlikely that a signalling bird can change or reduce the noise directly, they may indirectly decrease the level of noise by adjusting the timing or location of signalling to avoid or reduce masking effects. The most straightforward way to do this is to simply shift signalling to periods when environmental noise levels are low. Although this may seem a fairly simple solution, avoiding noise is not necessarily so easy. In many habitats, as discussed in Sect. 7.1.1, any individual signal may have to compete for acoustic space with a plethora of biotic and abiotic noise sources that vary in duration, frequency content, diurnal pattern, spectral complexity and amplitude. Consider a songbird singing in a forest: at any point during the day, he may confront potential masking noise from singing conspecifics, heterospecific birds, chorusing cicadas, chirping crickets, calling frogs, gusting wind, rustling leaves, rushing water or even airplanes flying overhead. So while, in theory, our singer might increase the signal-to-noise ratio of his broadcast by choosing silent "windows" within this acoustic din, in practice it is likely that finding enough opportunities to signal without competing noise is impossible. However, while it may be impractical to always wait for silence to send a signal, the temporal avoidance strategy can allow birds to avoid short-term periods of high-intensity noise. Indeed, there are now several studies that indicate that birds may adjust the timing of their signals to avoid masking by abiotic and biotic sounds. Lengagne and Slater (2002) found that tawny owls stopped calling during noisy rainy periods, although it is difficult to separate the effect of masking noise from other wet-weather effects such as decreased overall activity.

Urban noise follows a predictable diurnal pattern, and birds living in noisy cities may adjust the timing of their songs to minimise masking by traffic noise. European robins (*Erithacus rubecula*) in towns sing more during the night when traffic noise is low than robins in areas that are quieter during daytime hours (Fuller et al. 2007). Although changes in the diurnal pattern of song might also be attributed to light pollution effectively lengthening the daylight hours, Fuller et al. (2007) found that light levels showed a weaker correlation with changes in singing behaviour than ambient noise levels did.

Some birds may adjust the timing of their signalling on an even finer scale. In different forest communities, high numbers of birds sing at high rates at the same times of the day. Though the competition for acoustic space in such an environment is great, birds tend to avoid overlapping the songs of their nearest neighbours. This fine-tuning of the temporal presentation of song in order to avoid overlapping the songs of neighbours (be they heterospecific or conspecific) has been observed in many species. For example, common nightingales time the onset of their songs to fall between the songs of other species, or playback of other bird songs in the same frequency band as their own, thus avoiding potential masking of their own signals by those of other singers in their environment (Brumm 2006). Since many birds sing in a discontinuous pattern, with songs interspersed with silent pauses, a singing bird's best chance to avoid overlapping is to begin its song immediately after the song of a neighbouring singer ends. In natural (non-experimental) interactions, singing birds may adjust the timing of their peak song output to avoid the peak song activity of heterospecific neighbours (Cody and Brown 1969), and on a finer temporal scale, singers seem to avoid overlapping songs of nearby neighbours (Ficken et al. 1974; Popp et al. 1985). Whether the avoidance of song

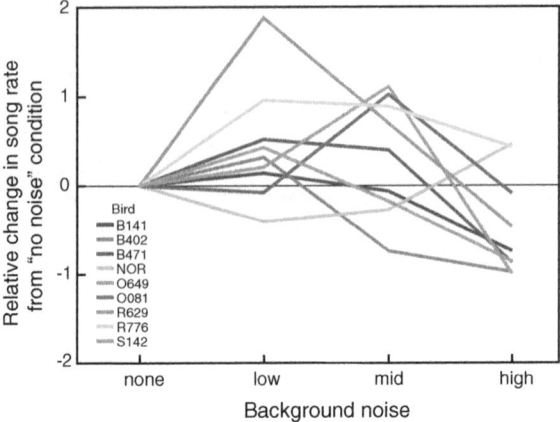

Fig. 7.4 Zebra finch song rate often increases with moderate levels of background noise (compared to song rate in a quiet room), but as noise levels approach 85 dB, song rates decline, and often song production ceases entirely. Song rates (motifs per hour) were normalised to relative change from song rate in no noise condition. Males were housed with a female in individual sound-attenuating chambers. We measured the number of song motifs produced during the first hour after lights turned on in the morning. Noise playback began 7 h prior to the song recording period and played during the night. Each day, birds were exposed to noise playback at one level (none ($\sim$33 dB), low (55–65 dB), moderate (70–80 dB), high (85–90 dB)). The level of noise playback for each bird each day was varied systematically, and recordings were made for 20 mornings in a row (a total of five days for each noise level). Unpublished data collected by Sue Anne Zollinger

overlapping with conspecifics is simply a way to limit masking interference or is a signal in itself is still a topic of some debate (Searcy and Beecher 2009, 2011; Naguib and Mennill 2010).

In domestic chickens *Gallus gallus domesticus*, call rate increased in response to increases in background noise levels up to around 70 dB, but then call rates began to decline as noises levels increased to even higher levels (Brumm et al. 2009). A similar pattern can be observed in zebra finches (Fig. 7.4): when exposed to noise at lower levels song rate often increases, or at least remains unchanged from song rate in silent recording boxes. However, as noise increases to sound pressure levels approaching 85 dB, song rate declines dramatically or ceases completely.

7.2.2.2 Increasing the Signal

It may be impractical or impossible to avoid signalling in noise, either because the timing of the signal is of critical importance, or the presence of noise is so ubiquitous that neither spatial nor temporal avoidance is possible. Yet, birds may still have a variety of ways to make themselves heard by adjusting the acoustic signal itself.

Increasing Signal Amplitude

100 years ago French otolaryngologist Étienne Lombard described for the first time the involuntary elevation of voice amplitude in human speakers in response to increases in background noise levels (Lombard 1911). This phenomenon, now known as the "Lombard effect", has been well documented in humans as well as in several species of non-human primates and other mammals (Chap. 9). In addition, laboratory studies reveal that the Lombard effect is also present in birds across a wide range of taxa including passerines, psittacids, galliformes, trochilids and the evolutionarily basal tinamiformes (Table 7.1). Given that the Lombard effect seems to be a shared trait of all extant birds, it is not surprising that it is independent of the ontogenetic origin of the vocalisation. It occurs in both vocalisations that are acquired through vocal production learning, such as the songs of song birds or the calls of parrots, as well as those that are not, such as the calls of quails and domestic chickens.

Manabe et al. (1998) demonstrated not only the Lombard effect in budgerigars *Melopsittacus undulatus* (Fig. 7.5a), they also showed in a series of experiments that temporal overlap between vocalisation and noise is not enough to induce an increase in call amplitude but that spectral overlap is necessary. In other words, the degree of masking is crucial for a Lombard response. In a later study, this finding was confirmed in a songbird, the common nightingale (Brumm and Todt 2002). These experiments suggest that birds assess the signal-to-noise ratio between their vocal output and background noise and adjust the amplitude of their vocalisation accordingly. Like humans, many songbirds depend on auditory feedback to control their vocalisations precisely (Woolley 2008). Thus, the Lombard effect is not only a means to maintain signal transmission but, at the same time, it can also be viewed as a feedback mechanism for vocal production.

The majority of studies on the Lombard effect in birds have been in the laboratory where acoustic conditions can be controlled and only a few studies have tested whether birds may exhibit the Lombard effect in the wild (Table 7.2). Correlations between vocal amplitude and background noise level have been shown in only three free-living birds as far as we are aware. Nightingales in areas within Berlin, Germany with higher levels of traffic noise sang louder than those in quieter areas (Fig. 7.5b). Male blue-throated hummingbirds (*Lampornis clemenciae*) near noisy brooks in the mountains of Arizona, U.S.A. call with higher amplitudes than males in quieter territories, and increase their call amplitude in response to transient increases in background noise (Pytte et al. 2003). On the other side of the globe, a similar trend was found in the alarm calls of noisy miners (*Manorina melanocephala*), an Australian honeyeater that is a successful coloniser of noisy urban habitats. Alarm call amplitude was positively correlated with traffic noise amplitude within an urban population of noisy miners, with individuals near busier arterial roads calling more loudly than birds near less busy roads (Lowry et al. 2012). These differences in vocal amplitude are likely to be the result of the Lombard effect, as suggested by the nightingale study (Brumm 2004), in which individual males varied the amplitude of their songs with the level of fluctuating traffic noise.

Table 7.1 Experimental tests of the Lombard effect in birds. While there have been only relatively few controlled, laboratory-based tests of the Lombard effect in birds to date, the species tested cover quite a broad range of taxa within the class. Empty cells indicate that a certain parameter was not measured (or not reported) in the study

Species	Change in amplitude?	Change in frequency?	Change in duration?	Change in redundancy or rate?	Reference
Tinamiformes—Tinamidae					
Elegant-crested tinamou (*Eudromia elegans*)	Yes	Yes	–	–	Schuster et al. (2012)
Galliformes—Phasianidae					
Japanese quail (*Coturnix japonica*)	Yes	–	–	Yes	Potash (1972)
Domestic chicken (*Gallus gallus domesticus*)	Yes	–	No	Increase in song rate at mid-level noise, but decrease with high noise	Brumm et al. (2009)
Psittaciformes—Psittaculidae					
Budgerigar (*Melopsittacus undulatus*)	Yes[1,2]	Yes[2]	Yes[2]	–	1. Manabe et al. (1998) 2. Osmanski and Dooling (2009)
Passeriformes—Hirundinidae					
Tree swallow (*Tachycineta bicolor*)	Yes	No (lab), yes (field)[a]	No (lab), yes (field)[a]	–	Leonard and Horn (2005)
—Muscicapidae					
Common nightingale (*Luscinia megarhynchos*)	Yes	–	No	–	Brumm and Todt (2002)
—Estrildidae					
Bengalese finch (*Lonchura striata domestica*)	Yes[b]	–	–	–	Kobayasi and Okanoya (2003)
Zebra finch (*Taeniopygia guttata*)	Yes[1,2]	–	No[2]	No[2]	1. Cynx et al. (1998) 2. Zollinger et al. (2011)

[a] Begging calls of nestlings experimentally exposed to noise in nest boxes in the field. In the laboratory, amplitude changed with noise exposure, but not frequency or duration
[b] Amplitude increase observed in undirected song, but not directed song

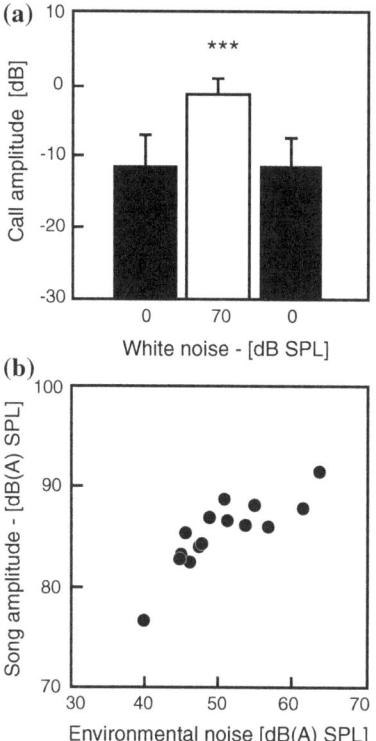

Fig. 7.5 The Lombard effect in birds. (**a**) Noise-dependent regulation of vocal amplitude in budgerigars ($N = 5$ birds). In the first phase of the experiment no noise was presented to the birds, then white noise with an amplitude of 70 dB(A) SPL was broadcast which elicited an increase in the amplitude of birds' contact calls. In the third phase, the noise was switched off again and in response the birds returned to their initial lower call amplitudes. (adapted from Manabe et al. 1998, used with permission). The Lombard Effect has now been demonstrated experimentally in nine species of birds across a wide range of taxa and is presumed to be a shared trait of the entire class. The Lombard effect is likely the cause of observed correlations between vocal amplitude and environmental noise levels in free-living birds such as common nightingales (**b**), which sing louder in noisier areas ($N = 15$ males, each data point gives the average values for one bird)(adapted from Brumm 2004, used with permission). All SPL values re. 20 µPa

This short-term regulation of vocal amplitude led the nightingales to sing more softly in the mornings of weekend days, when there was no commuting traffic, compared to working days when noisy cars made song transmission more difficult.

In addition to simply increasing the amplitude of a certain utterance, songbirds may be able to boost the amplitude of their songs in noisy habitats in a different way—by selectively singing the loudest song or element types from their repertoires. Any vocalising animal has a fixed range of frequencies and amplitudes that it can produce. Outside this range, phonation will be physiologically impossible due to constraints of the vocal organ, the respiratory system, body size or vocal tract. Voice range profiles describe the range of amplitudes and frequencies that

Table 7.2 Field studies of avian vocal responses to noise exposure. Empty cells indicate that a certain parameter was not measured

Species	Change in amplitude?	Change in frequency?	Change in duration?	Change in redundancy or rate?	Other changes?	Noise source	Reference
Columbiformes—Columbidae							
Eurasian collared dove (*Streptopelia decaocto*)	–	–	–	–	No change in onset of dawn chorus song	Normalised recordings of street traffic	Arroyo et al. (2013)
Apodiformes—Trochilidae							
Blue-throated hummingbird (*Lampornis clemenciae*)	Yes	–	–	–	–	Recordings of creek noise	Pytte et al. (2003)
Passeriformes—Paridae							
Black-capped chickadee (*Poecile atricapillus*)	–	Shifted in real-time to avoid masking tones	–	–	–	Tones targeted to match "bee" note frequency	Goodwin and Podos (2013)
Great tit (*Parus major*)	–	–	–	–	Adjustments to temporal song type switching behaviour	Low-frequency noise, white noise, and high-frequency noise	Halfwerk and Slabbekoorn (2009)
—Phylloscopidae							
Common chiffchaff (*Phylloscopus collybita*)	–	Higher minimum[a]	–	–	–	Recordings of highway noise	Verzijden et al. (2010)
—Sturnidae							
Spotless starling (*Sturnus unicolor*)	–	–	–	–	Earlier onset of dawn chorus song	Normalised recordings of street traffic	Arroyo et al. (2013)
—Muscicapidae							
European robin (*Erithacus rubecula*)	–	Higher minimum	Shorter songs	–	Reduced song complexity and reduced bandwidth	Low-pass filtered white noise	Montague et al. (2012)

(continued)

Table 7.2 (continued)

Species	Change in amplitude?	Change in frequency?	Change in duration?	Change in redundancy or rate?	Other changes?	Noise source	Reference
—Passeridae							
House sparrow (*Passer domesticus*)	–	–	–	–	Earlier onset of dawn chorus song	Recordings of normalised street traffic	Arroyo et al. (2013)
—Fringillidae							
European goldfinch (*Carduelis carduelis*)	–	–	–	–	No change in onset of dawn chorus song	Normalised recordings of street traffic	Arroyo et al. (2013)
European greenfinch (*Carduelis chloris*)	–	–	–	–	No change in onset of dawn chorus song	Normalised recordings of street traffic	Arroyo et al. (2013)
European serin (*Serinus serinus*)	–	–	–	–	No change in onset of dawn chorus song	Normalised recordings of street traffic	Arroyo et al. (2013)
House finch (*Carpodacus mexicanus*)	–	Higher minimum[b]	No change in song duration No change in syllable length[b]	–	–	Recordings of city traffic noise[b]	Bermudez-Cuamatzin et al. (2011)

(continued)

Table 7.2 (continued)

Species	Change in amplitude?	Change in frequency?	Change in duration?	Change in redundancy or rate?	Other changes?	Noise source	Reference
—Icteridae							
Red-winged blackbird (*Agelaius phoeniceus*)	–	Reduced spectral energy in higher frequencies of trills	No	–	Increased tonality (reduced entropy)	White noise low-pass filtered at 1830 Hz (below minimum frequency of blackbird trills)	Hanna et al. (2011)
—Emberizidae							
Reed bunting (*Emberiza schoeniclus*)	–	Higher minimum	Shorter songs	Lower song rate	–	Recordings of city traffic noise	Gross et al. (2010)

[a] Only for first 10 songs after onset of noise playback
[b] Wild-caught birds held temporarily in small cages during playback and recording

individuals are capable of producing. This range can be defined rather easily for humans, since a subject can simply be instructed to vocalise at the lowest and highest amplitude possible for any given frequency. In humans with healthy voices, higher frequencies can be produced louder than lower frequencies (Fig. 7.6a) (e.g. Heylen et al. 2002). So, for example, if you raise the frequency as well as the amplitude of your voice when yelling at someone across a football field, your shout can be louder than if you were to shout at a lower frequency. Determining the absolute upper and lower limits of frequency and amplitude for vocalising animals is not as simple, but it seems that a similar relationship between amplitude and frequency may occur in both birds and non-human primates. In chimpanzees and songbirds, average "voice range profiles" have been estimated from careful measurement of large numbers of recordings of different vocalisations produced in many different contexts and at many different amplitudes. While it will never be certain that an animal vocalised at the true extremes of its potential range, one may in this way delimit the range within which normal vocalisations occur. For example, in common chimpanzees, *Pan troglodytes*, the upper limits of

Fig. 7.6 **a** An average voice range profile for adult human females that use their voice professionally ($N = 46$ healthy school teachers, age range 22–51 years), maximum (*purple*) and minimum (*orange*) amplitude (*solid lines*) with 95 % confidence intervals (*dashed lines*). (adapted from Heylen et al. 2002, Figs. 2 and 3, used with permission). **b** Relationship between peak song frequency and amplitude in common blackbirds ($N = 12$ males recorded in sound-shielded chambers). Minimum peak frequency (*orange*); maximum peak frequency (*purple*) and mean peak frequency (*red*) curves are based on the weighted amplitude averages of all males). *Dashed lines* denote standard errors above and below these averages. (adapted from Nemeth et al. 2013, used with permission)

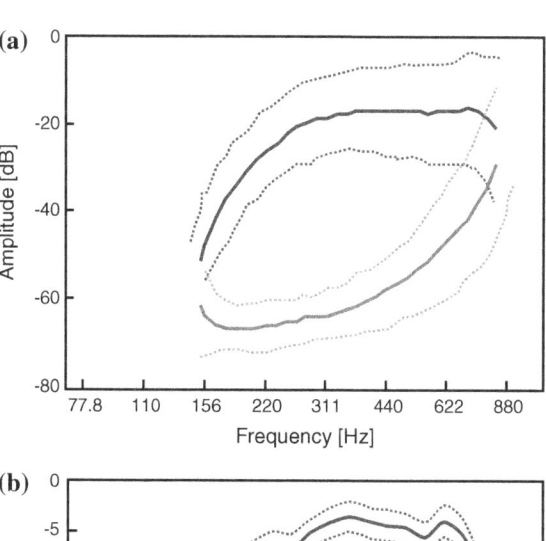

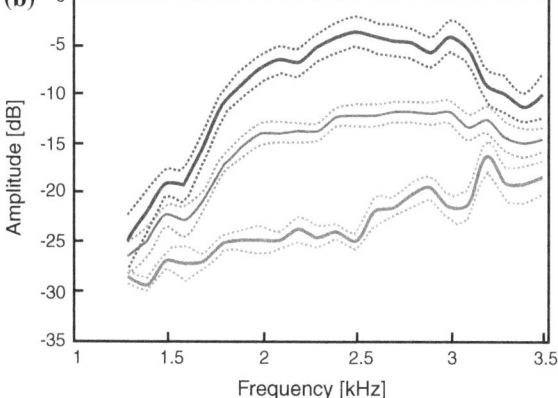

their vocal range have been suggested to be at (or just above) the point when the voice begins to "break apart", or in other words when nonlinear vocal phenomena begin to appear in call types that are normally free from such artefacts when produced at lower amplitudes or at lower frequencies (Riede et al. 2007). Using this method to chart the upper limits of vocal amplitude in chimpanzees, Riede et al. suggest that like humans, chimpanzees can produce higher frequency sounds at higher amplitudes than they can produce low-frequency ones. In common blackbirds (*Turdus merula*) and great tits, the relationship seems to be the same—higher frequency elements are usually produced at higher amplitudes than lower ones (Nemeth et al. 2012, 2013). Nemeth et al. (2013) used an averaged "voice range profile" from blackbirds recorded in the lab to show that urban birds boost the signal-to-noise ratio of their songs by switching to higher frequency song types (Fig. 7.6b). The most common motif elements sung by city birds at 2.3 kHz were on average 6 dB higher in amplitude than the most common song elements used by forest blackbirds, which were much lower in frequency at 1.8 kHz on average.

Increasing signal redundancy

The songs of many bird species are performed in a highly redundant manner, in which the same syllable or song type is repeated several times in sequence before the bird switches to a new type. In fluctuating noise conditions, birds can increase the probability that their signal is successfully transmitted by repeating the same information a number of times. This increased signal redundancy can considerably increase receiver performance (Wiley 2006). Such a noise-dependent increase in the redundancy of acoustic signals has been shown in at least three species of birds: Japanese quail (*Coturnix japonica*), king penguins (*Apenodytes patagonicus*) and common chaffinches (Tables 7.1 and 7.3). As the penguin and chaffinch studies report only correlations between redundancy and noise level, it is not possible to completely rule out that the observed increase in serial redundancy could be due to some other variable. However, the observation of this effect across such disparate clades as well as in frogs (Chap. 5) suggests that the increase in signal redundancy is indeed a widespread response to improve signal effectiveness in noise.

Other studies on songbirds living in areas with high levels of anthropogenic noise have reported that songs are more "hurried" in urban populations than in quiet forests. Common blackbirds in the city of Vienna, Austria, sing shorter, higher frequency songs with shorter between-song intervals, while in a quiet forest outside Vienna they sing slower, lower frequency songs with longer between-song intervals (Nemeth and Brumm 2009). Similar patterns of shorter, hurried songs in noisy areas have been reported in great tits (Slabbekoorn and den Boer-Visser 2006) as well as in black-capped chickadees (Proppe et al. 2011) but see Ríos-Chelén et al. (2012b) for different results in vermilion flycatchers *Pyrocephalus rubinus*. The pattern of shorter songs, with shorter intervals, in noisy habitats leads to higher song rates but this is not necessarily equivalent to an increase in signal redundancy, particularly if the bird switches between many different song types, as the blackbirds typically do. However, an increase in song rate may serve a similar

Table 7.3 Correlational field studies of avian vocal behaviour in noise. Listed in each column are differences that were positively correlated with increased levels of background noise. Empty cells indicate that a certain parameter was not measured or not reported

Species	Change in amplitude?	Change in frequency?	Change in duration?	Change in redundancy or rate?	Other changes?	Noise source	Reference
Sphenisciformes—Spheniscidae							
King penguin (Aptenodytes patagonicus)	–		Longer calls	Increased serial redundancy (greater number of calls and syllables)	–	Wind and conspecifics	Lengagne et al. (1999)
Apodiformes—Trochilidae							
Blue-throated hummingbird (Lampornis clemenciae)	Yes	–				Creek noise	Pytte et al. (2003)
Psittaciformes—Psittaculidae							
Rainbow lorikeet (Trichoglossus haematodus)	–	Higher minimum. No difference in dominant frequency	–	–	–	Street traffic	Hu and Cardoso (2010)
Eastern rosella (Platycercus eximius)	–	Higher minimum. No difference in dominant frequency	–	–	–	Street traffic	Hu and Cardoso (2010)
Passeriformes—Tyranni—Tyrannidae							
American grey flycatcher (Empidonax wrightii)	–	No	–	–	–	Gas extraction compressor	Francis et al. (2010)
Ash-throated flycatcher (Myiarchus cinerascens)	–	Higher peak	–	–	–	Gas extraction compressor	Francis et al. (2010)
Fork-tailed flycatcher (Tyrannus savana)	–	No (no difference in minimum, maximum or peak)	No		No difference in complexity (number of syllables or different syllable types)	Street traffic	Ríos-Chelén et al. (2012b)[a]
Great kiskadee (Pitangus sulphuratus)	–	Curitiba and Brasilia, Brazil—no (no difference in minimum, maximum or peak.) Manaus, Brazil—higher minimum and higher peak	No		No difference in complexity (number of syllables or different syllable types)	Street traffic	Ríos-Chelén et al. (2012b)[a]

(continued)

Table 7.3 (continued)

Species	Change in amplitude?	Change in frequency?	Change in duration?	Change in redundancy or rate?	Other changes?	Noise source	Reference
Small-billed elaenia (*Elaenia parvirostris*)	–	No (no difference in maximum, minimum or peak)	No	–	No difference in complexity (number of syllables or different syllable types)	Street traffic	Ríos-Chelén et al. (2012b)[a]
Tropical kingbird (*Tyrannus melancholicus*)	–	No (no difference in maximum, minimum or peak)	No	–	No difference in complexity (number of syllables or different syllable types)	Street traffic	Ríos-Chelén et al. (2012b)[a]
Vermillion flycatcher (*Pyrocephalus rubinus*)	–	No[1,2]	Longer songs[1]	More intro notes[1]	Reduced song complexity[1,b]	Street traffic	1. Ríos-Chelén et al. (2012a) 2. Ríos-Chelén et al. (2012b)[a]
Passeriformes–Tyranni––Furnariidae							
Olive spinetail (*Cranioleuca obsoleta*)	–	No (no difference in maximum, minimum or peak)	No	–	No difference in complexity (number of syllables or different syllable types)	Street traffic	Ríos-Chelén et al. (2012b)[a]
Rufous hornero (*Furnarius rufus*)	–	No (no difference in maximum, minimum or peak)	No	–	No difference in complexity (number of syllables or different syllable types)	Street traffic	Ríos-Chelén et al. (2012b)[a]
Passeriformes–Passeri––Meliphagidae							
Bell miner (*Manorina melanophrys*)	–	Higher dominant frequency, no difference in minimum	–	–	–	Street traffic	Hu and Cardoso (2010)
Noisy miner (*Manorina melanocephala*)	Higher amplitude with higher background noise[1]	No[2]	–	–	–	Street traffic	1. Lowry et al. (2012) 2. Hu and Cardoso (2010)
Red wattlebird (*Anthochaera carunculata*)	–	Higher minimum, No difference in dominant frequency	–	–	–	Street traffic	Hu and Cardoso (2010)
––Cracticidae							
Australian magpie (*Cracticus tibicen*)	–	No	–	–	–	Street traffic	Hu and Cardoso (2010)

(continued)

Table 7.3 (continued)

Species	Change in amplitude?	Change in frequency?	Change in duration?	Change in redundancy or rate?	Other changes?	Noise source	Reference
—Artamidae							
Grey butcherbird (*Cracticus torquatus*)	–	Higher dominant frequency, no difference in minimum	–	–	–	Street traffic	Hu and Cardoso (2010)
Pied currawong (*Strepera graculina*)	–	No	–	–	–	Street traffic	Hu and Cardoso (2010)
—Pachycephalidae							
Grey shrike-thrush (*Colluricincla harmonica*)	–	Higher dominant frequency of lowest tonal note	–	–	–	Street traffic	Parris and Schneider (2009)
—Vireonidae							
Grey vireo (*Vireo vicinior*)	–	Higher maximum frequency, but not higher minimum	Longer songs	–	–	Gas extraction compressor	Francis et al. (2011)
Plumbeous vireo (*Vireo plumbeus*)	–	Higher minimum, but not higher maximum	Shorter songs	–	–	Gas extraction compressor	Francis et al. (2011)
Red-eyed vireo (*Vireo olivaceus*)	–	No (no difference in maximum, minimum or peak)	No	–	No difference in complexity (number of syllables or different syllable types)	Street traffic	Ríos-Chelén et al. (2012b)[a]
Rufous-browed peppershrike (*Cyclarhis gujanensis*)	–	Higher minimum, No difference in maximum or peak	No	–	No difference in complexity (number of syllables or different syllable types)	Street traffic	Ríos-Chelén et al. (2012b)[a]
—Rhipiduridae							
Grey fantail (*Rhipidura albiscapa*)	.	No (no difference in dominant frequency of lowest tonal note)	–	–	–	Street traffic	Parris and Schneider (2009)
Willie wagtail (*Rhipidura leucophrys*)	–	No	–	–	–	Street traffic	Hu and Cardoso (2010)
—Monarchidae							
Magpie-lark (*Grallina cyanoleuca*)	–	No	–	–	–	Street traffic	Hu and Cardoso (2010)

(continued)

Table 7.3 (continued)

Species	Change in amplitude?	Change in frequency?	Change in duration?	Change in redundancy or rate?	Other changes?	Noise source	Reference
—Paridae							
Black-capped chickadee (*Poecile atricapillus*)	Amplitude ratio between "fee" and "bee" notes unchanged[2]	Higher peak[1,2]	Shorter songs[1]	–	–	Street traffic	1. Proppe et al. (2011) 2. Proppe et al. (2012)
Blue tit (*Cyanistes caeruleus*)	–	–	–	–	Earlier onset of dawn chorus song[c]	Street traffic	Bergen and Abs (1997)
Great tit (*Parus major*)	–	Higher minimum, no difference in peak, max or bandwidth[1,2,4]; Higher minimum, narrower bandwidth[3]	Shorter songs, with shorter within-song intervals (Leiden, Netherlands, and Cities across Europe)[1]; No difference in duration or song length (Cities across the UK)[2]; No difference in strophe length (Madrid, Spain)[3]; Longer songs (Tokyo, Japan)[4]	More phrases per song[4]	More "atypical" song types (with only one, or more than four notes per strophe)[1]; Earlier onset of dawn chorus song[5c]	Street traffic	1. Slabbekoorn and Peet (2003); Slabbekoorn and den Boer-Visser (2006) 2. Mockford and Marshall (2009) 3. Salaberria and Gil (2010) 4. Hamao et al. (2011) 5. Bergen and Abs (1997)
—Pycnonotidae							
Little greenbul (*Eurillas virens*)	–	Higher minimum, Higher maximum	–	–	–	Habitat—typical biotic and abiotic noise sources	Slabbekoorn and Smith (2002)
—Cettiidae							
Green hylia (*Hylia prasina*)	–	Lower peak[d]	–	–	–	Insects	Kirschel et al. (2009)
—Phylloscopidae							
Common chiffchaff (*Phylloscopus collybita*)	–	Higher minimum	Fewer syllables per song	–	–	Street traffic	Verzijden et al. (2010)

(continued)

Table 7.3 (continued)

Species	Change in amplitude?	Change in frequency?	Change in duration?	Change in redundancy or rate?	Other changes?	Noise source	Reference
—Zosteropidae							
Silvereye (*Zosterops lateralis*)	–	Higher minimum, no change in peak frequency or bandwidth	–	Syllable rate lower	–	Street traffic	Potvin et al. (2011)
—Troglodytidae							
Carolina wren (*Thryothorus ludovicianus*)	–	Higher minimum only at rural sites – interaction effect with amount of impervious surfaces	–	–	–	Street traffic	Dowling et al. (2012)
House wren (*Troglodytes aedon*)	–	No	–	–	–	Street traffic	Dowling et al. (2012)
Southern house wren (*Troglodytes musculus*)	–	Brasilia, Brazil—no Curitiba & Manaus, Brazil—higher minimum (no difference in maximum or peak)	No (all three populations)	–	Brasilia, Brazil—no difference in complexity Curitiba, Brazil—fewer different syllable types and reduced ratio between total syllables and different syllable types Manaus, Brazil—fewer different syllable types	Street traffic	Ríos-Chelén et al. (2012b)[a]
—Mimidae							
Gray catbird (*Dumetella carolinensis*)	–	Higher minimum	–	–	–	Street traffic	Dowling et al. (2012)
—Sturnidae							
Common myna (*Acridotheres tristis*)	–	No	–	–	–	Street traffic	Hu and Cardoso (2010)
—Turdidae							
American robin (*Turdus migratorius*)	–	No	–	–	–	Street traffic	Dowling et al. (2012)

(continued)

Table 7.3 (continued)

Species	Change in amplitude?	Change in frequency?	Change in duration?	Change in redundancy or rate?	Other changes?	Noise source	Reference
Black-billed thrush (*Turdus ignobilis*)	–	Lower maximum (no difference in minimum or peak)	Shorter songs	–	No difference in complexity (number of syllables or different syllable types)	Street traffic	Ríos-Chelén et al. 2012b[a]
Common blackbird (*Turdus merula*)	–	Higher peak (Austria)[1] (The Netherlands)[2] Higher minimum (The Netherlands)[2] Higher minimum, no difference in dominant frequency (Australia)[3]	Shorter songs[1]	–	Higher proportion of "twitter" elements vs. "motif" elements per song[2]	Street traffic	1. Nemeth and Brumm (2009) 2. Ripmeester et al. (2009) 3. Hu and Cardoso (2010)
Pale-breasted thrush (*Turdus leucomelas*)	–	Higher minimum (no difference in maximum or peak)	No	–	No difference in complexity (number of syllables or different syllable types)	Street traffic	Ríos-Chelén et al. 2012b[a]
Rufous-bellied thrush (*Turdus rufiventris*)	–	No (no difference in minimum, maximum or peak)	Shorter songs	–	Fewer total number of syllables	Street traffic	Ríos-Chelén et al. 2012b[a]
—Muscicapidae							
Common nightingale (*Luscinia megarhynchos*)	Higher amplitude	–	–	–	–	Street traffic	Brumm (2004)
European robin (*Erithacus rubecula*)	–	Higher minimum[2]	shorter songs[2]	Lower song rate[2]	Increased nocturnal song activity[1]	Street traffic	1. Fuller et al. (2007) 2. Montague et al. (2012)
—Fringillidae							
Common chaffinch (*Fringilla coelebs*)	–	Higher minimum[2]	–	Increased serial redundancy[2]	Earlier onset of dawn chorus song[c]	Street traffic[1] Waterfalls or torrents[2]	1. Bergen and Abs (1997) 2. Brumm and Slater (2006)
House finch (*Carpodacus mexicanus*)	–	Higher minimum[1,2]	Shorter songs[2]	–	–	Street traffic	1. Fernández-Juricic et al. (2005) 2. Bermúdez-Cuamatzin et al. (2009)
—Icteridae							
Red-winged blackbird (*Agelaius phoeniceus*)	–	No	No	–	Increased tonality (reduced entropy)	Street traffic	Hanna et al. (2011)

(continued)

Table 7.3 (continued)

Species	Change in amplitude?	Change in frequency?	Change in duration?	Change in redundancy or rate?	Other changes?	Noise source	Reference
—Emberizidae							
Dark-eyed junco (*Junco hyemalis*)	—	Higher minimum	—	—	—	Street traffic	Cardoso and Atwell (2011)
Reed bunting (*Emberiza schoeniclus*)	—	Higher minimum	No difference	Lower song rate	—	Street traffic	Gross et al. (2010)
Rufous-collared sparrow (*Zonotrichia capensis*)	—	No (no difference in minimum, maximum or peak)	Brasilia, Brazil—no Curitiba, Brazil—shorter songs	—	Brasilia, Brazil—fewer total number of syllables Curitiba, Brazil—fewer total number of syllables and Fewer different syllable types	Street traffic	Ríos-Chelén et al. (2012b)[a]
Song sparrow (*Melospiza melodia*)	Amplitude ratio between higher and lower frequency elements changes	Higher minimum[1] No difference in minimum or maximum[2]	—	—	—	Street traffic	1. Wood and Yezerinac (2006) 2. Dowling et al. (2012)
Yellow-browed sparrow (*Ammodramus aurifrons*)	—	No (no difference in maximum, minimum or peak)	No	—	No difference in complexity (number of syllables or different syllable types)	Street traffic	Ríos-Chelén et al. (2012b)[a]
—Thraupidae							
Blue-black grassquit (*Volatinia jacarina*)	—	Higher minimum (no difference in maximum or peak)	No	—	No difference in complexity (number of syllables or different syllable types)	Street traffic	Ríos-Chelén et al. (2012b)[a]
Blue-grey tanager (*Thraupis episcopus*)	—	Higher minimum (no difference in maximum or peak)	No	—	No difference in complexity (number of syllables or different syllable types)	Street traffic	Ríos-Chelén et al. (2012b)[a]
Saffron finch (*Sicalis flaveola*)	—	No (no difference in maximum, minimum or peak)	Longer songs	—	Increase in complexity (total number of syllables and different syllable types)	Street traffic	Ríos-Chelén et al. (2012b)[a]
Sayaca tanager (*Thraupis sayaca*)	—	No (no difference in maximum, minimum or peak)	No	—	No difference in complexity (number of syllables or different syllable types)	Street traffic	Ríos-Chelén et al. (2012b)[a]

(continued)

Table 7.3 (continued)

Species	Change in amplitude?	Change in frequency?	Change in duration?	Change in redundancy or rate?	Other changes?	Noise source	Reference
Yellow-bellied seedeater (*Sporophila nigricollis*)	–	No (no difference in maximum, minimum or peak)	No	–	No difference in complexity (number of syllables or different syllable types)	Street traffic	Ríos-Chelén et al. (2012b)[a]
—Cardinalidae							
Green-winged saltator (*Saltator similis*)	–	No (no difference in maximum, minimum or peak)	No	–	No difference in complexity (number of syllables or different syllable types)	Street traffic	Ríos-Chelén et al. (2012b)[a]
Northern cardinal (*Cardinalis cardinalis*)	–	Higher minimum	–	–	–	Street traffic	Dowling et al. (2012)

We have tried to include here every published study that correlates avian vocal behaviour with differences in environmental noise levels (including those with methodology prone to "false positive" measurement errors, see Zollinger et al. 2012). We apologise if we have overlooked any relevant studies; any such omissions were not intentional

[a] Including unpublished data from the same data set published in Ríos-Chelén et al. (2012a)

[b] May be an effect of time of day rather than noise level

[c] Earlier dawn chorus also correlated with higher levels of artificial light

[d] Lower peak frequency also correlated with higher elevation and lower tree cover levels

purpose as serial redundancy, in that it can increase the probability that each individual song is heard in an environment with rapidly fluctuating noise levels. Nonetheless, the "hurried" song of birds in high-traffic, high noise areas might be the result of some other factor associated with these habitats such as higher male density, differences in motivation, arousal, or stress. Controlled experimental investigations will be needed before these changes in behaviour can be causally linked to environmental noise.

Adjusting signal duration

A related mechanism by which a signaller might increase signal detectability is to extend the duration of individual syllables within the signal. In human speech, vowel lengthening is one of the adjustments made by speakers in response to increased noise (Junqua 1996). Indeed, in both mammals and birds signal lengthening of short sounds (less than 500 ms, depending on the frequency and the species) can greatly improve the detectability of acoustic signals in noise (Klump 1996; Brumm and Slabbekoorn 2005). The detection threshold for a signal goes down as the duration increases, thus increasing the duration of a brief signal in noise will make the signal more detectable. In birds, this connection between signal duration and detection thresholds has been reported for a number of parrot and songbirds species (Dooling 1979; Dooling and Searcy 1985; Klump and Maier 1989; Pohl et al. 2013)

So, do vocalising animals exploit this perception mechanism to make themselves heard in noise? Non-human primates, including common marmosets *Callithrix jacchus* (Brumm et al. 2004) and cotton-top tamarins *Saguinus oedipus* (Egnor and Hauser 2006), have indeed been shown to increase call duration in response to noise exposure. In birds, however, the evidence is mixed: budgerigars exposed to white noise in the laboratory increased the duration of their contact calls significantly with increasing noise level (Osmanski and Dooling 2009). Nestling tree swallows, *Tachycineta bicolour*, increase the duration of their begging calls in response to white noise playback in nest boxes in the field, although noise-exposed nestlings in the laboratory did not significantly increase call duration (Leonard and Horn 2005). In domestic chickens, no noise-dependent adjustment of signal duration was found (Brumm et al. 2009). Perhaps call duration cannot be modulated to communicate in noise in some bird species because this parameter is used to encode information, as is the case in some mammals (e.g. Schrader 1993; Le Roux et al. 2001; Swan and Hare 2008).

However, as with amplitude and frequency, adjustments of signal duration in noise must not be viewed in isolation from other signal parameters. On the sender's side, constraints imposed by vocal production mechanisms might link duration to amplitude or frequency. On the receiver's side, signal detection can be affected by the combination of different parameters. Since signal detection performance at threshold increases with both signal amplitude and signal duration, a given detection probability can be reached by different combinations of the two (Heil and Neubauer 2002). This means that a decrease in duration can be compensated for by an increase in amplitude, and vice versa.

Adjusting signal frequency

Another way to improve the signal-to-noise ratio of a signal is to shift the frequency content of the signal away from frequency bands containing the noise (or at least from the band with the highest intensity noise). These changes in signal frequency may occur as a result of natural selection when an animal lives in a habitat with a persistent and consistent noise source, such as next to noisy streams or waterfalls or, more recently in the evolutionary time scale, in areas exposed to intense anthropogenic noise such as alongside a busy motorway. A few examples of such microevolutionary changes in signal frequency are presented in Sect. 7.2.1 (Fig. 7.3). In these examples, it is presumed that calls or songs have been adapted over many generations to fit the acoustic niches that favour communication. Similar frequency shifts in response to environmental selection have been suggested for anurans (Chap. 5) and insects (Chap. 3).

It is also possible that individual birds may transiently adjust the frequency of their vocalisations on a shorter time scale in response to fluctuations in their acoustic environments, provided they possess the behavioural plasticity to allow such changes. Many studies have been published in the last decade reporting positive correlations between song frequency (peak frequency or minimum frequency) and level of urban or traffic noise (Table 7.3). The higher frequencies of urban bird songs—compared to the songs of neighbouring populations in quiet forests or less noisy sites—have been attributed to birds improving communication efficiency by moving the spectral content of their signals further from the loudest source of masking noise. As most of these studies have measured mean differences in populations of birds, rather than short-term changes in the vocalisations of individuals, it is not clear if the observed differences are the result of microevolutionary changes or of signal plasticity.

A handful of studies have examined responses of individual birds to fluctuations in environmental noise levels in the field. Bermúdez-Cuamatzin et al. (2011) exposed wild-caught house finches, *Carpodacus mexicanus*, to playback of recorded traffic noise and found real-time shifts in the minimum frequency of song during loud noise exposure (56–65 dB SPL). In this case, the birds did not switch to different song types composed of relatively higher notes, but rather increased the minimum frequency of individual notes within a song.

Halfwerk and Slabbekoorn (2009) exposed great tits to experimental playback of white noise, city-like noise (highest energy in frequencies below ∼4 kHz) or the inverse (highest energy in frequencies between 4 and 8 kHz) in 5 min intervals. Most of the birds (∼60 %) did not switch song types or adjust the frequency of the song types they were currently singing. However, when birds did change to a new song type during noise playback, they switched to songs with more high-frequency elements in the city-like noise and to songs with more low-frequency elements in the "inverse" noise. Reed buntings, *Emberiza schoeniclus*, (Gross et al. 2010) and chiffchaffs, *Phylloscopus collybita* (Verzijden et al. 2010) have both been reported to sing with higher minimum frequency in response to temporary playback of traffic-like noise, returning to lower minimum frequency songs

Fig. 7.7 Budgerigars call with increasing vocal amplitudes (**a**) as background noise increases (the Lombard effect). Budgerigars, like humans, also increase their call frequencies in noise (**b**) as vocal amplitude increases, even when these frequency shifts yield no release from masking (adapted from Osmanski and Dooling 2009, used with permission)

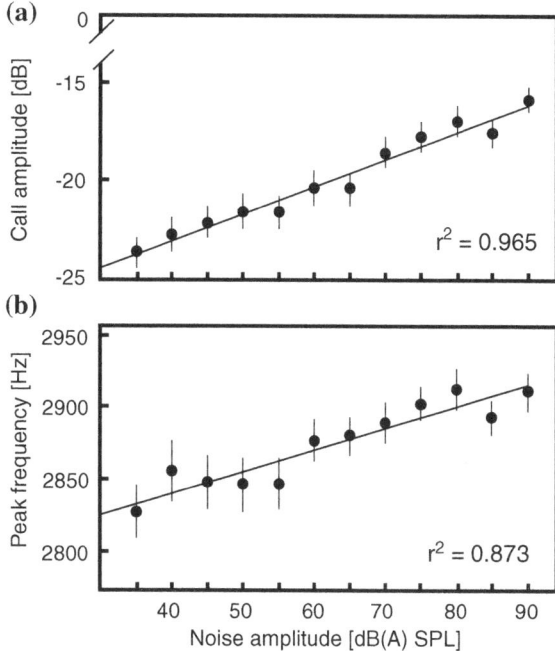

after cessation of the noise playback. While these studies rely on methods which are prone to potential measurement biases that could lead to false positive results (Zollinger et al. 2012), they potentially also provide evidence for real-time vocal plasticity in response to transient fluctuations in ambient noise conditions.

Interestingly, birds also increase their call frequencies in noise when this yields no release from masking (Osmanski and Dooling 2009; Schuster et al. 2012) (Fig. 7.7). While at first glance this may seem a counterintuitive phenomenon, it can probably be accounted for by a coupling of frequency and amplitude during vocal production in birds (Nelson 2000; Beckers et al. 2003; Elemans et al. 2008; Amador and Margoliash 2013), which predicts that both fundamental frequency and amplitude would increase when the system is driven with higher air flow rates. Thus, when the vocal amplitude increases due to the Lombard effect, a passive increase in frequency would be likely to occur in the absence of voluntary counter-measures to inhibit an increase in frequency. This also happens in mammalian vocalisations, including human speech, where the Lombard effect leads to an increase in fundamental frequency independent of any release from signal masking (Lu and Cooke 2008). Thus, considering all the evidence, it still remains an unresolved question whether the observed frequency shifts in urban birds have primarily arisen to evade masking by low-frequency traffic noise.

7.3 Conclusions

Noise has always presented a hurdle for signalling animals, and this is particularly true for vocalising birds. They use acoustic signals to exchange vital information, yet they have to deal with numerous noise sources in their habitats. Therefore, it is not very surprising that birds have evolved a number of ways to deal with the noise problem. These solutions may consist of microevolutionary adaptations, during which song characteristics get adapted to typical noise profiles in a given habitat, or individual vocal plasticity that can act during ontogeny or as real-time signal regulation. The Lombard effect appears to be the basic mechanism of vocal plasticity in birds for acoustic communication in noise. In addition, other forms of noise-related signal adjustments can be observed in some taxa, including changes in signal frequency, duration and redundancy.

Many birds acquire their vocalisations by copying those of other individuals, and it has been suggested that these individual learning processes accelerate the evolution of bird song (Lachlan and Servedio 2004). This may also apply to noise-dependent changes in song, since songs that are less audible in masking noise will be less likely to be copied. Thus, the cultural evolution of bird song may enhance environmental selection for songs that transmit well in noise. Intense masking noise that is present over considerable time periods in a habitat may even lead to the evolution of multimodal signals, in which additional signals in other modalities (mostly visual ones) facilitate information transfer by increasing the probability of signal detection (Chaps. 2, 5 and 6). Whether this is the case in birds awaits further study. Promising areas of future research might be the begging behaviour of nestlings that mutually mask their vocalisations or the displays of colony breeding birds that have to face the din produced by hundreds of individuals calling at the same time.

In many habitats, noise, especially high-intensity, low-frequency noise, is increasing because of human activity. Until recently, most of the studies examining vocal changes in environments with high levels of anthropogenic noise have dealt with frequency shifts as isolated effects (see Table 7.3). Although these first observations of frequency shifts in city birds relative to nearby populations in quieter forest or rural habitats were an important first step in identifying potential implications of urbanisation, we must now increase our efforts to examine the phenomenon from a more comprehensive and integrative perspective.

A disservice is performed by divorcing the work on avian vocal adjustment in noise from the literature on the Lombard effect. The same criticism can be made of studies addressing amplitude shifts in response to noise exposure, but which do not measure other vocal changes that often accompany Lombard shifts in amplitude. It would be surprising to find that birds exposed to high levels of anthropogenic noise increase the pitch of their songs but not the amplitude, even if the pitch shift is not directly a by-product of the Lombard effect. While accurate measurements of amplitude can be challenging from a technical point of view, we encourage future researchers to consider including these measurements as their inclusion can provide answers to many of the critical remaining questions about the effects of noise on the vocal behaviour in birds.

Acknowledgments We dedicate this review to our late friend and colleague Björn Siemers. Intially, Björn planned to write a chapter on the effects of noise on bat vocalisations for this volume, but then he had to cancel because he did not have the time to work on the manuscript. He did not know then that he did not have much time at all.

We are indebted to Tang Jun from China Bird Tours (www.chinabirdtour.com), Michelle and Peter Wong, Volker Deecke, and Stefan Greif for very generously allowing us to use their photographs in Fig. 7.3. Many thanks to Alejandro Ríos-Chelén and his co-authors for calculating the individual values for the many song parameters from many South American urban bird species in their data set, and for allowing us to publish those data in our table. Finally, we are grateful to Peter Slater for his thoughtful comments, which greatly improved this manuscript. Parts of this chapter are inspired by our own field and laboratory studies (as well as a foray into signal transmission modelling, which was strongly influenced by the work of Bob Dooling), and we thank the BBSRC, the DFG, and the MPG for supporting those research efforts.

References

Amador A, Margoliash D (2013) A mechanism for frequency modulation in songbirds shared with humans. J Neurosci 33:11136–11144

Arroyo AS, Segura JMC, Clemente MEF, Sánchez JLL, Slabbekoorn H (2013) Experimental evidence for an impact of anthropogenic noise on dawn chorus timing in urban birds. J Avian Biol 44:288–296

Aubin T, Jouventin P (2002) How to identify vocally a kin in a crowd ? The penguin model. Advances in the study of behaviour, vol 31. Academic Press, San Diego, pp 243–277

Barber JR, Crooks KR, Fistrup KM (2009) The cost of chronic noise exposure for terrestrial organisms. Trends Ecol Evol 25:180–189

Barrigon MJM, Gomez EV, Mendez SJA, Vilchez GR, Trujillo CJ (2002) An environmental noise study in the city of Caceres, Spain. Appl Acoust 63:1061–1070

Beckers GJL, Suthers RA, ten Cate C (2003) Mechanisms of frequency and amplitude modulation in ring dove song. J Exp Biol 206:1833–1843

Bergen F, Abs M (1997) Verhaltensökologische Studie zur Gesangsaktivität von Blaumeise (*Parus caeruleus*), Kohlmeise (*Parus major*) und Buchfink (*Fringilla coelebs*) in einer Großstadt. J Ornithol 138:451–467

Bermudez-Cuamatzin E, Rios-Chelen AA, Gil D, Garcia CM (2009) Strategies of song adaptation to urban noise in the house finch: syllable pitch plasticity or differential syllable use? Behaviour 146:1269–1286

Bermudez-Cuamatzin E, Rios-Chelen AA, Gil D, Garcia CM (2011) Experimental evidence for real-time song frequency shift in response to urban noise in a passerine bird. Biol Lett 7:36–38

Blumenrath SH, Dabelsteen T (2004) Degradation of great tit (*Parus major*) song before and after foliation: implications for vocal communication in a deciduous forest. Behaviour 141:935–958

Bocharov AA, Kolesnik AG, Soloviev AV (2012) Two-parametric model of the spectrum of traffic noise in Tomsk. Acoust Phys 58:718–724

Boersma HF (1997) Characterization of the natural ambient sound environment: measurements in open agricultural grassland. J Acoust Soc Am 101:2104–2110

Bolin K (2009) Prediction method for wind-induced vegetation noise. Acta Acust 95:607–619

Brainard MS, Doupe AJ (2000) Auditory feedback in learning and maintenance of vocal behaviour. Nat Rev Neurosci 1:31–40

Brenowitz EA (1982) The active space of red-winged blackbird song. J Comp Phys 147:511–522

Brumm H (2004) The impact of environmental noise on song amplitude in a territorial bird. J Anim Ecol 73:434–440

Brumm H (2006) Signalling through acoustic windows: nightingales avoid interspecific competition by short-term adjustment of song timing. J Comp Physiol A 192:1279–1285

Brumm H (2010) Anthropogenic noise: implications for conservation. In: Breed MD, Moore J (eds) Encyclopedia of animal behavior. Academic Press, Oxford, pp 89–93

Brumm H, Naguib M (2009) Environmental acoustics and the evolution of bird song. Advances in the study of behavior, vol 40. Academic Press, San Diego, pp 1–33

Brumm H, Ritschard M (2011) Song amplitude affects territorial aggression of male receivers in chaffinches. Behav Ecol 22:310–316

Brumm H, Schmidt R, Schrader L (2009) Noise-dependent vocal plasticity in domestic fowl. Anim Behav 78:741–746

Brumm H, Slabbekoorn H (2005) Acoustic communication in noise. Advances in the study of behavior, vol 35. Academic Press, San Diego, pp 151–209

Brumm H, Slater PJB (2006) Ambient noise, motor fatigue, and serial redundancy in chaffinch song. Behav Ecol Sociobiol 60:475–481

Brumm H, Todt D (2002) Noise-dependent song amplitude regulation in a territorial songbird. Anim Behav 63:891–897

Can A, Leclercq L, Lelong J, Botteldooren D (2010) Traffic noise spectrum analysis: dynamic modeling vs. experimental observations. Appl Acoust 71:764–770

Cardoso GC, Atwell JW (2011) Directional cultural change by modification and replacement of memes. Evolution 65:295–300

Catchpole CK, Slater PJB (2008) Bird Song, 2nd edn. Cambridge University Press, Cambridge

Chesson P (2000) Mechanisms of maintenance of species diversity. Ann Rev Ecol Syst 31:343–366

Cody ML, Brown JH (1969) Song asynchrony in neighbouring bird species. Nature 222:778–780

Cynx J, Lewis R, Tavel B, Tse H (1998) Amplitude regulation of vocalizations in noise by a songbird, *Taeniopygia guttata*. Anim Behav 56:107–113

Dooling RJ (1979) Temporal summation of pure tones in birds. J Acoust Soc Am 65:1058–1060

Dooling RJ, Searcy MH (1985) Temporal integration of acoustic signals by the budgerigar (*Melopsittacus undulatus*). J Acoust Soc Am 77:1920–1979

Dowling JL, Luther DA, Marra PP (2012) Comparative effects of urban development and anthropogenic noise on bird songs. Behav Ecol 23:201–209

Dubois A, Martens J (1984) A case of possible vocal convergence between frogs and a bird in Himalayan torrents. J Ornithol 125:455–463

Egnor SER, Hauser MD (2006) Noise-induced vocal modulation in cotton-top tamarins (*Saguinus oedipus*). Am J Primatol 68:1183–1190

Elemans CPH, Mead AF, Rome LC, Goller F (2008) Superfast vocal muscles control song production in songbirds. PLoS ONE 3(7):e2581

Ellinger N, Hoedl W (2003) Habitat acoustics of a neotropical lowland rainforest. Bioacoustics 13:297–321

Ey E, Fischer J (2009) The "acoustic adaptation hypothesis"—a review of the evidence from birds, anurans and mammals. Bioacoustics 19:21–48

Fegeant O (1999) Wind-induced vegetation noise. Part II: field measurements. Acustica 85:241–249

Fernandez-Juricic E, Poston R, De Collibus K, Morgan T, Bastain B, Martin C, Jones K, Treminio R (2005) Microhabitat selection and singing behavior patterns of male house finches (*Carpodacus mexicanus*) in urban parks in a heavily urbanized landscape in the Western U.S. Urban Habitats 3:49–69

Ficken RW, Ficken MS, Hailman JP (1974) Temporal pattern shifts to avoid acoustic interference in singing birds. Science 183:762–763

Francis CD, Ortega CP, Cruz A (2010) Vocal frequency change reflects different responses to anthropogenic noise in two suboscine tyrant flycatchers. Proc Roy Soc Lond B Bio 278:2025–2031

Francis CD, Ortega CP, Cruz A (2011) Different behavioural responses to anthropogenic noise by two closely related passerine birds. Biol Lett 7:850–852

Fuller RA, Warren PH, Gaston KJ (2007) Daytime noise predicts nocturnal singing in urban robins. Biol Lett 3:368–370

Funabiki Y, Funabiki K (2009) Factors limiting song acquisition in adult zebra finches. Dev Neurobiol 69:752–759

Funabiki Y, Konishi M (2003) Long memory in song learning. J Neurosci 23:6928–6935

Goodale E, Kotagama SW (2008) Response to conspecific and heterospecific alarm calls in mixed-species bird flocks of a Sri Lankan rainforest. Behav Ecol 19:887–894

Goodwin SE, Podos J (2013) Shift of song frequencies in response to masking tones. Anim Behav 85:435–440

Grant BR, Grant PR (2010) Songs of Darwin's finches diverge when a new species enters the community. Proc Nat Acad Sci USA 107:20156–20163

Greenfield MD (2005) Mechanisms and evolution of communal sexual displays in arthropods and anurans. Advances in the study of behavior, vol 35. Academic Press, San Diego, pp 1–62

Gross K, Pasinelli G, Kunc HP (2010) Behavioral plasticity allows short-term adjustment to a novel environment. Am Nat 176:456–464

Haff TM, Magrath RD (2013) Eavesdropping on the neighbours: fledglings learn to respond to heterospecific alarm calls. Anim Behav 85:411–418

Halfwerk W, Holleman LJM, Lessells CM, Slabbekoorn H (2011) Negative impact of traffic noise on avian reproductive success. J Appl Ecol 48:210–219

Halfwerk W, Slabbekoorn H (2009) A behavioural mechanism explaining noise-dependent frequency use in urban birdsong. Anim Behav 78:1301–1307

Hamao S, Watanabe M, Mori Y (2011) Urban noise and male density affect songs in the great tit, *Parus major*. Ethol Ecol Evol 23:111–119

Hanna D, Blouin-Demers G, Wilson DR, Mennill DJ (2011) Anthropogenic noise affects song structure in red-winged blackbirds (*Agelaius phoeniceus*). J Exp Biol 214:3549–3556

Heil P, Neubauer H (2002) A unifying basis of auditory thresholds based on temporal summation. Proc Nat Acad Sci USA 100:6151–6156

Henwood K, Fabrick A (1979) Quantitative-analysis of the dawn chorus-temporal selection for communicatory optimization. Am Nat 114:260–274

Heylen L, Wuyts FL, Mertens F, De Bodt M, Van de Heyning PH (2002) Normative voice range profiles of male and female professional voice users. J Voice 16:1–7

Holland J (2000) Song communication and degradation in the wren. PhD Thesis. Zoological Institute, University of Copenhagen, Copenhagen

Hu Y, Cardoso GC (2010) Which birds adjust the frequency of vocalizations in urban noise? Anim Behav 79:863–867

Hultsch H, Todt D (2008) Comparative aspects of song learning. In: Zeigler HP, Marler P (eds) Neuroscience of birdsong. Cambridge University Press, Cambridge, pp 201–216

Järvenpää M, Lindström K (2004) Water turbidity by algal blooms causes mating system breakdown in a shallow-water fish, the sand goby *Pomatoschistus minutus*. Proc Roy Soc Lond B Bio 271:2361–2365

Jilka A, Leisler B (1974) The relation between the frequency spectrum of the territorial songs of 3 reed warbler species acrocephalus schoenobaenus, acrocephalus scirpaceus acrocephalus arundinaceus and their respective habitats. J Ornithol 115:192–212

Junqua J-C (1996) The influence of acoustics on speech production: a noise-induced stress phenomenon known as the Lombard reflex. Speech Comm 20:13–22

Keast A (1994) Temporal vocalization patterns in members of a eucalypt forest bird community—the effects of weather on song production. Emu 94:172–180

Kight CR, Saha MS, Swaddle J (2012) Anthropogenic noise is associated with reductions in the productivity of breeding eastern bluebirds (*Sialia sialis*). Ecol Appl 22:1989–1996

Kight CR, Swaddle JP (2011) How and why environmental noise impacts animals: an integrative, mechanistic view. Ecol Lett 14:1052–1061

Kirschel ANG, Blumstein DT, Cohen RE, Buermann W, Smith TB, Slabbekoorn H (2009) Birdsong tuned to the environment: green hylia song varies with elevation, tree cover, and noise. Behav Ecol 20:1089–1095

Klump GM (1996) Bird communication in the noisy world. In: Kroodsma DE, Miller EH (eds) Ecology and evolution of acoustic communication in birds. Comstock Publishing Associates, Ithaca and London, pp 321–338

Klump GM, Maier EH (1989) Gap detection in the starling (*Sturnus vulgaris*), I: psychophysical thresholds. J Comp Physiol 164:531–539

Kobayasi KI, Okanoya K (2003) Sex differences in amplitude regulation of distance calls in bengalese finches. *Lunchula striata var. domestica*. Anim Biol 53:173–182

Krams I, Krama T (2002) Interspecific reciprocity explains mobbing behaviour of the breeding chaffinches, *Fringilla coelebs*. Proc Roy Soc Lond B Bio 269:2345–2350

Lachlan RF, Servedio MR (2004) Song learning accelerates allopatric speciation. Evolution 58:2049–2063

Le Roux A, Jackson TP, Cherry MI (2001) Does brants' whistling rat (*Parotomys brantsii*) use an urgency-based alarm system in reaction to aerial and terrestrial predators? Behaviour 138:757–773

Lengagne T, Aubin T, Lauga J, Jouventin P (1999) How do king penguins (*Aptenodytes patagonicus*) apply the mathematical theory of information to communicate in windy conditions? Proc Roy Soc Lond B Bio 266:1623–1628

Lengagne T, Slater PJB (2002) The effects of rain on acoustic communication: tawny owls have good reason for calling less in wet weather. Proc Roy Soc Lond B Bio 269:2121–2125

Leonard ML, Horn AG (2005) Ambient noise and the design of begging signals. Proc Roy Soc Lond B Bio 272:651–656

Lombard E (1911) Le signe de l'élévation de la voix. Ann Malad l'Oreille Larynx 37:101–119

Losos J (2010) Lizards in an evolutionary tree: ecology and adaptive radiation of anoles. University of California Press, Berkeley

Lowry H, Lill A, Wong BBM (2012) How noisy does a noisy miner have to be? Amplitude adjustments of alarm calls in an avian urban 'adapter'. PLoS ONE 7(1):e29960

Lu Y, Cooke M (2008) Lombard speech: effects of task and noise type. J Acoust Soc Am 123:3072

Luther D (2009) The influence of the acoustic community on songs of birds in a neotropical rain forest. Behav Ecol 20:864–871

Magrath RD, Pitcher BJ, Gardner JL (2007) A mutual understanding? Interspecific responses by birds to each other's aerial alarm calls. Behav Ecol 18:944–951

Manabe K, Sadr EI, Dooling RJ (1998) Control of vocal intensity in budgerigars (*Melopsittacus undulatus*): Differential reinforcement of vocal intensity and the Lombard effect. J Acoust Soc Am 103:1190–1198

Marler P (2004) Bird calls: a cornucopia for communication. In: Marler P, Slabbekoorn, H (eds) Nature's Music. Elsevier, San Diego, pp132–177

Martens J, Geduldig G (1990) Acoustic adaptations of birds living close to Himalayan torrents. In: Ornithologen-Gesellschaft D (ed) Proceedings of international 100th DO-G meeting. Verlag der Deutschen Ornithologen-Gesellschaft, Radolfzell, pp 123–131

Miller LN (1978) Sound levels of rain and of wind in the trees. Noise Control Eng J 11:101–109

Mockford EJ, Marshall RC (2009) Effects of urban noise on song and response behaviour in great tits. Proc Roy Soc Lond B Bio 276:2979–2985

Montague MJ, Danek-Gontard M, Kunc HP (2012) Phenotypic plasticity affects the response of a sexually selected trait to anthropogenic noise. Behav Ecol 24:343–348

Morton ES (1975) Ecological sources of selection on avian sounds. Am Nat 109:17–34

Naguib M, Mennill DJ (2010) The signal value of birdsong: empirical evidence suggests song overlapping is a signal. Anim Behav 80:E11–E15

Narins PM, Feng AS, Lin WY, Schnitzler HU, Denzinger A, Suthers RA, Xu CH (2004) Old world frog and bird, vocalizations contain prominent ultrasonic harmonics. J Acoust Soc Am 115:910–913

Nelson BS (2000) Avian dependence on sound pressure level as an auditory distance cue. Anim Behav 59:57–67

Nemeth E (2004) Measuring the sound pressure level of the song of the screaming piha *Lipaugus vociferans*: one of the loudest birds in the world? Bioacoustics 14:225–228

Nemeth E, Brumm H (2009) Blackbirds sing higher-pitched songs in cities: adaptation to habitat acoustics or side-effect of urbanization? Anim Behav 78:637–641

Nemeth E, Brumm H (2010) Birds and anthropogenic noise: are urban songs adaptive? Am Nat 176:465–475

Nemeth E, Pieretti N, Zollinger S, Geberzahn N, Partecke J, Miranda AC, Brumm H (2013) Bird song and anthropogenic noise: vocal constraints may explain why birds sing higher pitched songs in cities. Proc Roy Soc Lond B Bio 280:20122798

Nemeth E, Zollinger S, Brumm H (2012) Effect sizes and the integrative understanding of urban bird song. Am Nat 180:146–152

Oberweger K, Goller F (2001) The metabolic cost of birdsong production. J Exp Biol 204:3379–3388

Osmanski MS, Dooling RJ (2009) The effect of altered auditory feedback on control of vocal production in budgerigars (*Melopsittacus undulatus*). J Acoust Soc Am 126:911–919

Parris KM, Schneider A (2009) Impacts of traffic noise and traffic volume on birds of roadside habitats. Ecol Soc 14(1):29

Pohl N, Slabbekoorn H, Neubauer H, Heil P, Klump G, Langemann U (2013) Why longer song elements are easier to detect: threshold level-duration functions in the great tit and comparison with human data. J Comp Physiol A 199:239–252

Popp JW, Ficken RW, Reinartz JA (1985) Short-term temporal avoidance of interspecific acoustic interference among forest birds. Auk 102:744–748

Potash LM (1972) Noise-induced changes in calls of the Japanese quail. Psychonomic Science 26:252–254

Potvin DA, Parris KM, Mulder RA (2011) Geographically pervasive effects of urban noise on frequency and syllable rate of songs and calls in silvereyes (*Zosterops lateralis*). Proc Roy Soc Lond B Bio 278:2464–2469

Proppe DS, Avey MT, Hoeschele M, Moscicki MK, Farrell T, St Clair CC, Sturdy CB (2012) Black-capped chickadees *Poecile atricapillus* sing at higher pitches with elevated anthropogenic noise, but not with decreasing canopy cover. J Avian Biol 43:325–332

Proppe DS, Sturdy CB, St Clair CC (2011) Flexibility in animal signals facilitates adaptation to rapidly changing environments. PLoS ONE 6(9):e25413

Pytte CL, Rusch KM, Ficken MS (2003) Regulation of vocal amplitude by the blue-throated hummingbird, *Lampornis clemenciae*. Anim Behav 66:703–710

Quinn JL, Whittingham MJ, Butler SJ, Cresswell W (2006) Noise, predation risk compensation and vigilance in the chaffinch, *Fringilla coelebs*. J Avian Biol 37:601–608

Rainey HJ, Zuberbuhler K, Slater PJB (2004) Hornbills can distinguish between primate alarm calls. Proc Roy Soc Lond B Bio 271:755–759

Riede T, Arcadi AC, Owren MJ (2007) Nonlinear acoustics in the pant hoots of common chimpanzees (*Pan troglodytes*): vocalizing at the edge. J Acoust Soc Am 121:1758–1767

Ríos-Chelén A, Salaberria C, Barbosa I, Macias Garcia C, Gil D (2012a) The learning advantage: bird species that learn their song show a tighter adjustment of song to noisy environments than those that do not learn. J Evol Biol 25:2171–2180

Ríos-Chelén A, Quirós-Guerrero E, Gil D, Macías Garcia C (2012b) Dealing with urban noise: vermilion flycatchers sing longer songs in noisier territories. Behav Ecol Sociobiol 67:145–152

Ripmeester EAP, Mulder M, Slabbekoorn H (2009) Habitat-dependent acoustic divergence affects playback response in urban and forest populations of the European blackbird. Behav Ecol 21:876–883

Ryan MJ, Brenowitz EA (1985) The role of body size, phylogeny, and ambient noise in the evolution of bird song. Am Nat 126:87–100

Salaberria C, Gil D (2010) Increase in song frequency in response to urban noise in the great tit *Parus major* as shown by data from the Madrid (Spain) city noise map. Ardeola 57:3–11

Schrader L, Todt D (1993) Contact call parameters covary with social context in common marmosets (*Callitrix jacchus*). Anim Behav 46:1026–1028

Schroeder J, Nakagawa S, Cleasby I, Burke T (2012) Passerine birds breeding under chronic noise experience reduced fitness. PLoS ONE 7:e39200

Schuster S, Zollinger S, Lesku JA, Brumm H (2012) On the evolution of noise-dependent vocal plasticity in birds. Biol Lett 8:913–916

Searcy WA, Beecher MD (2009) Song as an aggressive signal in songbirds. Anim Behav 78:1281–1292

Searcy WA, Beecher MD (2011) Continued scepticism that song overlapping is a signal. Anim Behav 81:E1–E4

Seddon N (2005) Ecological adaptation and species recognition drives vocal evolution in neotropical suboscine birds. Evolution 59:200–215

Slabbekoorn H, den Boer-Visser A (2006) Cities change the songs of birds. Curr Biol 16:2326–2331

Slabbekoorn H, Peet M (2003) Birds sing at a higher pitch in urban noise. Nature 424:267–267

Slabbekoorn H, Smith TB (2002) Habitat-dependent song divergence in the little greenbul: an analysis of environmental selection pressures on acoustic signals. Evolution 56:1849–1858

Slater PJB (2003) Fifty years of bird song research: a case study in animal behaviour. Anim Behav 63:633–639

Slater PJB, Mann NI (2004) Why do the females of many bird species sing in the tropics? J Avian Biol 35:289–294

Swan DC, Hare JF (2008) The first cut is the deepest: primary syllables of richardson's ground squirrel, *Spermophilus richardsonii*, repeated calls alert receivers. Anim Behav 76:47–54

Tsai KT, Lin MD, Chen YH (2009) Noise mapping in urban environments: a Taiwan study. Appl Acoust 70:964–972

Tumer EC, Brainard MS (2007) Performance variability enables adaptive plasticity of 'crystallized' adult birdsong. Nature 450:1240–U1211

United Nations (2012) World urbanization prospects: the 2011 revision. Department of Economic and Social Affairs, United Nations, New York

Verzijden MN, Ripmeester EAP, Ohms VR, Snelderwaard P, Slabbekoorn H (2010) Immediate spectral flexibility in singing chiffchaffs during experimental exposure to highway noise. J Exp Biol 213:2575–2581

Ward S, Lampe HM, Slater PJB (2004) Singing is not energetically demanding for pied flycatchers, *Ficedula hypoleuca*. Behav Ecol 15:477–484

Ward S, Speakman JR, Slater PJB (2003) The energy cost of song in the canary, *Serinus canaria*. Anim Behav 66:893–902

Weir JT, Wheatcroft DJ, Price TD (2012) The role of ecological constraint in driving the evolution of avian song frequency across a latitudinal gradient. Evolution 66:2773–2783

Wheatcroft D, Price TD (2013) Learning and signal copying facilitate communication among bird species. Proc Roy Soc Lond B Bio 280:1–7

WHO (2011) Burden of disease from environmental noise—quantification of healthy life years lost in Europe. World Health Organization, Regional Office for Europe

Wiley RH (1991) Associations of song properties with habitats for territorial oscine birds of Eastern North-America. Am Nat 138:973–993

Wiley RH (2006) Signal detection and animal communication. In: Brockmann HJ, Slater PJB, Snowdon CT, Roper TJ, Naguib M, WynneEdwards KE (eds) Advances in the study of behavior, vol 36. Academic Press, San Diego, pp 217–247

Wiley RH, Richards DG (1978) Physical constraints on acoustic communication in atmosphere—implications for evolution of animal vocalizations. Behav Ecol Sociobiol 3:69–94

Wiley RH, Richards DG (1982) Adaptations for acoustic communication in birds: sound transmission and signal detection. In: Kroodsma DE, Miller EA (eds) Acoustic communication in birds. Academic Press, New York, pp 132–181

Wood WE, Yezerinac SM (2006) Song sparrow (*Melospiza melodia*) song varies with urban noise. Auk 123:650–659

Woolley SMN (2008) Auditory feedback and singing in adult birds. In: Zeigler HP, Marler P (eds) Neuroscience of birdsong. Cambridge University Press, Cambridge

Zannin PHT, Diniz FB, Barbosa WA (2002) Environmental noise pollution in the city of Curitiba, Brazil. Appl Acoust 63:351–358

Zevin JD, Seidenberg MS, Bottjer SW (2004) Limits on reacquisition of song in adult zebra finches exposed to white noise. J Neurosci 24:5849–5862

Zollinger SA, Goller F, Brumm H (2011) Metabolic and respiratory costs of increasing song amplitude in zebra finches. PLoS ONE 6:e23198

Zollinger SA, Podos J, Nemeth E, Goller F, Brumm H (2012) On the relationship between, and measurement of, amplitude and frequency in birdsong. Anim Behav 84:E1–E9

Chapter 8
Avian Sound Perception in Noise

Robert J. Dooling and Sandra H. Blumenrath

Abstract All environments are noisy, and auditory systems have evolved to cope with this noise. Indeed all sensory systems employ mechanisms that facilitate the separation of relevant signals from irrelevant noise. Interestingly, most of what we know about hearing comes from tests conducted in the near absolute quiet of an acoustic test booth. Because of their tractability in the laboratory, their complex vocal repertoires, and their elaborate acoustic communication systems, birds have proven valuable models for understanding the effects of noise on hearing and acoustic communication in part by bringing laboratory and field studies together. Noise can have at least four different kinds of effects occurring either alone or together. These four categories of effects are hearing damage and permanent threshold shift (PTS) from acoustic overexposure, temporary threshold shift (TTS) from acoustic overexposure, masking of acoustic communication signals (or other biologically relevant sounds), and a host of other physiological and behavioral responses including effects on attention. Here we consider masking as separate from these other effects of noise on hearing and acoustic communication. Furthermore, we take an 'auditory-centric' point of view and consider masking exclusively from the point of view of the listening bird. We review the behavioral and auditory strategies that birds use to maximize communication in a noisy environment and suggest an approach to assessing the risk posed by noise, whether natural or anthropogenic.

8.1 Introduction

Because of their tractability in the laboratory, their complex vocal repertoires, and their elaborate acoustic communication systems, birds are valuable models for understanding the effects of noise on hearing and acoustic communication. From an

R. J. Dooling (✉) · S. H. Blumenrath
Department of Psychology, Laboratory of Comparative Psychoacoustics,
University of Maryland, College Park, MD 20742, USA
e-mail: rdooling@umd.edu

H. Brumm (ed.), *Animal Communication and Noise*,
Animal Signals and Communication 2, DOI: 10.1007/978-3-642-41494-7_8,
© Springer-Verlag Berlin Heidelberg 2013

evolutionary and biological standpoint, there is a long interest in understanding the factors that determine the "active space" in which two birds can communicate effectively (e.g., Morton 1975; Marten and Marler 1977; Dooling 1982; Brenowitz 1982; Ryan and Brenowitz 1985; Dabelsteen et al. 1993; Klump 1996). Considered from a very different, practical standpoint, noise, especially anthropogenic noise, has been suspected to cause a variety of adverse effects on birds and other wildlife affecting, among other things, reproductive success, species diversity, and the structure of animal communities (Francis et al. 2009, 2011a, b; Slabbekoorn and Halfwerk 2009; Halfwerk et al. 2011; Schroeder et al. 2012). These effects include stress and physiological changes, auditory system damage from acoustic overexposure, and masking of communication and other important biological sounds (Ryals et al. 1999; Miller 1974; Forman et al. 2002; Brumm 2004; Foppen and Rejnen 2004; Brumm and Slabbekoorn 2005). A precise understanding of these effects is of interest to many groups including biologists, environmentalists, regulators, as well as roadway and construction engineers (see also Chap. 14). However, for a number of reasons, it has been difficult to reach a clear consensus on the causal relationships between noise levels and these adverse effects especially in natural environments. One reason is that there are surprisingly few studies in birds that can definitively identify noise alone as the principal source of stress or physiological effects. A second reason is that, while all humans have similar auditory capabilities and sensitivities, the same is not true for birds (Dooling 1982; Dooling et al. 2000). This means that noise that may be detrimental to one species may not be detrimental to another. Still another issue is separating out the various overlapping effects of noise. There are well-documented adverse consequences of elevated noise on humans including hearing loss, masking, stress, physiological and sleep disturbances, and changes in feelings of well-being (e.g., Miller 1974; Cohen et al. 1979; Smith 1991). In fact, recent studies and health reports have linked noise pollution to elevated cholesterol levels and increased risk of cardiovascular disease related deaths in humans (Babisch et al. 2003; WHO report 2011). It is therefore not surprising that similar behavioral and physiological effects of chronic noise have also been demonstrated in animals (Bedanova et al. 2010; Voslarova et al. 2011). In this chapter, we are concerned primarily with the auditory masking effects of noise.

8.2 The Bird Ear and Auditory System

The central and peripheral auditory system of birds is highly complex (Carr and Code 2000). The bird ear consists of an external membrane (tympanic membrane), a middle ear, and an inner ear. There is no external structure that resembles the mammalian outer ear flap, or pinnae (except in owls). The tympanic membrane is the outermost covering of the middle ear. The function of the bird tympanic membrane is to gather sound, as it does in mammals, and the middle ear acoustically couples air-borne sound to the fluids of the inner ear by impedance matching.

One factor that likely constrains the frequency range over which birds hear is the single-bone middle ear, the columella, a structure quite different in form than the three-bone middle bones (malleus, incus, and stapes) that are characteristic of mammals. A careful comparison of the function of the columellar systems of birds and reptiles demonstrate that, compared to the three-bone ossicular chain in mammals, these systems limit the hearing in most avian species to not much more than 10 kHz (Saunders et al. 2000).

The avian inner ear is similar to that of most vertebrates in having three semicircular canals to determine angular acceleration of the head and three otolith organs for detection of motion of the head relative to gravity. Birds have a cochlear duct which contains a basilar papilla upon which sit the sensitive sensory hair cells used for hearing. The basilar papilla is shorter and rather different in structure than that found in mammals (Tanaka and Smith 1978; Smith 1985; Gleich and Manley 2000), and some of these differences may account for the much narrower range of frequencies detected by birds as compared to mammals. There is a considerable amount of comparative data on the anatomy of the avian ear and on the behavior of hearing, showing a correlation between body mass, basilar membrane length, and features of hearing such as best frequency, bandwidth, and high frequency cut-off (Gleich et al. 2005). The basilar papillae of small birds such as the canary, budgerigar, and zebra finch are in the range of 1.6–3 mm and that of the barn owl is about 11 mm, while the size of the human basilar membrane is closer to 32 mm. Hair cells on the human basilar membrane are arranged as one inner row (primarily afferently innervated) and three outer rows (primarily efferently innervated). By contrast, there are many hair cells across the width of the bird basilar papilla ranging from tall hair cells on the neural side, which are predominantly afferently innervated, to short hair cells toward the abneural edge, which are primarily efferently innervated. Not only are there considerable species differences in the bird inner ear, but there is a remarkable complexity across the epithelium in the pattern of how the ciliary bundles are oriented, the shape of the hair cell bundle, and the number and height of stereovilli on each hair cell, suggesting several different mechanisms of frequency selectivity (Gleich and Manley 2000). Thus, in spite of its diminutive size, the bird ear is a highly specialized organ capable of supporting very fine auditory discrimination and perception which, in some cases, exceeds the acuity of many mammals, including humans (Dooling et al. 2000).

8.3 The Interrelated Effects of Noise

Depending on its level and spectral characteristics, noise exposure can have a variety of effects on birds and other animals including humans. It is useful to think about four overlapping categories of noise effects on an organism: hearing damage and permanent threshold shift (PTS) from acoustic overexposure, temporary threshold shift (TTS) from acoustic overexposure, masking of important biological sounds, and other physiological and behavioral responses. In all but the last case, these auditory effects depend strongly on the level of noise exposure which is highly correlated with the proximity of the bird to the noise source. These relations are schematically represented in Fig. 8.1. A noise source (left margin) can be any noise source (e.g., babbling brook, noisy waterfall, highway or roadway noise, etc.). The figure shows the conceptual relationships among noise level, distance of the bird from the noise source, and the different kinds of effects of noise on birds. This schematic illustrates that only under some conditions, primarily related to level, does noise have a single effect. Masking occurring some distance from the noise source would be one of these. In terms of areas around the noise source affecting birds, the four categories of effects are referred to as zones related to distances from the noise source.

Effects Expected in Zone 1: If a bird is close to an intense noise source, the potential effects include hearing loss, threshold shift, masking, and/or other behavioral and/or physiological effects. Laboratory evidence shows that continuous noise levels above 110 dB (A) SPL could result in physical damage of the auditory system and permanent threshold shift (Hashino et al. 1988, Hashino and Sokabe 1989; Dooling et al. 2008; Ryals et al. 1999). Birds present an especially interesting situation in terms of acoustic overexposure because they are more resistant to both temporary and permanent hearing loss and to hearing damage from acoustic overexposure than are humans and other mammals (Miller 1974).

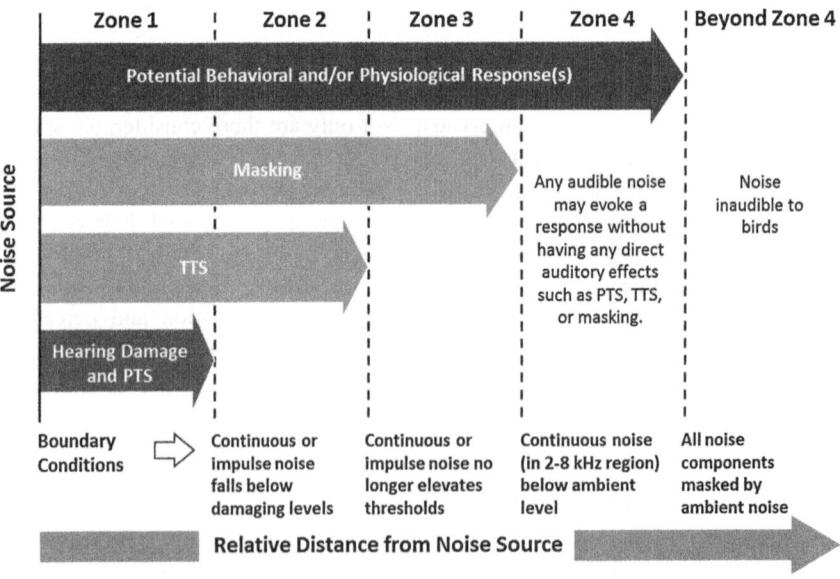

Fig. 8.1 Relation among noise level, distance, and potential effects on the exposed bird. The zones reflect distances within which a bird is exposed to a specific noise effect, such as hearing damage and permanent threshold shift (PTS) in Zone 1, or temporary threshold shift (TTS) in Zone 2

Moreover, unlike mammals, birds can regenerate the sensory cells of the inner ear, thereby providing a physiological mechanism for recovering from intense acoustic overexposure (Ryals et al. 1999; Adler et al. 1993; Dooling et al. 2008).

Effects Expected in Zone 2: At greater distances from the noise source, as noise from the source decreases in level, hearing damage and permanent threshold shift are less likely to occur. Laboratory data show that high noise levels above about 93 dB (A) SPL, though they do not cause permanent damage, might still cause a temporary elevation of a bird's hearing threshold and mask important communication signals. In nature, this could easily be seen to lead to other behavioral and/or physiological effects (Dooling et al. 2008; Saunders and Dooling 1974).

Effects Expected in Zone 3: At even greater distances from the noise source, noise levels of course would be expected to be even lower. Where the spectrum level of the noise is still at or above the natural ambient noise level, masking of communication or other biologically important signals can occur beyond the masking that already occurs from natural ambient noise. As far as we know, this range over which additional masking effects may occur in the wild has not been precisely and systematically investigated in the field. But it can be expected to be quite large in otherwise low noise environments. By contrast, in noisy environments, as for instance on a windy day, the masking effects from another point noise source will not extent as far out from the source, since masking from ambient noise will prevail. In either case, the range is bounded at the furthest distance from the noise source where the levels of noise from any particular prominent source such as a babbling brook falls below the ambient noise level due to other sources (Lohr et al. 2003; Langemann et al. 1998).

Effects Expected in Zone 4: Once the level of source noise falls below ambient noise levels in the critical frequencies for communication, masking of vocal signals is no longer an issue because the noise can no longer be heard by the receiver. However, faintly heard sounds falling outside the region of bird vocalizations, such as thunder from a distant storm, may still potentially cause other behavioral and/or physiological effects.

Effects Expected Beyond Zone 4: Beyond Zone 4, the level of noise from the source at all frequencies is completely inaudible to the bird (i.e., falls below the bird's masked threshold) and can be expected to have no effects of any kind on the bird.

While the above schematic representation may seem trivial, it provides a framework for conceptualizing the separate and integrated effects of noise on birds that can easily become conflated because they often occur together. This chapter focuses on masking occurring alone. This is useful because independent of other effects, masking of communication signals and other important biological sounds (e.g., sounds of an approaching predator) can potentially have significant adverse consequences for species' behavior (Brumm and Slabbekoorn 2005) and population viability (Chap. 14).

8.4 Effects of Masking Noise on Hearing

8.4.1 The Simple Laboratory Case: The Critical Ratio

Masking is the interference with the detection of one sound by another. Most commonly, masking refers to the increase in thresholds for detection or discrimination of sounds caused by the presence of another sound. The simplest kind of masking experiment in the laboratory is to measure the sound detection thresholds for pure tones in the presence of a flat, broadband noise. This ratio between the power in the pure tone at threshold and the power per Hertz (spectrum level) of the background noise is called the critical ratio (CR). Critical ratios reflect the operation of selectivity mechanisms of the vertebrate auditory system (e.g., Moore 2003). But they also have relevance for hearing under natural conditions including the perception of vocal signals, the range over which vocal signals may be heard, and the evolution of detection mechanisms. In a more ecological context, critical ratios might aid in understanding the selective pressures driving territory sizes and the extent to which anthropogenic noises may interfere with the acoustic communication of birds in their natural habitats. Signal-to-noise ratios (i.e., critical ratios) from masking measured in this way are now available for 13 different species of birds (Dooling et al. 2000), so we have a fairly good idea of how the typical bird hears in noise.

Figure 8.2 shows the median critical ratio functions for these 13 species of birds and the human (taken from Dooling et al. 2000). In the more well-studied humans and other mammals, CRs increase at a rate of about 3 dB/octave over a wide range of frequencies. This rate of increase is related to the mechanics of the peripheral auditory system and the logarithmic organization of the traveling wave's

Fig. 8.2 Comparison of critical ratios (CR) between birds and humans across frequencies. The dotted line represents the average rate of CR increase of 3 dB/octave in a bird or mammal (from Dooling et al. 2000)

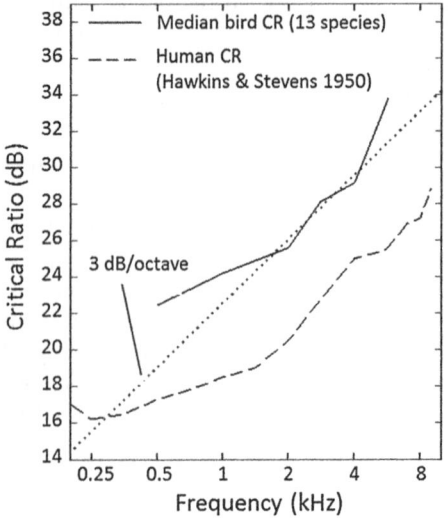

maximum displacement along the basilar membrane (Békésy 1960; Greenwood 1961a, b; Buus et al. 1995). For birds, the median CR function also shows an orderly increase of about 3 dB in critical ratio with each doubling of frequency but over a narrower frequency range of only about 2–3 octaves. The logarithmic organization of frequency along the papilla in birds is likely responsible for this, although there are clearly other sophisticated tuning mechanisms possible in the avian ear (Gleich and Manley 2000).

While, on average, birds show the same change in CR with frequency between 2 and 8 kHz as humans and other mammals, there are considerable species differences among birds. One can only speculate as to the adaptive value of these divergent critical ratio functions. The budgerigar (*Melopsittacus undulatus*), for instance, shows a very small CR at 2.8 kHz and larger CRs below and above that frequency (Dooling and Saunders 1975). Contact calls in budgerigars are used for long distance communication and recognition probably over great distances and often in large flocks. The energy in these calls falls exclusively in the frequency region with the smallest critical ratios suggestive of an auditory system with features for communication over long distances or under noisy conditions (Klump 1996). In great tits (*Parus major*), critical masking ratios are relatively flat over the entire range of hearing (Langemann et al. 1998). This means that at high frequencies, the great tit has unusually low critical masking ratios compared to other birds. Interestingly, the great tit produces an aerial predator alarm call that is pure tone-like and falls in the range of 8 kHz. Presumably, this call has been selected to be inaudible to the great tit's chief predator, the European sparrowhawk (*Accipter nisus*) (Klump et al. 1986). And in the barn owl, critical ratios are extremely small at 2 kHz (more like those of humans) and increase at the rate closer to 5 dB/octave rather than 3 dB/octave as in the average bird or mammals. High frequency hearing is used by the owls in localizing and capturing prey (Konishi 1973). Perhaps the unusually small critical ratios (for a bird) at these frequencies in the barn owl are related to this predatory lifestyle.

The difference between the CR of the typical bird and that of humans has considerable significance in the real world. The two functions parallel each other between about 2–8 kHz, with humans about 6 dB better (i.e., lower S/N ratio). What does this mean in detection terms? It means that humans can detect fainter sounds in noise than can the typical bird. In practical environmental terms, a human will be able to hear another bird vocalizing at twice the distance that another bird can. Conversely, noise from a waterfall or from an anthropogenic noise source that is barely audible to a human would be well below the threshold of detection for the typical bird. Indeed, simply by spherical spreading, the bird would have to be twice as close again to the noise source as a human to barely be able to hear it.

The effect of noise on detection thresholds is an important issue because most natural environments have a level of ambient noise sufficient to elevate an animal's pure tone thresholds far above what can be typically measured in the laboratory in an acoustic test booth. In other words, under natural conditions, a bird's

Fig. 8.3 Masked thresholds (*upper curve*) and background noise spectrum level (*lower curve*). Masked thresholds can be estimated from the noise spectrum and the listener's critical ratio. The thresholds shown here are based on average bird critical ratios for frequencies between 0.25 and 10 kHz (after Klump 1996)

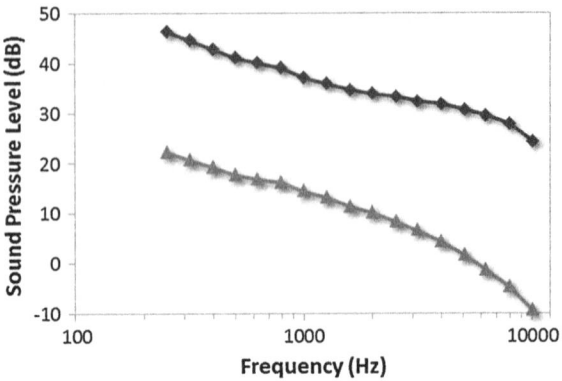

audiogram (as measured in the quiet of an acoustic test booth) does not define what the bird can hear. Rather, by definition, the bird's signal-to-noise ratio (i.e., CR) as a function of frequency—in combination with the spectrum and level of noise—defines what the bird can hear. Figure 8.3 taken from Klump (1996) nicely illustrates this point showing the spectrum and level of ambient noise in a wooded area and the function represented by values one CR above the spectrum of this ambient noise. The critical ratio function used by Klump was for the 'average' bird, which shows about a 3 dB/octave increase for frequencies above 2 kHz. The upper curve in this figure defines what the average bird would actually be able to detect in this environment. This is quite different from what one would expect based on the bird audiogram measured in the quiet.

8.4.2 Effect of Noise, Signal Spectrum, and Signal Level

Noises can be continuous or intermittent, broadband or narrowband, and predictable or unpredictable in time or space. All of these characteristics are important and have been widely studied in the laboratory under controlled conditions. Two of these characteristics, noise spectrum and signal spectrum, are less intuitive and have been particularly well studied. Most laboratory studies estimating the effects of noise on signal detection use continuous noises with precisely defined bandwidths, intensities, and spectral shapes. While this is rare in nature, it does occur, as for instance the noise from a babbling brook, a waterfall, sustained wind, etc. In terms of anthropogenic noise sources, traffic or highway noise on heavily traveled roads can also approximate these features (e.g., relatively continuous, relatively constant spectrum and intensity).

Laboratory masking studies in birds, humans, and other animals, show that the noise energy in the frequency region of a signal is most important in masking the signal; noise energy falling outside that frequency band contributes much less (e.g., Zwicker et al. 1957; Zwislocki 1963; Ehret 1975; Dooling et al. 2005;

Au and Moore 1990; Langemann et al. 1995; Dooling et al. 2000). The principle idea of the power spectrum model of hearing is that the peripheral auditory system acts like a series of overlapping bandpass filters laid end-to-end across the length of the basilar membrane or basilar papilla (in the case of birds). This fact is well established in the laboratory and is of major importance when trying to assess the effect of a masking noise on hearing. A listener detecting a signal in noise is using the bandpass filter nearest the signal. Noise in other frequency bands is far less important. Taking these values of signals and noise at threshold obtained from laboratory results, it is possible to estimate the effect of noise on various auditory behaviors of bird in the wild (e.g., communication distance).

In their study published in 2003, Lohr and his colleagues examined the relation between some of these key variables at play in considering masking of communication signals in natural environments as opposed to the masking of pure tones in the laboratory. In other words, can the masking of pure tones by noise in the laboratory (i.e., critical ratios) be used to predict the bird's ability to detect natural vocalizations in a background of noise? What is the effect of signal bandwidth on masking? Some vocalizations are more broadband than others. What is the effect of masker bandwidth on masking? As an example, typical highway noise has more energy below 1 kHz than above, and bird vocalizations generally contain more energy above 1 kHz than below, the masking effects of highway noise on bird vocalizations are less than would be expected from noise of the same level in the same frequency range of bird vocalizations (Lohr et al. 2003).

8.4.3 Hearing: Detection, Discrimination, and Recognition

Common sense and our own experience tell us that acoustic communication can be severely constrained if background noise is of a sufficient duration and level. Such noise can mask important signals and thereby reduce the animal's acoustic space (the combination of sound frequencies and levels that are audible). when we speak in these terms, we are almost always referring to the detection of a sound. Whether we are concerned with animal or human hearing, there is often a disconnect between what we typically measure in the laboratory and how we use our ears in the real world. This lack of precision in the way we use the term 'hearing' has real significance when we are concerned with the effects of masking noise. For instance, it is one thing to say that a speech stream can be heard (i.e., detected), or that we can tell one voice from another (i.e., discriminate), and quite another to be able to understand (i.e., recognize) what is being said. This distinction between detection of a sound and the recognition of a sound has long been familiar to field researchers who rely on playback studies and territorial responses of the receiver (Brenowitz 1982; Nelson and Marler 1990; Klump 1996). Klump frames this issue as a problem of signal detection and compares just noticeable differences (JNDs) with just meaningful differences (JMDs) for frequency change (Klump 1996).

The previously mentioned study by Lohr and his colleagues (2003) directly compared the detection and discrimination of contact calls by two different species of birds, the zebra finch (*Taeniopygia guttata*) and budgerigar, for conspecific calls and the calls of the other species. Zebra finch vocalizations are harmonic, broadband vocalizations while budgerigar contact calls are more tonal and narrowband. Spectra and time waveforms of these vocalizations are shown in Fig. 8.4. Differences in the spectral shapes of these vocalizations means, of course, that when these two types of vocalizations are presented in noise at the same overall signal level, the signal-to-noise ratio around 3 kHz will be higher for budgerigar vocalizations than for zebra finch vocalizations. We would expect then, when presented at the same overall level, budgerigar contact calls would be more easily detected than zebra

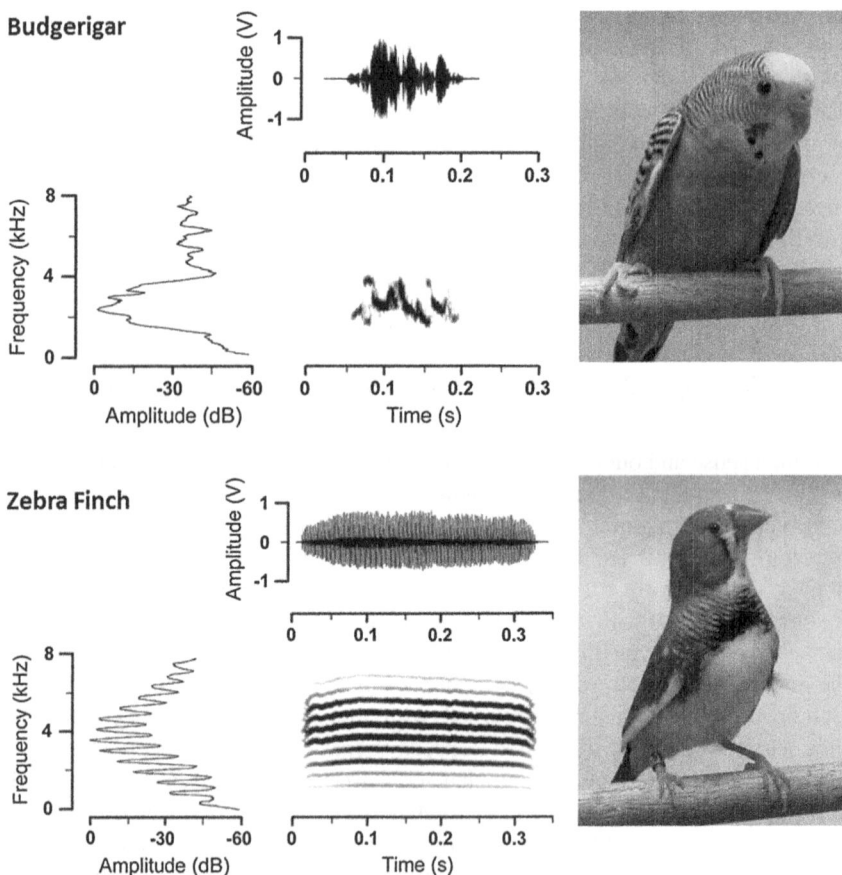

Fig. 8.4 Temporal and spectral patterns of representative examples of a budgerigar and zebra finch call. Zebra finch vocalizations are more harmonic and broadband than the rather tonal, narrowband budgerigar contact calls. Shown are the frequency spectrum (*left*), spectrogram (*bottom right*), and time waveform (*upper right*) of each call

finch calls. Moreover, this experiment was conducted with two different types of continuous noises of the same overall level: one that had a flat spectrum and one that had a sloping spectrum with more energy below 1 kHz and less energy in frequencies about 1 kHz. Such a sloping spectrum is characteristic of many anthropogenic noises such as noises form traffic and roadways. Results from these masking experiments with budgerigars and zebra finches tested on conspecific and other species calls are shown in Figs. 8.5 and 8.6. For both zebra finches and budgerigars, discrimination thresholds are about 2–5 dB higher than detection thresholds, whether the birds are being tested with conspecific or heterospecific calls (Lohr et al. 2003). Moreover, Fig. 8.6 shows that for masking noises of the same overall level, the noise having the spectral shape of traffic noise (with less energy in the region of the bird calls) caused less masking of vocalizations than did a flat spectrum noise for both species. Tests of the ability of these same birds to identify or recognize contact calls in noise show that signal-to-noise ratios are about 2–3 dB higher than that required for discrimination (Dooling, unpublished data). In

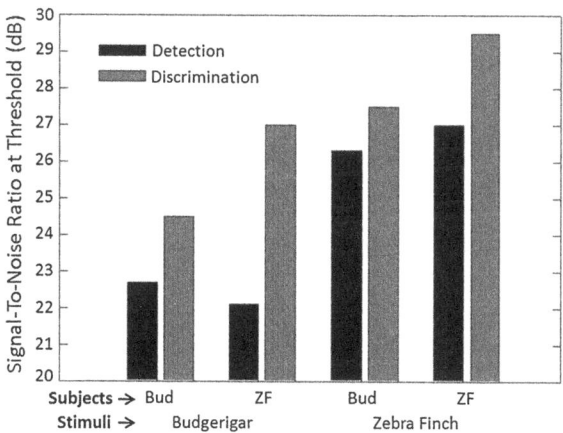

Fig. 8.5 Detection and discrimination thresholds for budgerigar and zebra finch calls. The signal-to-noise ratio (SNR) necessary to discriminate calls are about 2–5 dB higher than the SNR required for their detection

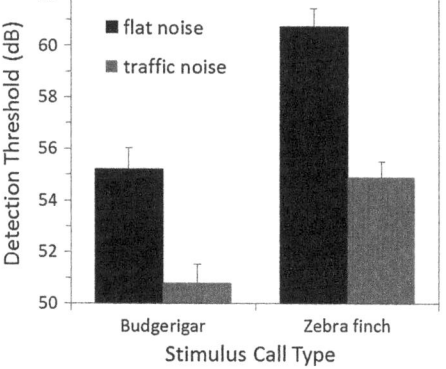

Fig. 8.6 Effect of the spectral shape of masking noise on budgerigar and zebra finch detection thresholds. Shown are average masked thresholds for birds tested with budgerigar and zebra finch contact calls in an overall noise level of 70 dB SPL ($N = 8$ birds/bar)

summary, the effect of background noise on hearing in birds and humans shows that signal discrimination requires a higher signal-to-noise ratio than detection; recognition requires a still higher signal-to-noise ratio than discrimination; and comfortable communication requires an even higher signal-to-noise ratio (Lohr et al. 2003; Freyaldenhoven et al. 2006). A signal-to-noise ratio that represents a comfortable communication level in animals is impossible to assess but quite easy to measure in humans, whose signal-to-noise ratio is about 15 dB (Freyaldenhoven et al. 2006). Armed with this information about the relation between the spectral levels of signals (vocalizations) and noise and a more precise definition of what we mean by hearing, helps us to understand the effects of these variables on acoustic communication between birds in the real world.

8.4.4 Sound Transmission and Masking Effects on Communication Behaviors

In nature, masking by noise can affect the detection of important biological signals. Because significant masking can occur without the permanent or temporary threshold effects resulting from more intense exposures, and because masking is almost always occurring in natural environments, it may be the most ubiquitous and insidious effect of noise. The study of the physical constraints on acoustic communication under natural environmental conditions is complex, including a variety of factors, of which some are linear and some nonlinear with distance (Wiley and Richards 1982). Two prominent effects that have been investigated with regard to the detection of bird vocalizations concern level changes due to spherical spreading (i.e., the inverse square law) and excess attenuation (Marten and Marler 1977; Marten et al. 1977; Dooling 1982). Under simple conditions, spherical spreading amounts to a quartering of power or halving of pressure with each doubling of distance (i.e., a sound pressure change of -6 dB/doubling of distance). In homogenous environments, there are also constant attenuation effects, which represent a deviation from the attenuation expected from the inverse square law. These are expected to be about 5 dB/100 m for a sound source 10 m above ground in an open field (e.g., Marten and Marler 1977).

A simple approach for predicting the effects of masking noise on communication distance in birds would include these variables as well as those described earlier, including the spectrum and level of the masking noise, the bird's hearing in quiet and in noise, and the spectrum and level of a signaling bird's vocalizations. The approach assumes that the spectrum and amplitude level of the noise and the signaler's vocalization are both known at the location of the receiver. These values can either be measured directly or they can be estimated by applying signal attenuation algorithms to both the noise source and the signals of the sender (Dooling 1982).

The challenge for the receiver in this scenario is to hear the signal in the presence of noise, which is primarily dependent on the species-specific auditory capabilities of the receiver in noisy conditions (i.e., its critical ratio) and the actual signal-to-noise ratio at the receiver's location. Taking into account that 'hearing,' as described earlier, can mean different things and each requires different signal-to-noise ratios, the four functions shown in Fig. 8.7 illustrate the effect of a 60 dB (A) SPL noise (stippled line) on four different auditory behaviors of birds. This example assumes both a median bird critical ratio function and that the calling bird is vocalizing at a peak sound pressure level of 100 dB through an open area with an excess attenuation of 5 dB/100 m beyond the loss due to spherical spreading. In this noise, a comfortable level of communication between two birds requires that the two birds be less than 60 m apart. Recognition and discrimination of a bird vocalization by the receiver, however, can still occur at greater inter-bird distances of up to about 220 and 270 m, respectively. And finally, simple detection of another bird's vocalization can occur at even greater distances of up to 345 m in this noise. These data can also be plotted as a set of concentric rings representing the different definitions of 'hearing' with the listening bird assumed to be in the center as shown in Fig. 8.8.

Of course, in real-world situations, the acoustic dynamics of signal transmission are highly variable, both spatially and temporally, depending upon distribution and character of habitat types, prevailing meteorological conditions, and the behaviors of both the caller and receiver in maximizing communication. Consequently, the shapes and sizes of the communication regions around the receiver will naturally

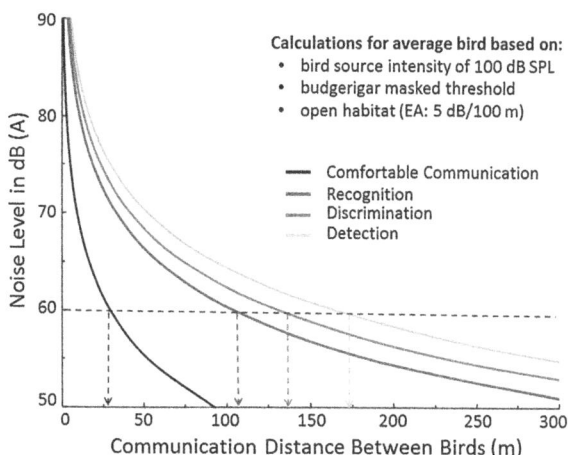

Fig. 8.7 Relationship between overall noise level and communication distances that allow detection, discrimination, recognition, and comfortable communication between birds. These curves are based on a mathematical model that takes into account the vocalization's source level (here 100 dB), the listener's masked threshold (here for the median bird), and the habitat's average excess attenuation (here open habitat: 5 dB/100 m)

Fig. 8.8 Average
communication distances
across bird species in low-
level noise (60 dB SPL).
Each shade of green
represents the range of
distances between two
communicating birds at
which the specified auditory
behavior, for instance
recognition, is possible

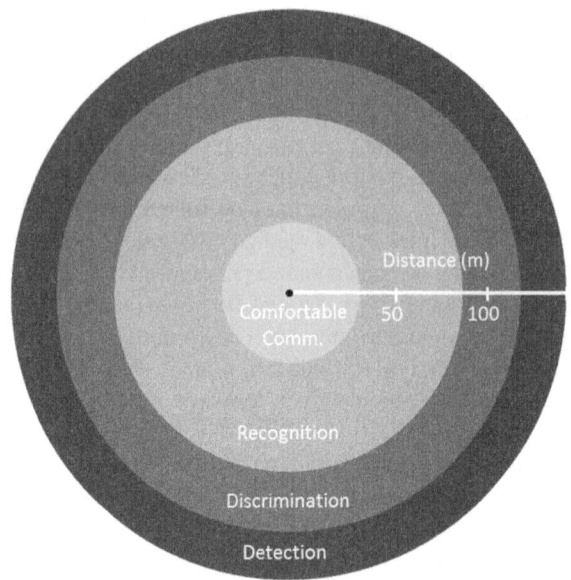

vary in accordance with the physical conditions of the area, the species-specific
hearing capabilities, the characteristics of the signal, and the strategies employed
in communicating acoustically.

Figure 8.9a illustrates the interplay of three of these variables: species differences in critical ratios, spectral characteristics of the signal, and different habitats—an open area with 5 dB/100 m excess attenuation versus a dense forest
habitat with up to 25 dB/100 m excess attenuation. The effect of increasing excess
attenuation (i.e., moving from an open area to a forested area) for the budgerigar,
on the other hand, has the effect of reducing communication (left).

In general, communication distances for birds closer to the noise source, or with
large critical ratios at 2 kHz, would be represented by smaller concentric circles.
Communication distances for birds further away from the noise source, or with
smaller critical ratios, would be represented by larger concentric circles. Figure 8.9b compares communication maps for budgerigars with canaries, which
have shown unusually large critical ratios at 2 kHz measured under laboratory
conditions. Taking such data into the field, we see that the communication range at
2 kHz is severely reduced for canaries (right).

To be sure, estimating the effective communication distance of acoustic communication signals between birds in the field is complicated by the many factors
that can contribute to the masking or release from masking of these signals. For
instance, noise in natural environments is rarely continuous (Klump 1996), and
birds (Chap. 7) as well as other animals (Chaps. 3 and 5) may take advantage of

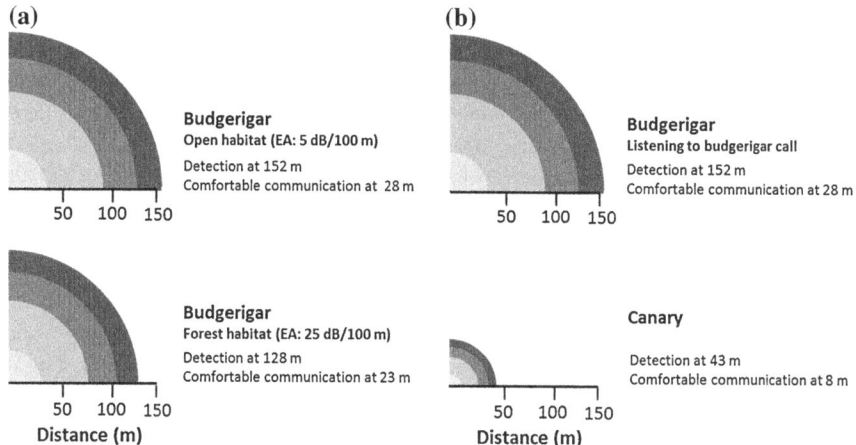

Fig. 8.9 Effect of low-level noise (60 dB SPL) on communication distance (**a**) for two different acoustic environments (open vs. forest habitat) in budgerigars, and (**b**) for two different species (budgerigars and canaries). Canaries, with much larger critical ratios, show a much restricted communication distance than budgerigars

gaps in noise to improve the signal-to-noise ratio. Birds also may use amplitude fluctuations affecting both signals and noise (Wiley and Richards 1982), to enhance their ability to detect signals in noise. Klump and his colleagues have shown that such spectrotemporal, or comodulation, Studies have shown that masking release can improve signal-to-noise ratios required for detection in birds by 10 dB or more (Klump and Langemann 1995; Dent et al. 1997).

Another mechanism to reduce masking is spatial release from masking, where the spatial separation of signal and noise source may be used to improve signal detection. Such spatial release from masking has been well studied in humans and other animals, including budgerigars (Saberi et al. 1991; Schwartz and Gerhardt 1989; Dent et al. 1997; Hine et al. 1994), and has been shown to similarly enhance signal detection abilities by 10–15 dB. In practical terms, this means that when the signal and the noise are coming from different locations, birds may be able to improve the signal-to-noise ratio in a natural environment simply by scanning or turning their heads.

Birds are also able to adjust the characteristics of their vocalizations in response to background noise (Chap. 7). The large-billed leaf-warbler (*Phylloscopus magnirostris*), which lives close to river torrents in the Himalayas, evades masking of its territorial songs by producing high-pitched notes in narrow frequency bands around 6 kHz (Dubois and Martens 1984). In fact, differences in song or call structure based on differences in habitat have been suspected and reported in a number of avian species (Douglas and Conner 1999; Slabbekoorn and Smith 2002; Slabbekoorn and Peet 2003; Nemeth and Brumm 2010; Mockford et al. 2011) such as for the songs of little greenbuls (*Andropadus virens*). However, it remains open at this point whether a given vocalization is adapted to environmental noise by evolutionary or by ontogenetic changes or both.

In response to increased noise levels, birds can adjust the amplitude of their vocalizations, a phenomenon also known as the Lombard Effect. A number of bird species have been shown to raise the level of their vocal output by as much as 10 dB in the presence of moderate background noise loud enough to affect the bird's perception of its own vocalizations (Potash 1972; Cynx et al. 1998; Manabe et al. 1998; Brumm and Todt 2002, 2003; Brumm 2009; Brumm and Naguib 2009; Brumm and Zollinger 2011; Schuster et al. 2012). Under exceedingly controlled conditions, laboratory birds trained to produce vocalizations for food while wearing headphones (Osmanski and Dooling 2006) raised the amplitude of vocal output by as much as 10 dB when noise was presented over headphones. Nightingale (*Luscinia megarhynchos*) males, too, sing louder in noisier territories, and birds in urban areas sing louder on work days than on weekend days when noise levels are reduced (Brumm 2004).

Location changes, which can often be observed both in senders and receivers, and changing the listening or signaling position can help counteract the effect of masking noise on acoustic communication. One strategy that will improve signal-to-noise ratio is to move to a position in the habitat in which the transmission pathway is better for the signal than the noise (Brumm and Slabbekoorn 2005). Thus, moving higher in the vegetation is one response that will improve the signal-to-noise ratio (Dabelsteen et al. 1993; Mathevon et al. 1996, 2005; Holland et al. 1998, Blumenrath and Dabelsteen 2004). For European blackbirds (*Turdus merula*) and great tits (*Parus major*), it is estimated that a receiver moving up from a low to a high perch at about 9 m results in an increase in audibility that is comparable to the receiver moving between one half to an entire territory diameter closer to the sender (Dabelsteen et al. 1993; Blumenrath and Dabelsteen 2004). Interestingly, this beneficial effect of perch elevation is more pronounced for receivers than for signalers.

A listener's ability to discriminate and recognize sounds is likely affected also by higher order cognitive processes, which relatively simple communication models as the one described above do not take into account. For instance, the pioneering work of Bregman and his colleagues (Bregman and Campbell 1971; Bregman 1990; Hartmann 1988; Bronkhorst 2000; Carlyon 2004) as well as many other human psychophysical studies have been pivotal in understanding auditory scene analysis as a solution to the so-called "cocktail party effect." They show that humans routinely segregate concurrent sounds into separate auditory objects, using a variety of cues such as common onset and offset times, common amplitude modulations, as well as similar frequency ranges, and spatial location (Bregman and Campbell 1971; Bregman 1990; Vliegen and Oxenham 1999). The formation of auditory objects is achieved by complex sound processing and cognitive integration of prior sensory experience and other simultaneous sensory input. The question of whether such an important phenomenon also occurs in nonhuman animals has been directly tested in starlings using both natural and synthetic stimuli. Hulse et al. (1997) showed that starlings could be trained to identify a sample of one species' bird song presented concurrently with a sample of another species' bird song. Moreover, these birds could learn to discriminate among many samples of the songs of two individual starlings and could maintain this discrimination when songs of a third starling were digitally added to both song sets and songs from additional starlings were added as further background distracters (Wisniewski and Hulse 1997; MacDougall-Shackleton et al. 1998). These results with starlings and results from many subsequent animal studies suggest that auditory scene analysis may also play an important role in auditory perception in birds and other nonhuman vertebrates that must parse the world into auditory objects (Fay 1998; Moss and Surlykke 2001; Hulse 2002; Izumi 2002; Barber et al. 2003; Bee and Klump 2004; Fishman et al. 2004; Bee and Klump 2005; Micheyl et al. 2005; Bee and Micheyl 2008; see also Chaps. 6 and 10).

While the reported thresholds with natural stimuli tested in the lab present something of an ideal case, they do go beyond traditional masking studies that use tones and white noise in providing data to estimate minimum distances over which calls can be transmitted in noise. An important caveat when considering the potential improvements in signal-to-noise ratio suggested above involves the environmental effects of the habitat on the signal. Vocalizations are altered when traveling through the habitat, and this acoustic degradation may change the characteristics important for evaluating detectability and discriminability (Wiley and Richards 1982; Dabelsteen et al. 1993; Naguib et al. 2000; Blumenrath and Dabelsteeen 2004). Signal degradation in the form of high-frequency attenuation, reverberation, and 'blurring' of amplitude and frequency patterns will alter a vocalization's spectral and temporal fine structure, affecting among other things the 'peakiness' of the waveform and thus its informational content (Wiley and Richards 1982; Dabelsteen et al. 1993; Dabelsteen 2005; Mockford et al. 2011). Some have suggested the Lombard Effect may have similar consequences (Brumm and Todt 2002; Brumm and Slater 2006). Using a nondegraded call in masked detection and discrimination tests may therefore provide a less accurate estimate of actual effective communication distance than a call re-recorded at biologically relevant distances in species-typical habitats. To fully understand the extent of effective communication ranges that allow vocally-mediated behaviors in animals, it is important to consider potentially synergistic effects of sound degradation and noise masking.

For instance, some of our most recent laboratory studies with small bird species that were tested with digitally reverberated stimuli suggest that discrimination of similar vocalizations from different individuals is significantly impaired when reverberation is paired with high abiotic noise levels, whereas neither reverberation nor noise alone had similarly detrimental effects (Blumenrath and Dooling 2011). Another experiment investigating how reverberation impacts the birds' ability to form auditory objects as described above showed that combining high levels of reverberation with biotic noise from simultaneously signaling conspecifics limit the birds' ability to segregate signals from multiple individuals (Blumenrath and Dooling 2012).

In summary, in order to predict whether and to what extent a given acoustic environment limits communication range or interferes with the detection, discrimination, and recognition of biologically important sounds it is important to consider effects caused by a combination of sound-altering habitat characteristics and species differences in both auditory sensitivity and cognitive processing.

8.5 Conclusions

The effects of masking noise on acoustic communication in the wild depends on: (1) the level of the noise but also its spectral composition, (2) the level and spectrum of the sender's vocalization at the receiver, and (3) the receiver's species-specific auditory capabilities. Noise within the spectral band of the signal, if it rises above ambient levels, can mask these communication signals thereby degrading or eliminating effective communication between individuals. In nature, the shape of the areas around the receiver demarcating different auditory effects as shown in this model would actually be irregular polygons reflecting habitat-specific differences in excess attenuation (e.g., ground effects and signal scattering in vegetation) as well as the relative locations of the two birds and the receiver's distance from the noise source. It is clear from this illustration that for birds communicating close to a noise source where noise levels are high, the area of the effective communication will be reduced. This approach of considering communication from the standpoint of the receiver may provide a useful metric for evaluating the actual noise impact on individuals, or collectively on populations, in areas subject to anthropogenic noise exceeding ambient levels. For instance, in determining risk to a species, the communication distances derived from this model might be considered in relation to other aspects of biology such as territory size.

Acknowledgments The work cited in this chapter was partially supported by NIH grants to RJD. We thank Marjorie Leek and Sue Anne Zollinger for comments on earlier versions of this chapter.

References

Adler HJ, Poje CP, Saunders JC (1993) Recovery of auditory function and structure in the chick after two intense pure tone exposures. Hearg Res 71:214–224

Au WW, Moore PW (1990) Critical ratio and critical bandwidth for the Atlantic bottlenose dolphin. J Acoust Soc Am 88:1635

Babisch W, Ising H, Gallacher JEJ (2003) Health status as a potential effect modifier of the relation between noise annoyance and incidence of ischaemic heart disease. Occup Environ Med 60:739–745

Barber J, Razak K, Fuzessery Z (2003) Can two streams of auditory information be processed simultaneously? Evidence from the gleaning bat Antrozous pallidus. J Comp Physiol A 189:843–855

Bee MA, Klump GM (2004) Primitive auditory stream segregation: a neurophysiological study in the songbird forebrain. J Neurophys 92:1088–1104

Bee MA, Klump GM (2005) Auditory stream segregation in the songbird forebrain: effects of time intervals on responses to interleaved tone sequences. Brain Behav Evolut 66:197–214

Bee MA, Micheyl C (2008) The cocktail party problem: what is it? How can it be solved? And why should animal behaviorists study it? J Comp Psychol 122:235–251

Bedanova I, Chloupek J, Chloupek P, Knotkova Z, Voslarova E, Pistekova V, Vecerek V (2010) Responses of peripheral blood leukocytes to chronic intermittent noise exposure in broilers. Reaktionen der peripheren Leukozyten auf die chronische Einwirkung von intermittierendem Lärm bei Broilern. Berl Münch Tierärztl 6:186–191

Békésy G (1960) Experiments in hearing. McGraw-Hill, New York

Blumenrath SH, Dabelsteen T (2004) Sound degradation before and after foliation: implications for acoustic communication in a deciduous forest. Behaviour 141:935–958

Blumenrath SH, Dooling RJ (2011) Communicating in Social Networks and Natural Environments: Effects of Reverberation and Noise. Abstract, 3rd symposium on acoustic communication by animals, Cornell University, Ithaca

Blumenrath SH, Dooling RJ (2012) Forming auditory objects in a reverberant cocktail party setting. Abstract, 10th International Congress of Neuroethology (ICN), International Society of Neuroethology, University of Maryland, College Park

Bregman AS (1990) Auditory scene analysis: the perceptual organization of sound. MIT Press, Cambridge

Bregman AS, Campbell JD (1971) Primary auditory stream segregation and perception of order in rapid sequences of tones. J Exp Psychol 89:244–249

Brenowitz EA (1982) The active space of red-winged blackbird song. J Comp Physiol A 147:511–522

Bronkhorst AW (2000) The cocktail party phenomenon: a review of research on speech intelligibility in multiple-talker conditions. Acustica 86:117–128

Brumm H (2004) The impact of environmental noise on song amplitude in a territorial bird. J Anim Ecol 73:434–440

Brumm H (2009) Song amplitude and body size in birds. Behav Ecol Sociobiol 63:1157–1165

Brumm H, Naguib M (2009) Environmental acoustics and the evolution of bird song. Adv Stud Behav 40:1–33

Brumm H, Slabbekoorn H (2005) Acoustic communication in noise. Adv Stud Behav 35:151–209

Brumm H, Slater PJ (2006) Animals can vary signal amplitude with receiver distance: evidence from zebra finch song. Anim Behav 72:699–705

Brumm H, Todt D (2002) Noise-dependent song amplitude regulation in a territorial songbird. Anim Behav 63:891–897

Brumm H, Zollinger A (2011) The evolution of the Lombard effect: 100 years of psychoacoustic research. Behaviour 148:1173

Buus S, Klump GM, Gleich O, Langemann U (1995) An excitation-pattern model for the starling (Sturnus vulgaris). J Acoust Soc Am 98:112–124

Carlyon RP (2004) How the brain separates sounds. Trends Cogn Sci 8:465–471

Carr CE, Code RA (2000) The central auditory system of reptiles and birds. In: Dooling RJ, Fay RR, Popper AN (eds) Comparative hearing: birds and reptiles. Springer, New York, pp 197–248

Cohen S, Glass DC, Phillips S (1979) Environment and health. In: Freeman H, Levine S, Reeder LG (eds) Handbook of medical sociology. Prentice-Hall, Englewood Cliffs, pp 134–149

Cynx J, Lewis R, Tavel B, Tse H (1998) Amplitude regulation of vocalizations in noise by a songbird, Taeniopygia guttata. Anim Behav 56:107–113

Dabelsteen T (2005) Public, private or anonymous? Facilitating and countering eavesdropping In: McGregor PK (ed) Animal communication networks. Cambridge University Press, Cambridge, pp 38–61

Dabelsteen T, Larsen ON, Pedersen SB (1993) Habitat-induced degradation of sound signals: quantifying the effects of communication sounds and bird location on blur ratio, excess attenuation and signal-to-noise ratio in blackbird song. J Acoust Soc Am 93:2206–2220

Dent ML, Larsen ON, Dooling RJ (1997) Free-field binaural unmasking in budgerigars (Melopsittacus undulatus). Behav Neurosci 111:590–598

Dooling RJ (1982) Auditory perception in birds. In: Kroodsma D, Miller E (eds) Acoustic Communication in Birds, vol 1. Academic Press, New York, pp 95–130

Dooling RJ, Dent ML, Lauer AM, Ryals BM (2005) Functional recovery after hair cell regeneration in birds. In: Salv RJ, Popper AN, Fay RR (eds) Hair cell regeneration, repair, and protection. Springer, New York, pp 117–140

Dooling RJ, Saunders JC (1975) Hearing in the parakeet (Melopsittacus undulatus): absolute thresholds, critical ratios, frequency difference limens, and vocalizations. J Comp Physiol Psych 88:1–20

Dooling RJ, Lohr B, Dent ML (2000) Hearing in birds and reptiles. In: Dooling RJ, Popper AN, Fay RR (eds) Comparative hearing: birds and reptiles. Springer, New York, pp 308–359

Dooling RJ, Dent ML, Lauer AM, Ryals BM (2008) Functional recovery following hair cell regeneration in birds. In: Salvi RJ, Popper AN, Fay RR (eds) Hair cell regeneration, repair and protection, vol 33. Springer Handbook of Auditory Research

Douglas Iii HD, Conner WE (1999) Is there a sound reception window in coastal environments? Evidence from shorebird communication systems. Naturwissenschaften 86:228–230

Dubois A, Martens J (1984) A case of possible vocal convergence between frogs and a bird in Himalayan torrents. J Ornithol 125:455–463

Ehret G (1975) Masked auditory thresholds, critical ratios, and scales of the basilar membrane of the housemouse (Mus musculus). J Comp Physiol A 103:329–341

Fay RR (1998) Auditory stream segregation in goldfish (Carassius auratus). Hearing Res 120:69–76

Fishman YI, Arezzo JC, Steinschneider M (2004) Auditory stream segregation in monkey auditory cortex: effects of frequency separation, presentation rate, and tone duration. J Acoust Soc Am 116:1656–1670

Foppen R, Reijnen R (2004) The effects of car traffic on breeding bird populations in woodland. II. Breeding dispersal of male willow warblers (Phylloscopus trochilus) in relation to the proximity of a highway. J Appl Ecol 31:95–101

Forman RTT, Reineking B, Hersperger AM (2002) Road traffic and nearby grassland bird patterns in a suburbanizing landscape. Environ Manage 29:782–800

Francis CD, Ortega CP, Cruz A (2009) Noise pollution changes avian communities and species interactions. Curr Biol 19:1415–1419

Francis CD, Ortega CP, Cruz A (2011a) Different behavioural responses to anthropogenic noise by two closely related passerine birds. Biol Lett 7:850–852

Francis CD, Ortega CP, Cruz A (2011b) Noise pollution filters bird communities based on vocal frequency. PLoS ONE 6:e27052

Freyaldenhoven MC, Smiley DF, Muenchen RA, Konrad TN (2006) Acceptable Noise Level: Reliability measures and comparison to preference for background sounds. J Am Acad Audiol 17:640–648

Gleich O, Dooling RJ, Manley GA (2005) Audiogram, body mass and basilar papilla length: correlations in birds and predictions for extinct archosaurs. Naturwissenshaften 92:595–598

Gleich O, Manley GA (2000) The hearing organ of birds and cocodilia. In: Dooling RJ, Popper AN, Fay RR (eds) Comparative hearing: birds and reptiles. Springer, New York, pp 70–138

Greenwood DD (1961a) Auditory masking and the critical band. J Acoust Soc Am 33:484–502

Greenwood DD (1961b) Critical bandwidth and the frequency coordinates of the basilar membrane. J Acoust Soc Am 33:1344–1356

Halfwerk W, Holleman LJ, Lessells CKM, Slabbekoorn H (2011) Negative impact of traffic noise on avian reproductive success. J Appl Ecol 48:210–219

Hartmann WM (1988) Pitch perception and the segregation and integration of auditory entities. In: Edelman GW, Gall E, Cowan WM (eds) Auditory function—neurobiological bases of hearing. Wiley, New York, pp 623–645

Hashino E, Sokabe M (1989) Kanamycin induced low-frequency hearing loss in the budgerigar (Melopsittacus undulatus). J Acoust Soc Am 85:289

Hashino E, Sokabe M, Miyamoto K (1988) Frequency-specific susceptivility to acoustic trauma in the budgerigar (Melopsittacus undulatus). J Acoust Soc Am 83:2450–2453

Holland J, Dabelsteen T, Pedersen SB, Larsen ON (1998) Degradation of wren Troglodytes troglodytes song: implications for information transfer and ranging. J Acoust Soc Am 103:2154–2166

Hulse SH (2002) Auditory scene analysis in animal communication. Adv Stud Behav 31:163–200

Hulse SH, MacDougall-Shackleton SA, Wisniewski AB (1997) Auditory scene analysis by songbirds: stream segregation of birdsong by European starlings (Sturnus vulgaris). J Comp Psychol 111:3–13

Hine JE, Martin RL, Moore DR (1994) Free-field binaural unmasking in ferrets. Behav Neurosci 108:196–205

Izumi A (2002) Auditory stream segregation in Japanese monkeys. Cognition 82:B113–B122

Klump GM (1996) Bird communication in the noisy world. In: Kroodsma DE, Miller EH (eds) Ecology and evolution of acoustic communication in birds. Cornell University Press, Ithaca, pp 321–338

Klump GM, Langemann U (1995) Comodulation masking release in a songbird. Hear Res 87:157–164

Klump GM, Windt W, Curio E (1986) The great tit's (Parus major) auditory resolution in azimuth. J Comp Physiol A 158:383–390

Konishi M (1973) How the owl tracks its prey. Am Sci 61:414–424

Langemann U, Klump GM, Dooling RJ (1995) Critical bands and critical-ratio bandwidth in the European starling. Hear Res 84:167–176

Langemann U, Gauger B, Klump GM (1998) Auditory sensitivity in the great tit: perception of signals in the presence and absence of noise. Anim Behav 56:763–769

Lohr B, Wright TF, Dooling RJ (2003) Detection and discrimination of natural calls in masking noise by birds: estimating the active space signal. Anim Behav 65:763–777

MacDougall-Shackleton SA, Hulse SH, Gentner TQ, White W (1998) Auditory scene analysis by European starlings (Sturnus vulgaris): Perceptual segregation of tone sequences. J Acoust Soc Am 103:3581

Manabe K, Sadr EI, Dooling RJ (1998) Control of vocal intensity in budgerigars (Melopsittacus undulatus): differential reinforcement of vocal intensity and the Lombard effect. J Acoust Soc Am 103:1190–1198

Marten K, Marler P (1977) Sound transmission and its significance for animal vocalization. I. Temperate habitats. Behav Ecol Sociobiol 2:271–290

Marten K, Quine D, Marler P (1977) Sound transmission and its significance for animal vocalization: II. Tropical forest habitats. Behav Ecol Sociobiol 2:291–302

Mathevon N, Aubin T, Dabelsteen T (1996) Song degradation during propagation: Importance of song post for the wren Troglodytes troglodytes. Ethology 102:397–412

Mathevon N, Dabelsteen T, Blumenrath SH (2005) Are high perches in the blackcap Silvia atricapilla song or listening posts? A transmission study. J Acoust Soc Am 117:442–449

Micheyl C, Tian B, Carlyon RP, Rauschecker JP (2005) Perceptual organization of tone sequences in the auditory cortex of awake macaques. Neuron 48:139–148

Miller JD (1974) Effects of noise on people. J Acoust Soc Am 56:729–763

Mockford EJ, Marshall RC, Dabelsteen T (2011) Degradation of rural and urban great tit song: testing transmission efficiency. PLoS ONE 6:e28242

Morton ES (1975) Ecological sources of selection on avian sounds. Am Nat 109:17–34

Moore BC (2003) An introduction to the psychology of hearing. Vol 4. San Diego: Academic press

Moss CF, Surlykke A (2001) Auditory scene analysis by echolocation in bats. J Acoust Soc Am 110:2207

Naguib M, Klump GM, Hillma E, Griessmann B, Teige T (2000) Assessment of auditory distance in a territorial songbird: accurate feat or rule of thumb? Anim Behav 59:715–721

Nelson DA, Marler P (1990) The perception of birdsong and an ecological concept of signal space. In: Stebbins WC, Berkley MA (eds) Comparative Perception, Vol 2: Complex Signals. Wiley, New York, pp 443–478

Nemeth E, Brumm H (2010) Birds and anthropogenic noise: are urban songs adaptive? Am Nat 176:465–475

Potash LM (1972) Noise induced changes in calls of the Japanese quail. Psychon Sci 26:252–254

Osmanski M, Dooling RJ (2006) Auditory feedback of vocal production in budgerigars using earphones. J Acoust Soc Am 119:3350

Ryals BM, Dooling RJ, Westbrook E, Dent ML, MacKenzie A, Larsen ON (1999) Avian species differences in susceptibility to noise exposure. Hear Res 131:71–88

Ryan MJ, Brenowitz EA (1985) The role of body size, phylogeny, and ambient noise in the evolution of bird song. Am Nat 126:87–100

Saberi K, Dostal L, Sadralodabai T, Bull V, Perrott DR (1991) Free-field release from masking. J Acoust Soc Am 90:1355–1370

Saunders JC, Duncan RK, Doan DE, Werner YL (2000) The middle ear of reptiles and birds. In: Dooling RJ, Popper AN, Fay RR (eds) Comparative hearing: birds and reptiles. Springer, New York, pp 13–69

Saunders JC, Dooling RJ (1974) Noise-induced threshold shift in the parakeet (Melopsittacus undulatus). Proc Natl Acad Sci USA 71:1962–1965

Schroeder J, Nakagawa S, Cleasby IR, Burke T (2012) Passerine birds breeding under chronic noise experience reduced fitness. PLOS ONE 7:e39200

Schuster S, Zollinger SA, Lesku JA, Brumm H (2012) On the evolution of noise-dependent vocal plasticity in birds. Biol Lett 8:913–916. doi:10.1098/rsbl.2012.0676

Schwartz JJ, Gerhardt HC (1989) Spatially mediated release from auditory masking in an anuran amphibian. J Comp Physiol A 166:37–41

Slabbekoorn H, Peet M (2003) Birds sing at a higher pitch in urban noise. Nature 424:267–269

Slabbekoorn H, Halfwerk W (2009) Behavioural ecology: noise annoys at community level. Curr Biol 19:R693–R695

Slabbekoorn H, Smith TB (2002) Habitat-dependent song divergence in the little greenbul: An analysis of environmental selection pressures on acoustic signals. Evolution 56:1849–1858

Smith A (1991) A review of the non-auditory effects of noise on health. Work Stress 5:49–62

Tanaka K, Smith CA (1978) Structure of the chicken's inner ear: SEM and TEM study. American Journal of Anatomy 153:251–271

Voslarova E, Chloupek P, Chloupek J, Bedanova I, Pistekova V, Vecerek V (2011) The effects of chronic intermittent noise exposure on broiler chicken performance. Anim Sci J 82:601–606

Vliegen J, Oxenham AJ (1999) Sequential stream segregation in the absence of spectral cues. J Acoust Soc Am 105:339

WHO Report 2011 Burden of disease from environmental noise - Quantification of healthy life years lost in Europe. ISBN: 978 92 890 0229 5 http://www.euro.who.int/__data/assets/pdf_file/0008/136466/e94888.pdf

Wisniewski AB, Hulse SH (1997) Auditory scene analysis in European starlings (Sturnus vulgaris): discrimination of song segments, their segregation from multiple and reversed conspecific songs, and evidence for conspecific song categorization. J Comp Psychol 111:337

Wiley RH, Richards DG (1982) Adaptations for acoustic communication in birds: sound transmission and signal detection. In: Kroodsma DE, Miller EH (eds) Acoustic communication in birds, vol 1. Academic Press, New York, pp 131–181

Zwicker E, Flottorp G, Stevens SS (1957) Critical band width in loudness summation. J Acoust Soc Am 29:548–557

Zwislocki J (1963) Analysis of some auditory characteristics. New York Lab of Sensory Communication, Syracuse University, New York

Chapter 9
Effects of Noise on Acoustic Signal Production in Marine Mammals

Peter L. Tyack and Vincent M. Janik

Abstract Marine mammals rely on sound for communication, orientation, and locating prey. Baleen whales use low-frequency sound, to frequencies below 10 Hz, to communicate over ranges of tens to hundreds of km. Toothed whales use clicks at center frequencies of 10–160 kHz to echolocate on targets at ranges of tens to a few hundreds of meters. Most marine mammals have sensitive enough hearing that they are limited by noise rather than the sensitivity of their auditory systems. Ocean noise is dominated by sounds of geological activity below about 20 Hz, by wind and waves above 200 Hz, but in the 20–200 Hz band, the dominant source of sound in the sea stems from a human source: the propulsion of ships. Other industrial and military activities also introduce very powerful, transient sounds into the oceans. As noise varies, the effective range for communication and echolocation would vary significantly if marine mammals did not have mechanisms to compensate for increased noise. Marine mammals have been shown to be able to compensate for noise by increasing the level of their own calls, by shifting their signal frequencies out of a noise band, by making their signals longer or more redundant, or by waiting to signal until noise is reduced. The mechanisms that involve modifying vocal output based upon auditory input have similarities with vocal production learning, and compensation for noise may have led to adaptations that close the neural loop between auditory input and vocal production. All of these mechanisms improve detection of signals in noise, but each is likely to incur costs, and it is not known whether they fully compensate for the effects of noise. At some levels of anthropogenic noise, animals leave an area near the source, reducing the amount of habitat available. As anthropogenic sound continues to increase in the ocean, the requirement for suitable conditions for communication means that effects of noise are one of the factors that must be monitored and regulated to maintain suitable environments for marine mammals.

P. L. Tyack (✉) · V. M. Janik
Scottish Oceans Institute, Sea Mammal Research Unit, School of Biology,
University of St Andrews, Fife KY16 8LB, UK
e-mail: plt@st-and.ac.uk

H. Brumm (ed.), *Animal Communication and Noise*,
Animal Signals and Communication 2, DOI: 10.1007/978-3-642-41494-7_9,
© Springer-Verlag Berlin Heidelberg 2013

9.1 Introduction

Noise can affect animals in a variety of ways. It can induce physiological and behavioral changes, and it can mask detection and recognition of vocal signals. Most receivers that evolved to detect distant signals are sensitive enough that they are limited by noise rather than the sensitivity of receptor organs. Many sensory systems have evolved remarkable sensitivity for detecting signals in noise (see Chaps. 3, 4, 6, 8, 10, 12, 13). If detection of a signal is noise-limited, then elevation of noise can mask detection of the signal, effectively reducing the range of communication or echolocation.

While the importance of noise for signal detection theory has been recognized in psychophysics for over half a century (Chap. 2), effects of noise have not been a central topic for behavioral ecology and ethology. This topic is not only an important area for basic research, but as researchers recognize how humans have elevated the ambient noise in many environments, it is becoming an important issue for conservation biology as well (Tyack 2008; Brumm 2010, see also Chap. 14). Within limits, vocalizing animals may be able to compensate for noise by increasing the level of their own calls, by shifting their signal frequencies out of the noise band, by making their signals longer or more redundant, or by waiting to signal until noise is reduced. Similarly, receivers have specific strategies to improve detection and recognition of sounds in noise (see Chap. 10). However, vocalizers as well as receivers may eventually leave areas of high noise levels. Ultimately, all of these changes are likely to incur costs and may not completely compensate for the noise.

There has been growing appreciation that terrestrial animals that use sound have developed mechanisms to compensate for noise (Chaps. 3, 5, 6, 7, 8), but because of the physics of sound propagation underwater, effects of noise are likely to have a greater impact on distance sensing and range of detection in aquatic environments. Sound propagates so much better than light in water that many aquatic animals have evolved ways to use sound as their primary distance sense to communicate and echolocate. Marine mammals have evolved mechanisms to use a wide range of frequencies of underwater sound. Not only does sound energy dilute as it spreads over a larger volume farther from the sound source, but the ocean also absorbs sound energy. Table 9.1 shows that the higher the frequency of sound, the shorter the range before half the sound energy is absorbed by passage through seawater. The lower absorption of low-frequency sound in the ocean has led large

Table 9.1 The distance it takes sounds of different frequencies to travel in the ocean before half of the sound energy is absorbed (from Tyack 1998)	Frequency (kHz)	Halving range (m)
	0.1	3×10^6
	1	3×10^4
	10	3,000
	40	300
	300	30

baleen whales (Mysticeti) to evolve mechanisms to produce and hear sound in the frequency range from less than 10 Hz to several hundred Hz. Absorption is trivial in the lower part of this frequency band. The low-frequency calls of whales can be detected at ranges of hundreds of kilometers, depending upon the ambient noise (Stafford et al. 1998). This supports a communication network of animals with no other means to maintain contact (Janik 2005). By contrast, some toothed whales (Odontoceti) have evolved specialized mechanisms to use high-frequency sounds to find and select prey with their sophisticated biosonar (Au 1993; Johnson et al. 2004; Madsen et al. 2005a). For sound to reflect energy efficiently from a rigid target, the wavelength of the sound must be less than or equal to the circumference of the target (Tyack 1998). Sound travels in water at nearly 1,500 m/s, which means that the wavelength to match a roughly 0.15 m circumference would have a corresponding frequency of 10 kHz or higher, that for a 0.015 m circumference would be 100 kHz or higher. Therefore, toothed whales must trade off the higher absorption of sound at high frequency against the greater efficiency of high-frequency sound as it reflects off of small targets. Toothed whales evolved high-frequency echolocation systems that can detect prey sized less than 1 m at ranges of tens to hundreds of meters (Madsen et al. 2007).

9.2 Mechanisms of Sound Production

Cetaceans share with other mammals a basic pneumatic mechanism for producing sound using air from the lungs. However, as diving mammals, they must conserve air; as they vocalize, air passes from the lungs (or a reservoir below the sound production organ—Wahlberg et al. 2005) through the sound production organ and is collected in sacs in the upper respiratory pathway, where it can be recycled back for further vocalization during the dive. As with most other mammals, baleen whales are thought to produce their sounds in the larynx. However, toothed whales are thought to produce their sounds as air passes through bony nasal nares. In both groups, some sounds are produced as air passes through a vibrating membrane that allows puffs of air through, creating oscillations in pressure that act as the source of the sound. The basic frequency of the sound produced by this mechanism is a function of how rapidly the air flows through the opening and closing of the constriction, which is influenced by the mass and tension or stiffness of the membrane and the pressure of the respiratory system causing the airflow. As in the syrinx of most birds and the larynx of most mammals, this mechanism can create sounds that are relatively tonal with harmonics, or sounds with more complex and "noisy" spectra.

The adaptations of odontocetes for high-frequency echolocation include hearing specialized for best frequencies in the 50–100 kHz region (see Chap. 10), and specialized organs for sound production. The sounds used for echolocation in odontocetes are short, high-frequency click sounds. These clicks are thought to be produced as air passes through "phonic lips" in the internal bony nares of the upper respiratory system and is collected in supracranial nasal air sacs between the nares and the blowhole (Cranford 2000). Odontocetes can produce individual pulses or series of pulses at rates of up to 800 clicks per second. Most odontocete species produce one or two typical echolocation pulse types, which are relatively stable as a function of water depth, even though some animals dive to depths of 150 atm or more, where the gas available for pneumatic sound production changes in density with a reduction in volume to 1/160th the original volume.

Pinniped sound production is less understood than mechanisms used by toothed whales. Generally, the larynx can be used to produce sounds, but animals often shift air between air sacs when producing sound underwater. It appears that the main sound production for underwater sound uses tracheal membranes and other parts of the respiratory tract. Walrus (*Odobenus rosmarus*) can also produce gong-like sounds by striking inflated pharyngeal sacs and can whistle by blowing through their lips (Tyack and Miller 2002).

9.3 Echolocation

Some acoustic characteristics of echolocation pulses of selected odontocete species are summarized in Table 9.2. The sperm whale (*Physeter macrocephalus*) in particular has evolved a powerful highly directional sonar for long distance echolocation. Some of the most intense and low-frequency toothed whale echolocation signals, such as those of the sperm whale, can detect echoes from larger targets such as the sea surface and seafloor at ranges of kilometers (Tyack 1997a; Zimmer et al. 2005b). More than one-third of the volume of the sperm whale is devoted to a sound production organ that has hypertrophied their head. Click

Table 9.2 Acoustic properties of echolocation clicks of selected odontocete species

Species	*Physeter catodon* regular	*Mesoplodon densirostris* regular	*Mesoplodon densirostris* buzz	*Tursiops truncatus*	*Phocoena phocoena*
Max. SL peak to peak (dB re 1 µPa @1 m)	236	213	–	228	157
Center freq (kHz)	15	38	51	120	128
Bandwidth (kHz)	5	25	55	30–60	16
Duration µs	52	271	104	50–80	150–300
Beamwidth -10 dB (degrees)	4	12	–	22	16
Ref.	Møhl et al.(2003)	Johnson et al. (2008); Schaffer et al. 2013	Johnson et al. (2008)	Au 1993	Au et al. (1999); Verboom and Kastelein (1997)

energy is produced by movement of air past phonic lips near the blowhole in the front of the head. This energy is primarily directed backwards through the spermaceti organ, where it reflects off of an air sac that overlies the skull, and then is directed forward in a beam only about 4° wide (Møhl et al. 2003). The peak source level of the on-axis click is about 230 dB re 1 μPa at 1 m, similar to that of a modern naval sonar.

The high-frequency components of clicks from most odontocete species project most of their energy forward in a directional beam, with half of the energy (−3 dB) within a beamwidth of <10°. Production of a directional sound allows for a higher effective source level in the direction of the signal. For example, if half of the energy of the sperm whale click is concentrated in a 4° radius, then the on-axis energy is 500 times higher than if the same energy were omnidirectional (Møhl et al. 2003). In addition, when the whale listens for echoes from the directional sonar, it only hears echoes from the direction of the target, reducing the interfering reverberation noise or clutter.

Sperm whales live in groups and often synchronize their deep foraging dives, so clicks of other whales may provide some of the most significant interference for this species. This may be one reason why sperm whales typically separate when they are foraging simultaneously at depth. Omnidirectional low-frequency components of these clicks are audible at ranges of 10 km or more, beyond the typical separation range. Sperm whales may thus be able to track one another's location and to reunite by eavesdropping on these clicks. These omnidirectional low-frequency components are probably below the frequencies used to detect prey, so may pose little interference for echolocation.

Other smaller odontocete species also use directional high-frequency signals for echolocation. The beaked whale species whose echolocation signals are known all produce several frequency modulated upsweeps each second with center frequencies about 40 kHz, bandwidth of 25kHz, duration of about 270 μs, and beamwidths of about 6° when they are searching for prey (Madsen et al. 2005a; Zimmer et al. 2005a). Once they switch from search mode to attempting to capture prey, Blainville's beaked whales (*Mesoplodon densirostris*) produce more rapid series of shorter clicks of ~100 μs duration, higher center frequency of 50 kHz, and broader bandwidth of 55 kHz. Dolphins of the delphinid family produce echolocation clicks with shorter durations and broad bandwidth similar to the rapid "buzz" clicks of beaked whales. Porpoises of the phocoenid family produce clicks that are longer in duration, higher in frequency, and narrower in bandwidth than those from any of the other odontocetes with the exception of delphinids of the genus *Cephalorhynchus* and the pygmy and dwarf sperm whales of the genus *Kogia* (Madsen et al. 2005b).

As discussed above, all toothed whales produce short sounds for echolocation, with durations on the order of tenths of a msec, and they are repeated with interclick intervals that are usually tens to hundreds of msec. This means that the duty cycle for vocalization is only about 1/100. It is generally thought that odontocetes time the next click to avoid any overlap between the echo from a target and their own emitted signal. As has been noted for bats (Schnitzler and

Kalko 2001), echolocating animals face significant problems separating the echo from a target from interfering signals. The short duration of odontocete clicks compared to the interclick interval can be seen as a mechanism for reducing self-noise and interference during echolocation.

Some bats are able to echolocate with a much higher duty cycle, in which the incoming echo from a target may overlap with the outgoing signal. They are able to avoid interference by producing narrow band signals, and adjusting the outgoing frequency so that the Doppler compensated echo falls in a very sensitive frequency for listening (Schnitzler and Kalko 2001). This kind of Doppler compensation has not been documented for odontocetes, but there is some evidence that odontocetes can shift the frequency of their echolocation signals to avoid noise that occurs in particular frequency bands. Echolocation signals from a beluga whale, *Delphinapterus leucas*, held captive in San Diego had center frequencies of 40−60 kHz. When this whale was moved to Hawaii to a site with loud noise from snapping shrimp, it increased the center frequencies of his clicks to 100−120 kHz where snapping shrimp noise is less intense (Au et al. 1985). When the beluga was trained to echolocate on a target in Hawaii, he also produced clicks on average 8.6 dB higher than when echolocating on the same target at the same range in San Diego. Thus, captive odontocetes can compensate for increased noise by increasing the level of their clicks and by shifting the clicks out of a noise band. There is a correlation between source level and frequency of echolocation clicks in some odontocetes (Au et al. 1995); so the increase in frequency might be a by-product of the increase in source level or vice versa (see Chap. 7 for a similar coupling in birds). Less is known about how echolocating odontocetes compensate for noise in the wild.

Many features of odontocete biosonar can be seen as mechanisms to reduce interference from noise in echolocation. These include highly directional signals (where the directionality also may correlate with source level and frequency, Kloepper et al. 2012), short durations with low duty cycles, and high frequencies that attenuate beyond the range of the targets for which the echolocation system is adapted.

9.4 Communication

Signals used for communication are under different design constraints than those used for echolocation. It is important to note that for a sound of a given level, the effective range of communication will be much higher than the range of echolocation. For a noise-limited receiver to detect a signal, the signal must just be above the noise level. In the case of echolocation, the sound travels to the target with some transmission loss, only a fraction of the sound energy impinging on the target is reflected back, and the returning sound has the same transmission loss. By contrast, in traveling from signaler to receiver a communication signal just undergoes the one-way transmission loss.

Some deep-diving species communicate using series of clicks that are similar to the clicks used for echolocation. For example, sperm whales produce rhythmic sequences of clicks, called codas, for communication (Watkins and Schevill 1977; Weilgart and Whitehead 1993), and Blainville's beaked whales produce rapid sequences of clicks, called rasps, to communicate at depth (Aguilar de Soto et al. 2012). Harbor porpoises (*Phocoena phocoena*) (Clausen et al. 2010), northern right whale dolphins (*Lissodelphis borealis*) (Rankin et al. 2007), and Hector's dolphins (*Cephalorhynchus hectori*) (Dawson 1991) are reported to produce stereotyped series of clicks in contexts that suggest they are used for communication. Here the timing of clicks differs from clicks used for echolocation and is critical for identifying the communication signals. Rather than having to wait to listen for echoes before producing the next click, an animal generating a stereotyped pattern of clicks for communication can produce more rapid series with little interference.

Other sounds used for communication by toothed whales are generally lower in frequency and less directional than those used for echolocation. Most delphinids, including river dolphins, and at least some ziphiid species produce frequency modulated narrow band harmonic signals called whistles. The fundamental frequency of whistles ranges from a few kHz to 30 kHz (Janik 2009), and duration ranges from about 0.1 s to several seconds. Delphinids also produce calls with broader short-term bandwidths, similar to the voiced calls of terrestrial mammals. The stereotyped calls of killer whales (*Orcinus orca*) are one of the best known examples. These calls include pulsed, broadband, and tonal whistle-like components (Ford 1989, 1991). Both whistles and stereotyped calls are thought to play a role in maintaining the cohesion of groups, especially when members of a group separate and need to reunite. The higher frequency components of both whistles and killer whale calls are more directional than the lower frequency components (Branstetter et al. 2012). Miller (2002) and Lammers and Au (2003) have hypothesized that by comparing the amount of energy in the high versus low components, a listener can determine whether the calling animal is moving toward or away. The effective range of these communication signals is thought to be in the range of several km to a few tens of kilometer (Janik 2000; Miller 2006), commensurate with the largest separations expected for conspecifics that share strong social bonds.

In contrast to the toothed whales, baleen whales have evolved communication signals that emphasize low frequencies. Baleen whales are the largest of animals, so have large enough sound production organs to generate sounds with long wavelengths and low frequencies. Baleen whales need low-frequency calls for long distance communication because they are both social and highly mobile. Many baleen whales have annual migrations of thousands of kilometer, and some species may disperse into low-latitude oceans during the breeding season. It is common for a migrating baleen whale to swim more than 100 km in a day (Mate et al. 1998). This puts a premium on the capability for long distance communication in these social oceanic animals, where sound is the only way to communicate at ranges greater than tens of meters.

Baleen whales can use low-frequency sound for long distance communication because of a specific feature of how sound propagates in the ocean. As sound passes through seawater some of the acoustic energy is absorbed, and the higher the frequency, the more sound energy is lost through absorption (Urick 1983). As Table 9.1 shows, a 100 Hz sound would have to travel more than 3,000 km before half of the sound energy was absorbed, while a 40 kHz sound would only have to travel about 300 m before the same halving of energy. This means that if a whale were communicating with another whale hundreds of km away, the lower the frequency, the less sound energy it would take to deliver the same level to the receiver.

The lowest frequency whale calls come from whales that disperse into low-latitude oceans during their winter breeding seasons (Tyack 1986). Blue whales (*Balaenoptera musculus*) produce calls with fundamental frequencies from 8 to 25 Hz that can last more than 10 s (Stafford et al. 1998), and fin whales (*Balaenoptera physalus*) produce calls with fundamental frequencies near 15−30 Hz and durations of near 1 s (Watkins et al. 1987). In a path-breaking paper, Payne and Webb (1971) used the standard theory of how sound propagates underwater (described for example by Urick 1983) to argue that these 20 Hz calls would have been audible at ranges of 1,000 or more kilometer, and in some propagation conditions, audible across whole ocean basins. While Northrop et al. (1968) used a bottom-mounted array of hydrophones to detect 20 Hz calls at ranges reported to be >160 km, there was some skepticism in the 1970s about the ability of whales to communicate over such huge ranges. More recently, use of the United States Navy sound surveillance system has routinely demonstrated detection of blue and fin whale calls at ranges of hundreds of km (Stafford et al. 1998; Watkins et al. 2000).

9.4.1 Does Shipping Noise Interfere with Communication by Marine Mammals?

An important point raised by Payne and Webb (1971) was that changes in the ambient noise in the ocean could have a significant effect on the range at which low-frequency whale calls could be detected. As humans introduced motorized shipping over the past century or so, the propulsion noise of ships has caused a remarkable change in the global ambient noise of the deep ocean. Figure 9.1 shows a set of typical levels for deep ocean ambient noise measured during the 1960s (Urick 1983). Above about 200 Hz, the ambient noise is affected mainly by sea state, which is driven by wind speed. For normal variations in wind speed, the ambient noise from 200 to 10,000 Hz can vary by at least 20 dB or a factor of 100 in terms of energy level. This natural variation was part of the acoustic environment in which marine animals evolved their hearing capacity. However, between 20 and 200 Hz, the ambient noise in the modern ocean is dominated by the propulsion sound of ships, and this level is increasing (Ross 2005; Chap. 14). The

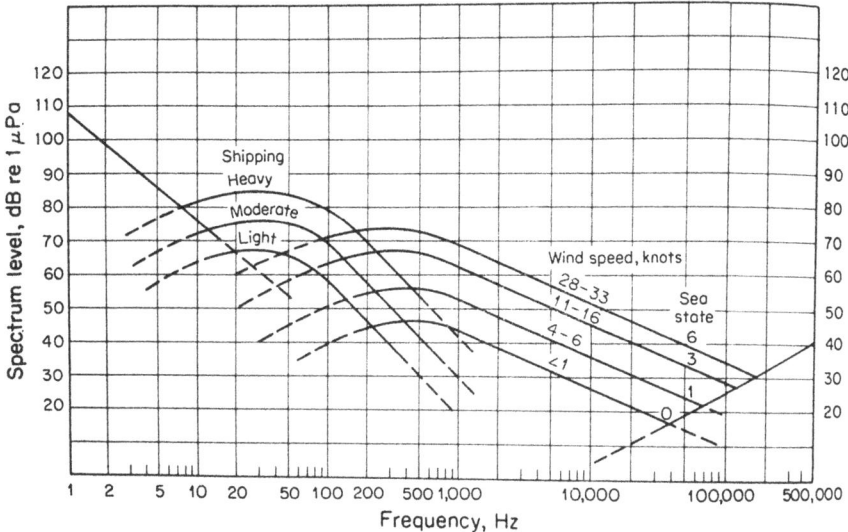

Fig. 9.1 Average deep sea noise levels as measured during the 1960s. The noise below about 20 Hz has natural causes. Shipping noise dominates the ambient noise from about 20–200 Hz. From 200 Hz to about 100 kHz, noise stems primarily from wind and waves. From Urick (1983)

various curves in this frequency region of Fig. 9.1 represent parts of the ocean with differing intensities of shipping.

Payne and Webb (1971) suggested that the introduction of shipping noise could reduce the range over which fin whales could communicate with their 20 Hz signals. They calculated a minimum detection range assuming poor sound propagation conditions of 90 km for a noise level at 20 Hz corresponding to moderate shipping in Fig. 9.1, and a range of about 280 km for a noise value corresponding to a pre-shipping ocean consistent with light shipping in Fig. 9.1, which is a relatively high value for natural ambient noise at 20 Hz judging by the figure.

One important condition for good sound propagation considered by Payne and Webb (1971) involves sound energy spreading in a duct in the deep ocean. Low-frequency sound can travel very efficiently when it refracts in this deep ocean sound channel. The deep ocean sound channel is supported by refraction in which sound bends toward a minimum sound speed at depth. Sound rays that are directed upward will often refract back downwards due to increasing speed caused by increasing temperature near the surface, and rays directed downward will often refract back upwards due to increasing speed caused by increasing pressure at depth. This means that the sound energy tends to concentrate at the depth of the minimum speed. The solid line in Fig. 9.2 illustrates how sound energy dissipates as it travels from a vocalizing whale. Imagine a whale is deep in the open ocean and it makes a sound that goes in all directions out to a range R. In this case, the sound energy dilutes as a $1/R^2$ function, which in decibel terms is—$10 \log R^2$ or $20 \log R$. Figure 9.2 plots this dilution from a rms source level of 180 dB re 1 µPa at

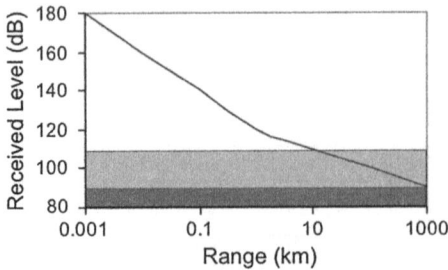

Fig. 9.2 The *solid line* illustrates how sound level reduces with range from a vocalizing fin or *blue* whale producing a call with a rms source level of 180 dB re 1 µPa at 1 m. The *dark rectangle* marks the estimated ambient noise level of the pre-industrial ocean. The line intersects this point at a range of 1,000 km, suggesting that whale calls could be detected in the pre-industrial ocean at ranges on the order of 1,000 km, the *lighter rectangle* marks how shipping noise has elevated the noise level in the frequency band of the calls of these whales by 20 dB, potentially reducing the detection range from 1,000 km to on the order of 10 km, where the higher noise level intersects the solid line

1 m from the whale out to 120 dB at 1,000 m from the whale. Now imagine that the sound switches from propagating in all three dimensions to concentrating energy in the sound duct, which can be approximated as two-dimensional or 1/R loss. In decibel terms, this is 10 log R, and Fig. 9.2 shows sound energy only decaying by 10 dB for every order of magnitude increase in range from this point on, down to a level of 90 dB at 1,000 km.

Figure 9.2 gives a generic illustration that the 180 dB whale call propagating as estimated by $1/R^2$ to 1 km and then 1/R loss beyond that would reach the 90 dB noise floor estimated for the pre-industrial ocean at a range of 1,000 km. Now if you raise the noise floor by 20 dB to the current estimated level of 110 dB, because this occurs in the zone of 1/R propagation loss, you reduce the estimated detection range from 1,000 to 10 km.

Payne and Webb (1971) used this kind of propagation modeling to calculate a detection range of 1,000 km for fin whale sounds in moderate shipping and 7,000 km in the ocean before motorized shipping. More modern acoustic models confirm reliable propagation of 20 Hz calls of fin whales well beyond 400 km (Spiesberger and Fristrup 1990). The minimum and maximum detection ranges calculated by Payne and Webb (1971) and illustrated in Fig. 9.2 are generic, and the actual detection range can be calculated with more accuracy for any specific set of propagation conditions (e.g., Spiesberger and Fristrup 1990). The general point is that the increase in ambient noise from shipping seems to have reduced the detectable range of low-frequency whale calls from many hundreds of km in the pre-propeller ocean down to tens of km in many settings today.

The noise estimates of Payne and Webb (1971) derive from the 1960s. Noise levels have continued to increase since the 1960s. Andrew et al. (2002) and McDonald et al. (2006) measured ambient sound from 1994 to 2004 at sites off California, and found that noise levels near 20 Hz were elevated by about

10−12 dB compared to the mid-60s. This elevation in ambient noise would reduce the minimum detection range estimated by Payne and Webb (1971) from 90 km in the 1960s to about 32 km now.

The primary uncertainty about the effect of this hypothesized reduction in range lies in our ignorance of the typical distance between a signaling whale and important receivers. Watkins and Schevill (1979) used an airplane to follow fin whales, and reported whales swimming 7−10 km to join a foraging group. Tyack and Whitehead (1983) reported a humpback whale (*Megaptera novaeangliae*) to stop singing and swim directly to a surface active group of whales 9 km away. It is possible, but was not demonstrated, that the approaching whales were responding to calls produced by the whales in these distant groups. While whales are likely to be able to detect calls at much greater ranges, we are not aware of any studies showing that whales communicate over ranges greater than this. However, while whales may not immediately react to distantly calling conspecifics, they could use such cues to find breeding or foraging grounds.

The costs associated with producing loud calls include the energy required for sound production and the risk that competitors, predators, or parasites may detect the call. These costs suggest that animals should be selected to produce sounds with source levels no higher than required for reliable communication over the ranges typical for important receivers. On the other hand, sexual selection may select for extreme values of advertisement displays. For acoustic displays, theory would suggest the possibility of selection for source levels much higher than required to detect the signal at typical distances of females monitoring song (Brackenbury 1979; Gil and Gahr 2002). These observations create problems for the argument that the required effective range for a signal must be the same as the actual range at which it can be detected by human acoustic sensors.

9.4.2 Do Marine Mammals Alter Their Vocal Behavior to Compensate for Noise?

As several chapters in this book argue, animals have evolved acoustic means to communicate and echolocate in the presence of natural ambient noise (see Chaps. 3, 4, 5, 6, 7, 8, 9, 10). Ambient noise in the ocean varies as a function of natural factors such as wind and waves or sounds of other animals, and anthropogenic factors such as shipping. We have just discussed a case where observed increases in noise could drastically reduce the range over which whales could communicate. This leads to the question of whether animals have evolved mechanisms to modify their vocal signals to compensate for changes in noise. Potential mechanisms for increasing the detectability of signals include waiting to call until noise decreases, increasing the rate of calling, increasing signal intensity, increasing the signal duration, and shifting signal frequency outside of the noise band. Most of these changes are likely to increase costs for signaling, so if animals show systematic use of compensation mechanisms, this would suggest that the noise is compromising effective communication sufficiently to make it worth the signaler incurring the cost. There is enough variation in natural sources of ambient noise that it is safe to assume that all animal communication systems evolved under conditions requiring adaptation to noise. For an aquatic example, Fig. 9.1 shows that wave noise varies as a function of wind speed, with nearly 30 dB of difference in the noise level at 1,000 Hz from calm seas to seas associated with 30 knot winds. Sounds of conspecifics or other animals can also dominate the ambient noise in areas where animals are likely to congregate. Fish, snapping shrimp, and marine mammals can elevate the ambient noise in coastal environments by 30 dB in specific frequency bands (Widener 1967; Cato and McCauley 2001). Conspecifics pose a particularly tough source of interference, as their signals usually overlap in frequency and some conspecifics may be actively competing with a signaler, attempting to reduce the salience of its calls or songs (Greenfield 1994).

The problem of communicating in a noisy channel is ubiquitous and important enough that it is likely to have created selection pressures for compensation mechanisms in most taxa that rely heavily upon sound for communication or echolocation. One of the simplest mechanisms involves waiting to signal until the noise level reduces, or timing vocalizations to minimize overlap with competing transient sounds. These mechanisms for timing calls with respect to interfering noise are well-developed in insects (Cade and Otte 1982, see also Chap. 3), anurans (Zelick and Narins 1983, see also Chap. 5) as well as birds (Ficken et al. 1974; Brumm 2006, see also Chap. 7) and terrestrial mammals (Egnor et al. 2007). Mechanisms for timing signals may have evolved particular sophistication when the signalers are competing for attention and the "noise" comprises competing signals from echoes or conspecific sounds (Greenfield 1994; Hall et al. 2006). As described above, echolocating toothed whales have evolved a low duty cycle system for echolocating, in which the outgoing signal is timed to minimize interference with capabilities for detecting the target echo. However, this effect has not been well-documented for communication among marine mammals. One example is the decrease in calling rate that we find when bottlenose dolphin (*Tursiops truncatus*) group sizes increases beyond 10 animals (Quick and Janik 2008). While calling rate would be expected to go up with more social interactions taking place in larger groups, individuals reduce their call rates when groups become very large. Whether marine mammals can time calls to minimize interference from intermittent noise is of practical importance for interpreting the potential effects of intermittent anthropogenic sound sources, such as airguns used for seismic survey. There is strong evidence that loud intermittent anthropogenic signals can cause some marine mammals to avoid a sound source at ranges of tens of km (Richardson et al. 1986; Morton and Symonds 2002), but we do not know whether marine mammals can reduce interference by timing calls so that they are received during quieter interludes. Harp seals (*Pagophilus groenlandicus*) (Terhune et al. 1979) appear to call less in the presence of vessel noise, and Blainville's beaked whales stop echolocating when exposed to sonar sounds (Tyack et al. 2011). However, it is unclear whether this is a strategy to optimize information transmission or a direct behavioral reaction to avoid detection in the presence of a threat. Similar reactions in response to killer whale sounds suggest that it might be the latter (Tyack et al. 2011). However, several cetacean species show the opposite pattern. Blue whales appear to call more during days that contain noise from seismic surveys (Di Iorio and Clark 2010). Long-finned pilot whales (*Globicephala melas*) whistled more during and immediately after exposure to low-level military sonar (Rendell and Gordon 1999). Bottlenose dolphins increase whistle rates when vessels approach (Buckstaff 2004). Groups of Pacific humpback dolphins (*Sousa chinensis*) that contained calves increased their whistling rates immediately after a boat passed within 1.5 km (van Parijs and Corkeron 2001). This effect was not found for groups without calves. Such an increase in calling could improve information transmission by introducing redundancy. Animals with social defenses against threats might be expected to call at a higher rate to increase cohesion when threatened, and the increased calling could also be viewed as a reaction to a perceived threat to facilitate group cohesion.

If the noise level is not changing rapidly enough, or if the animal cannot wait to get a signal through, then it can modify the acoustic structure of calls to compensate for the noise. One of the first such compensation mechanisms to be described is an increase in the source level of a vocalization as the noise level increases. This was described by Etienne Lombard in 1911 (Lombard 1911) and is known in psychophysics as the Lombard effect (Lane and Tranel 1971; Pick et al. 1989; Brumm and Zollinger 2011). More recent studies have demonstrated that several species of marine mammal in the wild, including beluga whales (Scheifele et al. 2005),West Indian manatees (*Trichechus manatus*–Miksis-Olds 2006) and right whales (*Eubalaena glacialis*—Parks et al. 2011) increase the source level of their calls when in the presence of elevated levels of shipping noise. Figure 9.3

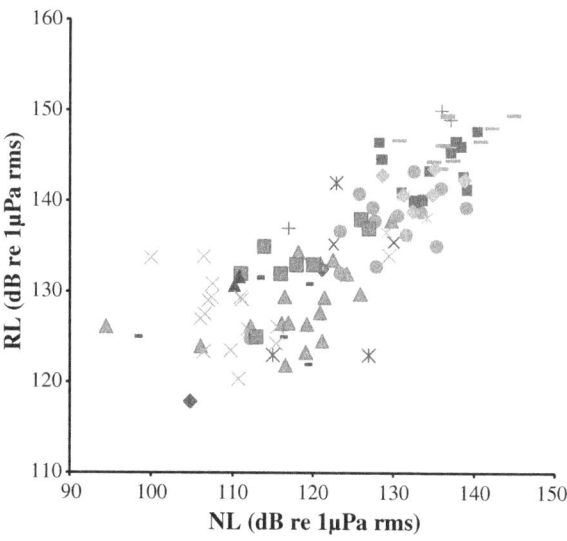

Fig. 9.3 Call received level versus noise level for 107 calls recorded from 14 tagged right whales. Data from each whale are presented with unique symbols (from Parks et al. 2011)

(from Parks et al. 2011) plots the source levels of calls of 14 right whales as a function of ambient noise measured at the same time. These data suggest that right whales modulate the level of each call they emit based upon the ambient noise level present at that time. Manatees are more likely to increase source level in noise when calves are present and when animals are dispersed, suggesting that they are particularly likely to incur more costly communication when they need to maintain contact with others with which they share a strong bond.

The observation in humans and several animal species that the Lombard effect is strongest for noise in the same frequency band as the vocalization frequency (e.g., Manabe et al. 1998) means that these species sense whether the interfering noise is in-band or not. Some animals respond to band-limited noise by changing the frequencies of their vocalizations to shift away from the noise. Just as some animals can wait to call until after a conspecific calls, avoiding interference in the time domain; so some animals can shift their call away from the frequency of a conspecific call, avoiding interference in the frequency domain. Some bats shift their echolocation calls away from the frequencies of conspecifics nearby; this is called a jamming-avoidance response (Ulanovsky et al. 2004). Serrano and Terhune (2002) have shown that when harp seals are calling at high rates during the breeding season, calls of different seals often overlap, and during these overlaps seals tend to produce calls that differ in frequency by more than one-third octave, the typical filter bandwidth of mammalian hearing. Terhune (1999) also proposed that Weddell seals (*Leptonychotes weddellii*) separate the frequency of their calls to avoid jamming.

This mechanism to avoid jamming by conspecific signals also appears to function for animals to avoid interference from band-limited noise. We discussed above a case when a beluga whale raised the frequency range and source level of its echolocation clicks effectively avoiding lower frequency noise (Au et al. 1985). Lesage et al. (1999) report an increase in the frequency of calls of beluga whales in the presence of low-frequency vessel noise. Bottlenose dolphins do not appear to change frequency or duration of their signature whistles when exposed to vessel noise (Buckstaff 2004). Since signature whistles carry identity information in their frequency modulation pattern (Janik et al. 2006), such changes would perhaps jeopardize the content of the signal.

Parks et al. (2007) document a remarkable long-term change in the frequency band of contact calls of North Atlantic right whales and South Atlantic right whales (*Eubalaena australis*), comparing low noise (1950s or South Atlantic) to high noise (present or North Atlantic) conditions (Fig. 9.4a). The average frequencies of these contact calls changed in the North Atlantic from 70 to 171 Hz in 1956 to 101−195 Hz in 2000−2004 and in the South Atlantic from 69 to 137 Hz in 1977 to 78−156 Hz in 2000. There was no significant difference between calls recorded in the North Atlantic in 1956 and the South Atlantic in 1977, but all other comparisons were highly significant (Fig. 9.4b). These results suggest that right whales have made long-term changes in the frequencies of their contact calls, apparently to compensate for increasing low-frequency shipping noise. Marine mammals have thus been demonstrated to have the capability to respond

Fig. 9.4 a Spectrograms of representative right whale contact calls from the South Atlantic (S. A.; *Eubalaena australis*) in 1977 and the North Atlantic (N. A.; *Eubalaena glacialis*) in 1956 and 2000. Notice the upwards shift in frequency in 2000 which represents almost a full octave change in start frequency. **b** Summary of start frequency differences between species and frequency differences over time for both species. Two asterisks (**) indicate $P < 0.001$; 2-way analysis of variance. From Parks et al. (2007)

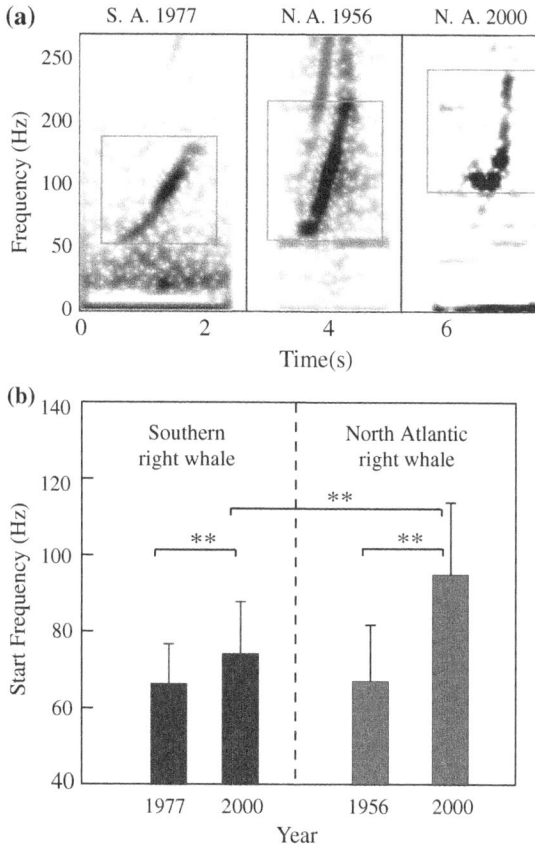

immediately to band-limited interference by shifting the frequency of their calls and also to gradually shift the frequency of a basic call type in the face of long-term changes in the spectrum of ambient noise.

Several animal taxa have been shown to increase the length of their calls in the presence of prolonged noise. Brumm et al. (2004) showed that a New World monkey, the common marmoset (*Callithrix jacchus*), lengthens its calls when exposed to white noise. Wieland et al. (2010) have shown that killer whales increased the length of some of their calls over the last 28 years, a period that coincided with an increase in noise caused by whale watching boats.

One of the predictions of communication theory (Shannon and Weaver 1949) is that the redundancy of signaling should increase as the channel becomes more noisy. As shown in Chap. 7, birds may increase the number of syllables in their calls or the bout duration of their songs with increasing noise. Examples for this increase in redundancy among birds include Japanese quail (*Coturnix coturnix japonica*—Potash 1972) and king penguins (*Aptenodytes patagonicus*—Lengagne et al. 1999). While nonhuman primates have not demonstrated such vocal flexibility (Brumm et al. 2004), humpback whales increased the repetitions of phrases in their songs when they were exposed to a low-frequency sonar (Miller et al. 2000; Fristrup et al. 2003). These responses have been interpreted as compensation to increase the ability of receivers to detect and classify signals in a noisy channel. Turnbull and Terhune (1993) have shown that a harbor seal (*Phoca vitulina*) can detect a regular series of calls at a lower signal to noise ratio than a single call alone, providing support on the receiver side for this interpretation.

9.5 Vocal Production Learning and Compensation for Noise

Some of the solutions to the problem of communication in varying noise, which is faced by all animals with sensitive hearing, may also involve vocal learning mechanisms. Janik and Slater (1997) and Tyack and Sayigh (1997b) argued that vocal learning in odontocetes most likely evolved in the context of individual recognition requirements in noisy environments. Vocal learning allows animals to create novel signals that are more diverse and distinctive than those in shared repertoires of nonlearners (Tyack 2000). Such diversity facilitates recognition of sounds in noise. Bottlenose dolphins appear to use vocal learning in such a context to create their individually distinctive signature whistles. These animals encode identity information in novel frequency modulation patterns (Janik et al. 2006) that they appear to develop by creating a variation on an existing signature whistle that they heard early in life (Fripp et al. 2005).

We have reviewed the evidence that animals have evolved mechanisms to compensate for varying ambient noise, including waiting to call until noise decreases, increasing the rate of calling, increasing signal intensity, increasing the signal duration, and shifting the frequency of a signal outside of the noise band. Even though the first two of these mechanisms indicate that auditory input modifies vocal behavior, they do not involve production learning by the definition of Janik and Slater (1997, 2000) because they do not involve modification of the acoustic features of vocalizations, but the last three, involving a shift in frequency out of the band of an auditory filter, or changing duration or intensity may qualify. A change in frequency would require control over the sound producing or phonatory system, a change in duration or intensity in its simplest form only requires control over the respiratory system (Janik and Slater 2000). While such changes using the respiratory system are relatively common in animals, control over the phonatory system to be able to produce novel calls is rare. By the traditional definition of vocal learning, however, our changes in response to noise would not qualify unless one could demonstrate that the signals have not appeared in this modified form beforehand. Vocal learning has been defined as requiring the acquisition of a novel signal through individual experience. Currently, we are unable to decide whether novelty occurs or not in these cases. However, even if the resulting signal is not novel, shifting parameters in response to noise could represent a pre-cursor to vocal learning.

The role of noise compensation mechanisms in the evolution of vocal learning may have been underestimated. It may also be that the genetic underpinnings for neural mechanisms linking vocal output to auditory input are widespread for ubiquitous problems such as changing signaling behavior to compensate for noise. Such relatively simple mechanisms have not normally been included in discussions of vocal production learning. But as long as they provide neural pathways to link auditory input with vocal motor output, they might form the substrate for evolution to work on in taxa that encounter niches with added uses for vocal learning.

9.6 Conclusions

Our review demonstrates that increased underwater noise causes marine mammals to alter the source level, frequency, duration, and redundancy of their signals. The evidence that marine mammals modify their calling behavior in response to anthropogenic noise also clearly suggests that it does interfere with their ability to echolocate and communicate. Several important questions follow from these observations: What are the costs of these compensation mechanisms? What are the limits of noise exposure beyond which animals cannot compensate? When does noise so degrade the usefulness of a habitat that animals leave? Can this level be predicted by the compensation behavior? What nonacoustic factors are important in predicting adverse effects of noise—e.g., what is the cost of missing a signal, are animals forced to change their distribution patterns in noise to maintain contact? Finally, very few data exist on how pinniped vocalizations change when they are exposed to noise. All of these are pressing questions that need to be addressed in the light of ever increasing noise levels in the oceans.

Acknowledgments Funding for the preparation of this review was provided by grants to PLT from the United States Office of Naval Research and support from the WHOI Marine Mammal Center. This work received funding from the MASTS pooling initiative (The Marine Alliance for Science and Technology for Scotland) and their support is gratefully acknowledged. MASTS is funded by the Scottish Funding Council (grant reference HR09011) and contributing institutions.

References

Aguilar de Soto N, Madsen PT, Tyack P, Arranz P, Marrero J, Fais A, Revelli E, Johnson M (2012) No shallow talk: cryptic strategy in the vocal communication of Blainville's beaked whales. Mar Mammal Sci 28:E75–E92
Andrew RK, Howe BM, Mercer JA (2002) Ocean ambient sound: comparing the 1960s with the 1990s for a receiver off the California coast. Acoust Res Lett 3:65–70

Au WWL (1993) The sonar of dolphins. Springer, Berlin

Au WWL, Carder DA, Penner RH, Scronce BL (1985) Demonstration of adaptation in beluga whale echolocation signals. J Acoust Soc Am 77:726–730

Au WWL, Pawloski JL, Nachtigall P, Blonz M, Gisiner RC (1995) Echolocation signals and transmission beam pattern of a false killer whale (*Pseudorca crassidens*). J Acoust Soc Am 98:51–59

Au WWL, Kastelein RA, Rippe T, Schooneman NM (1999) Transmission beam pattern and echolocation signals of a harbor porpoise (*Phocoena phocoena*). J Acoust Soc Am 106:3699

Brackenbury JH (1979) Power capabilities of the avian sound-producing system. J Exp Biol 78:163–166

Branstetter BK, Moore PW, Finneran JJ, Tormey MN, Aihara H (2012) Directional properties of bottlenose dolphin (*Tursiops truncatus*) clicks, burst-pulse and whistle sounds. J Acoust Soc Am 131:1613–1621

Brumm H (2006) Signalling through acoustic windows: nightingales avoid interspecific competition by short-term adjustment of song timing. J Comp Physiol A 192:1279–1285

Brumm H (2010) Anthropogenic noise: implications for conservation. In: Breed MD, Moore J (eds) Encyclopedia of animal behavior. Academic, Oxford, p. 89–93

Brumm H, Zollinger SA (2011) The evolution of the Lombard effect: 100 years of psychoacoustic research. Behaviour 148:1173–1198

Brumm H, Voss K, Köllmer I, Todt D (2004) Acoustic communication in noise: regulation of call characteristics in a New World monkey. J Exp Biol 207:443–448

Buckstaff KC (2004) Effects of watercraft noise on the acoustic behavior of bottlenose dolphins, *Tursiops truncatus*, in Sarasota Bay, Florida. Mar Mammal Sci 20:709–725

Cade WH, Otte D (1982) Alternation calling and spacing patterns in the Weld cricket *Acanthogryllus fortipes* (Orthopterea; Gryllidae). Can J Zool 60:2916–2920

Cato DH, McCauley RD (2001) Ocean ambient noise from anthropogenic and natural sources in the context of marine mammal acoustics. J Acoust Soc Am 110:2751–2751

Clausen KT, Wahlberg M, Beedholm K, Deruiter S, Madsen PT (2010) Click communication in harbor porpoises *Phocoena phocoena*. Bioacoustics 20:1–28

Cranford TW (2000) In search of impulse sound sources in odontocetes. In: Au WWL, Fay RR, Popper AN (eds) Hearing by Whales and Dolphins. Springer Handbook of Auditory Research, vol 12. Springer, Berlin, pp 109–155

Dawson SM (1991) Clicks and communication: the behavioural and social contexts of Hector's dolphin vocalizations. Ethology 88:265–276

Di Iorio L, Clark CW (2010) Exposure to seismic survey alters blue whale acoustic communication. Biol Lett 6:51–54

Egnor SER, Wickelgren JG, Hauser MD (2007) Tracking silence: adjusting vocal production to avoid acoustic interference. J Comp Physiol A 193:477–483

Ficken RW, Ficken MS, Hailman JP (1974) Temporal pattern shifts to avoid acoustic interference in singing birds. Science 183:762–763

Ford JKB (1989) Acoustic behavior of resident killer whales (*Orcinus orca*) off Vancouver Island, British Columbia. Can J Zool 67:727–745

Ford JKB (1991) Vocal traditions among resident killer whales *Orcinus orca* in coastal waters of British Columbia. Can J Zool 69:1454–1483

Fripp D, Owen C, Quintana-Rizzo E, Shapiro A, Buckstaff K, Jankowski K, Wells R, Tyack P (2005) Bottlenose dolphin (*Tursiops truncatus*) calves appear to model their signature whistles on the signature whistles of community members. Anim Cog 8:17–26

Fristrup KM, Hatch LT, Clark CW (2003) Variation in humpback whale (*Megaptera novaeangliae*) song length in relation to low-frequency sound broadcasts. J Acoust Soc Am 113:3411–3424

Gil D, Gahr M (2002) The honesty of bird song: multiple constraints for multiple traits. Trends Ecol Evol 17:133–141

Greenfield MD (1994) Cooperation and conflict in the evolution of signal interactions. Ann Rev Ecol Syst 25:97–126

Hall ML, Illes A, Vehrenkamp SL (2006) Overlapping signals in banded wrens: long-term effects of prior experience on males and females. Behav Ecol 17:260–269

Janik VM (2000) Source levels and the estimated active space of bottlenose dolphin *Tursiops truncatus* whistles in the Moray Firth, Scotland. J Comp Physiol A 186:673–680

Janik VM (2005) Acoustic communication networks in marine mammals. In: McGregor PK (ed) Animal communication networks. Cambridge University Press, Cambridge, pp 390–415

Janik VM (2009) Acoustic communication in delphinids. Adv Study Behav 40:123–157

Janik VM, Slater PJB (1997) Vocal learning in mammals. Adv Study Behav 26:59–99

Janik VM, Slater PJB (2000) The different roles of social learning in vocal communication. Anim Behav 60:1–11

Janik VM, Sayigh LS, Wells RS (2006) Signature whistle contour shape conveys identity information to bottlenose dolphins. Proc Natl Acad Sci USA 103:8293–8297

Johnson MP, Tyack PL (2003) A digital acoustic recording tag for measuring the response of wild marine mammals to sound. IEEE J Ocean Eng 28:3–12

Johnson MP, Madsen PT, Zimmer WMX, Aguilar de Soto N, Tyack PL (2004) Beaked whales echolocate on prey. Proc Roy Soc Lond B 271:S383–S386

Johnson M, Madsen PT, Zimmer WMX, Aguilar de Soto N, Tyack PL (2006) Foraging Blainville's beaked whales (*Mesoplodon densirostris*) produce distinct click types matched to different phases of echolocation. J Exp Biol 209:5038–5050

Johnson M, Hickmott LS, Soto NA, Madsen PT (2008) Echolocation behaviour adapted to prey in foraging Blainville's beaked whale (*Mesoplodon densirostris*). Proc Roy Soc B 275:133–139

Kloepper LN, Nachtigall PE, Donahue MJ, Breese M (2012) Active echolocation beam focusing in the false killer whale, Pseudorca crassidens. J Exp Biol 215:1306–1312

Lammers MO, Au WWL (2003) Directionality of the whistles of Hawaiian spinner dolphins (*Stenella longirostris*): a signal feature to cue direction of movement? Mar Mammal Sci 19:249–264

Lane H, Tranel R (1971) The Lombard sign and the role of hearing in speech. J Speech Hear Res 14:677–709

Lengagne T, Aubin T, Lauga J, Jouventin P (1999) How do king penguins (*Aptenodytes patagonicus*) apply the mathematical theory of information to communicate in windy conditions? Proc Roy Soc Lond B 266:1623–1628

Lesage V, Barrette C, Kingsley MCS, Sjare B (1999) The effect of vessel noise on the vocal behavior of belugas in the St. Lawrence River estuary. Canada Mar Mammal Sci 15:65–84

Lombard E (1911) Le signe de l'élévation de la voix. Annales des maladies de l'oreille, du larynx du nez et du pharynx 37:101–119

Madsen PT, Johnson M, Aguilar de Soto NA, Zimmer WMX, Tyack PL (2005a) Biosonar performance of foraging beaked whales (*Mesoplodon densirostris*). J Exp Biol 208:181–194

Madsen PT, Carder DA, Beedholm K, Ridgway S (2005b) Porpoise clicks from a sperm whale nose: convergent evolution of toothed whale echolocation clicks? Bioacoustics 15:195–206

Madsen PT, Wilson M, Johnson M, Hanlon RT, Bocconcelli N, Aguilar de Soto NA, Tyack PL (2007) Clicking for calamari: toothed whales can echolocate squid (*Loligo pealeii*)? Aquat Biol 1:141–150

Manabe K, Sadr EI, Dooling RJ (1998) Control of vocal intensity in budgerigars (*Melopsittacus undulatus*): Differential reinforcement of vocal intensity and the Lombard effect. J Acoust Soc Am 103:1190–1198

Mate BR, Gisiner R, Mobley J (1998) Local and migratory movements of Hawaiian humpback whales tracked by satellite telemetry. Can J Zool 76:863–868

McDonald MA, Hildebrand JA, Wiggins SM (2006) Increases in deep ocean ambient noise in the Northeast Pacific west of San Nicolas Island, California. J Acoust Soc Am 120:711–718

Miksis-Olds JL (2006) Manatee response to environmental noise levels. University of Rhode Island, Kingston, Ph.D. dissertation

Miller PJO (2002) Mixed-directionality of killer whale stereotyped calls: a direction of movement cue? Behav Ecol Sociobiol 52:262–270

Miller PJO (2006) Diversity in sound pressure levels and estimated active space of killer whale sounds. J Comp Physiol A 192:449–459

Miller PJO, Biassoni N, Samuels A, Tyack PL (2000) Whale songs lengthen in response to sonar. Nature 405:903

Møhl B, Wahlberg M, Madsen PT, Heerfordt A, Lund A (2003) The monopulsed nature of sperm whale clicks. J Acoust Soc Am 114:1143–1154

Morisaka T, Connor RC (2007) Predation by killer whales (*Orcinus orca*) and the evolution of whistle loss and narrow-band high frequency clicks in odontocetes. J Evol Biol 20:1439–1458

Morton AB, Symonds HK (2002) Displacement of *Orcinus orca* (L.) by high amplitude sound in British Columbia. Canada ICES J Mar Sci 59:71–80

Northrop J, Cummings WC, Thompson PO (1968) 20 Hz signals observed in the central Pacific. J Acoust Soc Am 43:383–384

Parks SE, Clark CW, Tyack PL (2007) Short and long-term changes in right whale calling behavior: the potential effects of noise on acoustic communication. J Acoust Soc Am 122:3725–3731

Parks SE, Johnson M, Nowacek D, Tyack PL (2011) Individual right whales call louder in increased environmental noise. Biol Lett 7:33–35

Payne RS, Webb D (1971) Orientation by means of long range acoustic signalling in baleen whales. Annal NY Acad Sci 188:110–141

Pick HLJ, Siegel GM, Fox PW, Gerber SR, Kearney JK (1989) Inhibiting the Lombard effect. J Acoust Soc Am 85:895–900

Potash LM (1972) Noise-induced changes in calls of the Japanese quail. Psychon Sci 26:252–254

Quick NJ, Janik VM (2008) Whistle rates of wild bottlenose dolphins: influences of group size and behavior. J Comp Psychol 122:305–311

Rankin S, Oswald J, Barlow J, Lammers M (2007) Patterned burst-pulse vocalizations of the northern right whale dolphin, *Lissodelphis borealis*. J Acoust Soc Am 121:1213–1218

Rendell LE, Gordon JCD (1999) Vocal response of long-finned pilot whales (*Globicephala melas*) to military sonar in the Ligurian Sea. Mar Mammal Sci 15:198–204

Richardson WJ, Würsig B, Greene CR (1986) Reactions of bowhead whales, *Balaena mysticetus*, to seismic exploration in the Canadian Beaufort Sea. J Acoust Soc Am 79:1117–1126

Ross D (2005) Ship sources of ambient noise. IEEE J Ocean Eng 30:257–261

Scheifele PM, Andrew S, Cooper RA, Darre M, Musiek FE, Max L (2005) Indication of a Lombard vocal response in the St. Lawrence River beluga. J Acoust Soc Am 117:1486–1492

Schnitzler HU, Kalko EKV (2001) Echolocation behavior of insect-eating bats. Bioscience 51:557–569

Serrano A, Terhune JM (2002) Antimasking aspects of harp seal (*Pagophilus groenlandicus*) underwater vocalizations. J Acoust Soc Am 112:3083–3090

Shaffer JW, Moretti D, Jarvis S, Tyack P, Johnson M (2013) Effective beam pattern of the Blainville's beaked whale (*Mesoplodon densirostris*) and implications for passive acoustic monitoring. J Acoust Soc Am 133:1770–1784. doi: 10.1121/1.4776177

Shannon CE, Weaver W (1949) The mathematical theory of communication. University of Illinois Press, Urbana

Spiesberger JL, Fristrup KM (1990) Passive localization of calling animals and sensing of their acoustic environment using acoustic tomography. Am Nat 135:107–153

Stafford KM, Fox CG, Clark DS (1998) Long-range acoustic detection and localization of blue whale calls in the northeast Pacific. J Acoust Soc Am 104:3616–3625

Terhune JM (1999) Pitch separation as a possible jamming-avoidance mechanism in underwater calls of bearded seals (*Erignathus barbatus*). Can J Zool 77:1025–1034

Terhune JM, Stewart REA, Ronald K (1979) Influence of vessel noises on underwater vocal activity of harp seals. Can J Zool 57:1337–1338

Turnbull SD, Terhune JM (1993) Repetition enhances hearing detection thresholds in a harbor seal (*Phoca vitulina*). Can J Zool 71:926–932

Tyack PL (1986) Population biology, social behavior and communication in whales and dolphins. Trends Ecol Evol 1:144–150

Tyack PL (1997a) Studying how cetaceans use sound to explore their environment. Persp Ethol 12:251–297

Tyack PL, Sayigh LS (1997b) Vocal learning in cetaceans. In: Snowdon C, Hausberger M (eds) Social influences on vocal development. Cambridge University Press, Cambridge, pp. 208–233

Tyack PL (1998) Acoustic communication under the sea. In: Hopp SL, Owren MJ, Evans CS (eds) Animal acoustic communication. Springer, Berlin Heidelberg New York, pp 163–220

Tyack PL (2000) Dolphins whistle a signature tune. Science 289:1310–1311

Tyack P (2008) Implications for marine mammals of large-scale changes in the marine acoustic environment. J Mammal 89:549–558

Tyack PL, Miller EH (2002) Vocal anatomy, acoustic communication and echolocation. In: Hoelzel AR (ed) Marine mammal biology: a evolutionary approach. Blackwell, Oxford, pp 142–184

Tyack PL, Whitehead H (1983) Male competition in large groups of male humpback whales. Behaviour 82:132–154

Tyack PL, Zimmer WMX, Moretti D, Southall BL, Claridge DE, Durban JW, Clark CW, D'Amico A, DiMarzio N, Jarvis S, McCarthy E, Morrissey R, Ward J, Boyd I (2011) Beaked whales respond to simulated and actual navy sonar. PLoS ONE 6:e17009

Ulanovsky N, Fenton MB, Tsoar A, Korine C (2004) Dynamics of jamming avoidance in echolocating bats. Proc Roy Soc Lond B 271:1467–1475

Urick RJ (1983) Principles of underwater sound. Peninsula Publishing, Los Altos

van Parijs SM, Corkeron PJ (2001) Boat traffic affects the acoustic behaviour of Pacific humpback dolphins, *Sousa chinensis*. J Mar Biol Assoc UK 81:533–538

Verboom WC, Kastelein RA (1997) Structure of harbour porpoise (*Phocoena phocoena*) click train signals. In: Read AJ, Wiepkema PR, Nachtigall PE (eds) The biology of the harbour porpoise. De Spil, Woerden, pp. 343–363

Wahlberg M, Frantzis A, Alexiadou P, Madsen PT, Møhl B (2005) Click production during breathing in a sperm whale (*Physeter macrocephalus*). J Acoust Soc Am 118:3404–3407

Watkins WA, Schevill WE (1977) Sperm whale codas. J Acoust Soc Am 62:1485–1490

Watkins WA, Schevill WE (1979) Aerial observation of feeding behavior in four baleen whales: *Eubalaena glacialis*, *Balaenoptera borealis*, *Megaptera novaeangliae*, and *Balaenoptera physalus*. J Mammal 60:155–163

Watkins WA, Tyack P, Moore K, Bird J (1987) The 20-Hz signals of finback whales (*Balaenoptera physalus*). J Acoust Soc Am 82:1901–1912

Watkins WA, George JE, Daher MA, Mullin K, Martin DL, Haga SH, DiMarzio NA (2000) Whale call data for the North Pacific: November 1995 through July 1999 occurrence of calling whales and source locations from SOSUS and other acoustic systems. Woods Hole Oceanographic Institution Technical Report 2000–2002

Weilgart L, Whitehead H (1993) Coda vocalizations in sperm whales off the Galapàgos Islands. Can J Zool 71:744–752

Widener MW (1967) Ambient-noise levels in selected shallow water off Miami, Florida. J Acoust Soc Am 42:904–905

Wieland M, Jones A, Renn SCP (2010) Changing durations of southern resident killer whale (*Orcinus orca*) discrete calls between two periods spanning 28 years. Mar Mammal Sci 26:195–201

Zelick RD, Narins PM (1983) Intensity discrimination and the precision of call timing in two species of neotropical treefrogs. J Comp Physiol A 153:403–412

Zimmer WMX, Johnson M, Madsen PT, Tyack PL (2005a) Echolocation clicks of Cuvier's beaked whales (*Ziphius cavirostris*). J Acoust Soc Am 117:3919–3927

Zimmer WMX, Tyack PL, Johnson MP, Madsen PT (2005b) Three-dimensional beam pattern of regular sperm whale clicks confirms bent-horn hypothesis. J Acoust Soc Am 117:1473–1485

Chapter 10
Effects of Noise on Sound Perception in Marine Mammals

James J. Finneran and Brian K. Branstetter

Abstract For marine mammals, auditory perception plays a critical role in a variety of acoustically mediated behaviors, such as communication, foraging, social interactions, and avoidance of predators. Although auditory perception involves many other factors beyond merely hearing or detecting sounds, sound detection is a required element for perception. As with many other processes, sound detection may be adversely affected by the presence of noise. This chapter focuses on two of the most common manifestations of the effects of noise on sound detection: auditory masking and noise-induced threshold shifts. The current state of knowledge regarding auditory masking and noise-induced threshold shifts in marine mammals is reviewed, and perceptual consequences of masking and threshold shifts are discussed.

10.1 Introduction

Auditory perception may be defined as the ability to detect, interpret, and attach meaning to sounds. For marine mammals, auditory perception plays a critical role in a variety of acoustically mediated behaviors, such as communication, foraging, social interactions, and avoidance of predators. Auditory perception can play an important role in detecting objects in the environment, discriminating between objects, and identifying the location of objects. Auditory perception is also a key component in auditory scene analysis—i.e., segregating a mixture of sounds from

J. J. Finneran (✉)
US Navy Marine Mammal Program, Space and Naval Warfare Systems Center Pacific, 53560 Hull St, San Diego, CA 92152, USA
e-mail: james.finneran@navy.mil

B. K. Branstetter
National Marine Mammal Foundation, 2240 Shelter Island Dr, San Diego, CA 92106, USA

a complex natural environment into "auditory streams" produced from individual sources and attending to those streams of interest (Bregman 1990).

Although perception involves many other factors beyond merely hearing or detecting sounds, sound detection is a required element for perception. As with many other processes, sound detection may be adversely affected by the presence of noise. Because auditory perception plays a key role in so many vital tasks, noise that adversely affects sound perception could ultimately result in fitness consequences to the individual.

This chapter focuses on two of the most common manifestations of the effects of noise on sound detection: auditory masking and noise-induced threshold shifts. Masking can be described as a reduction in the ability to hear a sound caused by the presence of another sound. A noise-induced threshold shift is a reduction in auditory sensitivity following a noise exposure. Both masking and threshold shifts have the effect of reducing an animal's auditory sensitivity over some frequency bandwidth, with the key distinction between the two that masking essentially occurs during the noise exposure, while a threshold shift persists after cessation of the noise. Because both processes are heavily influenced by the function of the peripheral auditory system, we begin with a brief overview of the anatomy and function of the ear in marine mammals, followed by individual discussions of masking and noise-induced threshold shifts. The relevant literature in each area is reviewed and synthesized to present the current understanding of these phenomena in marine mammals. Finally, some conclusions are presented and directions for future research proposed.

10.2 The Peripheral Auditory System in Marine Mammals

As in terrestrial mammals, the peripheral auditory system of marine mammals includes the external (outer) ear, middle ear, and inner ear. The external ear includes the pinnae (if present), the external auditory meatus (ear canal), and the tympanic membrane. The external ears of marine mammals exhibit a variety of adaptations from their terrestrial ancestors. The pinnae are absent in all cetaceans, and the external auditory meatus appears to be vestigial in most cetaceans (Ridgway 1999). The external ear pinna is small in otariid pinnipeds, but absent in phocids, odobenids, and sirenians (Nummela 2008b). For echolocating odontocetes, high frequency sounds are received through specialized fatty tissues in the lower jaws that offer a path to the ear (Ketten 2000; Nummela 2008b; Popov et al. 2008), thus these structures may also be considered as part of the external ear in these species. The ear of delphinoid cetaceans, unlike other species including Physeteridae, Kogiidae, and Ziphiidae, is suspended in the enlarged, air-filled peribullar space by fibrous bands with no bony connection to the skull (Ketten 2000). This suspension acoustically isolates each ear from the skull (McCormick et al. 1970).

The middle ear includes three small bones, the malleus, incus, and stapes that link the tympanic membrane to the fluid-filled cochlea of the inner ear. In odontocetes, the ossicular chain is more massive than in land mammals, but also stiffer, resulting in the middle ear apparatus being tuned to a higher frequency (Ketten 2000). In delphinoid cetaceans the malleus is not in direct contact with the tympanic membrane, but there is a large tympanic ligament that contacts the malleus. In mysticetes, the ossicles are also massive but apparently lack the stiffening elements, suggesting a lower frequency response (Ketten 2000). The middle ear ossicles are enlarged in sirenians, phocids, and odobenids; however, otariid middle ear ossicles are of similar size to terrestrial carnivores (Nummela 2008a, b).

Vibrations of the stapes are transmitted to the basilar membrane and organ of Corti located within the cochlea. The organ of Corti contains four rows of delicate mechanosensory hair cells: three rows of outer hair cells and one row of inner hair cells. Motion of the stapes causes fluid motion within the cochlea, which results in displacement of the basilar membrane, and deflection of the hair cell stereocilia. The inner hair cells generate neural impulses when their ciliary bundles are deflected, and thus provide the main neural output from the cochlea to the brain. In contrast, the outer hair cells have a motor function, and change their shape and stiffness in response to neural signals from the brain. The outer hair cells may therefore influence the mechanics of the cochlea, and form part of an active mechanical preamplifier which enhances the performance of the auditory system (de Boer and Nuttall 2010).

The mechanical properties of the basilar membrane vary along the length of the cochlea, from high stiffness near the base (where the stapes is attached), to lower stiffness at the apex. This results in a frequency-dependent vibration pattern of the basilar membrane with the basal portion responding best to high frequencies and the apical portion responding best to lower frequencies. For any specific location on the basilar membrane, there will be some frequency that produces a maximum vibration amplitude; lower frequencies will still displace the membrane (though with smaller amplitude) and higher frequencies will produce very little displacement at that location. Different populations of inner hair cells thus respond preferentially to different frequencies, depending on the physical position of the hair cell along the length of the basilar membrane. An inner hair cell is thus said to be "tuned" to a certain frequency, called the characteristic frequency, depending upon its location along the basilar membrane; hair cells near the cochlear base have higher characteristic frequencies than those located near the apex. The frequency-dependent basilar membrane motion and hair cell tuning therefore result in a frequency-to-place mapping within the cochlea. This mechanism is often referred to as the auditory filter, since, for a given nerve fiber, the cochlea performs band-pass filtering.

Hair cell tuning arises from two mechanisms: a passive component arising from the mechanical properties of the basilar membrane, and an active component that arises from outer hair cell motility. The passive component results in relatively broad tuning while the active component "sharpens" tuning by increasing the vibration amplitude over a narrow range of frequencies. As the received sound pressure level (SPL) increases, the relative contributions between the active and passive processes change, with the passive process becoming more dominant. The result is a broadening of hair cell tuning, or auditory filter width, at higher sound levels (Anderson et al. 1971; Moore and Glasberg 2003).

Compared to terrestrial species, the inner ears of marine mammals are functionally analogous, but differ in the contact with bones of the skull (fibrous suspension or bony connection), cochlear dimensions, basilar membrane length, thickness, and stiffness, hair cell densities, and innervation. This results in species-dependent parameters for the audible frequency range.

In summary, the sensation of hearing in marine mammals, results from sound conducted via the head to the cochlea. In many species the conduction chain is via the external and middle ear, while in delphinoid cetaceans experimental data suggest that transmission of sound is via the fat body of the lower jaw directly to the stapes or inner ear (McCormick et al. 1970). In all species, vibration of the basilar membrane causes deflection of the inner hair cell stereocilia and the generation of neural impulses. Although there are many species-specific differences and significant peripheral auditory system adaptations from land mammals, the inner ears of marine mammals are functionally analogous to those of land mammals, with the most substantial differences concerning the frequency range of hearing. As in land mammals, the complex, frequency-specific vibration patterns of the basilar membrane, the tuning characteristics of the hair cells, and the role of the outer hair cells in active cochlear amplification have a profound impact on the perception of sound. These factors also figure prominently in the discussion of auditory masking and noise-induced threshold shifts.

10.3 Auditory Masking

Auditory masking occurs when one sound (usually called noise) interferes with the detection, discrimination, or recognition of another sound (usually called the signal). Although well-studied in humans, only basic auditory masking studies related to signal detection have been performed on marine mammals due to animal availability and the difficulties associated with training an animal to perform a psychophysical hearing test. Of the few masking experiments performed on marine mammals, most are of the type where the animal is required to detect a tonal signal in the presence of another tone or broadband Gaussian noise. The results of these experiments can usually be explained within the framework of the power spectrum model of masking (described in detail below) and represent an important first step in understanding auditory masking in these animals. More recent experiments using complex and realistic sounds (both signal and noise) suggest that descriptions of auditory masking in marine mammals, like in humans, cannot be reduced to metrics exclusively related to frequency and SPL. At the very minimum, the temporal patterns of sounds, as well as the location of the sounds relative to each other, also play important roles in describing how two or more sounds are seg-regated in a complex auditory scene (for similar phenomena in anurans and birds, see Chaps. 6 and 8).

10.3.1 Signal Detection in Noise

10.3.1.1 Tone-on-Tone Masking

A bottlenose dolphin's (*Tursiops truncatus*) ability to detect a tonal signal (the "probe") in the presence of another tonal signal (the "masker") was first inves-tigated by Johnson (1971). In this experiment, behavioral thresholds for a 70 kHz probe tone were estimated in the presence of a masking tone where the frequency and SPL of the masker were independent variables. The masking pattern was similar to what is found in humans in that, (1) more masking occurred when the probe and masker frequencies were similar, (2) lower masker frequencies had a greater masking effect than higher masking frequencies, and (3) higher SPL noise masked a broader range of frequencies than lower SPL noise. As with humans, when the masker and probe frequencies were very similar, detection thresholds actually decreased rather than increased (Fig. 10.1). In humans, this threshold decrease was associated with the perception of "beats." Presumably, when both the probe and masker tones fall within a single auditory filter, listeners no longer perceive two tones, but instead, a single amplitude-modulated tone with a mod-ulation rate equal to the frequency difference between the tones. The dolphin, like humans, might have also perceived beats and used this cue for signal detection.

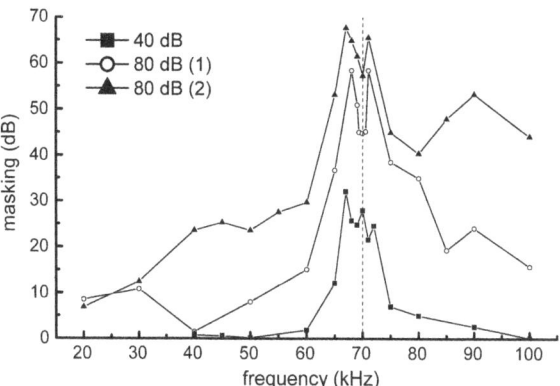

Fig. 10.1 Two tone masking [adapted from Johnson (1971)]. The vertical line indicates the frequency of the probe tone. Symbols indicate the threshold of the probe tone in the presence of the masker tone at various frequencies and SPLs. The 80 dB re 1 µPa masker was repeated with different results, apparently this difference reflected learning by the dolphin

Neurophysiological techniques have also been used to measure frequency tuning curves in a number of odontocetes using the tone-on-tone masking paradigm (Popov et al. 1996; Supin and Popov 1986). In these studies, the addition of a tonal masker was found to suppress the evoked response to a tonal probe much in the same way that tonal maskers affect the detectability of tones in psychophysical experiments. For short duration tone-pip stimuli, masker frequencies below the tone-pip frequency produced a tuning curve with an average slope of 52 dB/octave. For masker frequencies higher than the tone-pip frequency, the average slope of the tuning curve was 96 dB/octave, almost twice as steep as that of lower frequency maskers. A common feature of the above studies is that lower frequency maskers appear to have a greater masking effect on higher frequency tones than vice versa. This result is directly related to basilar membrane mechanics discussed earlier. When the basilar membrane is excited by two or more tones of different frequencies, the traveling wave of the lower frequency tone will propagate through the higher frequency regions thus causing a greater masking effect on the higher frequency even when the frequency separation is relatively large.

10.3.1.2 Critical Bands and Critical Ratios

Fletcher (1940) conducted a series of seminal experiments with human listeners that have been repeated with several animal species including a few odontocetes and pinnipeds. Using a band-widening paradigm, Fletcher discovered that thresholds for a tonal signal centered in band-limited Gaussian noise increased proportionally with the bandwidth of noise, but only up to a certain "critical bandwidth." Noise bandwidths beyond this critical bandwidth no longer contributed to the masking of the signal. To account for this result, Fletcher envisioned the auditory system behaving as a series of continuously overlapping band-pass filters, where masking only occurred if the signal and the masker were within a common auditory filter or *critical bandwidth* (CB). Because of this relationship, the bandwidth of a hypothetical auditory filter can be estimated by simply measuring tonal thresholds in broadband noise, since only the noise within an auditory filter centered on the signal will effectively mask the signal. If the power spectral density of the noise, N, and the power of the signal at threshold, S_{th}, are known, the CB is given by

$$\Delta F_{CB} = S_{th}/(K \cdot N), \qquad (10.1)$$

where ΔF_{CB} is the CB and K is a constant. If K is assumed to equal 1, the equation simplifies to

$$\Delta F_{CR} = S_{th}/N, \qquad (10.2)$$

where ΔF_{CR} is called the *critical ratio* (CR). The CR expressed as a frequency level, in dB re 1 Hz, is calculated by subtracting the noise pressure spectral density

Fig. 10.2 Critical ratios
measured in different
odontocete species

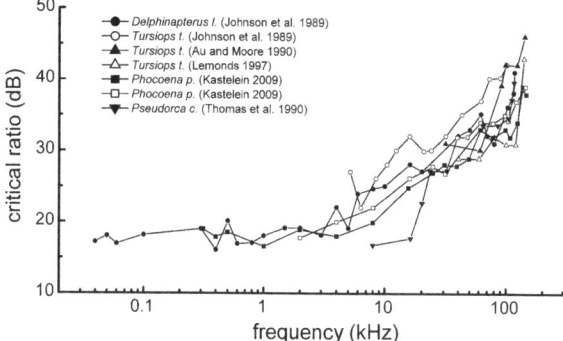

level (L_N, in dB re 1 μPa^2/Hz) from the signal SPL at threshold (L_S, in dB re
1 μPa):

$$L_{CR} = L_S - L_N. \tag{10.3}$$

For example, a CR of 20 dB re 1 Hz (equivalent to 100 Hz) states that the
signal must be 20 dB greater than the noise spectral density level of the masker to
be detected. This simple metric is most commonly used to predict masking effects
of noise found in a marine mammal's environment (e.g., anthropogenic noise, see
Chap. 14). Compared to the band-widening technique used to estimate CBs, CRs
require only a fraction of the time and effort with respect to data collection. As a
result, CRs have become a standard first step at understanding auditory masking in
many marine mammal species.

Critical ratios for several odontocete cetaceans demonstrate a similar pattern of
masking in which more masking occurs at high frequencies, presumably because
of the increasing bandwidth of auditory filters at higher frequencies (Fig. 10.2).
CRs appear flat for signal frequencies of 1 kHz and below. Critical ratios for
pinnipeds also demonstrate an increase as a function of signal frequency for both
underwater and airborne sounds (Fig. 10.3). CRs and CBs for both odontocetes

Fig. 10.3 Critical ratios
from different pinniped
species

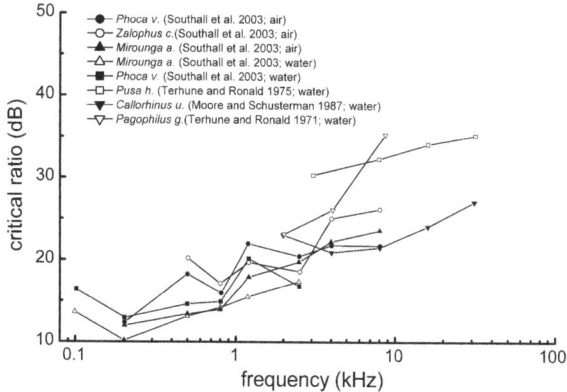

and pinnipeds suggest that auditory filter bandwidths increase as a function of the center frequency of the filter.

10.3.1.3 The Power Spectrum Model of Masking and the Auditory Filter

Fletcher's (1940) original concept of an auditory filter bank developed into what is now referred to as the power spectrum model (PSM) of auditory masking (Patterson and Moore 1986). The model makes the following assumptions:

(1) The auditory system can be modeled as a series of continuously overlapping band-pass filters.
(2) Only the spectral components of a noise masker that are within a filter centered on the signal frequency will effectively mask the signal.
(3) Signal detection is accomplished by monitoring an energy detector at the output of the filter centered on the signal. More energy will be present in a signal-plus-noise interval than a noise-alone interval.
(4) Signal thresholds are proportional to the noise power that passes through a single auditory filter. Noise is represented by its long-term spectrum.

Formally, the PSM can be expressed as:

$$P_s = K \int_{-\infty}^{\infty} N(f)W(f)\mathrm{d}f, \qquad (10.4)$$

where P_s is the power of the signal at threshold, $N(f)$ is the noise power spectral density and $W(f)$ is a weighting function described by the shape of the auditory filter. Auditory filter shapes have been derived for bottlenose dolphins (Finneran et al. 2002a; Lemonds 1999) and a beluga (*Delphinapturus leucas*, Finneran et al. 2002a) using a behavioral response, notched-noise masking paradigm (Patterson 1976). An assumption is made that the auditory filter shape can be estimated by a simple-rounded exponential function (roex) with a limited number of free parameters. In both Finneran et al. (2002a) and Lemonds (1999) a two-parameter, roex (p,r) function was used:

$$W(g) = (1 - r)(1 + pg)\,e^{-pg} + r \qquad (10.5)$$

where g is the normalized frequency deviation [$g = |f-f_o|/f_o$, where f is frequency and f_0 is the signal frequency], and p and r are adjustable parameters. Common features of the auditory filters are that bandwidths increase with both increased noise level and increased center frequency. The relationship between bandwidth and center frequency of the filter can be described by the quality factor, Q:

$$Q = f_o/\Delta f, \qquad (10.6)$$

Fig. 10.4 Roex auditory filter banks for **a** *Tursiops truncatus*, **b** *Delphinapterus leucas*, and **c** *Phocoena phocoena*

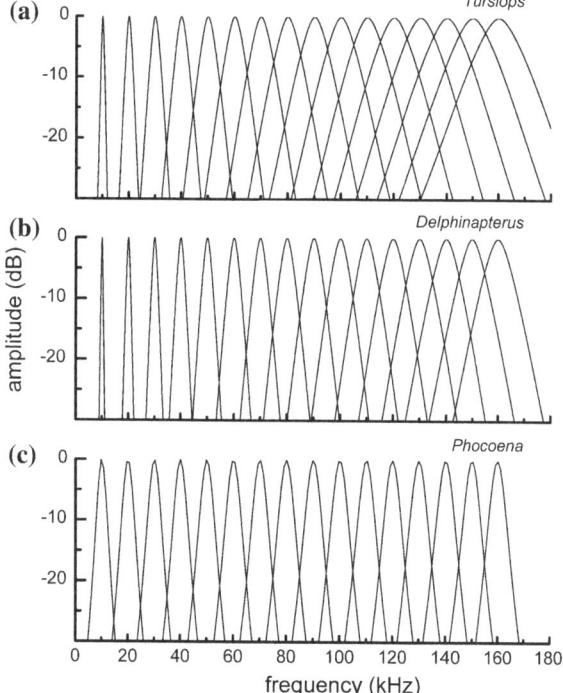

where f_o is the frequency of the signal and Δf is the filter bandwidth. For many mammals, the entire auditory periphery can be reasonably approximated using the same value for Q (constant-Q filters). Auditory filter Q values tend to vary depending on the methodology used to estimate thresholds. For example, Q values of 2.2 and 12.3 were estimated for a bottlenose dolphin using CB and CR techniques, respectively (Au and Moore 1990).

High Q values reflect narrow filter bandwidths which result in enhanced frequency resolution, with the trade-off of compromised temporal resolution. Auditory filter banks for bottlenose dolphins and belugas have properties where frequency resolution is best at lower frequencies while temporal resolution is better at higher frequencies (Fig. 10.4). This may not be the case for smaller porpoises. Tuning curves derived from electrophysiological measurements suggest at least two species of porpoises (*Phocoena phocoena* and *Neophocaena phocaenoidis asiaeorientalis*) have auditory filter banks with relatively constant bandwidths across frequencies (Popov et al. 2006). Such a filter bank may allow for enhanced frequency resolution at the cost of compromised temporal resolution. A recent re-evaluation of critical ratio data suggests that the auditory filter bank of the bottlenose dolphin might be better modeled as a constant-Q filter bank for frequencies below 40 kHz and a constant bandwidth filter bank for frequencies above 40 kHz (Lemonds et al. 2011).

Modeling the auditory periphery proves useful not only for describing auditory masking, but the auditory filter banks can be used to model other hearing phenomena such as discrimination and recognition abilities during passive hearing and echolocation (Au et al. 2009; Branstetter et al. 2007; Roitblat et al. 1993). Figure 10.4 displays roex(p,r) auditory filter banks constructed for three odontocete species: bottlenose dolphins (Lemonds 1999), belugas (Finneran et al. 2002a), and harbor porpoises (Popov et al. 2006). Filter bandwidths for these three species predict that critical ratios at higher frequencies should be highest for the dolphin and lowest for the harbor porpoise, which is consistent with the empirical findings in Fig. 10.3.

10.3.2 Masking with Complex Stimuli

10.3.2.1 Comodulation Masking Release

The use of simple but well-defined stimuli in masking experiments has proven useful in elucidating the underlying mechanisms of the auditory system. For example, the power spectrum model of masking, which is based almost exclusively on experiments using pure tones and Gaussian noise stimuli, can adequately describe most of the masking results discussed thus far in this chapter. This is not surprising since most of these experiments were conducted using pure tones and Gaussian noise. However, sounds marine mammals encounter in their natural environment are likely to be more complex than pure tones and Gaussian noise. Models derived from simple stimuli may be limited in their ability to generalize to environmental noise. For example, one of the primary assumptions of the PSM is that only noise within a CB centered on a signal contributes to the masking of that signal. However, if the noise is coherently amplitude modulated (comodulated noise) across frequency regions, a release from masking relative to a Gaussian masker of the same pressure spectral density occurs for noise bandwidths greater than a CB; i.e., more total noise power results in less masking. This phenomenon is known as comodulation masking release (CMR) and has been demonstrated in anurans (Chap. 6), birds (Chap. 8), and several mammalian species (Bee et al. 2007; Nelken et al. 2001; Pressnitzer et al. 2001), including humans (Hall et al. 1990) and the bottlenose dolphin (Branstetter and Finneran 2008). (For a discussion of potential CMR in insects see Chap. 3). Figure 10.5 displays masked threshold patterns for both Gaussian and comodulated noise within a standard band-widening paradigm (Fletcher 1940). Consistent with the PSM, thresholds for Gaussian noise increase up to a specific bandwidth (the CB) and then asymptote because noise at frequencies beyond the CB no longer contributes to the masking of the signal. A similar pattern emerges for comodulated noise for masker bandwidths less than the CB. However, there is a monotonic decrease in thresholds for

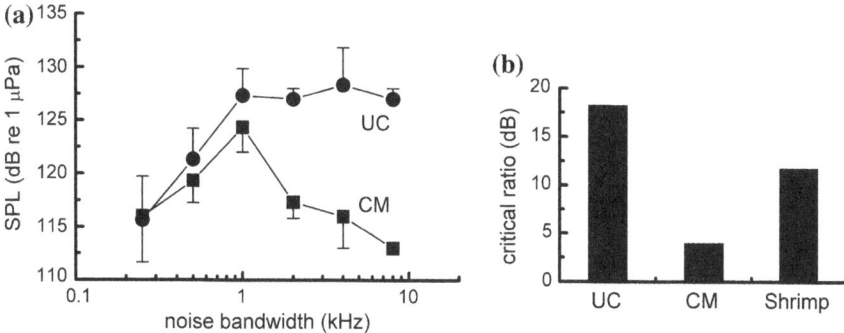

Fig. 10.5 **a** Masking patterns for Gaussian and comodulated noise (adapted from Branstetter and Finneran 2008) and **b** critical ratios from three different noise types (data calculated from Trickey et al. 2011)

masker bandwidths greater than the CB. The release from masking is substantial (17 dB at the largest bandwidth) and is beyond the capability of the PSM to explain. Although several explanations for CMR have been proposed, numerous studies suggest that the auditory system compares temporal envelopes between an auditory filter centered on the signal and flanking auditory filters (Hall et al. 1984; McFadden 1988). The addition of a tonal signal to comodulated noise decreases the modulation depth in the signal channel, thus reducing the envelope correlation between the signal and flanking bands. The presence or absence of a tonal signal can be determined by comparing envelope correlation across frequency channels (Hall et al. 1984).

The extent to which ocean noise is comodulated has not been fully investigated; however, at least two studies suggest CMR may play a role in auditory masking for environmental noise that marine mammals encounter. Erbe (2008) estimated detection thresholds for a beluga using pure tones and beluga vocalization signals with Gaussian, ice-cracking, underwater bubble generator, and propeller noise types. Thresholds for ice-cracking noise, which is comodulated, were at least 6 dB lower than the other uncomodulated noise types (Erbe 2008). A similar release from masking was found for bottlenose dolphins detecting a 10 kHz pure-tone in snapping shrimp noise (Trickey et al. 2011), which is also comodulated. CRs from Gaussian noise overestimated masked thresholds using snapping shrimp noise, primarily because CRs assume that only noise within a single auditory filter contributes to masking.

Additional studies, initially using realistic signals and maskers and then using controlled stimuli, are needed to determine not only the masking patterns for realistic sounds, but also the mechanisms that govern these masking patterns. If environmental noise is similar to Gaussian noise, the PSM can provide accurate predictions. However, if natural noise is not Gaussian, additional mechanisms yet unknown will need to be determined before accurate predictions can be made.

10.3.2.2 Spatially Separated Sound Sources

In realistic acoustic environments with multiple sound sources, detecting a biologically relevant signal in noise depends not only on the physical attributes of the
signal and noise, but also on the location of the signal and noise relative to each
other and to the listener's position and orientation. In humans, where research on
this topic is more extensive, the relative position of sound sources can act as one of
the most salient cues in segregating multiple sounds in a complex auditory scene
(Bregman 1990), and can lead to a spatial release from masking (SRM). Many
types of ocean noise (e.g., boat vessel noise, industrial sites) are emitted from
directional sources that can be well off-axis from a biologically relevant signal. In
such situations, masking predictions based only on the CR may over-estimate the
amount of actual masking.

Au and Moore (1984) measured hearing thresholds for pure tones emitted from
an on-axis transducer while Gaussian noise was emitted from a second transducer
that varied in position in both the horizontal and vertical planes. Although the
authors intended to measure the dolphin's receiving beam pattern, their data are
also an example of a spatial release from masking. Figure 10.6 displays threshold
values relative to when the noise source was directly in front of the animal (i.e., the
position where most masking occurs).

Levels at off-axis positions represent the amount of SRM. Off-axis noise
positions produced less masking and the effect was stronger at higher frequencies.
Au and Moore (1984) were interested in the receiving beam pattern for processing
echolocation signals, and as a result, only tested frequencies of 30 kHz and above
and only at angles in front of the animal. Lower frequencies associated with

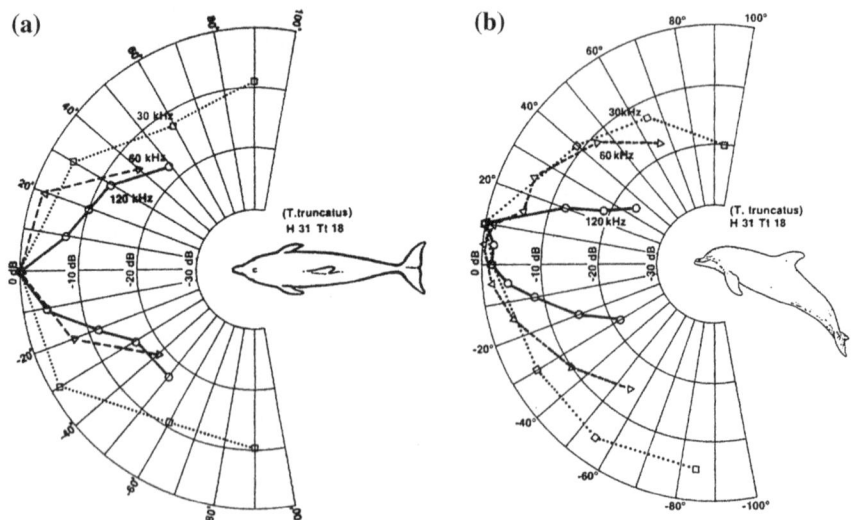

Fig. 10.6 Spatial release from masking (i.e., receiving beam patterns) for the bottlenose dolphin
(adapted from Au and Moore 1984)

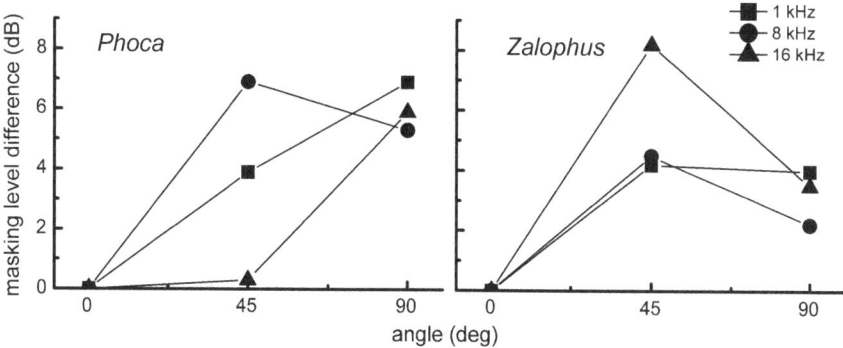

Fig. 10.7 Masking level differences for the harbor seal (*Phoca vitulina*) and the California sea lion (*Zalophus californianus*)

communication were not tested, although if the trend that lower frequencies exhibit less SRM holds true, communication signals will likely be more susceptible to masking than sonar signals. Furthermore, noise locations behind the animal will likely result in even a larger SRM. Additional studies using lower frequencies are therefore warranted.

SRM for airborne sounds has been studied with a harbor seal (*Phoca vitulina*) and California sea lion (*Zalophus californianus*) using a similar approach as Au and Moore (1984), except that the noise transducer's position was held constant at the on-axis position and the position of the signal transducer varied in the horizontal plane (Fig. 10.7, Holt and Schusterman 2007). Because detection thresholds will vary as a function of position even without masking noise, Holt and Schusterman (2007) used a metric called the masking level difference (*MLD*) to account for unmasked thresholds differences:

$$MLD = \left(M_q - M_0\right) - \left(U_q - U_0\right), \qquad (10.7)$$

where U_0 and U_q are the unmasked threshold at $0°$ and $q°$, respectively, and M_0 and M_q are masked thresholds at $0°$ and $q°$, respectively. Overall, the results suggest that signals are better detected when they are separated in spatial location from the noise, although the relationships between threshold, frequency, and noise angular position were inconsistent across these two species. The difference in MLD patterns may be related to differences in external ear (i.e., pinnae) morphology between these species or to individual differences between the subjects.

10.3.3 Echolocation

Of all the marine mammals, only odontocete cetaceans have conclusively demonstrated the ability to echolocate. Although their detection, discrimination, and recognition abilities have been well-studied, very little research has been conducted on their ability to echolocate in the presence of noise. What is known is that odontocetes appear to have the capability to modify their echolocation signal to compensate for noise levels. This was demonstrated when echolocation discrimination tasks were conducted in both San Diego Bay, California and Kaneohe Bay, Hawaii with the same beluga (Au et al. 1988). The ambient noise in both locations is dominated by snapping shrimp, although the noise spectral density levels in Kaneohe Bay were typically 15–20 dB greater than those of San Diego Bay. Beluga clicks recorded in San Diego bay typically had peak–peak (p–p) source levels between 201 and 202 dB re 1 μPa, with peak frequencies typically between 40 and 60 kHz. However, in Kaneohe Bay, which possessed higher ambient noise levels, the beluga clicks had p–p source levels between 210 and 214 dB re 1 μPa, with peak frequencies between 100 and 120 kHz. Apparently, the animal increased the level and peak frequency of its incident signal to compensate for the increased ambient noise in Kaneohe Bay. It is unclear, however, if the animal intentionally shifted the peak frequency of its signals to the higher end of the spectrum to avoid low-frequency masking. Odontocete echolocation signals show a strong positive correlation between amplitude and peak frequency (Au 1980), suggesting the frequency shift may have simply been a by-product of increasing the source level (see Chap. 7 for a similar discussion for bird songs).

10.3.4 Consequences of Auditory Masking

The most obvious consequence of auditory masking is a reduction in the distance at which an animal could detect a sound of interest. Because sound absorption is frequency-dependent, with low frequencies traveling farther than higher frequencies, low-frequency noise has the potential to affect marine mammals at larger distances compared to higher frequency noise. Consequently, the communication ranges of mysticetes that rely on very low-frequency sounds have likely been reduced (compared to preindustrial ranges), thus compromising the biological functions of these signals (Clark et al. 2009). Communication ranges of other marine mammals (e.g., odontocetes and pinnipeds) that utilize higher frequency sounds may be affected by auditory masking by higher frequency noise sources such as small boat engines and marine construction. For specific scenarios involving Gaussian-like noise sources, knowledge or estimates of the hearing threshold and CR for a species, along with the signal and noise properties, can be used to estimate the resulting detection range (e.g., Clark et al. 2009; Janik 2000). For more complex noise sources that may be comodulated, simple estimates based on Gaussian noise and the PSM will tend to over-estimate the masking effects of noise and under-estimate the range at which a particular signal can be detected.

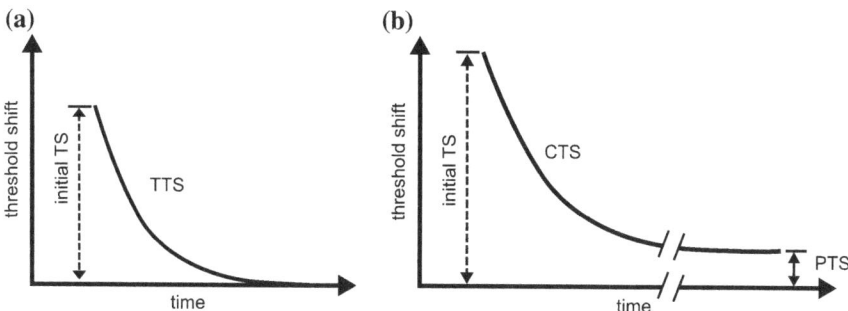

Fig. 10.8 Distinctions between TTS, PTS, and CTS

Simple models for masking and animal communication range also typically do not include the compensatory mechanisms that animals use to communicate in suboptimal environments. For example, when humans communicate in noisy environments, we often increase speech amplitude, move closer together, read lips, turn our backs toward a noisy sound source, or simply leave the noisy area. Marine mammals appear to employ similar strategies but little is known about their effectiveness or cost. If an animal is able to leave, or avoid an area of potential masking there may be associated metabolic costs that are yet to be determined. In many circumstances, leaving a zone of auditory masking may not be an option (e.g., pervasive low-frequency shipping noise). Some areas may be too important to leave such as feeding and breeding grounds. In these cases, an animal may attempt to compensate for the noise by increasing its signal amplitude while communicating (Holt et al. 2008; Parks et al. 2011), shifting signal frequencies (McDonald et al. 2009), or increasing its repetition rate or duration (Miller et al. 2000). Again, compensation may come with a cost and the effectiveness is unknown. In other cases, consequences may be unavoidable and may include a decreased ability to maintain group cohesion, decreased ability to detect predators and prey, and decreased foraging and breeding success.

Detection of a sound only implies that the sound registered in the listener's auditory system. If an animal can detect a signal but is unable to recognize or make sense of the information (e.g., humans detecting speech but not understanding it because of noise) the signal's utility will be lost. The harmonic structure of odontocete whistles has a direction-dependant pattern (Branstetter et al. 2012) that has been hypothesized to convey information on location and direction of travel of the signaler (Lammers and Au 2003; Miller 2002). If odontocetes use the whistle harmonic structure to monitor the direction of travel of group members, masking may reduce the animal's ability to maintain group cohesion when separated at larger distances. The potential effect would be to limit the distance between group members, and thus reduce the area covered during cooperative behaviors such as foraging.

10.4 Noise-Induced Threshold Shifts

Most adults living in industrialized countries have experienced a loss of hearing sensitivity, and eventual recovery, after exposure to high intensity sound at concerts, while operating firearms, or in the presence of industrial machinery or power tools. This phenomenon is called a noise-induced threshold shift (NITS), and is characterized as an increase in auditory threshold (loss of sensitivity) over some frequency range, that persists after the cessation of a noise exposure. The magnitude of a NITS generally decreases with increasing time after the noise exposure. If the hearing threshold returns to normal after some period of time, the NITS is called a temporary threshold shift (TTS). If, however, thresholds remain elevated after some extended period of time (typically 30 days), then the remaining amount of NITS is called a permanent threshold shift (PTS). The term compound threshold shift (CTS) is used to describe an initial NITS that only partially recovers, leaving some residual PTS; i.e., a CTS represents some combination of TTS and PTS (Ward 1997). Figure 10.8 illustrates the relationships between TTS, PTS, and CTS.

A NITS may result from a variety of mechanical and biochemical processes, including physical damage or distortion of the tympanic membrane and cochlear hair cell stereocilia, hair cell death resulting from oxidative stress, changes in cochlear blood flow, and swelling of cochlear nerve terminals from glutamate excitotoxicity (Henderson et al. 2006; Kujawa and Liberman 2009). Although the outer hair cells are the most prominent target for noise effects, severe noise exposures may also result in inner hair cell death and loss of auditory nerve fibers (Henderson et al. 2006). Recent studies in mice have also revealed that a TTS near the limits of reversibility, e.g., a 40 dB maximum TTS, measured 24 h after exposure via auditory brainstem response and compound action potential, may result in acute loss of afferent nerve terminals, delayed cochlear nerve degeneration, and permanently attenuated suprathreshold neural responses, despite complete recovery of auditory thresholds (Kujawa and Liberman 2009). These data suggest that there may be progressive consequences to noise exposure not revealed by conventional threshold testing.

A great deal of work has been done to characterize TTS and PTS in humans and other terrestrial mammals (rev Clark 1991; Henderson and Hamernik 1986; Kryter 1973; Melnick 1991; Miller 1974; Quaranta et al. 1998; Ward 1997; see Chaps. 4 and 8 for reviews of TTS and PTS in fish and birds, respectively). The primary emphasis of these efforts has been to predict and mitigate human occupational hearing loss, thus the particular exposure conditions have focused on those conditions most often encountered in industrial or military settings: multi-hour exposure to broadband noise and exposure to impulse and impact noise. A goal of early human work was to relate the amount of TTS experienced at the end of an 8 h work day to the amount of PTS that would be experienced after many years of comparable daily exposures (e.g., Nixon and Glorig 1961). Although these efforts were not completely successful, and no clear predictive relationship has been found between TTS and PTS, much has been learned about the relationships between threshold shifts and exposure parameters such as SPL, duration, frequency, and duty cycle. It is also clear that larger exposures are necessary to produce PTS compared to TTS, thus knowledge of TTS-inducing exposure levels can be used to mitigate the occurrence of PTS. For example, terrestrial mammal data have shown that a NITS less than 40 dB, measured 2–4 min after exposure, is not likely to result in PTS (e.g., Kryter et al. 1966).

TTS and PTS data from humans and terrestrial mammal models have been used to define safe limits for occupational noise exposure. For steady-state (i.e., non-impulsive) noise exposures, current US regulations prescribe a maximum permissible exposure SPL of 90 dBA for an 8-h period; for each halving of exposure time, the permissible SPL increases by 5-dB, called a 5 dB exchange rate (29CFR1910.95 2009). The maximum permissible exposure to impulsive or impact noise is 140 dB re 20 μPa peak SPL (29CFR1910.95 2009).

Despite the wealth of knowledge accumulated via human and terrestrial mammal studies, the applicability of these data to marine mammals is limited. There are significant differences between the peripheral auditory systems of marine and terrestrial mammals and the sound transduction mechanisms in air and water, thus direct extrapolation of human noise exposure criteria to marine mammals is not practical. Also, the types of noise exposures most relevant for people (e.g., 8-h exposure to broadband noise) may not be relevant to marine mammals exposed to shorter duration, intermittent sources such as military sonars, pile driving, and seismic airguns. For these reasons, a number of TTS measurements have been conducted with marine mammals to determine noise exposure conditions necessary for TTS, and to predict those capable of causing PTS, in these animals.

10.4.1 Measuring NITS in Marine Mammals

Studies of NITS in marine mammals have focused on measuring TTS after exposure to relatively long duration, broadband noise (Kastak et al. 1999, 2005, 2007; Kastak and Schusterman 1996; Kastelein et al. 2011; Mooney et al. 2009a; Nachtigall et al. 2003, 2004; Popov et al. 2011), relatively short duration tones (Finneran et al. 2005, 2007c, 2010a, b; Finneran and Schlundt 2010; Mooney et al. 2009b; Ridgway et al. 1997; Schlundt et al. 2000), and single underwater impulses (Finneran et al. 2000, 2002b, 2003; Lucke et al. 2009). Subjects have consisted of bottlenose dolphins, belugas, a harbor porpoise (*Phocoena phocoena*), Yangtze finless porpoises (*Neophocaena phocaenoides asiaeorientalis*), California sea lions (*Zalophus californianus*), a harbor seal (*Phoca vitulina*), and a Northern elephant seal (*Mirounga angustirostris*).

The experimental approaches for TTS measurements in marine mammals are analogous to those used to measure TTS in terrestrial mammals. Tests begin with a pre-exposure hearing threshold measurement at one or more frequencies. This is followed by the fatiguing sound exposure—the sound that may cause TTS. Finally, post-exposure hearing thresholds are measured at one or more frequencies. The NITS at each frequency is typically defined as the difference (in decibels) between the post-exposure and pre-exposure thresholds at that frequency, though some studies (e.g., Mooney et al. 2009a, b) have used an average "baseline" threshold instead of the pre-exposure threshold. To assess the recovery of hearing after a NITS, and to verify that the shift was in fact temporary, post-exposure thresholds are typically measured several times, over a period that may extend for several days.

There have been no designed studies of PTS in marine mammals; however, Kastak et al. (2008) reported incomplete recovery of a 50-dB initial threshold shift in a harbor seal, resulting in 7–10 dB of PTS measured about 2 months after exposure.

10.4.2 Predicting the Onset of NITS

One of the goals of marine mammal TTS research has been to identify exposure levels that are just-sufficient to cause a TTS. These exposure levels are often referred to as "onset TTS" levels, and have been widely used in environmental analyses to estimate the numbers of animals that may be adversely affected by human-generated noise (e.g., US Navy 2008). The first controlled TTS experiments in marine mammals used a 6-dB criterion to identify a measurable TTS (Ridgway et al. 1997; Schlundt et al. 2000); for this reason, a noise exposure sufficient to induce 6 dB of TTS has often been taken as the onset-TTS exposure level.

The onset of PTS in marine mammals has been estimated by assuming that a TTS greater than 40 dB has the potential to result in some PTS. Exposures sufficient to induce 40 dB of TTS are estimated from onset-TTS exposure levels and TTS growth rates (see Southall et al. 2007).

10.4.3 Parameters that Affect NITS

The major findings to arise from marine mammal TTS experiments parallel findings from terrestrial mammal experiments. As in terrestrial mammals, the most significant factors that affect hearing loss are the exposure SPL, exposure duration, exposure frequency, temporal pattern, and recovery time. In addition to those factors that affect the actual function of the subject's auditory system, some additional parameters affect the amount of TTS that is measured. For example, the amount of TTS varies with frequency, so the specific hearing test frequency will influence the amount of TTS that is observed. Also, the methodology used to perform the hearing test has been found to affect the amount of TTS observed. The following sections discuss each of these factors individually and provide example data to illustrate what is currently known about TTS in marine mammals.

10.4.3.1 Hearing Test Method

Marine mammal hearing assessments are conducted using behavioral (i.e., psychophysical) or electrophysiological methods. For behavioral methods, subjects are trained to perform a specific action, such as vocalizing or pressing a paddle, in response to hearing test tones. Tone SPLs are manipulated and the subject's responses tracked to estimate the threshold. Most TTS studies have used adaptive staircase paradigms, where the tone SPL is reduced after each detection and increased following a nondetection (Cornsweet 1962; Levitt 1971). The threshold is then estimated from the reversal points, where the tone SPL changes from increasing to decreasing or vice versa. During behavioral approaches it is also important to feature signal-absent trials, so that any changes to the subject's response bias can be identified. Behavioral methods are straightforward to implement and the resulting data are easy to interpret. The amount of time required to obtain a behavioral threshold depends on the specific experimental paradigm. With a staircase procedure and multiple stimulus presentations within each reinforcement interval, behavioral thresholds can be obtained in as little as 2–4 min (Finneran et al. 2005; Schlundt et al. 2000); however, regardless of the specific behavioral test paradigm, initial subject training typically requires several months.

Electrophysiological approaches use passive electrodes placed on the head (Fig. 10.9) to record changes in the electroencephalogram (EEG) that are synchronized with the onset of a sound stimulus. These small voltages, on the order of microvolts, are called auditory evoked potentials (AEPs). To measure AEPs, relatively short duration (typically tens of milliseconds) stimuli are presented hundreds or thousands of times, and the resulting AEPs synchronously averaged, to reduce residual physiological background noise caused by breathing, head movement, eye movement, etc. Marine mammal TTS measurements have

Fig. 10.9 A bottlenose dolphin participating in an AEP-based hearing test. The electrodes are embedded in suction cups attached to the head, back, and dorsal fin

generally used amplitude modulated stimuli to produce a steady-state, harmonic AEP called the auditory steady-state response (ASSR) or envelope following response (EFR). The ASSR amplitude at the stimulus modulation rate is recorded as the stimulus SPL is manipulated. Thresholds are based on the lowest detectable response (e.g., Finneran et al. 2007c) or by fitting a curve to the ASSR-stimulus SPL graph and extrapolating to the zero-crossing point (e.g., Nachtigall et al. 2004). The most appropriate modulation rates vary across species; for odontocetes, frequencies around 1 kHz are optimal (e.g., Dolphin et al. 1995; Finneran et al. 2007b, 2009; Nachtigall et al. 2005, 2008; Popov et al. 2005; Schlundt et al. 2011; Supin and Popov 1995), while in pinnipeds, frequencies near 150–200 Hz have worked well (Mulsow and Reichmuth 2007, 2010; Mulsow et al. 2011a, b). Evoked potential thresholds may be obtained as quickly as behavioral thresholds and are not limited by the requirements to train subjects for behavioral testing; however, AEP methods, and especially the ASSR technique, tend to work better at relatively high frequencies. For dolphins, the ASSR method is most effective at frequencies of ∼8 kHz and above; in sea lions, the ASSR has been successfully used at frequencies of 500 Hz and above.

It is important to keep in mind that ASSR thresholds and behavioral thresholds are not equivalent. Behavioral testing is a cognitive task—the subject must hear the sound stimulus and make a decision whether to respond. The signal processing chain includes the auditory cortex and centrally located processing centers in the brain. In contrast, at the modulation rates typically employed in marine mammal threshold testing, the ASSR is composed of summed neuronal activity from many individual generators at locations ranging from the auditory nerve to the brainstem. In this sense, ASSR and behavioral thresholds provide different glimpses of the function of the auditory system. There is no reason to expect behavioral and ASSR thresholds to perfectly agree—and they normally do not, with ASSR thresholds typically 5–15 dB higher than behavioral thresholds (e.g., Finneran et al. 2007a; Mulsow et al. 2011b; Mulsow and Reichmuth 2010; Schlundt et al. 2007, 2008; Yuen et al. 2005). TTS results obtained with the two techniques may also differ. In the only direct comparison between TTS obtained from behavioral and ASSR

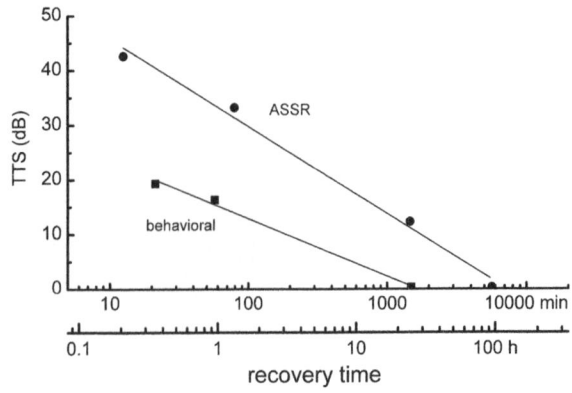

Fig. 10.10 Comparison of TTS recovery, from the same exposure, measured using ASSR and behavioral methods (adapted from Finneran et al. 2007c)

threshold measurements (Finneran et al. 2007c), the ASSR technique consistently resulted in larger amounts of TTS and longer recovery times (Fig. 10.10). These data caution against pooling TTS data obtained with behavioral and ASSR methods and show that even after recovery of behavioral thresholds, some functions of the auditory system may still be adversely affected. This suggests that the ASSR technique may be a more sensitive indicator of auditory damage compared to psychophysical threshold testing.

10.4.3.2 Hearing Test Frequency

The specific hearing test frequency will also affect the amount of TTS that is observed. Studies of dolphins and belugas exposed to tones have shown that the maximum TTS does not occur at the exposure frequency, but normally at frequencies one-half to one octave above the exposure frequency (Finneran et al. 2007c; Schlundt et al. 2000). The spread of TTS from tonal exposures can thus extend over a broad frequency range; i.e., narrowband exposures can produce broadband (greater than one octave) TTS (Fig. 10.11). These findings match those from human and terrestrial mammal studies (e.g., McFadden 1986; Ward 1962). For octave band noise exposures, the upward spread of TTS, or "half-octave shift," has not always been observed, with some pinniped studies showing the maximum TTS near the center frequency of the exposure (Kastak et al. 2005), and dolphin experiments showing the maximum TTS one-half octave above the center of the noise band (Mooney et al. 2009a). This result is also consistent with terrestrial mammal data, where the half-octave shift is most commonly associated with tonal noise exposures. The failure for broadband noise to result in an upward spread of TTS may also be related to the TTS magnitudes induced; as the exposure level increases, the activation area on the basilar membrane spreads more toward the basal end of the cochlea and thus affects higher frequencies to a greater extent (McFadden and Plattsmier 1983). At lower amounts of TTS, the activation pattern tends to be more symmetrical about the noise center frequency.

Fig. 10.11 Influence of hearing test frequency on the amount of TTS that is observed. For tonal exposures, the maximum TTS normally occurs one-half to one octave above the exposure frequency (adapted from Finneran et al. 2007c)

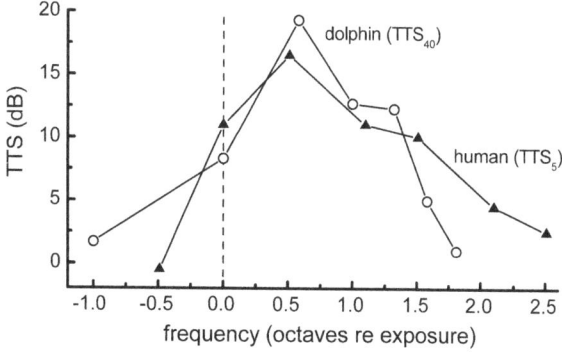

10.4.3.3 Recovery Time

Since TTS is a temporary phenomenon, the amount of TTS observed will be a function of the recovery time—the amount of time that has elapsed since the cessation of the noise exposure. For this reason, numeric subscripts are normally used to indicate the recovery time associated with a specific TTS measurement; i.e., TTS_4 indicates a TTS measured 4 min after the exposure.

The amount of TTS normally decreases with increasing recovery time; however, the relationship is not necessarily monotonic, and it is common to see examples of delayed recovery, where the TTS may remain nearly constant for some time after the exposure (e.g., Finneran et al. 2007c; Popov et al. 2011). In many cases the recovery function is not linear with time, but approximately linear with the logarithm of time. In these cases, the recovery rates are often described by the slope of the recovery function; for dolphins, recovery rates between 1.5 and 2 dB per doubling of time have been measured when the initial shifts were ~ 5–15 dB (Finneran et al. 2007c; Mooney et al. 2009a; Nachtigall et al. 2004). For larger amounts of TTS, up to ~ 40 dB, recovery rates of 4–6 dB per doubling of time have been measured in a dolphin (Finneran et al. 2007c). For a sea lion, recovery rates from TTS_{12} of ~ 20–35 dB were ~ 2.5 dB per doubling of time (Kastak et al. 2007). Complex TTS recovery patterns have been observed in dolphins after exposure to 3-kHz tones (Finneran et al. 2010a). These curves often contained regions where TTS was linear with the logarithm of time, but also often contained regions with varying slopes. Double exponential functions used to fit human TTS recovery data (Keeler 1968; Patuzzi 1998) fit the dolphin recovery data and, for 3-kHz exposures with durations from 1 to 128 s, the recovery functions were described using TTS_4 and recovery time only; i.e., recovery functions did not depend on the specific SPL and duration but only on the resulting TTS_4 (Fig. 10.12; Finneran et al. 2010a). The extent to which this result may be extrapolated to other exposure conditions is unknown.

10.4.3.4 Noise Sound Pressure Level

As in many other animal groups, the amount of TTS generally increases with the noise SPL; however, the relationship is neither monotonic nor linear. Ward (1976) defined "effective quiet" as the highest SPL that would not produce a significant TTS or affect recovery from a TTS produced by a prior, higher level exposure. For humans, effective quiet for octave band noise with center frequencies from 250 to 4,000 Hz is around 68–76 dBA (Ward et al. 1976). To date, there have been no studies performed to measure effective quiet in a marine mammal; however, we can estimate the upper limit for effective quiet by examining the lowest noise exposure SPLs that have resulted in measurable amounts of TTS. For dolphins, effective quiet must be less than 155–160 dB re 1 μPa, since this SPL produced

Fig. 10.12 TTS recovery after 3 kHz exposures, as a function of TTS$_4$ and the logarithm of post-exposure time (in min). Symbols indicate the experimentally measured values for four dolphins. The *color bar* indicates TTS in dB (adapted from Finneran et al. 2010a)

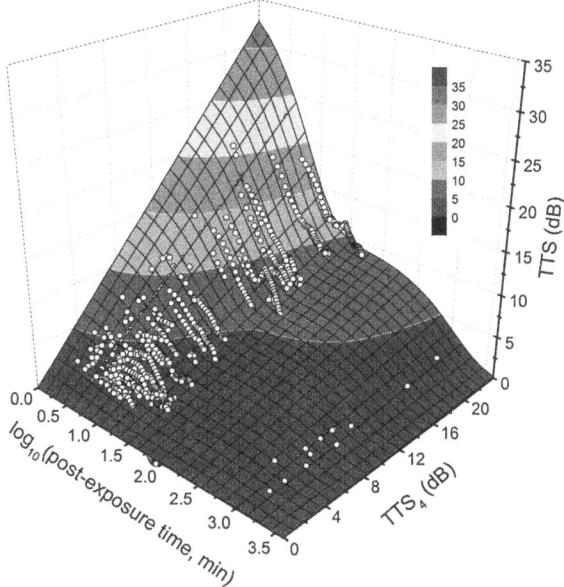

TTS after only 30 min of exposure to broadband noise centered around 6–7 kHz (Mooney et al. 2009a; Nachtigall et al. 2004). For sea lions, harbor seals, and Northern elephant seals, effective quiet must be less than 80 dB re 1 μPa, which produced TTS at 2.5 kHz after 22-min underwater exposures to octave band noise centered at 2.5 kHz (Kastak et al. 2005). For sea lions in air, effective quiet must be less than 94 dB re 20 μPa, which produced ∼5 dB of TTS after only 25 min exposures to 2.5-kHz, octave band noise (Kastak et al. 2007).

At exposure levels above effective quiet, the amount of TTS increases with SPL in an accelerating fashion. This is illustrated in Fig. 10.13, which shows the increase, or growth, of TTS$_4$ with increasing SPL in a dolphin exposed to short duration, 3-kHz tones (Finneran et al. 2010a). At low exposure SPLs, the amount of TTS is small and the growth curves have shallow slopes. At higher SPLs, the growth curves become steeper and approach linear relationships with the noise SPL. TTS growth curves for dolphins, harbor seals, sea lions, and northern elephant seals have been successfully fit by equations with the form:

$$y(x) = a \log_{10}\left[1 + 10^{(x-b)/10}\right], \tag{10.8}$$

where y is the amount of TTS, x is the exposure level, and a and b are fitting parameters (Finneran et al. 2005, 2010a; Kastak et al. 2005, 2007). This particular function has an increasing slope when $x < b$ and approaches linearity for $x > b$ (Maslen 1981). The linear portion of the curve has a slope of $a/10$ and an x-intercept of b. TTS growth curves for dolphins have been shown to be frequency-dependent, with growth rates at 3-kHz of approximately 0.2–0.7 dB/dB, while

Fig. 10.13 Growth of TTS$_4$ as a function of SPL for a bottlenose dolphin exposed to 3 kHz tones. *Vertical error bars indicate SD for the mean TTS$_4$ in each exposure group. Horizontal error bars indicate the SD for the mean exposure SPLs in each group. The solid lines are functions with the form of Eq. (10.8) fit to the data (adapted from Finneran et al. 2010a)*

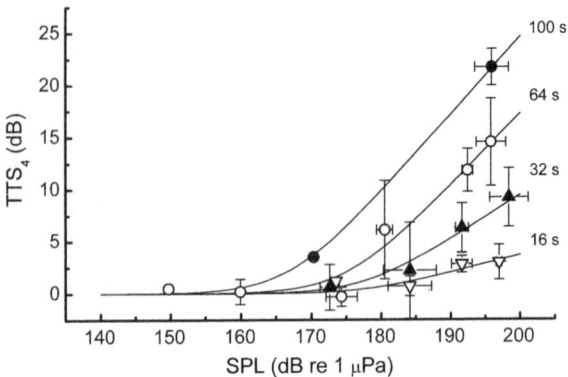

those at higher frequencies have steeper slopes, such as 1.2 dB/dB at 20 kHz (Finneran et al. 2010a; Finneran and Schlundt 2010). The growth rate for a California sea lion tested in air was ~ 2.5 dB/dB at 2.5 kHz (Kastak et al. 2007).

10.4.3.5 Noise Duration

TTS also generally increases with noise duration; however, as with SPL the growth functions are nonmonotonic. Growth functions relating TTS to the exposure duration are also accelerating functions, where the slope is shallow at low amounts of TTS (e.g., less than 10 dB) and becomes increasingly steep as the duration (and amount of TTS) increase. At low amounts of TTS, the functions for TTS growth with increasing exposure duration appear roughly linear (Finneran et al. 2010a; Mooney et al. 2009a), but approach linear behavior with the logarithm of time as the exposure duration and resulting amount of TTS increase. TTS growth functions based on exposure duration, up to about 20 dB of TTS$_4$, have been successfully fit by Eq. (10.8) (Finneran et al. 2010a).

Terrestrial mammal data have shown that if the noise SPL is fixed and the exposure duration continually increased, the amount of TTS will eventually reach a plateau, where further increases in exposure duration do not result in additional threshold shift. This region is called asymptotic threshold shift (ATS). ATS has been hypothesized to represent the upper bound of PTS that could be produced by noise of a specific SPL, regardless of duration (Mills 1976). Exposure durations sufficient to induce ATS in terrestrial mammals have generally been at least 4–12 h (Mills 1976; Mills et al. 1979), much longer than the maximum exposure durations used with marine mammal testing (less than 1 h). As a result, ATS has not been observed in any marine mammals; however, given the similarities in cochlear function it is likely that similar patterns of TTS growth would be found in marine mammals, including regions of ATS. When ATS is taken into account, TTS growth with exposure duration is best described using exponential functions (Keeler 1968; Mills et al. 1979).

10.4.3.6 Sound Exposure Level and the "Equal Energy Rule"

Sound exposure is an "energy-like" metric, defined as the time integral, over the duration of the exposure, of the instantaneous sound pressure-squared (American National Standards Institute 1994); the term *sound exposure level* (SEL) refers to the sound exposure expressed in decibels, referenced to 1 $\mu Pa^2 \cdot s$ in water or (20 $\mu Pa)^2 \cdot s$ in air (American National Standards Institute 2011). For multiple or intermittent exposures, the cumulative SEL, defined as the total SEL calculated over the "on-time" of the noise exposure, is often used to characterize the exposure. SEL is linearly related to the SPL and logarithmically related to the exposure time, meaning that SEL will change on a 1:1 basis with SPL, and change by 3 dB for each doubling/halving of exposure time. For plane progressive waves, sound exposure is proportional to sound energy flux density, so the use of SEL is often described as an "equal-energy" rule, whereby exposures of equal energy are assumed to produce equal amounts of NITS, regardless of how that energy is distributed over time. Since the SEL changes by 3 dB for each doubling or halving of exposure duration, the use of SEL or an equal energy rule can also be described as a "3-dB exchange rate" for acoustic damage risk criteria. This means that the permissible noise exposure SPL will change by 3 dB with each doubling or halving of exposure time; e.g., an equal energy rule means that if the permissible exposure limit is 90 dB re 1 μPa for an 8-h exposure, the limit for a 4-h exposure would be 93 dB re 1 μPa.

Because threshold shifts depend on both the exposure SPL and duration, it has become convenient to use SEL as a single numeric value to characterize a noise exposure and to predict the amount of NITS. SEL has been shown to be an effective predictor of TTS, and has been useful in establishing acoustic damage risk criteria for marine mammals (Finneran et al. 2010a, 2005; Kastak et al. 2007, 2005; Mooney et al. 2009a). However, the marine mammal studies, like terrestrial mammal studies, have shown that the equal energy rule has limitations, and is most applicable to single, continuous exposures. As the exposure duration increases, the relationship between TTS and SEL also begins to break down. Specifically, duration has a more significant effect on TTS than what would be predicted on the basis of SEL alone (Finneran et al. 2010a; Kastak et al. 2005; Mooney et al. 2009a). This means that if two exposures have the same SEL but different durations, the exposure with the longer duration will tend to produce more TTS. For this reason, recent models for TTS in marine mammals have begun to treat TTS as a function of both exposure SPL and duration, representing TTS growth as a surface rather than a curve (e.g., Fig. 10.14; Finneran et al. 2010a; Kastak et al. 2007; Mooney et al. 2009a).

The marine mammal data serve to emphasize that the equal energy rule is an over-simplification. The temporal pattern of noise exposure is known to affect the resulting threshold shift. It is also well-known that the equal energy rule will over-estimate the effects of intermittent noise, since the quiet periods between noise exposures will allow some recovery of hearing compared to noise that is continuously present with the same total SEL (Ward 1997). However, despite its

Fig. 10.14 TTS$_4$ as a function of SPL and duration for 3-kHz tone exposures. Symbols represent individual TTS$_4$ values measured in four dolphins. The *color bar* indicates TTS$_4$ in dB (adapted from Finneran et al. 2010a)

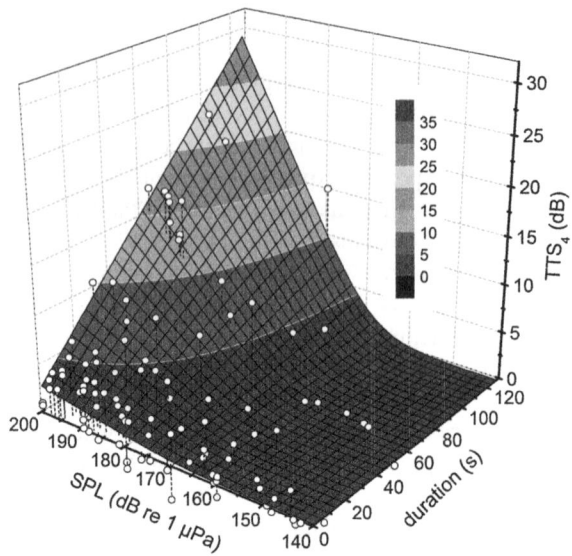

simplistic nature and obvious limitations, the equal energy rule continues to be a useful concept, since it highlights the need to consider both the noise amplitude and duration when predicting auditory effects. Early efforts to mitigate the effects of noise on marine mammals often neglected the noise duration and predicted zones of hearing loss based on the SPL alone. Predictive models have significantly advanced since, and the use of SEL, while clearly not perfect, is simple, allows the effects of multiple noise sources to be combined in a meaningful way, and is accurate, especially when applied to a limited range of noise durations. The use of cumulative SEL for intermittent exposures also errors on the side of caution since it will always over-estimate the effects of intermittent sources.

10.4.3.7 Noise Frequency

For humans, TTS increases with increasing noise frequency, at least up to 2–6 kHz, which is near the range of best hearing sensitivity (Elliott and Fraser 1970; Miller 1974). Because of the similarities in inner ear structure/function, it seems logical that marine mammals would respond in a similar fashion; i.e., that animals would be more susceptible to TTS at frequencies where auditory sensitivity is higher. Most marine mammal TTS data, however, have been collected at relatively low frequencies, generally between 1 and 10 kHz. This frequency range contains some of the most intense anthropogenic sources, but is below the region of best sensitivity for many species. Early TTS data obtained at multiple noise frequencies in dolphins did not reveal significant differences in TTS onset at 3, 10, and 20 kHz, perhaps because of inter-subject differences in susceptibility or because TTS values were based on masked hearing thresholds (Schlundt et al.

2000). As a result, most acoustic impact criteria have used similar numeric thresholds for the onset of TTS, regardless of exposure frequency (e.g., Southall et al. 2007). More recent data, however, have revealed large differences (~ 15 dB) between TTS onset at 3 kHz compared to 20 kHz (Finneran and Schlundt 2010; Finneran et al. 2007c). TTS growth rates in dolphins have also been shown to increase with exposure frequency above 3 kHz, with the maximum growth rate, and lowest threshold for the onset of TTS, occurring near 14–28 kHz in dolphins (Finneran 2010; Finneran and Schlundt 2010). The occurrence of maximum TTS in an odontocete at a few tens of kilohertz, but not at the frequency of maximum sensitivity, is also supported by the data of Popov et al. (2011), who found higher susceptibility in the Yangtze finless porpoises at 32 kHz compared to higher frequencies.

These data demonstrate the need for frequency-specific criteria for noise susceptibility. For humans, susceptibility to noise across frequency is handled through the use of auditory weighting functions. Weighting functions describe a series of frequency-specific correction factors, or "weights" that are added to noise levels to increase the calculated noise dose at frequencies where individuals are more susceptible, and to decrease the noise dose at frequencies where individuals are less sensitive. Human auditory weighting functions were derived from equal loudness contours and measures of subjective loudness level, not auditory sensitivity. For marine mammals, equal loudness levels have only recently been measured (Finneran and Schlundt 2011), and only in a single bottlenose dolphin. Auditory weighting functions derived from the equal loudness contours agree remarkably well with TTS onset values in dolphins exposed to short duration tones (Finneran 2010), and suggest that, in the absence of equal loudness level data for other species, the use of auditory sensitivity curves as weighting functions may provide a reasonable alternative.

10.4.3.8 Temporal Pattern of Noise

Most marine mammal TTS experiments have featured single, continuous noise, or single impulses, and there have been only two studies designed to examine the effects of intermittency and temporal pattern on TTS (Finneran et al. 2010b; Mooney et al. 2009b). These studies have shown that TTS can accumulate across multiple exposures, but the resulting TTS will be less than the TTS from a single, continuous exposure with the same total SEL. This result is not surprising, since the equal energy rule is known to over-estimate the effects of intermittent noise because it does not account for recovery that may occur in the quiet intervals between noises. Finneran et al. (2010b) found that the modified power law model (Humes and Jesteadt 1989) fit the growth of TTS across multiple, short duration tonal noise exposures; however, it is unknown to what extent this method would fit other test conditions.

10.4.3.9 Impulse Noise

The term "impulse noise" is generally used to denote any short duration, high amplitude sound with relatively broad frequency content and relatively fast rise time. Common examples of impulsive sound sources would include impact pile driving, explosions, and seismic air guns. Terrestrial mammal studies of the auditory effects of impulse noise have revealed that impulse noise may be particularly hazardous to hearing, and that the variability associated with NITS measurements is higher when using impulsive fatiguing sources (Henderson and Hamernik 1986). In addition to the factors affecting NITS listed above, the rise time and number of impulses will also affect the resulting amount of NITS (Henderson and Hamernik 1986).

Very few TTS studies have been conducted with marine mammals exposed to impulsive noise sources. Finneran et al. (2000) exposed dolphins and a beluga to single impulses from an array of underwater sound projectors designed to produce pressure signatures resembling underwater explosions, but found no TTS after exposure to the highest level the device could produce (SEL $= 179$ dB re 1 $\mu Pa^2 \cdot s$). Similarly, no TTS was found in two California sea lions exposed to single impulses from an arc-gap transducer with SELs of 161–163 dB re 1 $\mu Pa^2 \cdot s$ (Finneran et al. 2003). Finneran et al. (2012, 2011) also reported preliminary data showing no behavioral TTS in three bottlenose dolphins exposed to a sequence of 10 impulses, produced from a seismic air gun at an interval of 10 s/impulse. The cumulative SEL for the 10 impulses was ~ 176 dB re 1 $\mu Pa^2 \cdot s$. One of the three dolphins had also been exposed to 10 impulses with cumulative SEL of ~ 195 dB re 1 $\mu Pa^2 \cdot s$ with no TTS (Finneran et al. 2011).

For impulse noise studies, measurable TTS has only been observed in a single beluga exposed to an impulse from a seismic watergun (Finneran et al. 2002b), and a single harbor porpoise exposed to an impulse from a seismic air gun (Lucke et al. 2009). The SEL necessary for the onset of TTS in the beluga was 186 dB re 1 $\mu Pa^2 \cdot s$, 9 dB lower than that required for TTS after exposure to a 1-s tone (Schlundt et al. 2000), which supports the idea that impulsive noise exposures are more hazardous than nonimpulsive exposures with the same energy. The exposure SEL required for onset TTS in the harbor porpoise was ~ 164 dB re 1 $\mu Pa^2 \cdot s$; however, the impulsive data are the only TTS data available at present for harbor porpoises, so there can be no impulsive/nonimpulsive comparison. At present, the relationship between exposure frequency content and the occurrence and frequency spread of impulse noise TTS is unclear. The TTS in the beluga and the harbor porpoise exposed to single impulses occurred at frequencies above the predominant energy in the exposures, suggesting an upwards shift in TTS as one would expect based on terrestrial mammal data (Finneran et al. 2002b; Lucke et al. 2009). It is also possible that the failure of air gun impulses to produce TTS in a dolphin at cumulative SELs higher than those producing TTS in a beluga exposed to a single impulse may be related to the frequency content of the exposures (Finneran et al. 2012, 2011).

10.4.4 Perceptual Consequences of NITS

Exposures required for the onset of TTS are relatively large; e.g., for dolphins exposed to short duration tones at 3-kHz, the SEL required for TTS is about 195 dB re 1 $\mu Pa^2 \cdot s$ (Finneran et al. 2005; Schlundt et al. 2000). This means that for short or moderate duration exposures, relatively high SPLs are generally required to induce TTS in marine mammals. This in turn results in relatively small areas around a sound source where received levels may reach sufficient values to induce TTS, and even smaller regions where a PTS may occur. From this standpoint, a NITS may not be as significant to marine mammal populations as other potential effects, such as masking, which may occur at lower received SPLs and thus within larger areas around a sound source. However, from an individual animal's perspective, a NITS could be a serious consequence, since the loss of hearing sensitivity associated with PTS is permanent and that associated with TTS could last for hours to days after the cessation of the noise. During this time, any activities that depended upon the animal's hearing ability would be compromised to a degree determined by the extent and character of the hearing loss.

The consequences of a NITS will vary depending on the extent and frequency regime of the loss, the amount of time required for recovery, and the particular hair cell populations that are affected. For humans, the severity of hearing loss is normally described categorically as normal (0–15 dB hearing loss), slight (16–25 dB), mild (26–40 dB), moderate (41–55 dB), moderately severe (56–70 dB), severe (71–90 dB), and profound (91 dB or more) (Clark 1981). Although this scale is for humans, it gives an idea of the significance of various amounts of hearing loss to an animal; i.e., a NITS of 10 dB is a small amount of hearing loss, while 70 dB could be considered severe.

The most obvious consequence of a NITS is an increase in absolute threshold, which may arise from loss or damage to inner and/or outer hair cells. Elevated hearing thresholds would result in reduced detection ranges for sounds within the frequency range of loss, potentially affecting communication, navigation, and echolocation detection ranges during foraging. Damage or loss of outer hair cells would also reduce the active cochlear processes and cause a reduction in the compressive nonlinearity in the basilar membrane motion and a loss of frequency selectivity, which would broaden the excitation pattern along the basilar membrane (Moore 1998). Reduced frequency selectivity can in turn affect loudness perception, frequency discrimination, and the perception of complex sounds (Moore 1998). Abnormal frequency selectivity may also cause masking effects to be more pronounced in hearing-impaired listeners, especially when the masker and signal frequencies differ (Moore 1996). Hearing loss is often accompanied by a phenomenon called loudness recruitment, where the growth rate of loudness is higher in impaired ears compared to normal ears. This can cause an exaggerated sense of dynamic fluctuations in sounds, since the apparent loudness would change more dramatically than for a normal listener (Moore et al. 1996). Loudness recruitment could also result in an exaggerated sense of how fast a sound source is approaching or receding, since recruitment would result in a higher rate of a change of loudness compared to an unimpaired ear. Unilateral hearing loss can result in abnormal binaural or spatial hearing, leading to difficulties in localizing sound sources and using spectral cues to identify sound sources within background noise, and making it more difficult to spatially separate the locations of sound sources amidst background noise (Moore 1996, 1998). Hearing loss can also affect temporal resolution, making it more difficult to follow the temporal structure of time-varying sounds.

10.5 Conclusions

Much progress has been made in understanding the function of the auditory system in marine mammals and the potential adverse effects of noise on the hearing of these animals. Much of the resulting data have shown that marine mammal ears are very much analogous to those of terrestrial mammals, a result of their possessing inner ears very similar to terrestrial mammals. Detection of tones and complex sounds in Gaussian and comodulated noise, and measures of TTS have revealed that auditory masking and noise-induced threshold shifts in marine mammals behave similarly as those in humans and terrestrial mammals. The most significant differences concern the specific noise exposures required for masking and threshold shift effects in the various marine mammal species, and the frequency patterns of those effects.

Almost all of our information concerning the effects of noise on marine mammal perception has come from controlled experiments on captive animals. In many cases, the studies involved complex psychoacoustic tasks with "expert" subjects—animals for whom much time and effort have been spent in behavioral conditioning for specific experimental paradigms. Although conducting psychophysical tasks with captive subjects is a time consuming process which limits the maximum number of subjects for whom data can be obtained, many of the questions regarding perceptual effects of noise can only be answered in this fashion, and the degree to which stimuli can be controlled and manipulated cannot typically be matched in field studies.

Despite the progress made in understanding masking and noise-induced threshold shifts in marine mammals, many gaps in our understanding of how marine mammals perceive sound in noisy environments still remain. Aside from extrapolations based on anatomical data, information on mysticete hearing is almost completely lacking. Almost no data on masking and echolocation exist, even though all odontocetes rely on echolocation to capture prey, navigate, and potently detect predators. Almost all masking studies have employed Gaussian noise, assumed masking was restricted to a single auditory filter, and that the noise could be represented by its spectral density. This metric ignores temporal fluctuations that appear to play a significant role in an animal's ability to segregate a signal from noise. Identifying the proper noise metrics and a better understanding of auditory mechanisms that govern masking will help aid in making more accurate predictions about the effects of noise on communication. Data on noise-induced threshold shifts in marine mammals are available for only very few species, and few individuals within these species. There also remain significant questions regarding the effects of exposure frequency, the rate of TTS growth and recovery after exposure to intermittent noise, the effects of single and multiple impulses, and the extent and manner in which TTS data can be extrapolated to other species.

Acknowledgments The preparation of this paper was supported by the International Association of Oil and Gas Producers Joint Industry Programme (JIP) on Exploration & Production Sound and Marine Life, the US Navy Chief of Naval Operations (N45) Living Marine Resources Program, and the US Office of Naval Research Marine Mammal S&T Program.

References

29CFR1910.95 (2009) Occupational noise exposure. Occup Safety Health Stand 1910

American National Standards Institute (2011) ANSI S1.8-1989 (R2011) American national standard reference quantities for acoustical levels, vol ANSI S1.8-1989 (R2011). Acoustical Society of America, New York

Anderson DJ, Rose JE, Hind JE, Brugge JF (1971) Temporal position of discharges in single auditory nerve fibers within the cycle of a sine-wave stimulus: frequency and intensity effects. J Acoust Soc Am 49(4):1131–1139

ANSI S1.1-1994 (R 2004) (1994) American national standard acoustical terminology. Acoustical Society of America, New York

Au WWL (1980) Echolocation signals of the Atlantic bottlenose dolphin (*Tursiops truncatus*) in open waters. In: Busnel RG, Fish JF (eds) Animal sonar systems. Plenum, New York, pp 251–282

Au WWL, Moore PWB (1984) Receiving beam patterns and directivity indices of the Atlantic bottlenosed dolphin (*Tursiops truncatus*). J Acoust Soc Am 75(1):255–262

Au WWL, Moore PWB (1990) Critical ratio and critical bandwidth for the Atlantic bottlenose dolphin. J Acoust Soc Am 88(3):1635–1638

Au WWL, Carder DA, Penner R, Scronce BL (1988) Demonstration of adaptation in beluga whale echolocation. J Acoust Soc Am 93:1–14

Au WWL, Branstetter BK, Benoit-Bird KJ, Kastelein RA (2009) Acoustic basis for fish prey discrimination by echolocating dolphins and porpoises. J Acoust Soc Am 126(1):460–467

Bee MA, Buschermohle M, Klump GM (2007) Detecting modulated signals in modulated noise: (II) neural thresholds in the songbird forebrain. Eur J Neurosci 26(7):1979–1994

Branstetter BK, Finneran JJ (2008) Comodulation masking release in bottlenose dolphins (*Tursiops truncatus*). J Acoust Soc Am 124(1):625–633

Branstetter BK, Mercado III E, Au WWL (2007) Representing multiple discrimination cues in a computational model of the bottlenose dolphin auditory system. J Acoust Soc Am 122 (4):2459–2468

Branstetter BK, Moore PW, Finneran JJ, Tormey MN, Aihara H (2012) Directional properties of bottlenose dolphin (*Tursiops truncatus*) clicks, burst-pulse, and whistle sounds. J Acoust Soc Am 131(2):1613–1621

Bregman AS (1990) Auditory scene analysis: the perceptual organization of sound. The MIT Press, Massachusetts

Clark JG (1981) Uses and abuses of hearing loss classification. ASHA 23(7):493–500

Clark WW (1991) Recent studies of temporary threshold shifts (TTS) and permanent threshold shift (PTS) in animals. J Acoust Soc Am 90(1):155–163

Clark CW, Ellison WT, Southall BL, Hatch L, Van Parijs SM, Frankel A, Ponirakis D (2009) Acoustic masking in marine ecosystems: intuitions, analysis, and implication. Mar Ecol Prog Ser 395:201–222

Cornsweet TN (1962) The staircase method in psychophysics. Am J Psych 75:485–491

de Boer E, Nuttall AL (2010) Cochlear mechanics, tuning, non-linearities. In: Fuchs PA (ed) The ear, vol 1., The Oxford handbook of auditory scienceOxford University Press, New York

Dolphin WF, Au WW, Nachtigall PE, Pawloski J (1995) Modulation rate transfer functions to low-frequency carriers in three species of cetaceans. J Comp Physiol A 177(2):235–245

Elliott DN, Fraser WR (1970) Fatigue and adaptation. In: Tobias JV (ed) Foundations of modern auditory theory, vol I. Academic Press, New York, pp 117–155

Erbe C (2008) Critical ratios of beluga whales (*Delphinapterus leucas*) and masked signal duration. J Acoust Soc Am 124(4):2216–2223

Finneran JJ (2010) Auditory weighting functions and frequency-dependent effects of sound in bottlenose dolphins (*Tursiops truncatus*) (trans: 322 C). Marine Mammals and biological oceanography annual reports: FY10. Office of Naval Research (ONR), Washington, DC

Finneran JJ, Schlundt CE (2010) Frequency-dependent and longitudinal changes in noise-induced hearing loss in a bottlenose dolphin (*Tursiops truncatus*). J Acoust Soc Am 128(2):567–570

Finneran JJ, Schlundt CE (2011) Subjective loudness level measurements and equal loudness contours in a bottlenose dolphin (*Tursiops truncatus*). J Acoust Soc Am 130(5):3124–3136

Finneran JJ, Schlundt CE, Carder DA, Clark JA, Young JA, Gaspin JB, Ridgway SH (2000) Auditory and behavioral responses of bottlenose dolphins (*Tursiops truncatus*) and a beluga whale (*Delphinapterus leucas*) to impulsive sounds resembling distant signatures of underwater explosions. J Acoust Soc Am 108(1):417–431

Finneran JJ, Schlundt CE, Carder DA, Ridgway SH (2002a) Auditory filter shapes for the bottlenose dolphin (*Tursiops truncatus*) and the white whale (*Delphinapterus leucas*) derived with notched noise. J Acoust Soc Am 112(1):322–328

Finneran JJ, Schlundt CE, Dear R, Carder DA, Ridgway SH (2002b) Temporary shift in masked hearing thresholds (MTTS) in odontocetes after exposure to single underwater impulses from a seismic water gun. J Acoust Soc Am 111(6):2929–2940

Finneran JJ, Dear R, Carder DA, Ridgway SH (2003) Auditory and behavioral responses of California sea lions (*Zalophus californianus*) to single underwater impulses from an arc-gap transducer. J Acoust Soc Am 114(3):1667–1677

Finneran JJ, Carder DA, Schlundt CE, Ridgway SH (2005) Temporary threshold shift (TTS) in bottlenose dolphins (*Tursiops truncatus*) exposed to mid-frequency tones. J Acoust Soc Am 118(4):2696–2705

Finneran JJ, Houser DS, Schlundt CE (2007a) Objective detection of bottlenose dolphin (*Tursiops truncatus*) steady-state auditory evoked potentials in response to AM/FM tones. Aquat Mammals 33(1):43–54

Finneran JJ, London HR, Houser DS (2007b) Modulation rate transfer functions in bottlenose dolphins (*Tursiops truncatus*) with normal hearing and high-frequency hearing loss. J Comp Physiol A 193:835–843

Finneran JJ, Schlundt CE, Branstetter B, Dear RL (2007c) Assessing temporary threshold shift in a bottlenose dolphin (*Tursiops truncatus*) using multiple simultaneous auditory evoked potentials. J Acoust Soc Am 122(2):1249–1264

Finneran JJ, Houser DS, Mase-Guthrie B, Ewing RY, Lingenfelser RG (2009) Auditory evoked potentials in a stranded Gervais' beaked whale (*Mesoplodon europaeus*). J Acoust Soc Am 126(1):484–490

Finneran JJ, Carder DA, Schlundt CE, Dear RL (2010a) Growth and recovery of temporary threshold shift (TTS) at 3 kHz in bottlenose dolphins (*Tursiops truncatus*). J Acoust Soc Am 127(5):3256–3266

Finneran JJ, Carder DA, Schlundt CE, Dear RL (2010b) Temporary threshold shift in a bottlenose dolphin (*Tursiops truncatus*) exposed to intermittent tones. J Acoust Soc Am 127(5):3267–3272

Finneran JJ, Trickey JS, Branstetter BK, Schlundt CE, Jenkins K (2011) Auditory effects of multiple underwater impulses on bottlenose dolphins (*Tursiops truncatus*). J Acoust Soc Am 130:2561(A)

Finneran JJ, Branstetter BK, Trickey JS, Schlundt CE, Jenkins K (2012) Temporary threshold shift in bottlenose dolphins exposed to multiple air gun impulses. Paper presented at the joint industry programme on E&P sound and marine life programme review meeting II, Washington, DC

Fletcher H (1940) Auditory patterns. Rev Mod Phys 12:47–65

Hall JW, Haggard MP, Fernandes MA (1984) Detection in noise by spectro-temporal pattern analysis. J Acoust Soc Am 76:50–56

Hall JW, Grose JH, Haggard MP (1990) Effects of flanking band proximity, number, and modulation pattern on comodulation masking release. J Acoust Soc Am 87(1):269–283

Henderson D, Hamernik RP (1986) Impulse noise: critical review. J Acoust Soc Am 80(2):569–584

Henderson D, Bielefeld EC, Harris KC, Hu BH (2006) The role of oxidative stress in noise-induced hearing loss. Ear Hear 27(1):1–19

Holt MM, Schusterman RJ (2007) Spatial release from masking of aerial tones in pinnipeds. J Acoust Soc Am 121(2):1219–1225

Holt MM, Noren DP, Veirs V, Emmons CK, Veirs S (2008) Speaking up: killer whales (*Orcinus orca*) increase their call amplitude in response to vessel noise. J Acoust Soc Am 125(1):EL27–EL32

Humes LE, Jesteadt W (1989) Models of the additivity of masking. J Acoust Soc Am 85(3):1285–1294

Janik VM (2000) Source levels and the estimated active space of bottlenose dolphin (*Tursiops truncatus*) whistles in the Moray Firth Scotland. J Comp Physiol A 186(7–8):673–680

Johnson CS (1971) Auditory masking of one pure tone by another in the bottlenosed porpoise. J Acoust Soc Am 49 (4 (part 2)):1317–1318

Kastak D, Schusterman RJ (1996) Temporary threshold shift in a harbor seal (*Phoca vitulina*). J Acoust Soc Am 100(3):1905–1908

Kastak D, Schusterman RJ, Southall BL, Reichmuth CJ (1999) Underwater temporary threshold shift induced by octave-band noise in three species of pinniped. J Acoust Soc Am 106(2):1142–1148

Kastak D, Southall BL, Schusterman RJ, Kastak CR (2005) Underwater temporary threshold shift in pinnipeds: effects of noise level and duration. J Acoust Soc Am 118(5):3154–3163

Kastak D, Reichmuth C, Holt MM, Mulsow J, Southall BL, Schusterman RJ (2007) Onset, growth, and recovery of in-air temporary threshold shift in a California sea lion (*Zalophus californianus*). J Acoust Soc Am 122(5):2916–2924

Kastak D, Mulsow J, Ghoul A, Reichmuth C (2008) Noise-induced permanent threshold shift in a harbor seal. Acoustics 2008

Kastelein R, Gransier R, van Mierlo R, Hoek L, de Jong C (2011) Temporary hearing threshold shifts and recovery in a harbor porpoise (*Phocoena phocoena*) and harbor seals (*Phoca vitulina*) exposed to white noise in a 1/1 octave band around 4 kHz. J Acoust Soc Am 129:2432 (A)

Keeler JS (1968) Compatible exposure and recovery functions for temporary threshold shift-mechanical and electrical models. J Sound Vib 2:220–235

Ketten DR (2000) Cetacean ears. In: Au W, Popper AN, Fay RR (eds) Hearing by whales and dolphins. Springer handbook of auditory research, 1st edn. Springer, New York, pp 43–108

Kryter KD (1973) Impairment to hearing from exposure to noise. J Acoust Soc Am 53(5):1211–1234

Kryter KD, Ward WD, Miller JD, Eldredge DH (1966) Hazardous exposure to intermittent and steady-state noise. J Acoust Soc Am 39(3):451–464

Kujawa SG, Liberman MC (2009) Adding insult to injury: cochlear nerve degeneration after "temporary" noise-induced hearing loss. J Neurosci 29(45):14077–14085

Lammers MO, Au WWL (2003) Directionality in the whistles of Hawaiian spinner dolphins (*Stenella longirostris*): a signal feature to cue direction of movement? Mar Mammal Sci 19(2):249–264

Lemonds DW (1999) Auditory filter shapes in an Atlantic bottlenose dolphin (*Tursiops truncatus*). University of Hawaii

Lemonds DW, Kloepper LN, Nachtigall PE, Au WWL, Vlachos SA, Branstetter BK (2011) A re-evaluation of auditory filter shape in delphinid odontocetes: evidence of constant-bandwidth filters. J Acoust Soc Am 130(5):3107–3114

Levitt H (1971) Transformed up-down methods in psyhcoacoustics. J Acoust Soc Am 49:467–477

Lucke K, Siebert U, Lepper PA, Blanchet M-A (2009) Temporary shift in masked hearing thresholds in a harbor porpoise (*Phocoena phocoena*) after exposure to seismic airgun stimuli. J Acoust Soc Am 125(6):4060–4070

Maslen KR (1981) Towards a better understanding of temporary threshold shift of hearing. Appl Acoust 14:281–318

McCormick JG, Wever EG, Palin J, Ridgway SH (1970) Sound conduction in the dolphin ear. J Acoust Soc Am 48(6):1418–1428

McDonald MA, Hildebrand JA, Mesnick S (2009) Worldwide decline in tonal frequencies of blue whale songs. Endanger Species Res 9:13–21

McFadden D (1986) The curious half-octave shift: evidence for a basalward migration of the traveling-wave envelope with increasing intensity. In: Salvi RJ, Henderson D, Hamernik RP, Coletti V (eds) Basic and applied aspects of noise-induced hearing loss, vol 111. Proceedings of a NATO advanced studies institute on applied and basic aspects of noise-induced hearing loss, held September 23–29, 1985, in Lucca. NATO ASI Series A, Life Sciences edn. Plenum, New York, pp 295–312

McFadden D (1988) Comodulation masking release: effects of varying the level, duration, and time delay of the cue band. J Acoust Soc Am 80:1658–1672

McFadden D, Plattsmier HS (1983) Frequency patterns of TTS for different exposure intensities. J Acoust Soc Am 74(4):1178–1184

Melnick W (1991) Human temporary threshold shift (TTS) and damage risk. J Acoust Soc Am 90(1):147–154

Miller JD (1974) Effects of noise on people. J Acoust Soc Am 56(3):729–764

Miller PJO (2002) Mixed-directionality of killer whale stereotyped calls: a direction of movement cue? Behav Ecol Sociobiol 52:262–270

Miller PJO, Biassoni N, Samuels A, Tyack PL (2000) Whale songs lengthen in response to sonar. Nature 405(6789):903

Mills JH (1976) Threshold shifts produced by a 90-day exposure to noise. In: Henderson D, Hamernik RP, Dosanjh DS, Mills JH (eds) Effects of noise on hearing. Raven Press, New York, pp 265–275

Mills JH, Gilbert RM, Adkins WY (1979) Temporary threshold shifts in humans exposed to octave bands of noise for 16 to 24 hours. J Acoust Soc Am 65(5):1238–1248

Mooney TA, Nachtigall PE, Breese M, Vlachos S, Au WWL (2009a) Predicting temporary threshold shifts in a bottlenose dolphin (*Tursiops truncatus*): the effects of noise level and duration. J Acoust Soc Am 125(3):1816–1826

Mooney TA, Nachtigall PE, Vlachos S (2009b) Sonar-induced temporary hearing loss in dolphins. Biol Lett 5(4):565–567

Moore BC (1996) Perceptual consequences of cochlear hearing loss and their implications for the design of hearing aids. Ear Hear 17(2):133–161

Moore BCJ (1998) Cochlear hearing loss. Whurr Publishers Ltd, London

Moore BCJ, Glasberg BR (2003) Behavioural measurement of level-dependent shifts in the vibration pattern on the basilar membrane at 1 and 2 kHz. Hear Res 175:66–74

Moore BCJ, Wojtczak M, Vickers DA (1996) Effect of loudness recruitment on the perception of amplitude modulation. J Acoust Soc Am 100(1):481–489

Mulsow J, Reichmuth C (2007) Electrophysiological assessment of temporal resolution in pinnipeds. Aquat Mammals 33(1):122–131

Mulsow JL, Reichmuth C (2010) Psychophysical and electrophysiological aerial audiograms of a Steller sea lion (*Eumetopias jubatus*). J Acoust Soc Am 127(4):2692–2701

Mulsow J, Reichmuth C, Gulland FMD, Rosen DAS, Finneran JJ (2011a) Aerial audiograms of several California sea lions (*Zalophus californianus*) and Steller sea lions (*Eumetopias jubatus*) measured using single and multiple simultaneous auditory steady-state response methods. J Exp Biol 214:1138–1147

Mulsow JL, Finneran JJ, Houser DS (2011b) California sea lion (*Zalophus californianus*) aerial hearing sensitivity measured using auditory steady-state response and psychophysical methods. J Acoust Soc Am 129(4):2298–2306

Nachtigall PE, Pawloski J, Au WWL (2003) Temporary threshold shifts and recovery following noise exposure in the Atlantic bottlenosed dolphin (*Tursiops truncatus*). J Acoust Soc Am 113(6):3425–3429

Nachtigall PE, Supin AY, Pawloski J, Au WWL (2004) Temporary threshold shifts after noise exposure in the bottlenose dolphin (*Tursiops truncatus*) measured using evoked auditory potentials. Mar Mammal Sci 20(4):673–687

Nachtigall PE, Yuen MML, Mooney TA, Taylor KA (2005) Hearing measurements from a stranded infant Risso's dolphin, *Grampus griseus*. J Exp Biol 208:4181–4188

Nachtigall PE, Mooney TA, Taylor KA, Miller LA, Rasmussen MH, Akamatsu T, Teilmann J, Linnenschmidt M, Vikingsson GA (2008) Shipboard measurements of the hearing of the white-beaked dolphin *Lagenorhynchus albirostris*. J Exp Biol 211:642–647

Navy US (2008) Southern California Range Complex: Final Environmental Impact Statement/Overseas Environmental Impact Statement. Department of the Navy, Washington, DC

Nelken I, Jacobson G, Ahdut L, Ulanovsky N (eds) (2001) Neural correlates of co-modulation masking release in auditory cortex of cats. Physiological and psychophysical basis of auditory function. Shaker Publishing

Nixon JC, Glorig A (1961) Noise-induced permanent threshold shift at 2000 cps and 4000 cps. J Acoust Soc Am 33(7):904–908

Nummela S (2008a) Hearing. In: Perrin WF, Wursig B, Thewissen JGM (eds) Encyclopedia of marine mammals, 2nd edn. Academic Press, Burlington, pp 553–561

Nummela S (2008b) Hearing in aquatic mammals. In: Thewissen JGM, Nummela S (eds) Sensory evolution on the threshold. University of California Press, Berkeley, pp 211–224

Parks SE, Johnson M, Nowacek D, Tyack PL (2011) Individual right whales call louder in increased environmental noise. Biol Lett 7:33–35

Patterson RD (1976) Auditory filter shapes derived with noise stimuli. J Acoust Soc Am 59(3):640–654

Patterson RD, Moore BCJ (1986) Auditory filters and excitation patterns as representations of frequency resolution. In: Moore BCJ (ed) Frequency selectivity in hearing. Academic, London, pp 123–127

Patuzzi R (1998) Exponential onset and recovery of temporary threshold shift after loud sound: evidence for long-term inactivation of mechano-electrical transduction channels. Hear Res 125:17–38

Popov VV, Supin AY, Klishin VO (1996) Frequency tuning curves of the dolphin's hearing: envelope-following response study. J Comp Physiol A 178(4):571–578

Popov VV, Supin AY, Wang D, Wank K, Xiao J, Li S (2005) Evoked-potential audiogram of the Yangtze finless porpoise *Neophocaena phocaenoides asiaeorientalis* (L). J Acoust Soc Am 117(5):2728–2731

Popov V, Supin A, Wang D, Wang K (2006) Nonconstant quality of auditory filters in the porpoises, *Phocoena phocoena* and *Neophocaena phocaenoides* (Cetacea, Phocoenidae). J Acoust Soc Am 119(5):3173–3180

Popov VV, Supin AY, Klishin VO, Tarakanov MB, Pletenko MG (2008) Evidence for double acoustic windows in the dolphin *Tursiops truncatus*. J Acoust Soc Am 123(1):552–560

Popov VV, Supin AY, Wang D, Wang K, Dong L, Wang S (2011) Noise-induced temporary threshold shift and recovery in Yangtze finless porpoises *Neophocaena phocaenoides asiaeorientalis*. J Acoust Soc Am 130(1):574–584

Pressnitzer D, Meddis R, Winter IM (2001) Physiological correlates of comodulation masking release in the mammalian ventral cochlear nucleus. J Neurosci 21(16):6377–6386

Quaranta A, Portalatini P, Henderson D (1998) Temporary and permanent threshold shift: an overview. Scand Audiol 48:75–86

Ridgway SH (1999) The cetacean central nervous system. In: Adelman G, Smith BH (eds) Elsevier's Encyclopedia of Neuroscience, pp 352–357

Ridgway SH, Carder DA, Smith RR, Kamolnick T, Schlundt CE, Elsberry WR (1997) Behavioral responses and temporary shift in masked hearing thresholds of bottlenose dolphins, *Tursiops truncatus*, to 1-second tones of 141–201 dB re 1 μPa. Naval Command, Control, and Ocean Surveillance Center, RDT&E Division, San Diego

Roitblat HL, Moore PWB, Helweg DA, Nachtigall PE (1993) Representation and processing of acoustic information in a biomimetic neural network. In: Meyer J-A, Roitblat HL, Wilson SW (eds) Animals to animats 2: stimulation of adaptive behavior. MIT press, pp 1–10

Schlundt CE, Finneran JJ, Carder DA, Ridgway SH (2000) Temporary shift in masked hearing thresholds of bottlenose dolphins, *Tursiops truncatus*, and white whales, *Delphinapterus leucas*, after exposure to intense tones. J Acoust Soc Am 107(6):3496–3508

Schlundt CE, Dear RL, Green L, Houser DS, Finneran JJ (2007) Simultaneously measured behavioral and electrophysiological hearing thresholds in a bottlenose dolphin (*Tursiops truncatus*). J Acoust Soc Am 122(1):615–622

Schlundt CE, Finneran JJ, Branstetter BK, Dear RL, Houser DS, Hernandez E (2008) Evoked potential and behavioral hearing thresholds in nine bottlenose dolphins (*Tursiops truncatus*). J Acoust Soc Am 123:3506(A)

Schlundt CE, Dear RL, Houser DS, Bowles AE, Reidarson T, Finneran JJ (2011) Auditory evoked potentials in two short-finned pilot whales (*Globicephala macrorhynchus*). J Acoust Soc Am 129(2):1111–1116

Southall BL, Bowles AE, Ellison WT, Finneran JJ, Gentry RL, Greene CR Jr, Kastak D, Ketten DR, Miller JH, Nachtigall PE, Richardson WJ, Thomas JA, Tyack PL (2007) Marine mammal noise exposure criteria: initial scientific recommendations. Aquat Mammals 33(4):411–521

Supin AY, Popov VV (1986) Tonal hearing-masking curves in bottlenosed dolphins. Doklady Akademii Nauk SSSR 289:242–246

Supin AY, Popov VV (1995) Envelope-following response and modulation transfer function in the dolphin's auditory system. Hear Res 92:38–46

Trickey JS, Branstetter BB, Finneran JJ (2011) Auditory masking with environmental, comodulated, and Gaussian noise in bottlenose dolphins (*Tursiops truncatus*). J Acoust Soc Am 128(6):3799–3804

Ward WD (1962) Damage-risk criteria for line spectra. J Acoust Soc Am 34(10):1610–1619

Ward WD (1997) Effects of high-intensity sound. In: Crocker MJ (ed) Encyclopedia of acoustics. Wiley, New York, pp 1497–1507

Ward WD, Cushing EM, Burns EM (1976) Effective quiet and moderate TTS: implications for noise exposure standards. J Acoust Soc Am 59(1):160–165

Yuen MML, Nachtigall PE, Breese M, Supin AY (2005) Behavioral and auditory evoked potential audiograms of a false killer whale (*Pseudorca crassidens*). J Acoust Soc Am 118(4):2688–2695

Part III
Optical, Electric, and Chemical Signals

Chapter 11
Noise in Visual Communication: Motion from Wind-Blown Plants

Richard A. Peters

Abstract Animals demonstrate with their signalling strategies that they are sensitive to signal efficacy. Signallers can choose favourable conditions or alter the structure of their signals at times of increased noise. The nature of these adjustments has provided important insights into how signal evolution is constrained by the noise landscape. Only recently, have we shown that the structure of movement-based visual signals depends on ambient motion noise caused by wind-blown plants, but our depth of understanding has been constrained by our limited knowledge of motion noise. We therefore need to understand in detail how plants move. In this chapter, I outline how and why plant interactions with wind will vary according to plant species, plant geometry, microhabitat structure, and the light environment. Ultimately, we will need to consider signal and noise together to truly determine the masking effect of plant motion. With this in mind, I conclude by suggesting that a fresh look at movement-based signals and plant motion noise is needed.

11.1 Introduction

To explain the evolution of signals, it is crucial to understand the conditions in which signalling takes place. Signals must be transmitted through a given environment and their structure must enable reliable detection and efficient processing by receiver sensory systems. A signal that does not achieve these outcomes is ineffective (Guilford and Dawkins 1992) and would be, in time, replaced by signals that fulfil these requirements (Endler and Basolo 1998). Therefore, signals need to be designed to minimise corruption by transmission channel properties

R. A. Peters (✉)
Department of Zoology La Trobe University, Bundoora, VIC 3086, Australia
e-mail: richard.peters@latrobe.edu.au

H. Brumm (ed.), *Animal Communication and Noise*,
Animal Signals and Communication 2, DOI: 10.1007/978-3-642-41494-7_11,
© Springer-Verlag Berlin Heidelberg 2013

(e.g. attenuation, scattering and reflections) and to remain salient in the ambient noise background. Not all sources of signal corruption and background noise are predictable and animals, therefore, can be expected to make use of any available feedback regarding the efficacy of their signals and adjust them accordingly. The ways in which senders change their signalling strategy in noisy conditions are cases in point (Chaps. 3, 4, 5, 7 and 9). Clearly, consideration of signal structure thus must go hand-in-hand with the careful analysis of the structure and dynamics of noise. This is more important because noise not only affects communication systems, but also other aspects of animal behaviour. For example, urban noise not only causes changes in signal structure (Chap. 7), it also might influence species distributions and foraging activities (Chap. 14).

Aside from recent contributions humans have made to the 'noise landscape', animal signals have evolved to compete with noise from biotic and abiotic sources. Biotic noises are those emanating from living organisms. The signals from sympatric animals can interfere with reliable signal transmission and must be overcome. Wind and running water are common sources of abiotic noise. For example, even a few drops of rain can adversely affect the signalling environment of insects that generate vibrations in host plants to communicate with conspecifics (Barth et al. 1988). Moderate rainfall leads to a decreasing signal-to-noise ratio for these insects and heavy rain is likely to make efficient signalling impossible (Cocroft and Rodriguez 2005).

Wind can also have a direct effect on the environment, which in turn influences signalling. For example, surface wave breaking in strong winds causes humpback whales to favour surface-generated acoustic signals like breaching and pectoral fin slapping over underwater vocalisations (Dunlop et al. 2010). Another major environmental impact of wind is the effect it has on plants. Indeed, wind affects vibratory, acoustic and visual communication strategies. It possibly has the most significant influence on vibratory communication because the vibrations it generates in plants will influence how and when signalling can take place (Cocroft and Rodriguez 2005; Mcnett et al. 2010). Importantly, plant responses to wind vary depending on the transmission properties that differ between plant species (Barth et al. 1988). These vibratory niches are likely to affect other aspects of animal behaviour (see Cocroft and Rodriguez 2005).

The effect of wind on plants also affects animals that communicate using movement and is the focus for this chapter. There have been relatively few empirical studies on the structure of plant motion noise in the context of animal communication. In the following, I will demonstrate the importance of wind–plant interactions for movement-based animal communication systems by:

- Reviewing movement-based signalling and the masking effect of plant motion;
- Highlighting variation in wind–plant interactions;
- Considering microhabitats as distinct image motion environments;
- Suggesting future directions: virtual microhabitats.

11.2 Movement-Based Visual Signals and Evidence for the Masking Effect of Plant Motion

Animals use movement to communicate. The image motion that animal move-ments generate is one of the most salient features in the visual world of animals: image motion sensing is crucial for prey capture, predator avoidance, camouflage breaking and navigation. Unlike sounds, however, the detection of motion is highly dependent on viewer orientation. Adaptations to overcome this constraint, like sampling most of the visual field (Smolka and Hemmi 2009) or high con-centrations of motion sensitive cells for peripheral vision (Stein and Gaither 1981, 1983), facilitate the detection of salient motion information and make the gener-ation of motion cues a credible option for signalling.

Movement-based visual signals are indeed quite common. They have been documented in diverse taxonomic groups ranging from mammals (e.g. Thomson's gazelles, *Eudorcas thomsonii*: Caro 1995; squirrels, *Spermophilus beecheyi*: Rundus et al. 2007) to invertebrates (e.g. fiddler crabs, *Uca* sp.: How et al. 2009; jumping spiders, *Habronattus dossenus*: Elias et al. 2006), and are used in a variety of contexts including opponent assessment (Peters and Ord 2003), mate choice (How et al. 2008) and antipredator signalling (Leal 1999, Rundus et al. 2007). Movement-based signals are diverse in interesting ways. A nice illustration of this can be seen in the fiddler crabs (Genus: *Uca*), which raise and lower their enlarged claw to generate a conspicuous signal (Crane 1975). Even for such a simple motor pattern, crabs exhibit inter- and/or intra-specific structural variation due to tidal cycle (Crane 1975), age (Hyatt 1977), ambient temperature (Doherty 1982), receiver distance (How et al. 2008), as well as individual identity, geographic location and social context (How et al. 2007, 2009).

Fiddler crabs exhibit a surprisingly diverse range of signals yet it remains unclear whether the signals vary in response to signalling conditions (e.g. How et al. 2009; Milner et al. 2008). To examine how movement-based signals are affected by signalling conditions we shift our focus to lizards (Order: Squamata). Lizards from the Iguanid and Agamid families in particular have been widely studied in the context of movement-based signalling (see Ord and Blumstein 2002; Ord et al. 2001, 2002 and references therein). Early work indicated that Iguanid lizards typically have a single species-specific display or choose from a limited repertoire of displays (Crews 1975; Jenssen 1975; Rothblum and Jenssen 1978; Stamps and Barlow 1973). More recent analyses have suggested that signal diversity in lizard displays relates to habitat structure, with differences in signal complexity between species influenced by ecological factors such as home range size and diet (Ord et al. 2002). In addition, the plant environment is emerging as an ecological parameter of note because of the changing background it creates depending on prevailing wind conditions. The potential masking effect of plant movement for the detection of movement-based signals was first considered in detail in work undertaken by Leo Fleishman (reviewed in Fleishman 1992). Fle-ishman (1986) conducted a series of elegant experiments in which he considered the relative effectiveness of artificial lures at eliciting orienting responses in *Anolis auratus*. Fleishman varied the motion characteristics of these lures, as well as simulating plant motion in the background. Performance declined when the lure moved at similar frequencies to the oscillating plant background. I extended this idea to the detection of movement-based signals in a recent playback study with Jacky lizards (*Amphibolurus muricatus*; Fig. 11.1a). Employing a radio-controlled flicking tail within a planted habitat, I demonstrated that plant motion lengthened the time taken to detect a (simulated) movement-based signal (Fig. 11.1b; Peters 2008). My conclusions were much like those posited by Fleishman 20 years earlier that detection difficulties might be "due to direct interference with the sensory mechanism (the background acting as 'noise') such that the motion detection apparatus cannot distinguish the motion of the stimulus" (Fleishman 1986, p 718). Clearly, the spatiotemporal properties of movement-based signals and plant motion are sufficiently similar to make signal detection more difficult.

The masking effect of noise in any communication system is perhaps best demonstrated by careful consideration of what animals do in response to changing signalling conditions. A number of recent studies have now demonstrated that plant motion noise is an important sensory constraint for the reception of movement-based signals (Ord et al. 2007; Ord and Stamps 2008; Peters et al. 2007). Ord et al. (2007) were the first to report a relationship between the structure of movement signals and that of background noise. By analysing separately the image motion generated by plant motion and the displays of Puerto Rican lizards, *Anolis cristatellus* and *A. gundlachi,* we identified a strong correlation between the maximum speeds of displays and surrounding plant motion. Peters et al. (2007) subsequently reported the first experimental evidence for modification of movement-based signals due to environmental conditions in any taxonomic group. Rather than adjusting display speed, Australian Jacky lizards significantly lengthened the duration of tail flicking in conditions of greater plant motion (Fig. 11.1c). Tail flicking in Jacky lizards represents the introductory portion of the territorial display of these lizards (Peters and Ord 2003), and because of its alerting function it is the motor pattern most likely to be influenced by prevailing noise levels. The lizards also changed the structure of their signal by switching to intermittent tail flicking in windy conditions. Interestingly, the behaviour of Jacky lizards in adverse conditions matches nicely the predictions from an earlier video playback study (Peters and Evans 2003b). Using animated tail flicking signals in constant noise conditions, flicking intermittently for long durations was the most effective at attracting attention of lizard receivers compared with other strategies like faster speeds or larger amplitudes (see Chap. 7 for examples of longer song and call durations by birds in response to noise). Increasing the efficacy of a movement-based signal in noise can also be achieved by adding a novel and conspicuous display component (Ord and Stamps 2008). *Anolis gundlachi* were found to add a motor pattern at the start of their display when signalling conditions were less conducive to signal transmission. Using robotic playback techniques, Ord and Stamps (2008) also demonstrated that the facultative addition of this motor pattern significantly shortened detection times compared to the same signal

Fig. 11.1 **a** The Australian Jacky lizard (*Amphibolurus muricatus*). **b** Mean (±S.E.) response latencies of Jacky lizards in detecting a mechanical lizard at 1 and 3 m viewing distances during playback experiments. Responses were significantly delayed during windy conditions when plant motion was stronger (*Source* Peters 2008). **c** Jacky lizards adjusted the duration of introductory, attention-grabbing tail flicking in windy conditions (*Source* Peters et al. 2007)

(a)

(b)

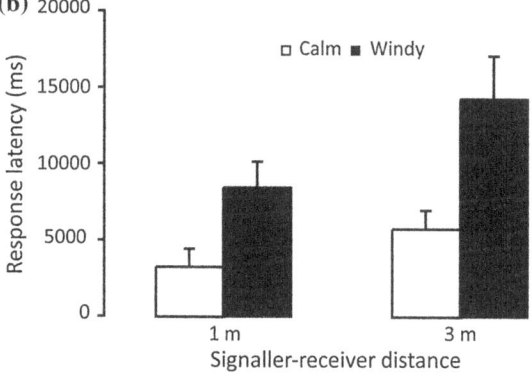

(c)

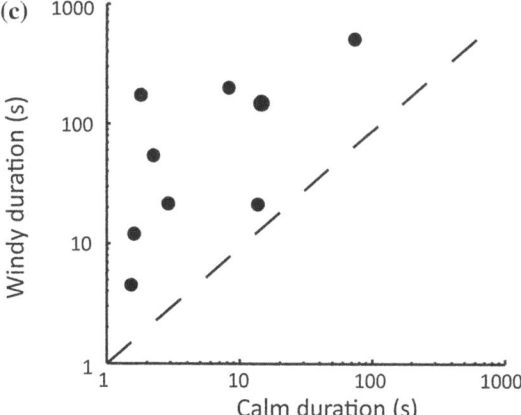

without the added component. A final, yet obvious, strategy for overcoming plant motion noise is to time signalling to coincide with lulls in prevailing wind conditions (Ord et al. 2011).

The studies described above provide a clear demonstration that plant motion acts as noise for movement-based signals and that animals are sensitive to the likely reduction in efficacy of certain signal characteristics. Given what we know from the acoustic domain (see Chaps. 3, 4, 5, 6, 7, 8, 9, 10), we still have much to learn about the relationship between movement-based signals and plant motion. From the outset, it will be worthwhile to consider the circumstances that have led to the evolution of alternative strategies for offsetting the masking effect of plant motion. One path of investigation must consider evolutionary starting points giving rise to morphological and/or physiological differences between species (Ord et al. 2011), but as the acoustic literature demonstrates, understanding different signalling environments is of fundamental importance to understanding signal structure. Therefore, within the context of movement-based animal signalling, we must understand in detail *how plants move*.

11.3 Wind–Plant Interactions

Quantifying the way plants move has received only modest attention in the animal communication literature. Fleishman (1986, 1988) was possibly the first to address the topic. Using Fourier analysis to characterise the movement of single blades of grass, he showed that plant motion varied as a function of species and the prevailing wind conditions (Fleishman 1988). I extended Fleishman's early work some years later by analysing the motion of whole plants (Peters and Evans 2003a) and microhabitats (Fig. 11.2a; Peters et al. 2008) using computational motion analysis that once again revealed plant species differences in movement characteristics and the importance of prevailing wind, and hinted at the masking effect of plant movement (Fig. 11.2b). However, there is a sizeable literature dealing with the response of plants to wind, predominantly because of the important role wind plays in biomass production (food and materials) and therefore the economic consequences of wind. de Langre (2008) presents an excellent review of the literature outlining the breadth of interest, motivations for research and levels of analysis. Published models of plant motion under wind vary in their focus from canopies and individual trees (Py et al. 2005; Sellier et al. 2006) to branches and leaves (Staelens et al. 2003; Watt et al. 2005). I have no intention of summarising here the details of such models and refer interested readers to thorough and technical descriptions provided by others (e.g. de Langre 2008; Coutts and Grace 1995; Niklas 1992). Rather, my goal is to outline why microhabitats are likely to reflect distinct image motion environments for movement-based signalling systems (Peters et al. 2008).

Environments feature a variety of plant species and multiple exemplars of the same species, yet each will move differently in response to wind as a consequence

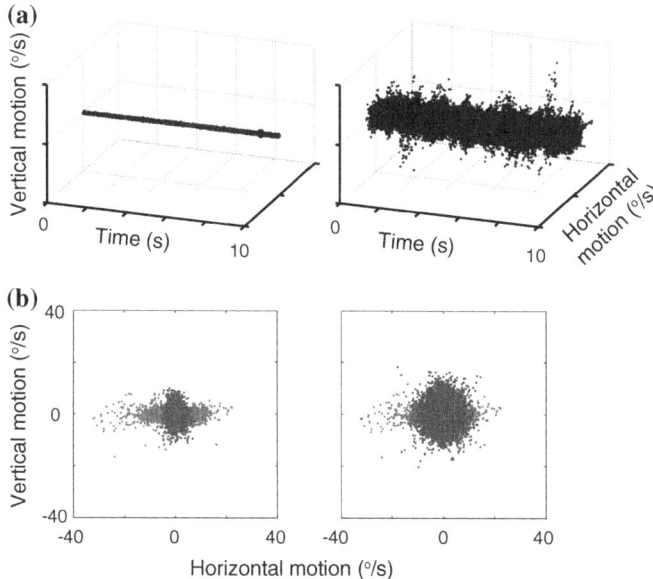

Fig. 11.2 **a** Prevailing wind generates stronger plant movements leading to greater image motion. This is illustrated for 10 s clips of *Acacia longifolia* during wind-still conditions (*left*) and when the prevailing wind fluctuated between 2 and 3 m/s (*right*). In these charts, scatter plots of the velocity field summarise the horizontal and vertical motion measured at each spatial location using gradient detectors (Adapted from Peters et al. 2008). **b** Image motion generated during separate 8 s clips of wind-blown *Lomandra longifolia* (*red dots*) and a lizard display (*blue dots*) in calm (*left*) and windy (*right*) conditions reveal the potential masking effect of plant motion that will likely influence the detection of movement-based animal signals in certain conditions (Adapted from Peters et al. 2007)

of the structure and geometry of individual plants. In order to illustrate the scope for variation in wind–plant interactions I will focus on coastal heath environments of south-eastern Australia where I study the Australian Jacky lizard (*A. muricatus*; Fig. 11.1a), and in particular, an area around Point Hicks (37°48′ S 149°16′ E) within Croajingolong National Park, Victoria. Plant species found in this region are similar to those seen at other coastal heath habitats along the Australian south-east coast. Such coastal plant communities thrive in salt rich substrates and atmospheres (sea spray) and can cope with variable wind conditions (Costermans 2005). The harsh environment results in plants with particular adaptations to suit the conditions, however, plant movement in response to wind is not homogenous and noise profiles vary due to variation in plant structure and geometry. I consider several representative monocotyledons and dicotyledons separately because plants from each botanical subdivision respond to prevailing wind in similar ways.

Monocotyledons that contribute to observable plant motion around Point Hicks include a variety of native grasses (Poeceae), rushes (Juncaceae) and the grass tree *Xanthorrhoea resinosa*. Wind generates motion in leaves as well as stems and inflorescences (seed head) and structural variation between specimens produces different motion characteristics. I have not quantified the characteristic movement of these plants but take this opportunity to speculate on general movement patterns based on my observations and the published literature. To begin with, movement will be constrained by the plant's active degrees of freedom (Fig. 11.3a). In addition, a number of key parameters affect wind-induced plant movement that apply both between and within species. Figure 11.4 sketches four local species to help introduce these parameters and identify likely sources of variation: *Poa poiformis*, *Themeda triandra*, *Lomandra longifolia* and *X. resinosa*. Variation in plant movement will be influenced by:

- *Height/length*. Generally speaking, larger amplitude movements are more likely to be generated by plants with taller stems and/or longer leaves. The flowering part of *X. resinosa* is the longest structure of the four species in Fig. 11.4 and can reach 2.7 m in height (scape + spike). The sweep area of this structure will be relative large, however, it also has a limited range of motion due to other parameters like stiffness.
- *Stiffness*. Movement amplitudes will be constrained by the structure's ability to resist bending forces and is applicable to the plant stem and leaves. The shape of the structure as well as the material that defines it determines stiffness. Furthermore, as these types of plants are fixed at the base, movement might also be constrained by stiffness at the root base.
- *Weight at the top* of the stem affects the mechanical behaviour of a plant such that greater deflection angles are likely with increased weights (Niklas 1992). Inflorescent structures for each of the four examples in Fig. 11.4 will affect movement.
- *Leaf width and shape*. Plant movement occurs because of the drag force of the wind. Differences in the area exposed to the wind will lead to observable

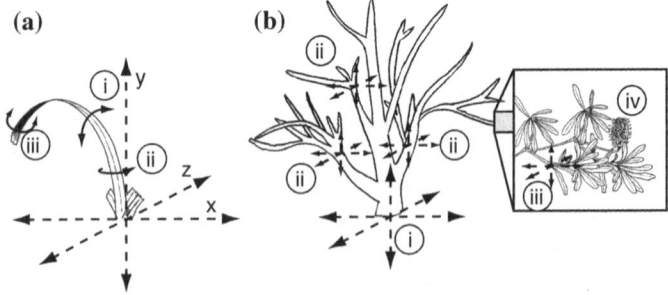

Fig. 11.3 **a** Degrees of freedom of monocotyledons illustrated for *Lomandra longifolia*. Movement of a given leaf, fixed at the base, includes up and down movement along the *x–y* plane (*i*), extending into the *z*-plane (*ii*) as well as rotational movement of the leaf itself (*iii*). **b** Dicotyledons contain multiple fixation points. These include: (*i*) movement of the trunk of the plant that will generate more global movement, (*ii*) relatively independent movements at branching points and (*iii*) leaves, as well as movement of (*iv*) flowering parts of the plant

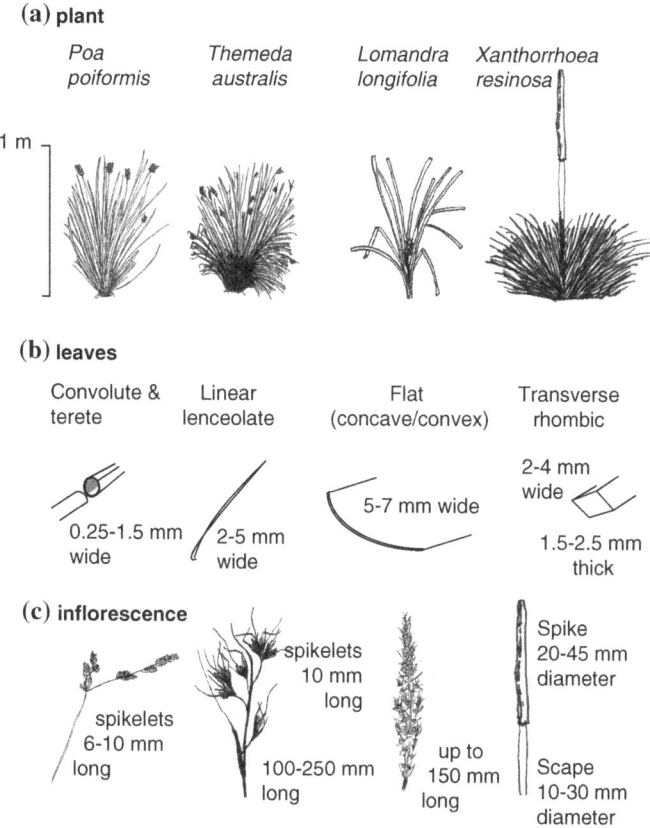

(a) plant

Poa poiformis   *Themeda australis*   *Lomandra longifolia*   *Xanthorrhoea resinosa*

1 m

(b) leaves

Convolute & terete | Linear lenceolate | Flat (concave/convex) | Transverse rhombic

0.25-1.5 mm wide

2-5 mm wide

5-7 mm wide

2-4 mm wide

1.5-2.5 mm thick

(c) inflorescence

spikelets 6-10 mm long

spikelets 10 mm long

100-250 mm long

up to 150 mm long

Spike 20-45 mm diameter

Scape 10-30 mm diameter

Fig. 11.4   Representative monocotyledons found in coastal heath at Point Hicks, Croajingolong National Park, Victoria, Australia. Although superficially similar in their overall size and shape (**a**) there are considerable differences in leaf structure (**b**) and inflorescence structure (**c**)

differences in plant motion. Leaf structural differences between *P. Poiformis* (terete), *L. longifolia* (flat or concave/convex) and *X. resinosa* (transverse rhombic) may well contribute to variation in response to wind. Of the four examples, *L. longifolia* tends to dominate the plant motion landscape and a major contributing factor may indeed be leaf shape. The leaves of this species are relatively wide and coupled with its flat shape will generate greater drag and hence more conspicuous movements. In comparison, younger specimens of *L. longifolia* and the other grasses (*T. triandra* and *P. poiformis*) have differently shaped leaves that result in lower drag forces and hence do not seem to generate large amplitude movements.

The monocotyledons are mostly single stem plants, which makes quantifying their response to wind more straightforward as modellers can draw heavily from engineering theories of bending (Wood 1995). Branching found in dicotyledons introduces a degree of complexity to modelling wind-induced plant movements (Fig. 11.3b). Depending on the characteristics of the prevailing wind, movement in shrubs and trees can vary from simple flutters of the leaves to small areas of localised movement of one or more branches to coordinated movement of the whole plant (Fig. 11.3b). All of these types of movement are relevant for movement-based animal signalling. I have identified six plants common in the study area to illustrate how plant motion noise might vary from one location to the next: coastal wattle (*Acacia longifolia* var *sophorae*), coastal banksia (*Banksia integrifolia*), coastal tea-tree (*Leptospermum laevigatum*), coastal beard-heath (*Leucopogon parviflorus*), scented paper bark (*Melaleuca squarrosa*) and coastal rosemary (*Westringia fruticosa*).

Wind-induced plant movement is still apparent on relatively wind-still days, although the effect is largely limited to fluttering of the leaves. Differences between species under the same conditions may be attributable to variation in leaf structure and arrangement. Figure 11.5 illustrates variation in the six common

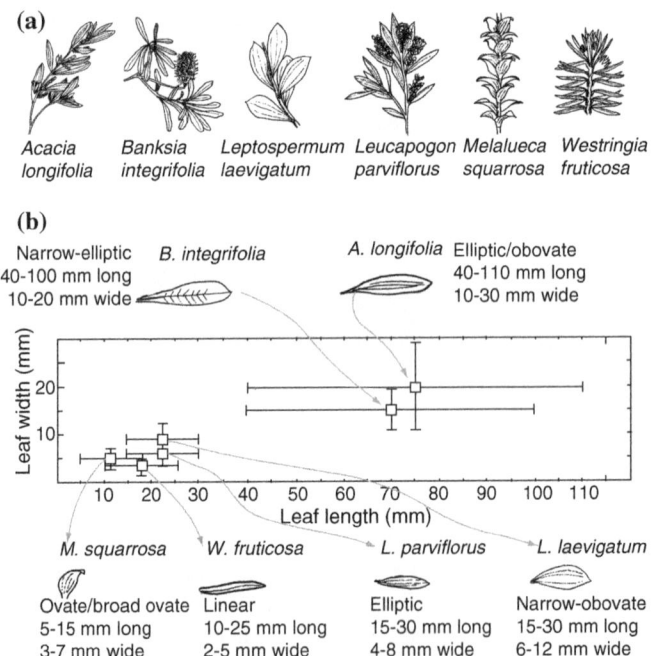

Fig. 11.5 **a** Leaf structure of six representative dicotyledons found in coastal heath at Point Hicks, Croajingolong National Park, Victoria, Australia. These plants are all suited to the environmental conditions of the coastal region but still exhibit diversity in structure. **b** The response of plants to wind will vary in many ways as described in the text, including the shape and size of leaves. Four of these plants (*M. squarrosa*, *W. fruticosa*, *L. parviflorus* and *L. Laevigatum*) have relatively short and narrow leaves but are also relatively densely packed around its host twig/branch. In contrast, the other two (*B. integrifolia* and *A. longifolia*) are considerably longer and wider and extend further from its host twig/branch

dicotyledons in the study area. Four species exhibit short and narrow leaves, densely arranged such that light wind will likely generate only small amplitude movements (*L. laevigatum*, *L. parviflorus*, *M. squarrosa* and *W. fruticosa*). In contrast, *A. longifolia* var *sophorae* and *B. integrifolia* exhibit considerably longer and wider leaves and therefore the capacity to capture more wind and generate greater movement amplitudes. Nevertheless, these species do show marked differences in response to wind that is likely to be a consequence of host branches. For example, *B. integrifolia* branches are noticeably stiffer than that of *A. longifolia* var *sophorae* and movement tends to be less pronounced.

Leaf motion is also expected to be strongly influenced by the underlying branch structure of plants (Diener et al. 2006). Branching determines the overall shape of plants and will vary considerably within species (Fig. 11.6a). As each branching element of a plant will have its own frequency of oscillation determined by its length and stiffness (Niklas 1992), branching plants exhibit a multimodal response to wind. Modelling wind–plant interactions in woody plants requires an alternative approach than the engineering principles relevant to single stem monocotyledons (Rodriguez et al. 2008). Modal analysis techniques, which are used for measuring vibrations in mechanical structures, have proven to be useful for quantifying complex plant motion by decomposing the plant into a set of vibration nodes (Rodriguez et al. 2012; Diener et al. 2008; Rodriguez et al. 2008). Rodriguez et al. (2008) have demonstrated that vibration frequencies increase from the trunk through second-order and higher order branches (Fig. 11.6b). As a consequence, plants with several orders of branching have a range of responses to wind. This will lead to some regions of the plant showing clear movement, in the leaves for instance, while other regions remain motionless.

Thus far, I have outlined the dynamic response of plants to wind forces, treating each stem or branch as an isolated unit. However, inertia and the damping processes that dissipate mechanical energy of the plant are also major contributors to plant movement (Niklas 1992). A plant's resistance to changing momentum is highly dependent on plant architecture and the mass of the plant unit (leaves, branches, etc.). Conversely, damping in plants is brought about by the inherent aerodynamics of plants and contact with other structures (de Langre 2008). Small trees, branches and leaves can effectively reduce the amount of drag through reconfiguration in response to initial wind forces (Vogel 1989). Similarly, leaf fluttering and twisting reduces drag by decreasing the area subjected to wind forces (Niklas 1992). Dissipation also occurs as a result of interactions with neighbouring stems, leaves and branches and will be greater for dense plants with small distances between other structures.

To conclude this section, it is worth mentioning that the adaptive growth hypothesis suggests that plants grow as strong as they need to be to resist the forces they experience during their growth history (Wood 1995). It is thus intriguing to consider the consequences for plant motion of altered environments in which previously sheltered plants are exposed to the elements.

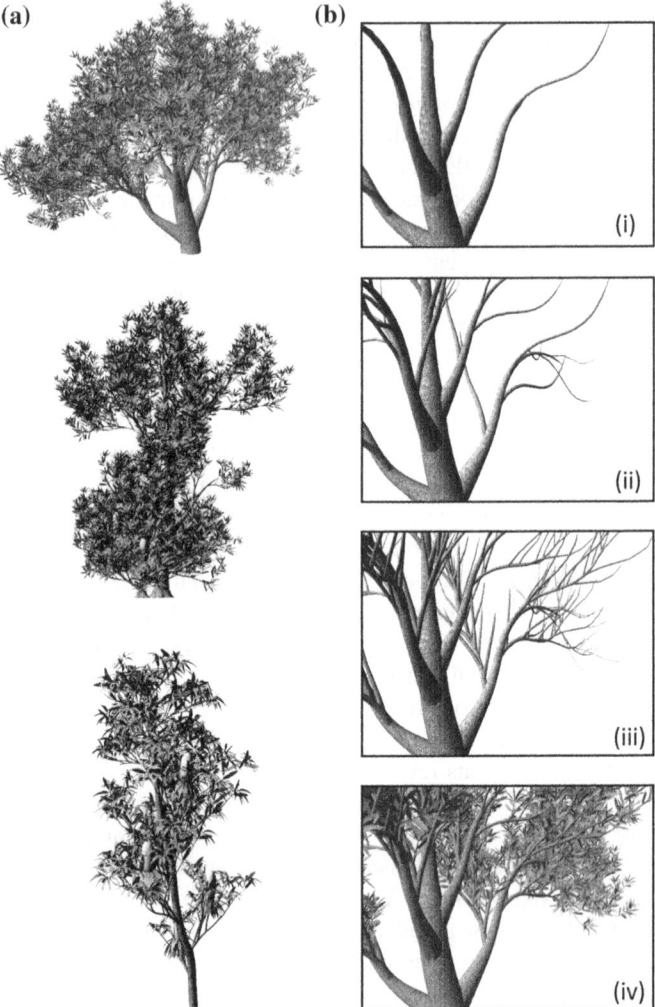

Fig. 11.6 **a** Plant shape can vary considerably from one plant to the next even within the same species, as shown in this *Banksia* sp. **b** Dicotyledons exhibit multiple orders of branching from the main trunk. This sequence shows: (*i*) first-order branching, (*ii*) first-, second- and some third-order branching, (*iii*) first through higher order branching and (*iv*) leaf detail

11.4 Microhabitats as Distinct Image Motion Environments

It should now be clear that wind-induced plant movements are highly variable both within and between species. From the bending behaviour of monocotyledons to the multimodal response of branching shrubs and trees, movement patterns will be determined by the physical characteristics of a given plant. Furthermore, habitat location and topography will affect relative exposure to wind (Hannah et al. 1995), while the presence of other plants in the environment will affect the characteristics of wind including eddy size and frequency (see de Langre 2008). The combined effect of plant characteristics, location and habitat structure will inevitably result in vastly different plant motion habitats. Quantifying the extent to which habitats vary in this regard is important for understanding the noise characteristics influencing movement-based signals.

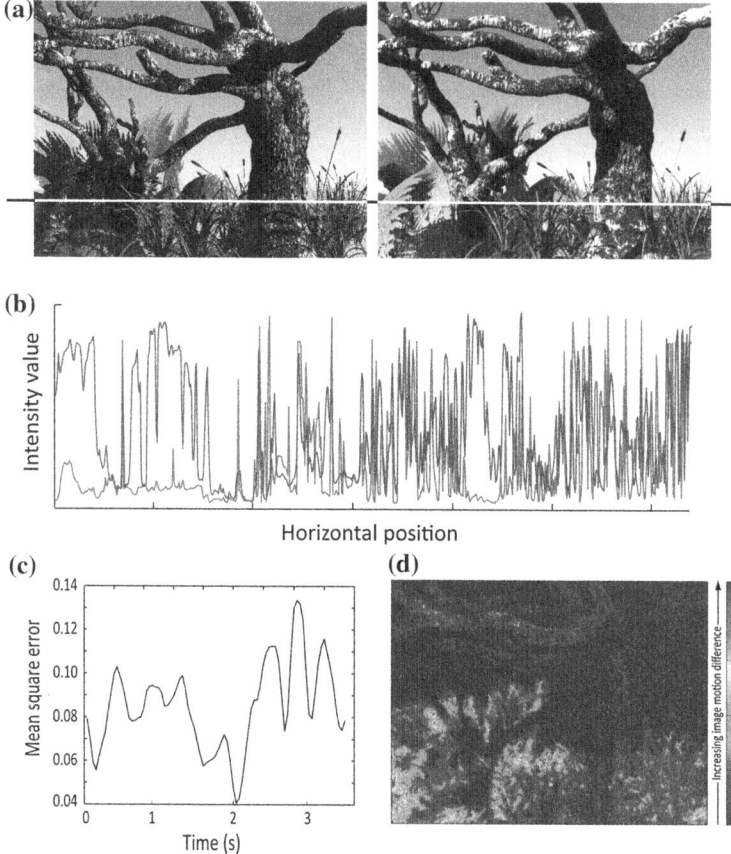

Fig. 11.7 The light environment will influence the measurement of image motion. **a** Frames from two animation sequences featuring identical movement characteristics but simulating different times of the day (*sun position*). **b** Each frame shows very different intensity profiles after the colour information is converted to grey scale values (*blue line* representing the image on the *right*). **c** A gradient detector algorithm was used to quantify image frame for each sequence and the difference in measured motion between corresponding frames was determined by computing the mean square error over time. **d** Summing across time the absolute difference between frames in the otherwise identical sequences identifies the spatial locations where different light conditions influenced perceived motion

In order for plant movement to become 'noise' it must stimulate the same sensory mechanisms as movement-based signals. Briefly, mechanisms of (first-order) motion perception rely on luminance differences between an object and its background but movement is not explicitly represented at the sensory (photoreceptor) level. Visual motion must be computed from correlated changes in brightness in neighbouring photoreceptors and two prominent models of this computation are the Reichardt or correlation-type detectors and gradient detectors. Many animals have been shown to employ correlation-type motion detectors, despite the fact that their output depends strongly on the contrast and the spatial structure of natural scenes. The ecological implications are that the perception of movement is most reliable for high-contrast visual objects. Most of the highest contrast features in natural scenes are created by shadows. Shadows result when the sun's rays are partially blocked by other plants, for instance, and depend on the celestial position of the sun. The shadows themselves will generate a simplified, silhouetted version of plant movement. In theory, shadow movement could be distracting even when the plant *per se* is not in the field of view of signal receivers. The salience of shadows in a given environment will vary with time of day as the position of the sun changes (Zeil et al. 2003). Similarly, the position of the sun will modify the intensity profile of scenes (Fig. 11.7). This will not add to the noise environment in the way that shadows do, but will influence the salience of plant image motion as perceived by an observer. Figure 11.7c, d illustrate that exactly the same plant movements will generate different motion profiles at the eye of the receiver depending simply on the time of the day. This is a consequence of the distinct intensity profiles produced at these times (Fig. 11.7a, b). It is possible that the shadow structure and light profiles of environments will vary predictably with time of day on clear days with no clouds. The addition of clouds in the atmosphere, however, will probably alter the predicted patterns considerably.

11.5 Future Directions: Virtual Microhabitats

I have emphasised in this chapter the need to consider in more detail the way plants move because of its relevance to animal signalling. A variety of techniques have been used to measure the physical movement of plants including strain gauges, displacement transducers, accelerometers and approaches using optical target monitoring (see de Langre 2008 and references therein). Physical measurements of plant movements alone will not be sufficient for quantifying motion noise, as they do not reflect, in a straightforward way, how it affects the motion detection filters of receivers. A solution is needed that allows for simultaneous consideration of the physical movements of plants, microhabitat structure, environmental variables and animal signals while also enabling us to systematically manipulate one or more of these variables. To achieve this in nature would be time consuming and impractical. However, simulating nature in virtual animation environments is achievable and potentially a powerful tool for exploring the evolutionary constraints on movement-based signalling. I have been making use of animation technology to generate lizard displays for a few years (e.g. Peters and Evans 2003b, 2007) and I was excited to discover that animated film and game production industries are pushing the development of models of plant motion (Diener et al. 2006, 2008).

Creating and experimenting with virtual habitats may offer exciting experimental opportunities and has the potential to demonstrate how the creative arts can contribute to answering fundamental questions in animal biology. To conclude this chapter, I describe the main components of such an approach and the type of questions that can now be addressed in detail. At the heart of the approach is what I call virtual microhabitats. Realistic 3D models of individual plants arranged to reproduce convincing natural scenes have been possible for some time. Reproducing plant motion to the same high levels of realism has been problematic and time consuming because models had to be manually animated. However, recent work demonstrates that wind–plant interactions can be simulated with limited user intervention by relying on statistical models of plant motion (Diener et al. 2006, 2008). Consequently, known microhabitats can be reconstructed in detail complete with variable environmental conditions. In addition to simulating variable wind conditions, the celestial position of the sun can be varied thereby reproducing different effects of shadows and variable intensity profiles as illustrated in Fig. 11.7. Accurately modelled signalling animals can be incorporated into the scene (see New and Peters 2010). After simulation of such a dynamic habitat, environmental conditions and signals, the scene can be exported as video or image sequence and analysed using established techniques such as elementary motion detector networks (Fleishman and Pallus 2010) or saliency analysis (Peters 2010).

A virtual lab approach recreates nature, which saves considerable time and effort in capturing the range of environmental conditions we would need to understand fully motion signalling systems. Grounded in realistic simulations of how plants move in response to wind, it will allow for the consideration of movement-based signalling in unprecedented detail and with extraordinary control. A number of important questions can now be tackled in a systematic manner, including the effects of fluctuating wind conditions on the distribution of image motion signals:

- *Plant species.* A central tenet of this chapter is that plant species vary in response to wind and this can now be explored explicitly.
- *Depth structure of environment.* Perceived motion speeds vary as a function of viewing distance (Peters et al. 2008). The spatial layout of plants will ensure complex distributions of image motion, while the position of the signaller relative to surrounding plants is also predicted to be important (Peters 2010; Peters et al. 2008).
- *Light conditions.* The potential distracting effect of shadows and variation in the light environment due to sun position and/or clouds (Fig. 11.7) can be explored with unparalleled control.

- *Positions of signaller and receiver.* Plant motion noise is not distributed evenly in space, with many microhabitats showing regions of high and low noise at the same time (Peters et al. 2008). Clearly then the position of the receiver is crucial in determining whether the signal is dominated by noise. The virtual world allows us to model such a scenario, as both the location of the signaller in the scene and the location of the 'camera', which records the scene and represents the position of the receiver, can be manipulated (Fig. 11.7). It is intriguing to consider whether signals can be effective even if surrounded by strong plant motion, and to explore how signal and noise might be viewed by potential predators that have very different perspectives compared to conspecific receivers of such signals (Fig. 11.8b).

(a)

(b)

Fig. 11.8 Animation will be a powerful tool for understanding variation in the efficacy of movement-based signals. We should begin by recreating actual scenes involving planted environments and a signalling animal (**a**). The power of animation can then allow us to explore, for example, how signal structure varies depending on the position of the receiver (**b**): at the same height as the signaller (*top left*), below the signaller (*bottom left*) or above and at a distance from the signaller (*top right*), which is where potential predators could be watching

11.6 Concluding Remarks

I have endeavoured to describe the potential significance of plant motion on the evolution of movement-based animal signals. It is clear from the animals we have studied thus far that plant motion is a major factor in determining signal structure. There is now clear evidence that animals adjust their signals to compensate for changing plant motion conditions. However, as the responses of plants to wind will vary in many ways, we should not assume plant motion would be homogenous across sites. Only after careful analysis of microhabitat motion noise might we be able to identify similarities across sites, or determine the properties of motion vision systems that effectively filter any variation.

With such scope for variation in the image motion environment it will be exciting to see the range of strategies that animals might use to deal with plant motion noise. At a general level, animals may be aware of variations in wind conditions and exploit daily and/or seasonal patterns in plant motion. For example, wind speeds along the south-east coast of Australia, near to sites considered in the present chapter, increased predictably throughout the day (see Fig. 11.2 in Peters and Evans 2003a, b), although the signalling patterns of lizards in these areas are not known. Conversely, Ord (2008) identified clear dawn and/or dusk peaks in display behaviour as a function of time of day in four species of *Anolis* lizard in Jamaica, but in this case, it is unknown whether wind conditions match these peak activity patterns. In both study systems, however, signalling still occurs throughout the day and signallers must deal with prevailing wind and concomitant motion noise. Rather than seeking to contrast with plant movements, signallers could, of course, simply wait for lulls in wind and exploit the decreased motion noise conditions. Signalling for longer durations, as seen in Jacky lizards (Peters et al. 2007), might simply improve the chance of tail flicks occurring in lulls in wind conditions.

As some animals are believed to adopt movement patterns that are characteristic of plant movements to avoid detection (e.g. Fleishman 1985), it is logical to expect that the converse is also true such that a reliable strategy to enhance detection of signals is to move inherently *unlike* a plant. For example, plant image motion can be strongly directional for short periods of time in conditions of strong wind (Peters et al. 2008). Detection might be optimised, therefore, if the dominant motion direction of the signal strongly contrasted with that of plant motion. Similarly, plants do not abruptly start and stop moving. Animals that use such a strategy for signalling will not only contrast with plant motion noise but should generate conspicuous peaks in the motion vision systems of receivers, as large onset transients are characteristic of the responses of motion sensitive neurons (Ibbotson and Clifford 2001). Indeed Fleishman and Pallus (2010) nicely demonstrated the potential for generating strong neural signals with abrupt up and down head movements in the presence of plant motion noise. Intriguingly, one of the five lizard species in their study, *Anolis stratulus*, does not use this characteristic movement pattern and their signals are duly swamped by motion noise. The authors suggested that this species, which occupies the forest canopy, must use an alternative strategy for making their display visible. Fleishman and Pallus (2010) concluded that the complexity of the image motion environment at the canopy, relative to the forest floor, has lead to such a divergence in signalling strategy but the mechanisms of this was not immediately obvious.

In conclusion, quantifying the structure of plant motion noise will generate new insights about the evolution of motion signalling strategies and document the importance of habitat structure for signal efficacy. Furthermore, we will gain insight into the likely impact of altered habitats on the motion signalling behaviour of lizards that might have important ramifications for other functional behaviour.

Acknowledgments I would very much like to thank Henrik Brumm for the invitation to submit this chapter, and for providing me with the opportunity to organise a collection of thoughts and ideas about why we should care about plant movements. I am also indebted to a number of individuals with whom I have enjoyed many discussions on the topic including Jan Hemmi, Jochen Zeil, Leo Fleishman and Terry Ord. I am also grateful to Tom Chandler for helping me recognise the potential of animation for understanding motion signals, and to Henrik Brumm, Leo Fleishman, Haven Wiley and Jochen Zeil for their thoughtful and constructive comments on earlier versions of this chapter.

References

Barth F, Bleckmann H, Bohnenberger J, Seyfarth E (1988) Spiders of the genus Cupiennius Simon 1981 (Arameae, Ctenidae). 2. On the vibratory environment of a wandering spider. Oecologia 77:194–201

Caro TM (1995) Pursuit-deterrence revisited. Trends Ecol Evol 10:500–503

Cocroft R, Rodriguez R (2005) The behavioral ecology of insect vibrational communication. Bioscience 55:323–334

Costermans L (2005) Native trees and shrubs of South-Eastern Australia. Reed New Holland, Sydney

Coutts M, Grace J (1995) Wind and trees. Cambridge University Press, Cambridge

Crane J (1975) Fiddler crabs of the world. Ocypodidae *Genus Uca*. Princeton University Press, Princeton

Crews D (1975) Inter- and intraindividual variation in display patterns in the lizard, *Anolis carolinensis*. Herpetologica 31:37–47

de Langre E (2008) Effects of wind on plants. Annu Rev Fluid Mech 40:141–168

Diener J, Reveret L, Fiume E (2006) Hierarchical retargetting of 2D motion fields to the animation of 3D plant models. Eurographics/ACM SIGGRAPH symposium on computer animation, 2006, pp 187–195

Diener J, Rodriguez M, Baboud L, Reveret L (2008) Wind projection basis for real-time animation of trees. Computer graphics forum, pp 533–540

Doherty J (1982) Stereotypy and the effects of temperature on some spatio-temporal subcomponents of the 'courtship wave' in the fiddler crabs *Uca minax* (Le Conte) and *Uca pugnax* (Smith) (Brachyura, Ocypodidae). Anim Behav 30:352–363

Dunlop RA, Cato DH, Noad MJ (2010) Your attention please: increasing ambient noise levels elicits a change in communication behaviour in humpback whales (*Megaptera novaeangliae*). Proc R Soc Lond B 277:2521–2529

Elias D, Land B, Mason A, Hoy R (2006) Measuring and quantifying dynamic visual signals in jumping spiders. J Comp Physiol A 192:785–797

Endler JA, Basolo AL (1998) Sensory ecology, receiver biases and sexual selection. Trends Ecol Evol 13:415–420

Fleishman LJ (1985) Cryptic movement in the vine snake *Oxybelis aeneus*. Copeia 1985:242–245

Fleishman LJ (1986) Motion detection in the presence or absence of background motion in an *Anolis* lizard. J Comp Physiol A 159:711–720

Fleishman LJ (1988) Sensory and environmental influences on display form in *Anolis auratus*, a grass anole of Panama. Behav Ecol Sociobiol 22:309–316

Fleishman LJ (1992) The influence of the sensory system and the environment on motion patterns in the visual displays of anoline lizards and other vertebrates. Am Nat 139(Supplement):S36–S61

Fleishman LJ, Pallus AC (2010) Motion perception and visual signal design in Anolis lizards. Proc R Soc Lond B 277:3547–3554

Guilford T, Dawkins MS (1992) Understanding signal design: a reply to Bloomberg and Alberts. Anim Behav 44:384–385

Hannah P, Palutikof J, Quine C (1995) Predicting windspeeds for forest areas in complex terrain. In: Coutts M, Grace J (eds) Wind and trees. Cambridge University Press, Cambridge, pp 113–129

How M, Hemmi JM, Zeil J, Peters RA (2008) Claw waving display changes with receiver distance in fiddler crabs, *Uca perplexa*. Anim Behav 75:1015–1022

How M, Zeil J, Hemmi J (2009) Variability of a dynamic visual signal: the fiddler crab claw-waving display. J Comp Physiol A 195:55–67

How M, Zeil J, Hemmi JM (2007) Differences in context and function of two distinct waving displays in the fiddler crab, *Uca perplexa*. Behav Ecol Sociobiol 62:137–148

Hyatt GW (1977) Quantitative analysis of size dependent variation in the fiddler crab wave display Uca-pugilator Brachyura Ocypodidae. Mar Behav Physiol 5:19–36

Ibbotson M, Clifford C (2001) Characterising temporal delay filters in biological motion detectors. Vision Res 41:2311–2323

Jenssen TA (1975) Display repertoire of a male *Phenacosaurus heterodermus* (Sauria: Iguanidae). Herpetologica 31:48–55

Leal M (1999) Honest signalling during prey-predator interaction in the lizard *Anolis cristatellus*. Anim Behav 58:521–526

Mcnett GD, Luan LH, Cocroft RB (2010) Wind-induced noise alters signaler and receiver behavior in vibrational communication. Behav Ecol Sociobiol 64(12):pp 2043–2051

Milner R, Jennions M, Backwell P (2008) Does the environmental context of a signalling male influence his attractiveness? Anim Behav 76:1565–1570

New S, Peters R (2010) A framework for quantifying properties of three-dimensional movement-based signals. Current Zoology 56:327–336

Niklas KJ (1992) Plant Biomechanics: an engineering approach to plant form and function. University of Chicago Press, Chicago, p 607

Ord T, Peters R, Clucas B, Stamps J (2007) Lizards speed up visual displays in noisy motion habitats. Proc R Soc Lond B 274:1057–1062

Ord T, Stamps JA (2008) Alert signals enhance animal communication in "noisy" environments. Proc Natl Acad Sci U S A 105:18830–18835

Ord TJ (2008) Dawn and dusk "chorus" in visually communicating Jamaican anole lizards. Am Nat 172:585–592

Ord TJ, Blumstein DT (2002) Size constraints and the evolution of display complexity: why do large lizards have simple displays? Biol J Linn Soc 76:145–161

Ord TJ, Blumstein DT, Evans CS (2001) Intrasexual selection predicts the evolution of signal complexity in lizards. Proc R Soc Lond B 268:737–744

Ord TJ, Blumstein DT, Evans CS (2002) Ecology and signal evolution in lizards. Biol J Linn Soc 77:127–148

Ord TJ, Charles GK, Hofer RK (2011) The evolution of alternative adaptive strategies for effective communication in noisy environments. Am Nat 177(1):54–64

Peters R (2008) Environmental motion delays the detection of movement-based signals. Biol Lett 4:2–5

Peters R (2010) Movement-based signalling and the physical world: modelling the changing perceptual task for receivers. In: Tosh C, Ruxton G (eds) Modelling perception with artificial neural networks. Cambridge University Press, Cambridge, pp 269–292

Peters R, Evans C (2007) Active space of a movement-based signal: response to the Jacky dragon (*Amphibolurus muricatus*) display is sensitive to distance, but independent of orientation. J Exp Biol 210:395–402

Peters R, Hemmi J, Zeil J (2007) Signalling against the wind: modifying motion signal structure in response to increased noise. Curr Biol 17:1231–1234

Peters R, Hemmi J, Zeil J (2008) Image motion environments: background noise for movement-based animal signals. J Comp Physiol A 194:441–456

Peters R, Ord T (2003) Display response of the Jacky Dragon, *Amphibolurus muricatus* (Lacertilia : Agamidae), to intruders: a semi-Markovian process. Austral Ecol 28:499–506

Peters RA, Evans CS (2003a) Design of the Jacky dragon visual display: signal and noise characteristics in a complex moving environment. J Comp Physiol A 189:447–459

Peters RA, Evans CS (2003b) Introductory tail-flick of the Jacky dragon visual display: signal efficacy depends upon duration. J Exp Biol 206:4293–4307

Py C, de Langre E, Moulia B, Hemon P (2005) Measurement of wind-induced motion of crop canopies from digital video images. Agr Forest Meteorol 130:223–236

Rodriguez M, Langre ED, Moulia B (2008) A scaling law for the effects of architecture and allometry on tree vibration modes suggests a biological tuning to modal compartmentaliza-tion. Am J Bot 95:1523–1537

Rodriguez M, Ploquin E, de Langre E, Moulia B (2012) The multimodal dynamics of a walnut tree: experiments and models. J Appl Mech 79:044505

Rothblum L, Jenssen TA (1978) Display repertoire analysis of *Sceloporus undulatus hyacin-thinnus* (Sauria: Iguanidae) from South-Western Virginia. Anim Behav 26:130–137

Rundus AS, Owings DH, Joshi SS, Chinn E, Giannini N (2007) Ground squirrels use an infrared signal to deter rattlesnake predation. Proc Natl Acad Sci USA 104:14372–14376

Sellier D, Fourcaud T, Lac P (2006) A finite element model for investigating effects of aerial architecture on tree oscillations. Tree Physiol 26:799–806

Smolka J, Hemmi JM (2009) Topography of vision and behaviour. J Exp Biol 212:3522–3532

Staelens J, Nachtergale L, Luyssaert S, Lust N (2003) A model of wind-influenced leaf litterfall in a mixed hardwood forest. Can J Forest Res 33:201–209

Stamps JA, Barlow GW (1973) Variation and stereotypy in the displays of *Anolis aeneus* (Sauria: Iguanidae). Behaviour 47:67–94

Stein BE, Gaither NS (1981) Sensory representation in reptilian optic tectum: some comparisons with mammals. J Comp Neurol 202:69–87

Stein BE, Gaither NS (1983) Receptive-field properties on reptilian optic tectum: some comparisons with mammals. J Neurophysiol (Bethesda) 50:102–124

Vogel S (1989) Drag and reconfiguration of broad leaves in high winds. J Exp Bot 40:941–948

Watt MS, Moore JR, Mckinlay B (2005) The influence of wind on branch characteristics of Pinus radiata. Trees 19:58–65

Wood C (1995) Understanding wind forces on trees. In: Coutts M, Grace J (eds) Wind and trees. Cambridge University Press, Cambridge, pp 133–164

Zeil J, Hofmann M, Chahl J (2003) Catchment areas of panoramic snapshots in outdoor scenes. J Opt Soc Am A: 20:450–469

Chapter 12
Neural Noise in Electrocommunication: From Burden to Benefits

Jan Benda, Jan Grewe and Rüdiger Krahe

Abstract Weakly electric fish generate an electric field, called electric organ discharge (EOD), that they use for active electrosensation. This system is used for both object localisation and electrocommunication. Both, objects that are close to the fish and the EODs of other nearby electric fish, modulate the amplitude of a fish's EOD. Localisation signals are low in amplitude and frequency whereas electrocommunication signals are large amplitude signals with higher frequencies. Electroreceptor neurons are tuned to the frequency of the fish's own EOD. This tuning, however, is rather broad to allow for the reception of EODs of other fish with different frequencies. This is the basis for electrocommunication. Spike trains of electroreceptor afferents are surprisingly noisy even in the absence of any external signal. From theoretical studies it is known that in populations of spiking neurons such internal noise can improve the information carried about a common input signal in comparison to the noiseless case. In particular, the processing of high-frequency signals benefits from internal noise and the convergence of large populations of neurons. The target neurons of the electroreceptor afferents, the pyramidal cells in the electrosensory lateral line lobe, are organised in three distinct maps of the electroreceptive body surface that are characterised by different receptive field sizes, i.e. the number of afferents that converge on them, and frequency tuning. The properties of these three maps can be understood based on the differential impact of the noise in the electroreceptor afferent spike trains on the processing of the distinct types of signals arising in localisation and

J. Benda (✉) · J. Grewe
Institute for Neurobiology, University of Tübingen,
Auf der Morgenstelle 28 E 72076 Tübingen, Germany
e-mail: jan.benda@uni-tuebingen.de

J. Grewe
e-mail: jan.grewe@uni-tuebingen.de

R. Krahe
Department of Biology, McGill University, 1205 Docteur Penfield,
Montreal, QC H3A 1B1, Canada
e-mail: rudiger.krahe@mcgill.ca

H. Brumm (ed.), *Animal Communication and Noise*,
Animal Signals and Communication 2, DOI: 10.1007/978-3-642-41494-7_12,
© Springer-Verlag Berlin Heidelberg 2013

communication contexts. Further, the noise in the electroreceptors allows for the discrimination of synchronous spikes from all spikes fired by the afferent population. The level of synchrony seems particularly important for encoding high-frequency communication signals. The electrosensory system is thus a showcase for demonstrating how neural systems actually use noise to enhance processing of signals.

12.1 Introduction

Considering that the idea of an electric sense seems quite exotic to most of us, it may come as a surprise to hear that the earliest vertebrates presumably had the capability to sense external electric fields (Zupanc and Bullock 2005). Electrosensation was then lost, but re-evolved in several lineages, among them elasmobranchs and teleost fish, and is found today even in the bill of the platypus (Pettigrew 1999). Several groups of fishes, including the South American gymnotiformes, the African mormyriformes, siluriform catfish and elasmobranchs, possess a passive, ampullary, electrosense that allows them to detect the weak and low-frequency electric fields generated by the muscle activity of aquatic organisms. Gymnotiform and mormyriform fishes have a second electrosensory system, the tuberous system that responds to the discharges of their own electric organ (EOD) and to the EODs generated by conspecifics and other species. Their combined electrogeneration and electrosensation system is a dual-use system: on the one hand, it allows the fish to detect objects and navigate their habitat based on perturbations of the electric field caused by objects whose electrical properties differ from those of the surrounding aquatic medium. Because the animal itself provides the energy used to sense its environment, electrolocation can be considered an active sense comparable to echolocation in bats and marine mammals (bats: Schnitzler et al. 2003, Chaps. 9, 10 on marine mammals). On the other hand, the EOD and modulations of its discharge frequency serve a communication function in various contexts ranging from aggressive encounters between males to courtship. It has even been proposed that EOD signals are used to coordinate pack-hunting behaviour in the mormyrid *Mormyrops anguilloides* (Arnegard and Carlson 2005).

Weakly electric fish come in two types, wave-type and pulse-type (Fig. 12.1). Wave-type weakly electric fish generate a quasi-sinusoidal EOD by discharging their electric organ periodically with the pauses between EOD pulses being of similar duration as the pulses themselves. Pulse-type fish, on the other hand, generate brief pulses separated by longer pauses that, in many species, are of variable duration (Moller 1995). This chapter will focus on gymnotiform wave-type weakly electric fish, whose processing of electrosensory information has been investigated in more detail than that of any other group of electric fish (but see Sawtell and Bell 2008; Kawasaki 2005). Of particular appeal has been the

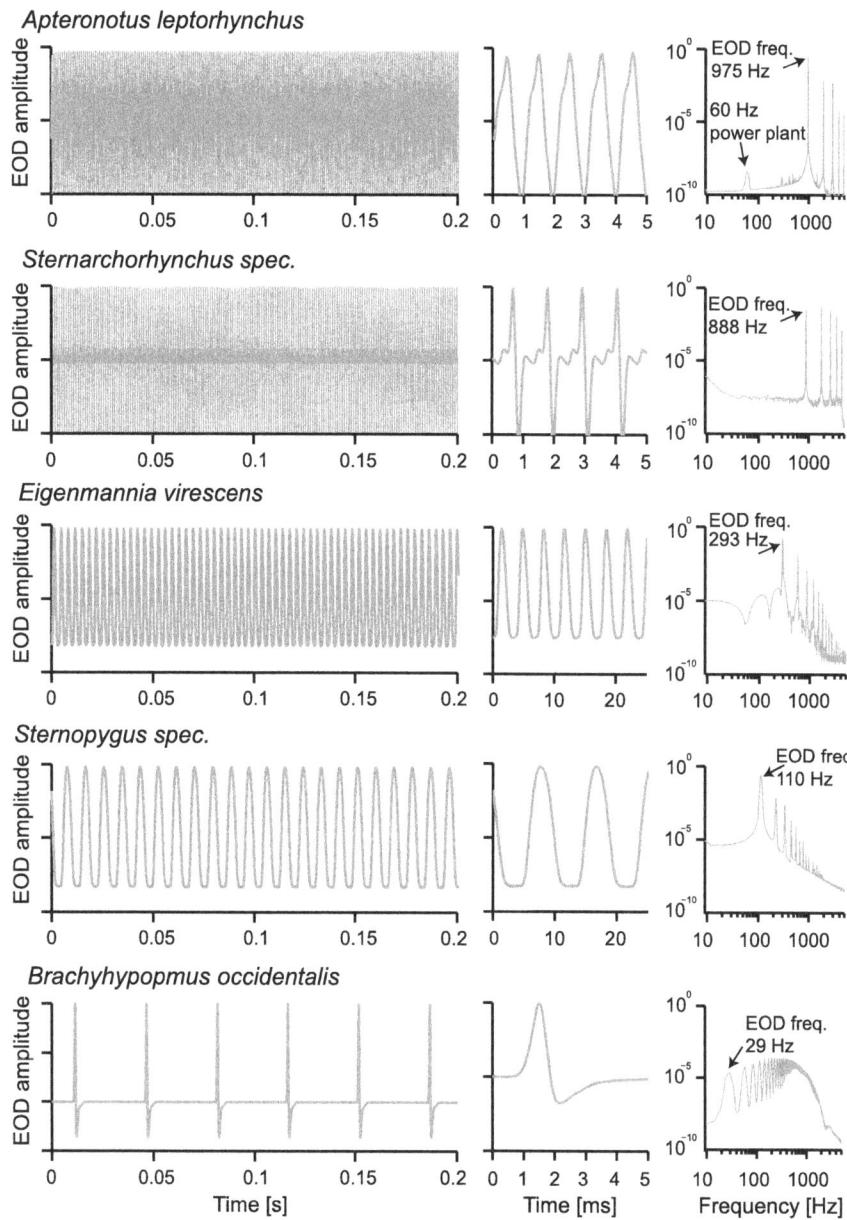

Fig. 12.1 EOD characteristics of five species of Gymnotiform electric fish. *Brachyhypopomus occidentalis* (*bottom*) is a pulse-type electric fish, the four other fish are wave-type. *Left column* A 200 ms sequence of EOD recordings. Amplitudes are normalised to the largest positive amplitude value. *Middle column* A close up of the EODs illustrating individual discharges of the electric organ. *Right column* Power spectra of the EOD traces shown in the *left column*

excellent experimental accessibility of the electrosensory system in animals that display normal electric behaviours. As any other sensory system, the electrosense functions in the face of noise that originates from both external and internal sources. Remarkably, compared to other communication channels, such as hearing and vision, electrocommunication in these fish may be relatively little affected by external sources of noise. And with respect to internal sources of noise, the electrosensory system may be a showcase for demonstrating how neural systems actually use noise to enhance information processing.

12.2 Electrocommunication

The EOD signals of wave-type weakly electric fish can be largely described by their fundamental frequency and amplitude. As outlined in the following sections, both of these as well as their modulations appear to be used by the fish for communication purposes.

12.2.1 EOD Frequency Identifies Species and Individuals

The most basic communication signal of a wave-type weakly electric fish is its EOD frequency, which is individual specific and astoundingly constant; at a time scale of seconds to hours, the standard deviation of the EOD cycle periods is in the sub-microsecond range and the coefficient of variation of the EOD cycle period is as low as 10^{-4}, making the EOD generating mechanism the most regular biological oscillator known (Moortgat et al. 1998). While the baseline EOD frequency of a given fish is quite stable, different individuals of the same species use different frequencies within a certain range that is characteristic for that species (Fig. 12.2). For example, one of the most intensely studied species, *Apteronotus leptorhynchus*, the brown ghost knifefish, occupies a frequency range from approximately 700–1,100 Hz, while the EOD frequency of *Sternopygus macrurus* can take values between 40 and 200 Hz. If fish evaluate EOD frequency for communication purposes, then overlapping frequency ranges of different, co-occurring species could constitute a source of error. Such overlapping frequency ranges have been reported for sympatric species of gymnotiforms (Kramer et al. 1981); there is, however, a lack of data demonstrating syntopy of such species during the breeding season (occurring within the same microhabitat, which makes physical interaction likely, whereas sympatry only implies occurrence in the same general area). Given that the EOD waveforms of many species are quite different due to differences in harmonic content (Crampton and Albert 2006; Turner et al. 2007), and that waveform sensitivity has been demonstrated in *Eigenmannia* (Kramer and Otto 1991), it is conceivable that these fish use waveform information rather than EOD frequency to distinguish their own from other species. Surprisingly, experiments

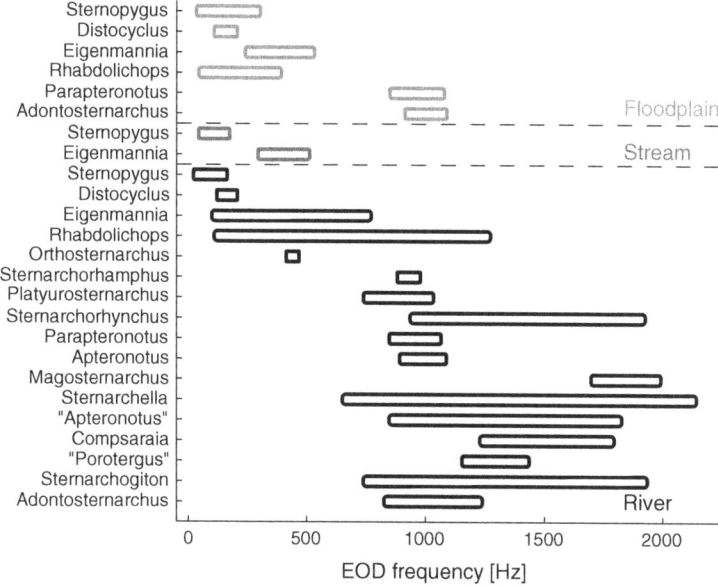

Fig. 12.2 Ranges of EOD frequencies of different species in different habitats. *Bars* indicate the distribution of EOD frequencies observed in various genera of gymnotiform fish and how they are distributed across different habitats. The *quotation marks* indicate that the respective group of species is not a well defined genus and should be considered a species group. Redrawn from Crampton and Albert (2006)

on *A. leptorhynchus* suggest the exact opposite: fish showed behavioural responses to playback stimuli as long as the carrier frequency was in the right range, whereas the quite distinct EOD waveforms of various species that were tested had no effect on the responses (Fugère and Krahe 2010). Interestingly, discrimination between individuals based on waveform has been demonstrated in one species of pulse-type gymnotiform fish (McGregor and Westby 1992).

EOD frequency is not only species specific in its range, it can also carry information about age and/or size, sex and dominance. Gymnotiform fish begin to produce a measurable EOD within the first 10 days after hatching (Kirschbaum 1983; Kirschbaum and Westby 1975; Meyer et al. 1987). EOD frequency is initially low and, in *Eigenmannia*, reaches adult values within 20 days (Kirschbaum and Westby 1975), whereas in *Apteronotus*, it takes up to 1 year for EOD frequency to increase from initial values around 300 Hz to between 600 and 800 Hz (Kirschbaum 1983; Meyer et al. 1987). A correlation between body size and EOD frequency has also been observed in adult male *A. leptorhynchus* studied in the laboratory (Dunlap 2002; Triefenbach and Zakon 2003) and in a related apteronotid species, *Sternarchorhynchus sp.*, observed in its natural habitat (Fugère et al. 2011). Behavioural experiments on the latter species demonstrated that EOD frequency carries information on dominance status and is used by these fish to decide contests about access to hiding places (Fugère et al. 2011). EOD frequency has also been found to be sexually dimorphic in many wave-type species (Crampton and Albert 2006). In the family Sternopygidae, which includes *Eigenmannia*, mature males occupy the lower end of a given species' range, whereas mature females occupy the upper end (Hopkins 1972; Hagedorn and Heiligenberg 1985). In many species of the family Apteronotidae, which comprises the electric fishes with the highest EOD frequencies, males occupy the high range, whereas at least in *A. albifrons*, this sexual dimorphism appears to be reversed (Dunlap et al. 1998; Crampton and Albert 2006). Not surprisingly, EOD frequency is under control of steroid hormones (e.g. Meyer et al. 1987; Zakon et al. 1991; Dunlap 2002; Cuddy et al. 2011, see below).

Besides developmental and hormonal effects, EOD frequency is strongly correlated with water temperature (Dunlap et al. 2000). Although the temperature of freshwater bodies in the lowland tropical rainforest varies a lot less through the year than that of streams and lakes in the temperate zone, it can fluctuate by several degrees in the course of a day and through the year (Kramer 1978; Flecker et al. 1991). At present, there is no evidence that temporal or spatial variations in temperature affect communication interactions between wave-type electric fish.

How is the frequency of one wave-type fish perceived by another? Because wave-type EODs are almost sinusoidal, the superposition of two EODs will lead to a beat (Fig. 12.3a), that is, a combined signal that oscillates in amplitude and phase at a frequency equal to the difference in EOD frequencies between the two fish. When two animals of the same sex interact, the beat frequency will be relatively low, when two fish of opposite sex interact, the beat frequency can reach values of up to 400 Hz depending on the species. The strength of the beat modulation experienced by a given animal is a function of its distance from the conspecific, because EOD amplitude drops steeply with distance from the source (Knudsen 1975; Chen et al. 2005). In natural groups of fish, multiple EOD signals can interact, leading to multiple simultaneous beats as well as to interactions between the beat frequencies themselves ("beat of beats"; Partridge and Heiligenberg 1980; Tan et al. 2005; Stamper et al. 2010). Depending on the frequency composition of the group, such beats of beats can cause periodic low-frequency envelopes of the original beat signal (Fig. 12.3b). Interestingly, recordings from groups of *Apteronotus* in streams in Ecuador suggest that the fish adjust their EOD frequencies to avoid low-frequency envelopes (Stamper et al. 2010).

12.2.2 EOD Amplitude is Related to Body Size

It is conceivable that electric fish not only evaluate the frequency of each other's EOD, but also its amplitude. Based on the strong positive correlation between EOD amplitude and body size (Knudsen 1975), they might assess a neighbour's body size by measuring the strength of its electric field. This relationship is complicated by the decrease in electric field strength with distance as pointed out in the preceding section and by the fact that in pulse-type gymnotiforms and in

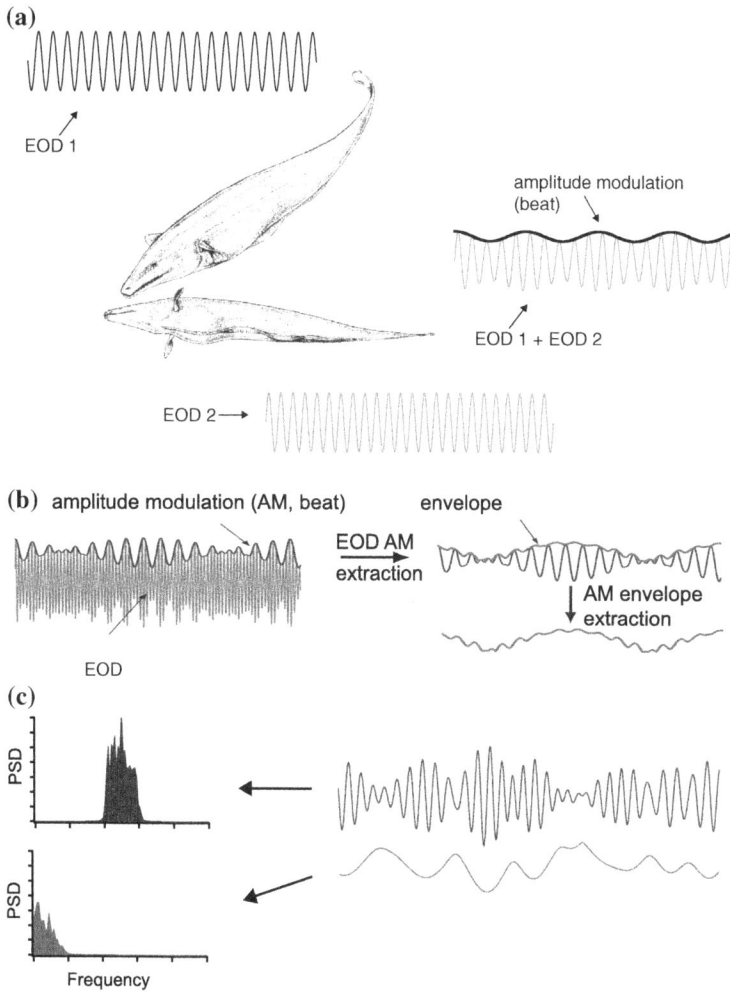

Fig. 12.3 Beats and envelopes. **a** If two fish are in proximity of each other the two individual waveforms will interact. Each fish receives an interference of the two EODs. Both fish perceive their own EOD amplitude modulated by the field of the other one. The resulting amplitude and phase modulation is referred to as the beat. Here, we will only consider the amplitude component (AM, *black line*) of the beat. Its frequency is given by the frequency difference of the individual EODs. The modulation depth of it depends on the relative amplitudes of the individual EODs. For example, since the amplitude of the EODs decline with distance from the fish, at the position of the lower fish the EOD amplitude of the distant upper fish is smaller compared to the lower fish's EOD amplitude. If the fish come closer the modulation depth or contrast of the beat increases. Drawings of *A. leptorhynchus* taken from Hagedorn and Heiligenberg (1985). **b** If the EODs of multiple fish interfere, higher-order AMs occur. These are called envelopes (*red line* in the *top row*). **c** Spectral power that is contained in the beat and envelope. One can observe that the higher-order AMs (i.e. the envelope) have lower frequency power (*bottom* traces and power spectra) than the beat. Modified from Middleton et al. (2006)

sternopygid wavefish EOD amplitude follows a circadian rhythm (reviewed in Stoddard et al. 2006; Markham et al. 2009). The reduction in amplitude during the daylight hours when the fish are usually hiding, may serve the dual purpose of being less conspicuous to electrosensory predators and of saving energy (Markham et al. 2009; Salazar and Stoddard 2008; Reardon et al. 2011). At night, signal amplitude is boosted, which is expected to increase the range of electrolocation and the signal-to-noise ratio of the electric images on the skin created by the object-induced perturbation of the fish's electric field (Assad et al. 1998). Stronger electric fields will also make fish more conspicuous to conspecifics and to electroreceptive predators. The hypothesis that amplitude plays a role in social signalling in these fish is supported by the observation that male pulse-type fish, *Brachyhypopomus gauderio*, show larger amplitude increases at night than females and that exposure to conspecifics has a boosting effect on EOD amplitude in both *B. gauderio* and *Sternopygus macrurus* (Stoddard et al. 2006; Markham et al. 2009; Salazar and Stoddard 2008).

12.2.3 Electrocommunication by EOD Frequency Modulations

12.2.3.1 Jamming Avoidance Response

Wave-type weakly electric fish use several kinds of frequency modulations of their EOD in interactions with conspecifics. The most intensely studied frequency modulation is the jamming avoidance response (JAR) which was first described in *Eigenmannia virescens* (Watanabe and Takeda 1963; Bullock 1969; Bullock et al. 1972a, b). When a fish encounters a conspecific whose EOD frequency is close to its own, the resulting low-frequency beat interferes with the animal's ability to detect nearby objects (Heiligenberg 1973, 1991; Bastian 1987a, b). The fish show extraordinary sensitivity in determining whether their own EOD is of higher or lower frequency than that of their neighbour and, as a consequence, lower or raise their frequency away from that of their neighbour's (Fig. 12.4, also Kawasaki 1997). By increasing the frequency difference, the fish free up the frequency range most critical for electrolocation, hence the term jamming avoidance. The JAR can be viewed as a noise avoidance behaviour, because the low-frequency beat modulation, which affects the entire body surface of the fish and interferes with the animal's prey detection mechanisms, can be seen as strong background noise whose frequency range overlaps with the frequency of object-induced perturbations of the fish's electric field (Nelson and MacIver 1999). In addition to avoiding low-frequency noise, the higher-frequency beats resulting from the JAR have been found to enhance directional selectivity for object movements of neurons in the midbrain (Ramcharitar et al. 2006). The enhancement of directional selectivity is

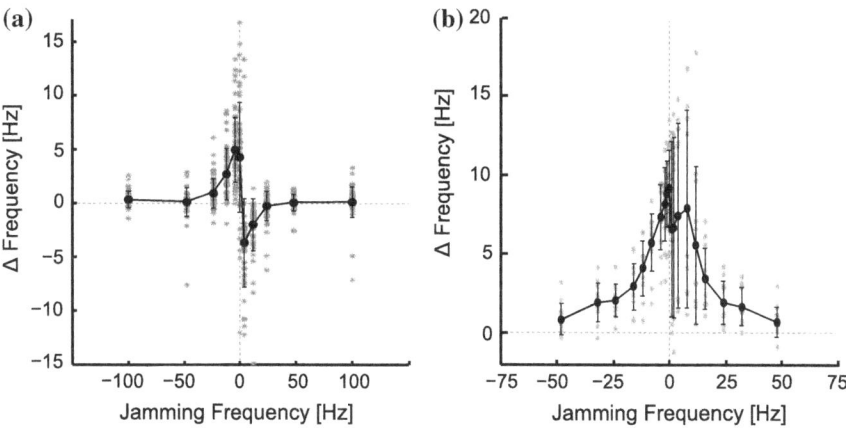

Fig. 12.4 The jamming avoidance response (JAR). **a** The JAR of *E. virescens*. In response to a jamming beat of various frequencies the fish shifts its EOD frequency to higher or lower values. The strength of the JAR is given relative to the "resting" condition (*n* = 56). *Grey asterisks* are individual measurements, *black dots* are averages across all recordings. *Error bars* are standard deviations. **b** Same as (**a**) but for *Apteronotus leptorhynchus* (*n* = 6). *A. leptorhynchus* shows only a positive JAR and appears to be unable to lower its EOD frequency actively (Heiligenberg et al. 1996). There is experimental evidence that *A. leptorhynchus* actively uses jamming as an aggressive signal (Tallarovic and Zakon 2005)

caused by frequency-dependent synaptic depression that is most pronounced in the gamma frequency range corresponding to the beat frequencies found in natural groups of *Eigenmannia*.

12.2.3.2 Chirps

More rapid frequency modulations have been described in a number of wave-type species, some of them with a duration of hundreds of milliseconds and others, called chirps, that are as short as 10 ms (e.g. Hagedorn and Heiligenberg 1985; Zakon et al. 2002; Zupanc et al. 2006; Turner et al. 2007). Chirps are produced mostly by male fish during aggressive and courtship encounters and have been studied most thoroughly in *A. leptorhynchus* (e.g. Zakon et al. 2002; Hupé and Lewis 2008; Triefenbach and Zakon 2008). In an interaction between two fish, chirps always occur on the background of a beat. The rapid frequency increase and return to baseline causes a phase advance of the beat-related amplitude modulation (AM), experienced by the fish (Fig. 12.5). The most commonly observed chirps are the so-called "small", or type-2, chirps, which show frequency increases between 60 and 200 Hz and are produced mostly in response to EOD frequencies similar to the fish's own frequency (Engler and Zupanc 2001). Large chirps, whose frequency excursions can reach several hundred Hertz, are observed more rarely and are produced mostly in response to large EOD frequency differences, that is,

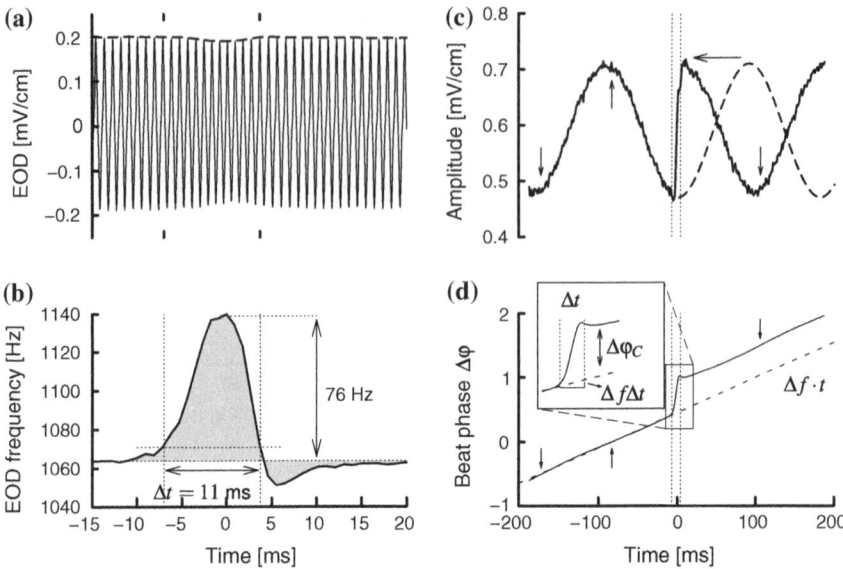

Fig. 12.5 Chirps and their impact on beats. **a** The EOD of a chirping fish. Small vertical bars indicate beginning and end of a small chirp. During the chirp the EOD frequency is increased and the EOD amplitude is slightly reduced. **b** EOD frequency of the chirping fish shown in a as a function of time. The chirp is characterised by its frequency excursion (here about 76 Hz) and duration Δt. **c** Amplitude of the resulting beat as a function of time (*solid line*). *Dashed line* indicates the undisturbed beat. The chirp, occurring at time 0 induces a phase advance of the beat. **d** beat phase $\Delta \phi$ as a function of time. The chirp occurs at time 0 and induces the phase shift ($\Delta \phi$ C). Figure modified from Benda et al. (2005)

on a background of high-frequency beats. They are thought to play a role in courtship and spawning (Hagedorn and Heiligenberg 1985; Bastian et al. 2001), but have also been proposed to serve as submissive signals of low-frequency males towards dominant males with a high EOD frequency (Cuddy et al. 2011). Similar to communication signals in other species and modalities (e.g. Goodson and Bass 2001; Albers et al. 2002; Gutzler et al. 2011; Allee et al. 2008; Pasch et al. 2011), chirping has been shown to be regulated by steroid hormones (Dunlap et al. 1998, 2002, 2011; Dunlap 2002), serotonin (Telgkamp et al. 2007; Smith and Combs 2008), and also by arginine-vasotocin (Bastian et al. 2001).

12.3 External Sources of Noise

The electrosensory system is exposed to noise from several sources, but their impact on electrocommunication may be small compared to noise effects on other communication channels. Noise of abiotic origin includes the electric fields caused by lightning. As discussed by Hopkins (1973), wave-type weakly electric fish are expected to be relatively immune to lightning-related electrical noise due to the quasi-sinusoidal nature of their EOD that contrast with the randomly occurring brief lightning pulses. A more serious, anthropogenic problem for some wave-type species may be electropollution from power plants and power lines, which emit electric fields at 60 Hz and its higher harmonics that can be picked up by simple electrodes over large distances (R. Krahe, personal observation). It is conceivable that "line noise", due to its harmonic structure, interferes with the EOD signals of a number of wave-type species (van der Sluijs et al. 2011). It is still unknown if species avoid producing EOD frequencies at 60 Hz and its harmonics in the vicinity of settlements and power lines or if some of these species even disappear from heavily polluted areas because of electrical interference.

The weak and local AMs caused by objects in the vicinity of an electric fish (much less than 1 % modulation in amplitude for a typical prey item close to the skin; (Nelson and MacIver 1999; Chen et al. 2005) can also be obscured by large-scale AMs due to the fish's own body movements. Chief among those may be tail bending, which changes the geometry of the electric field (Heiligenberg 1975). The signal driving the electroreceptors in the skin, the transdermal potential difference, will increase in amplitude on the side of the body to which the tail is bent, and decrease on the opposite side (Bastian 1995; Chen et al. 2005). The change in transdermal potential amplitude caused by tail bending is in the range of several percent of the baseline value (Heiligenberg 1975; Chen et al. 2005). The undulations of the anal fin that propel the fish forward and backward may also lead to modulations of the animal's own electric field, but the effect has not been quantified so far.

When two or more fish interact, the frequency of the beat modulations is determined by the difference in EOD frequency between the animals (Fig. 12.3). The contrast, i.e., the strength of AM during the beat, depends on the distance between fish, because the strength of the electric field of a fish falls off steeply with distance (Knudsen 1975; Rasnow and Bower 1996; Chen et al. 2005). Therefore, contrast will be modulated as fish move relative to each other. These second-order AMs are also called envelopes. Similar, but periodic, envelopes of beat signals are created when three or more fish interact (Middleton et al. 2006; Stamper et al. 2010); this is because the beats between a fish and its neighbours will themselves interfere and create a "beat of beats". Although these envelopes may contain important information for a fish, such as the distance from conspecifics and the EOD frequency relationships in groups, it is conceivable that envelopes can interfere with the fish's ability to electrolocate objects, and may thus constitute noise in the context of foraging. This is supported indirectly by the finding that EOD recordings of groups of *Apteronotus* in their natural habitat did not contain envelope power at very low frequencies (Stamper et al. 2010).

In the following sections, we will first discuss various ways in which the electrosensory system deals with noise that may be considered as a burden for detecting and processing important sensory signals. The last section will take a different, and complementary, approach by looking at the benefits of noise for neural processing in general and electrosensory processing of communication signals in particular.

12.4 Neural Tuning

In order to understand how different sources of noise can affect the encoding of electrosensory information we will now discuss how the nervous system processes electric signals in gymnotiform weakly electric fish (Fig. 12.6). After a short introduction of the electrosensory periphery we first consider the frequency tuning of the electrosensory system, before we describe the role of neuronal noise in sensory processing.

The circuitry involved in electrosensation and electric behaviour has been studied in great detail and an enormous amount of information is available on the neuroanatomy of the system, its neurotransmitters and neuromodulators, and the transmitter receptors and ion channels (for reviews, see Berman and Maler 1999; Bell and Maler 2005; Maler 2009b). This extensive knowledge forms an excellent foundation for in-depth studies of neural processing, which have been facilitated by the experimental accessibility of the electrosensory system under *in vivo* conditions and the persistence of electric behaviours, such as the JAR and chirping, in immobilised preparations.

The first point to note about the electrosensory system is that its circuitry takes up a large proportion of the brain volume (Maler et al. 1991), which can be seen as an indirect argument for its dominant role among the senses of electric fish. Sensing electric fields starts with the electroreceptor organs which are distributed

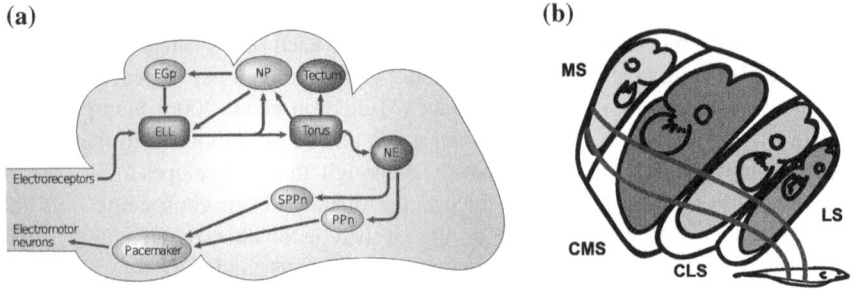

Fig. 12.6 The electrosensory system. **a** Overview of the brain of the weakly electric fish illustrating the main components of the electrosensory circuitry. *ELL* electrosensory lateral line lobe, *NP* nucleus praeeminentialis dorsalis, *EGp* eminentia granularis pars posterior, *NE* nucleus electrosensorius, *PPn* prepacemaker nucleus, *SPPn* subleminiscal prepacemaker nucleus. Figure taken from Rose (2004). **b** The *ELL* is separated in four segments/maps. *MS* medial segment, *CMS* centromedial segment, *CLS* centrolateral segment and *LS* lateral segment. Figure taken from Krahe et al. (2008)

all over the skin, and of which two basic types exist: the first one, ampullary electroreceptors are sensitive to low-frequency electric fields generated by muscle activity of other organisms and, at least in *Eigenmannia*, to EOD interruptions, which in this genus contain low-frequency power and are produced in communication interactions (Metzner and Heiligenberg 1991). The second type of electroreceptors consist of the so-called tuberous electroreceptors, which sense the fish's own EOD and its perturbations. Each receptor organ is composed of a number of electroreceptors, which are innervated by the electroreceptor afferents that carry electrosensory information to the hindbrain via the octavolateralis nerve (Zakon 1986a). The skin of an adult *A. albifrons* contains approximately 15,000 tuberous receptor organs compared with approximately 700 ampullary organs and roughly 300 neuromasts for the mechanosensory lateral line (Carr et al. 1982). Within the population of tuberous primary afferents, two sub-populations are observed, so-called P-units and T-units (Scheich et al. 1973). T-units respond in a precisely phase locked, one-to-one fashion to the fish's own EOD pulses, while the P-units fire probabilistically, and their response probability for a given EOD pulse depends on its amplitude.

12.4.1 Matched Tuning of Electroreceptors to EOD Frequency

The active electrosensory system is used for the two purposes introduced above: electrolocation and electrocommunication. In both cases the AM of the carrier, the EOD, contains information that needs to be extracted. The AM sets the probability with which a P-unit emits a spike during an EOD cycle. Thus, the P-unit afferents encode the time course of the AMs in their firing rate. As in other AM coding systems, such as the auditory system (see Chaps. 3, 4, 6, 8 and 10), the electroreceptors as well as their afferents are specifically tuned to the fish's own EOD frequency (Hopkins 1976; Viancour 1979a, b; Knudsen 1974; Fig. 12.7). The tuning of receptors and receptor afferents is interpreted to constitute a matched filter (Hopkins 1976), which will be discussed below.

12.4.1.1 Emergence of Tuning

Evidence from newly regenerating electroreceptors, which initially show a rather broad, coarse tuning and eventually are adjusted to the fish's EOD frequency, indicates that the tuning is only in part genetically defined (Zakon 1986b). Fine tuning appears to be achieved via hormonal influence. Steroid hormones affect the EOD frequency as well as the receptor tuning (Meyer et al. 1987; Keller et al. 1986). In *Apteronotus* treatment with 5-α-*dihydrotestosterone* (DHT) increases and 17-β-*estradiol* decreases the EOD frequency. Interestingly the tuning of the

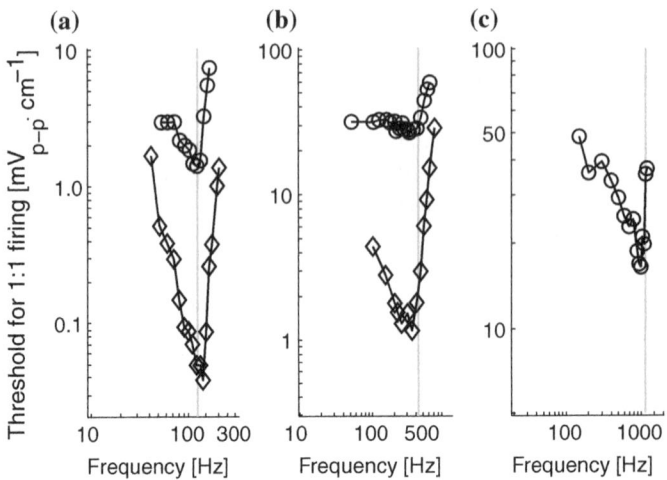

Fig. 12.7 Tuning of electroreceptors. Example tuning curves of tuberous T-units (*diamonds*) and P-units (*circles*) of three different species of weakly electric fish. **a** *Sternopygus macrurus*, **b** *E. virescens* and **c** *Apteronotus albifrons* (only P-units). The tuning curves show the field intensity (in peak-to-peak amplitude per cm) required for a 1:1 firing (one action potential per EOD cycle) of the cell at various frequencies of an artificial electric field. This is the upper limit of the operating range, and, for P-units, therefore not representative of the behaviourally relevant range of field intensities. The tuning curves are matched to the fish's own EOD frequency (indicated by the *grey lines*). Redrawn from Hopkins (1976)

electroreceptors shifts accordingly. A further line of evidence in this direction comes from the correlation of the EOD frequency with the maturation of the animals. In *Apteronotus* the EOD frequency of newly hatched fish starts at 300–400 Hz and increases with maturation (Meyer et al. 1987).

12.4.1.2 Tuning Mechanism

The question arises by which mechanisms the tuning of the cells is established. At the best frequency the electroreceptors show a resonance behaviour (e.g. Viancour 1979b; Keller et al. 1986; Meyer et al. 1987; Zakon 1986b) which is also known from auditory hair cells. In these, Ca^{2+}-activated K^+ ion channels are responsible for oscillatory behaviour (e.g. bullfrog hair cells, Lewis and Hudspeth 1983). The close relationship between hair cells and electroreceptors suggests that the same mechanism may apply here (Keller et al. 1986). Changing the density of these channels via hormonal influence can affect the resonance behaviour and thus the tuning. A modelling study by Koch (1984) demonstrated that a population of such Ca^{2+}-activated K^+ channels can render the membrane *quasi-active* inducing resonance behaviour. The density of such channels defines the resonance frequency.

As any other biological mechanism, the tuning is also influenced by temperature. For both the EOD and the receptor tuning similar effects with $Q10$ values in the range if 1.4–1.6 have been observed (e.g. Enger and Szabo 1968; Hopkins 1976; Dunlap et al. 2000). Because temperature affects EOD frequency and receptor tuning in similar ways, changes in water temperature are not expected to interfere with the coding of localisation and communication information.

12.4.1.3 Matched Filtering in a Multi-Purpose System

The close match between the receptor's best frequency (frequency with lowest threshold, Fig. 12.7) and the frequency of the fish's own EOD suggests that the filter characteristics constitute a matched filter turning electrosensation into a private channel of information acquisition. This notion, however, needs to be discussed in some more detail.

If only the carrier frequency is of interest, a sharply tuned matched filter would reject all frequencies that are not relevant, e.g. those due to biotic or abiotic noise sources. The tuning of electroreceptors and of auditory hair cells has been interpreted as a mechanism for reducing noise and increasing response specificity (e.g. Hopkins 1976; Ricci et al. 2005, see also Chaps. 8 and 10). Similar mechanisms are found in various other systems. For example, neurons in the auditory system of certain species of bats exhibit a very sharp tuning to the *reference frequency*, i.e. the frequency of their call's echo (e.g. Suga 1965; Pollak and Bodenhamer 1981). But how does the level of specificity found in the frequency tuning of the electroreceptors relate to their role in electrosensation? Many other sensory systems consist of an array of receptors that, as a population, but not individually, cover the relevant stimulus space. In weakly electric fish the electrosensory receptors serve two purposes each with its own requirements, but there is only a single population of similarly tuned P-units that has to cover the behaviourally relevant frequency range. For electrolocation (auto-communication) a sharp tuning should be advantageous because all information is contained in the AMs of the fish's own EOD, and contamination from other frequencies should be avoided.

Communication with conspecifics, however, would not be possible with an extremely sharp receptor tuning for two reasons. First: when two, or more, individuals are in proximity, their EODs interfere, resulting in a beat (Fig. 12.3). This beat carries information about the nature of the interaction (same-sex or opposite-sex encounter; interacting with a dominant or subordinate animal). Accordingly, the production of the different chirp types depends on the context of the encounter, i.e. the beat frequency (Engler and Zupanc 2001). To sense a beat, the receiving fish must be sensitive not only to its own EOD but also to the frequency of the other fish. Because encounters in *Apteronotus* can easily produce beats up to 400 Hz, the tuning must be wide enough to permit sensing such signals, especially when considering that large EOD frequency differences, i.e. high beat frequencies, are likely to occur in mating contexts. Second: The communication signals themselves (chirps) are characterised by frequency excursions of up to several hundred Hertz (Fig. 12.5). The various chirp types are produced in different contexts (agonistic, submissive, mating) and a sharp receptor tuning would not permit to discriminate chirps which have frequency excursions shifting the beat to frequency ranges outside the range that can be sensed. The tuning of the electroreceptors is thus constrained by these two purposes. An optimal tuning can only be a compromise between noise rejection and sensitivity bandwidth. Whether the tuning width matches the frequency range that naturally occurs for a given species remains unknown.

12.4.1.4 Electrocommunication Interferes with Electrolocation

The multi-purpose usage of the broadly tuned electroreceptors for both electrolocation and -communication signals makes the electrosensory system vulnerable to interferences among the two types of signals. Low-amplitude electrocommunication signals can be obscured by distortions of the electric field by nearby objects and non-conducting boundaries, such as the water surface. Vice versa, strong jamming signals from conspecifics, that overwhelm electrolocation signals, affect electrolocation on the behavioural as well as the physiological level (Matsubara and Heiligenberg 1978; Bastian 1987a, b). By means of the JAR (see above, Bullock 1969) the fish actively move the jamming beats out of the AM frequency range of electrolocation signals to higher frequencies. The frequency shifts due to the JAR, however, are relatively small compared to the width of the tuning curves (compare Figs. 12.4 and 12.7). Further separation between electrolocation and -communication is left to the neural system. For example, the generation of a "negative image" removes predictable low-frequency distortions (see below).

12.4.1.5 Receptor Tuning as a Species-Specific Filter

The tuning seen in the electroreceptors restricts the band of frequencies that affect the P-unit responses. With this, the perception of other wave-type species that share the same habitat, but use different frequency bands (Fig. 12.2), is greatly reduced. In that way, a species-specific, rather than an individual-specific, frequency channel is established. The broader tuning resembles to some extent the tuning of ascending neurons in bush-crickets which show species-specific tuning (e.g. Stumpner 2002, see also Chap. 3) interpreted to act as a frequency filter for the songs of conspecifics.

Still, the match between the individual EOD frequency and the individual receptor's best frequency makes the fish most sensitive to their own carrier frequency. This fits experimental results on mormyrids showing that electrolocation is more robust against disturbances than electrocommunication (Schief et al. 1971). The EODs of pulse-type electric fish, however, pose a problem for syntopic wave-type fish. These pulses have spectral power in a very broad range (Fig. 12.1, last row, right column) and thus will contaminate the field even of very sharply tuned receptors. The discharge rates of pulse-type fish are quite low and thus contamination from EOD pulses is limited to occasional and short time windows.

The tuning of the electroreceptors thus appears to be a compromise between the constraints imposed by electrolocation and electrocommunication. Optimal noise rejection is sacrificed for an increased bandwidth of the communication channel. This trade-off shares some similarity with olfactory communication in moths. There, male moths respond to a wider variation of pheromones than are usually produced by female moths. The wider acceptance window, or tuning, is of advantage if failure to respond is more expensive than a false alarm (see Chap. 13).

12.4.2 Temporal Tuning of Primary Afferents for AMs

Besides the frequency tuning discussed above, P-units have temporal response characteristics which act as frequency filters on the AM signal. Measurements of the P-unit sensitivity to AMs of increasing frequency suggest that they have high-pass characteristics that attenuate low frequencies (Bastian 1981; Nelson et al. 1997; Chacron et al. 2005; Benda et al. 2005). For electrolocation and prey detection this seems to be counterproductive because the AMs induced by moving objects and prey have mainly low-frequency power (Bastian 1981). Still, the fish are well able to detect and successfully capture prey (Nelson and MacIver 1999). Nelson and MacIver (1999) suggest that the high-pass characteristics of the P-unit afferents may assist prey capture by serving as a predictive filter pointing to the location at which the prey will be next. By applying information theoretic measures Chacron et al. (2005) could show that despite the high-pass characteristics, low frequencies are similarly well represented in the responses as higher frequencies. This is attributed to reduced low-frequency noise due to correlations between consecutive interspike intervals (Chacron et al. 2001, 2004, 2005). In the context of electrocommunication, i.e., the encoding of chirps in the P-unit responses, the high-pass behaviour arising from spike-frequency adaptation was concluded to be advantageous by separating the responses to the transient chirps from low-frequency background modulations (Benda et al. 2005).

12.4.3 Spatio-Temporal Tuning in the ELL

Upon entering the hindbrain, each primary afferent fibre trifurcates, so that three somatotopic maps of the body surface are formed in the electrosensory lateral line lobe (ELL) of the hindbrain (Carr et al. 1982; Heiligenberg and Dye 1982). Based on their arrangement in the ELL, these maps are called lateral segment (LS), centrolateral segment (CLS) and centromedial segment (CMS). A fourth map, the medial segment (MS) is dedicated to processing ampullary information (Fig. 12.6b). T-units synapse onto spherical cells, which send axons to the torus semicircularis of the midbrain, where EOD phase information is compared between different parts of the body surface for computations involved in the generation of the JAR (Heiligenberg 1991). P-units, on the other hand, synapse onto pyramidal cells in the ELL as well as on local interneurons.

The pyramidal cells of the ELL come in two types, E-cells and I-cells, analogous to the ON and OFF cells of the lateral geniculate nucleus (Krahe and Gabbiani 2004). E-cells respond with excitation to increases in EOD amplitude, whereas I-cells respond with excitation to decreases in amplitude. Similar to thalamic relay neurons, pyramidal cells have antagonistically organised spatial receptive fields with excitatory centre and inhibitory surround (E-cells) or inhibitory centre and excitatory surround (I-cells) (Maler et al. 1981; Shumway 1989; Bastian et al. 2002). E- and I-cells can each be further subdivided into superficial, intermediate and deep cells according to the location of their somata in the pyramidal cell layer of the ELL and the length of their apical dendrites (Bastian and Courtright 1991). These six pyramidal cell types have been found to be arranged in columns containing one cell of each type (Maler 2009a).

A common feature in nervous systems is the increasing stimulus specificity from the periphery to higher processing stages in the brain which extract behaviourally relevant information form the continuous flow of sensory information. This is precisely what is seen in the electrosensory system of the weakly electric fish. While the electroreceptor afferents reliably code for AMs of the carrier (e.g. Chacron et al. 2005; Gabbiani 1996) the pyramidal cells in the ELL respond much more specifically (Gabbiani 1996). The ELL pyramidal cells in the different maps show different temporal coding properties ranging from high-pass in the lateral segment (LS) to low-pass characteristics in the CMS, tuning them to different temporal aspects of the sensory input (Krahe et al. 2008). High-pass tuning itself is induced by the increased expression of SK (small-conductance potassium) channels in the LS that suppresses the responses to low frequencies (Ellis et al. 2007). Measurements of these *in vivo* tuning properties match those found under *in vitro* conditions (Mehaffey et al. 2008).

The spiking activities of electroreceptor afferents are independent from each other (Chacron et al. 2005; Benda et al. 2006). Pyramidal cells integrate convergent information from the number of afferents within their receptive fields. Across the ELL maps the receptive field size changes from small (CMS) to large in the LS (Maler 2009a) making them better suited for electrolocation or electrocommunication, respectively (see below).

Dendritic mechanisms can lead to bursts of action potentials in pyramidal neurons (Doiron et al. 2001). The role of these bursts is different in the different maps. In the CMS bursts increase the representation of prey-like signals. In the LS, on the other hand, fast, transient AMs as induced by chirps lead to bursting (Marsat et al. 2009). Bursting makes the coding of communication signals more robust against noise (Ávila-Åkerberg et al. 2010).

12.4.4 Feedback Removes Predictable Information

A striking similarity between ELL pyramidal cells and thalamic relay neurons is the extensive feedback both types of neurons receive on their apical dendrites (Krahe and Gabbiani 2004). Two feedback pathways arise from the isthmic nucleus praeeminentialis dorsalis (NP, Fig. 12.6a), which receives electrosensory input from a subset of pyramidal cells. The so-called direct pathway projects from NP directly back to the ELL. It has been discussed to be involved in a sensory searchlight mechanism originally proposed by Francis Crick for corticothalamic circuitry (Crick 1984; Bratton and Bastian 1990). Another set of output neurons of NP projects to the eminentia granularis posterior (EGp) of the cerebellum, from where parallel fibres innervate the molecular layer of the ELL which contains the apical dendrites of the pyramidal cells. This indirect feedback pathway is thought to be involved in gain control and the cancellation of redundant sensory input (Berman and Maler 1999).

Sensory systems in general are set up as change detectors. Anything that is novel could be important, whereas sensory input that is predictable does not carry much information and should be disregarded. Predictable inputs to the electrosensory system arise, for example, from movements of the fish's tail, which change the geometry of the electric field, and from periodic AMs due to interaction with nearby conspecifics. These signals can also be viewed as highly structured narrowband noise sources. Superficial and intermediate ELL pyramidal cells display a pronounced plasticity in their responses to such predictable electrosensory signals; their responses decline gradually thanks to an active mechanism that learns to counterbalance the feed-forward input from primary afferents by generating a "negative image" of the predictable input (Bastian 1995). Such negative image mechanisms that cancel redundant sensory input have been proposed by Curtis Bell to be a primary function of cerebellum-like structures in general (reviewed in Bell 2002). Interestingly, the ELL, which is located directly ventral of the eminentia granularis posterior of the cerebellum, is itself a cerebellum-like structure. A massive set of parallel fibres originating from cerebellar granule cells courses through the molecular layer of the ELL, where they interact with the large apical dendrites of ELL pyramidal cells (Berman and Maler 1999). Proprioceptive and electrosensory feedback provided by the parallel fibres has been shown to mediate the cancellation of predictable input due to tail bending and beats. Also, this mechanism potentially improves the signal of an object in front of background water plants, roots or rocks as well (Babineau et al. 2007).

An interesting problem for such a mechanism is that the plastic neurons that learn to disregard predictable input cannot themselves provide information on the redundant signals to higher brain centres. Thus, another, non-plastic, pathway is needed whose activity can be used by cerebellar feedback to cancel the sensory responses of the plastic neurons. This non-plastic pathway, which continues to respond to predictable signals, was found to consist of deep pyramidal cells in the

ELL, which project to the nucleus praeeminentialis dorsalis, from where information is relayed to the cerebellum (Bastian et al. 2004).

The suppression of predictable sensory responses due to tail bending, or the continued presence of a conspecific, is expected to increase the signal-to-noise ratio of responses to novel signals, such as prey or chirps (change detection). This has indeed been shown through in vivo electrophysiological experiments in *A. leptorhynchus* (Marsat et al. 2009), where cerebellar feedback to plastic pyramidal cells supports the firing of bursts in response to chirps riding on a background beat.

12.4.5 Beyond the ELL

The next stage of electrosensory processing is the torus semicircularis where information from the time-coding (T-unit) system and the amplitude-coding pathway converges (Heiligenberg and Rose 1985) and input from the tuberous and ampullary maps is integrated (Metzner and Heiligenberg 1991). In addition, many toral neurons show direction-selective responses to objects moving along the fish (Ramcharitar et al. 2005). From the torus, one pathway leads to the optic tectum, which processes information on the motion of objects (Bastian 1982), and another to the nucleus electrosensorius. The latter contains neurons exquisitely sensitive to the sign of the frequency difference between a fish's EOD and that of its neighbour, a computation that has been shown to be instrumental for the JAR (Heiligenberg 1991). The nucleus electrosensorius provides input to two prepacemaker nuclei, the mesencephalic sublemniscal prepacemaker and the diencephalic prepacemaker (Heiligenberg et al. 1996), which, in turn, controls the medullary pacemaker nucleus that determines EOD frequency (Fig. 12.6a).

The following discussion of the role of noise in electrosensory processing will focus on the amplitude-coding pathway, specifically P-units and ELL pyramidal cells.

12.5 Noisy Neurons

Apart from the extrinsic sources of noise discussed above, electrosensory processing is also affected by intrinsic neuronal noise. This section first describes the noise observed in individual electroreceptor afferents and later reviews the effects of noise in populations of spiking neurons. In many cases, intrinsic noise is seen as a problem that is destructive and needs to be eliminated by averaging over many neurons. In populations of spiking neurons, however, the presence of noise and reliable information transmission are not necessarily contradictory.

12.5.1 Neuronal Noise Introduced by Primary Afferents

The frequency tuning of the electroreceptors leads to a high degree of temporal coupling to the fish's EOD (Fig. 12.8a, b). This imposes a temporal structure on the responses. Under baseline conditions, i.e. when no external signal is applied, P-units exhibit highly irregular response patterns, shown for *A. leptorhynchus* (e.g. Nelson et al. 1997) as well as *E. virescens* (Fig. 12.8). This is expressed in the broad multimodal distribution of the interspike intervals (ISI) shown in Fig. 12.8c, (see also Nelson et al. 1997). The response regularity can be described by the coefficient of variation (CV_{ISI}) relating the standard deviation of the observed ISIs to their mean. Perfect regularity would lead to a CV_{ISI} close to zero, while random Poisson-like firing of action potentials results in *CV*s around one.

The *CV*s observed in *Apteronotus* and *Eigenmannia* P-units are rather large for primary afferents (0.54 ± 0.23 and 0.37 ± 0.17 for *Apteronotus* and *Eigenmannia* P-units, respectively, Fig. 12.8d). Despite the apparently high degree of response variability the internal noise in the P-units was shown to be too small to distinctly affect the coding of AM signals (Kreiman et al. 2000). Interspike-interval corre-lations, as found in P-units, were further shown to reduce low-frequency response variability and increase information transfer (Chacron et al. 2001, 2004).

In addition to the high degree of response irregularity the population of elec-troreceptor afferents exhibits a considerable level of heterogeneity that can be viewed as a static noise source. Both, the CV_{ISI} (Fig. 12.8d) as well as the baseline rates (Wessel et al. 1996; Gussin et al. 2007) are widely distributed (Fig. 12.8e). It is, however, not clear (i) what the reason for these differences is, nor (ii) what the consequences for the coding of electrosensory information are. Heterogeneity found on higher processing levels, however, was concluded to be beneficial for coding of electrosensory information (Marsat and Maler 2010; Ávila-Åkerberg et al. 2010).

Why are the P-units so noisy, even in the unperturbed baseline condition? To answer this question we first introduce some general coding properties of popu-lations of noisy neurons in the following section.

12.5.2 Noise in Populations of Spiking Neurons

In contrast to graded neurons that more or less operate linearly on an input signal, the generation of action potentials is a highly non-linear process that endows spiking neurons with interesting computational possibilities. For example, the all-or-none property of action potentials is the basis for decision processes. Either an input signal was able to trigger an action potential or not. Making decisions or classifications is not at all possible with linear systems. Similarly, noise in spiking neurons can have effects that are impossible in linear systems. In the following paragraphs, we introduce some very general and fundamental concepts on the role

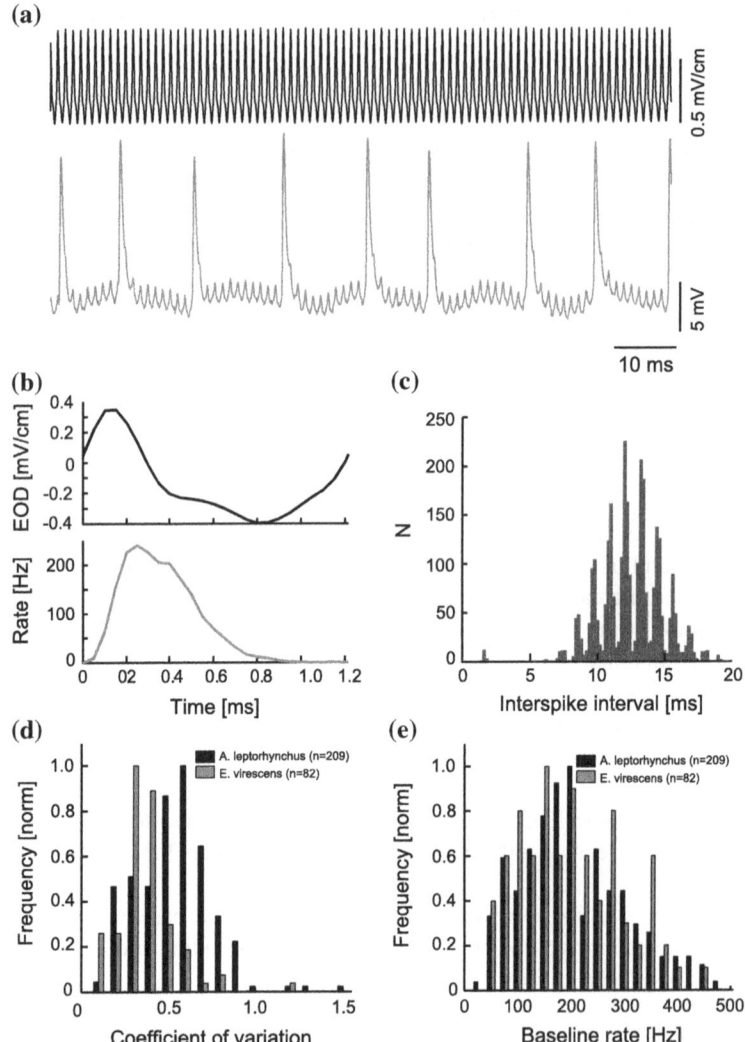

Fig. 12.8 P-unit baseline characteristics. **a** Recording of P-unit baseline activity (*bottom trace, grey*) and the EOD (*top trace, black*). Action potentials couple to the EOD. **b** Close up to the coupling. The *top trace* shows a single EOD cycle. *Bottom trace* PSTH averaged across all recorded EOD cycles. **c** Interspike-interval (ISI) histogram. Coupling to EOD leads to the dented structure of the ISI histogram. The response regularity is summarised with the CV_{ISI} of the ISIs. **d** Population heterogeneity is represented by the broad distributions of baseline rates of *Apteronotus leptorhynchus* and *E. virescens*. **e** Distribution of CVs observed in *Apteronotus leptorhynchus* and *E. virescens*

of noise in spiking neurons. These concepts are in no way specific to the electrosensory system, they apply to all neural systems. We need these concepts to be able to understand why the electroreceptor afferents are so noisy.

Best known for a possible beneficial role of noise in neural systems is the phenomenon of sub-threshold stochastic resonance (e.g. Bulsara and Zador 1996, Jaramillo and Wiesenfeld 1998). A weak sub-threshold input signal does not trigger any action potentials in a noiseless neuron. The addition of some intrinsic noise once in a while pushes the membrane potential over the firing threshold. This random generation of action potentials is more likely whenever the signal amplitude was high (and thus the membrane potential was closer to threshold) and less likely whenever the signal amplitude was low. In this way the number of action potentials per time contains some information about the amplitude of the sub-threshold input signal. However, if the noise level is too strong the noise itself dominates the generation of action potentials and less information about the signal is conveyed. Thus, there is an optimal non-zero noise level for a given signal amplitude, for which most information about the signal is encoded in the resulting spike train. There is a resonance in information transmission with respect to the noise level. Therefore, the term "stochastic resonance".

This example of sub-threshold stochastic resonance already illustrates three important aspects: (i) A non-linearity, here the firing threshold, is needed for the noise to have a beneficial effect. (ii) Because of the non-linearity the noiseless case does not perform optimally. In the example, the information transmitted about the sub-threshold input signal is even zero. (iii) Not the absolute noise level determines the performance of the system but rather its relation to the signal amplitude, i.e. the signal-to-noise ratio (SNR).

Let us now turn to the case of suprathreshold signals that drive even a noiseless neuron sufficiently so that the neuron responds with some mean firing rate that is modulated according to the input signal (Fig. 12.9b). In this case a single neuron "samples" the input signal with its instantaneous rate (inverse interspike interval) and similar to the Nyquist theorem can maximally transmit signal frequencies up to about half its firing rate (Knight 1972; Pressley and Troyer 2011). Given two successive action potentials with interspike interval T there is no way to figure out whether these two spikes have been generated by a sinusoidal signal of period T or by one with half the period (twice the frequency, Fig. 12.9a). Only if there are at least two spikes per signal period, i.e. the firing rate is at least twice the maximum frequency of the input signal, one can imagine that the signal waveform can be inferred from the action potentials of a single neuron (Fig. 12.9b). Because the maximum firing rate of neurons is limited this sets a limit to the maximum frequency of an input signal that can be encoded by a single neuron.

The situation changes for a population of noisy neurons. Consider many identical but independent neurons (no lateral connections), a so-called homogeneous population, all receiving the same input. A downstream target neuron reads out the action potentials from all neurons in the population (Fig. 12.9c). Such a convergent feed-forward network is a common and basic network motif in neural systems.

If the neurons in the population were all noiseless, they would all fire at the same times (given they are forgetful, i.e. current leaks through the membrane). The responses are highly redundant and the population carries just the same information

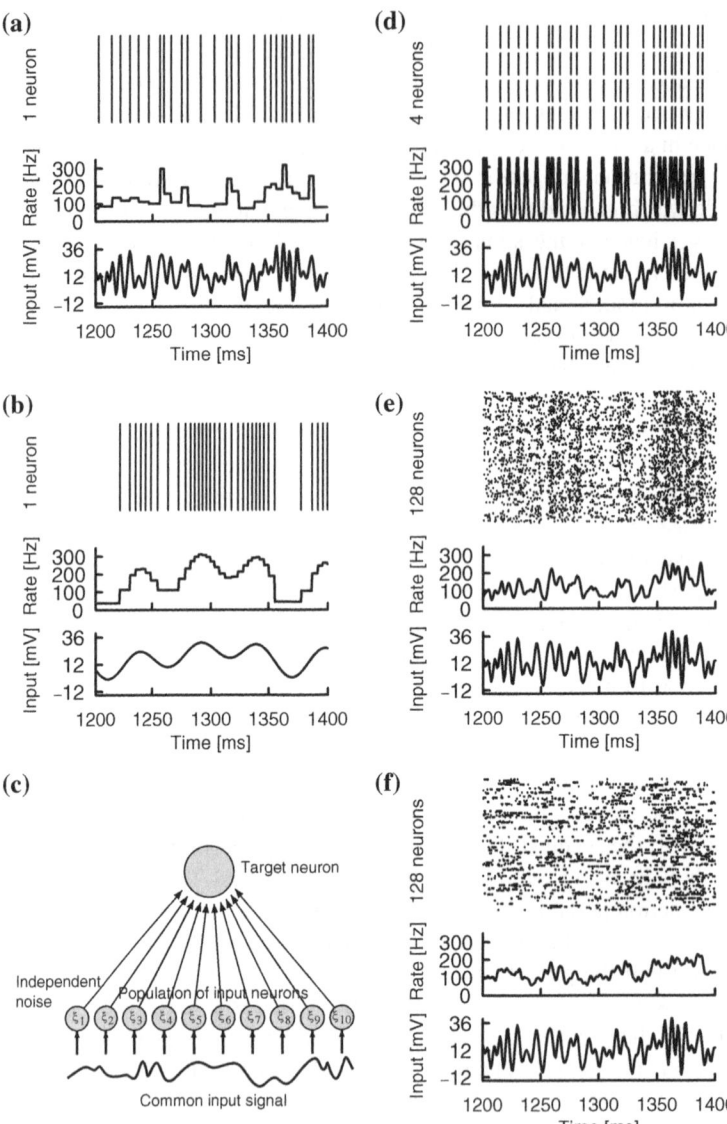

Fig. 12.9 Noise improves signal representation in populations of spiking neurons. Shown are spike rasters (*top panels*), firing rates (*middle panels*) and input signals (*bottom panels*) obtained from simulations of standard leaky integrate-and-fire neurons with firing threshold at 10 mV, reset voltage at 0 mV, membrane time constant of 10 ms, input resistance of 1 (therefore the current stimulus is in mV and the noise intensity in mV^2/Hz) and additive white noise. **a** The instantaneous rate of a single noiseless neuron cannot resolve fluctuations of the input signal faster than the interspike intervals. **b** Input signals that are much slower than the interspike intervals are well represented by the instantaneous rate of noiseless neurons. **c** In a feed-forward convergent network the input neurons are independent and all receive the same input signal. **d** A population of noiseless neurons is identical to a single neuron. The population rate does not capture the signal waveform. **e** With the right amount of noise, the population rate follows also fast fluctuations of the signal well. **f** Too much noise deteriorates the population response

as a single neuron. So, nothing would be gained by having more than a single neuron representing the input signal (Fig. 12.9d). With a sufficient amount of intrinsic noise, however, each of the neurons in the population will fire at more or less different times. This way, the population samples the common input signal at any point in time (Fig. 12.9e). When the signal is weak only a small fraction of the population will fire, and when the signal is strong a much larger fraction will fire an action potential. Thus, the population rate $r(t)$, defined as the fraction of neurons in a population that fires a spike within a small period of time, will nicely represent the input signal (Knight 1972; Manwani et al. 2002). This is the firing rate that is usually measured as the PSTH from multiple trials recorded in a single neuron. The independent intrinsic noise sources decorrelate the responses of the neurons and thus reduce redundancy. Each neuron contributes slightly different information about the input signal.

In particular, even signal frequencies that are much higher than the firing rate can be transmitted by such a population of neurons (Fig. 12.9e). How exactly is described by power law functions that depend on the specific dynamics of the spike generator (Fourcaud-Trocmé et al. 2003). Contrary to common sense, the membrane time constant does not set any limit to the information carried about a suprathreshold signal (Pressley and Troyer 2011).

Although the intrinsic noise reduces the information about the input signal that is carried by a single neuron, summing up or averaging the responses of the whole population of noisy neurons can result in a much better representation of the signal than a single noiseless neuron can provide (Fig. 12.10a, Stocks and Mannella 2001). The common more pessimistic view on this characteristic is that, because of the noisy responses, upstream neurons have to average over many neurons to get rid of the noise. This view neglects the fact that by means of this averaging over a population of noisy spiking neurons potentially much more information about the input signal can be gained than in the noiseless case. Note also that in the context of computing Bayesian inference the variability of neural responses can be directly used for encoding the uncertainty of the estimated mean (Knill and Pouget 2004). How much noise in a population can enhance the mutual information depends on the properties of the signal and is in particular strong for high-frequency signals. In addition, the higher the noise level, the more the neurons are decorrelated, and the more information can be transmitted, provided the number of neurons in the population is large enough. The more noise is in the system, the more neurons are needed for achieving the same fidelity of the population code (Fig. 12.10b).

Noise in such homogeneous populations of neurons with a common input signal is thus potentially beneficial for signal representation. Furthermore, for a given number of neurons and a given class of input signals there is an optimal noise level that maximises the mutual information between signal and population response (Fig. 12.10c), because too much noise will eventually deteriorate the code (Fig. 12.9f, suprathreshold stochastic resonance, Stocks and Mannella 2001). Experimentally, however, this optimality of the noise level has not been shown yet.

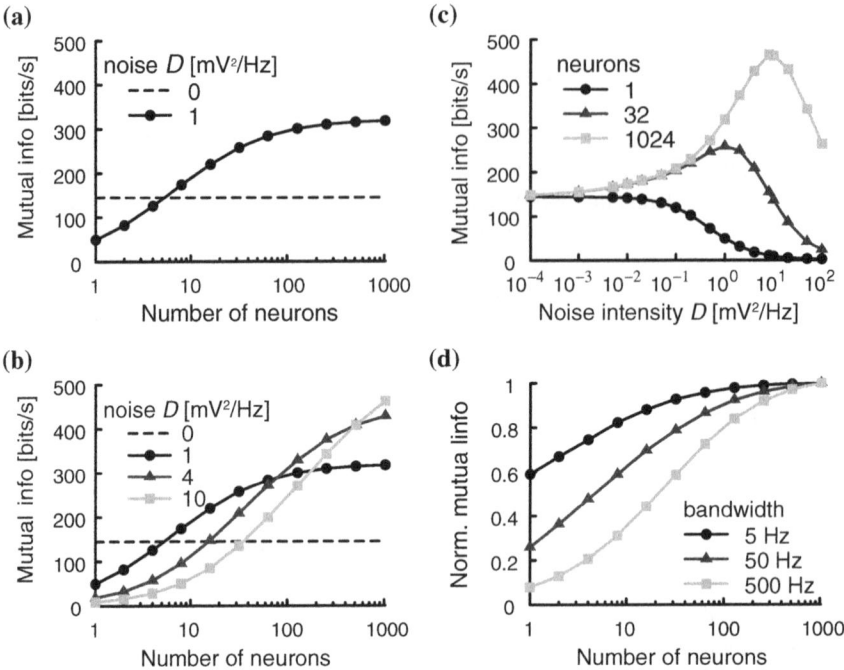

Fig. 12.10 Mutual information between input signal and population response. **a** The mutual information of a population of noisy neurons (*solid line*) increases with the number of neurons and can get much higher than the mutual information of noiseless neurons (*dashed line*). **b** The stronger the noise, the more neurons are needed to achieve the same mutual information. However, with even more neurons the mutual information can be increased even further. **c** For a given population size the mutual information is maximised by a particular non-zero noise level. **d** The higher the bandwidth of the signal (high-pass filtered Gaussian white noise with cutoff-frequency as indicated) the more neurons are needed before the mutual information saturates. Shown are simulations of the same leaky integrate-and-fire neuron as in Fig. 12.9. Mutual information was estimated from the coherence between input signal and spiking activity. The mean firing rate was 75 Hz

Usually the intrinsic noise of a neuron is thought to arise from ion channel noise, i.e. the stochastic opening and closing of voltage- or ligand-gated ion channels (White et al. 2000); and thus is a fixed quantity. Note that in higher brain centres the noise level could also be controlled by the amount of balanced excitatory and inhibitory synaptic input (Chance et al. 2002). What a peripheral sensory system in general cannot control is the amplitude of the sensory signal in relation to the fixed noise level. However, the size of the neural population can be adapted as well as the convergence ratios of the target neurons (on long, probably evolutionary time scales). For a given signal amplitude and frequency content the mutual information between the population response and the signal first increases with the number of neurons in the population, but eventually saturates (Fig. 12.10a). The minimum number of neurons needed for maximum mutual information depends on the noise level, the signal amplitude and the frequency content of the signal. The higher the signal amplitude (higher SNR) or the lower its frequency the fewer neurons are needed for a maximum information transmission at a given noise level (Fig. 12.10d).

To summarise, let us highlight four aspects of coding common input signals in populations of noisy neurons; (i) Intrinsic noise exerts its beneficial role only in spiking neurons, not in graded-potential neurons. (ii) Populations of independent neurons are required—in single neurons noise in general deteriorates the coding quality (however, under certain assumptions even in single neurons noise can be beneficial, as for sub-threshold stochastic resonance discussed above). (iii) The relevant quantity that determines the fidelity of the representation of a common signal in a population of spiking neurons is not the absolute noise level, but the signal-to-noise ratio. For a two-fold increase in signal amplitude one needs twice as much intrinsic noise in order to achieve the same coding performance (Stocks and Mannella 2001). (iv) The frequency content of the signal is also an important factor. For lower maximum frequencies of the signal fewer neurons are sufficient.

In addition to the intrinsic noise discussed above, another kind of noise results from the heterogeneity of neurons in a given population. Neurons are not identical in their response properties; they have different thresholds, sensitivities, baseline firing rates, membrane time constants, adaptation strength, etc. Even without any intrinsic current noise this heterogeneity has a similar effect in that it decorrelates the responses of the neurons and thus results in an enhanced representation of the signal. A heterogeneous population has in addition the possibility to cover much wider ranges of input intensities by having different sensitivities.

In light of these fundamental properties of neural populations, the high intrinsic variability of P-units and their strong heterogeneity is not so surprising any more. By averaging over many noisy P-units the pyramidal cells in the ELL can obtain much more information about a signal than from a population of homogeneous and noiseless P-units. Let us now return to the active electrosensory system and discuss the characteristics of the natural signals the system has to process and how the information carried by the P-units is integrated and processed at the next level of the electrosensory pathway, the ELL.

12.5.3 Decoding Population Codes in Three Different Maps of the ELL

As described above the electrosensory system has to deal with several different classes of electric signals that differ in amplitude, frequency content and spatial extent and thus require different processing strategies in the electrosensory pathway. In fact, the receptor afferents of the active electrosensory system project onto three different types of pyramidal cells that are arranged in three separate maps, the CMS, CLS and LS (Fig. 12.6b) that differ in receptive field size and temporal tuning properties.

For the task of object localisation the fish have to extract the three coordinates for the relative position of the object as well as its size from the information available in the three ELL maps. Lewis and Maler (2001) and Maler (2009b) recognised that this is a multi-parameter estimation problem and investigated it in simulations using Fisher-information on the spike count. They conclude that the parameters can be successively obtained from the three maps (Lewis and Maler 2001) and that this parameter estimation is heavily influenced by possible synaptic scaling that normalises for receptive field sizes (Maler 2009b). The spike-count responses of the pyramidal cells are modelled with Gaussian functions describing their receptive fields plus additive noise. The noise strength was fixed across cells and the same for all the maps. However, this simple assumption models only the additional intrinsic noise of the pyramidal cells themselves and neglects the noise from the P-unit input. The latter, however, depends on the interacting effects of population size, signal amplitude and signal frequency and thus determines the fidelity and thus the noisiness of the representation of a signal by the population of P-unit afferents as explained in the previous section.

Therefore, we here qualitatively evaluate the role of the three ELL maps by summarising the different types of electric signals in terms of the properties that are important in the context of population codes introduced above (Fig. 12.11a, b). The relevant quantities are signal amplitude (in relation to the intrinsic noise level), signal frequencies and spatial extent. The latter defines the number of electroreceptor afferents receiving the same input signal. We then compare these requirements with the known properties of the target neurons, in particular, their frequency tuning and receptive field size, i.e. the number of receptor afferents converging onto a single pyramidal cell (Fig. 12.11c, d).

Let us first discuss the signals arising from distortions of the EOD by nearby objects that are used by the fish for electrolocation (Lissmann and Machin 1958, red regions in Fig. 12.11). If the objects are "far" away (at most a few centimetres, Nelson and MacIver 1999; MacIver et al. 2001), the electric images they cast on the body of the fish are faint in amplitude and blurred. They cover several centimetres on the body surface and thus provide weak input to many (hundreds) of P-units. If the fish approaches an object the electric image gets stronger and more focused. This is called a "local" stimulus that excites only a few (tens) receptors. How the shape and amplitude of electric images look in detail has been studied both experimentally and theoretically (Heiligenberg 1975; Bastian 1981; Bacher 1983; Rasnow 1986; Chen et al. 2005; Babineau et al. 2006). The temporal properties of these signals are mainly determined by the speed (typical 9 cm/s) of the fish swimming past an object (MacIver et al. 2001). Temporal frequencies are low and range from a few Hertz for distant objects to about 25 Hz for nearby prey (Nelson and MacIver 1999).

Thus, for electrolocation we expect the target neurons of the electroreceptor afferents, the pyramidal cells in the ELL, to be sensitive to low signal frequencies. This excludes pyramidal cells of the LS because of their high-frequency tuning (Krahe et al. 2008) and leaves the CMS and CLS as the only segments that are sensitive to low-frequency signals (Shumway 1989; Krahe et al. 2008). Because of

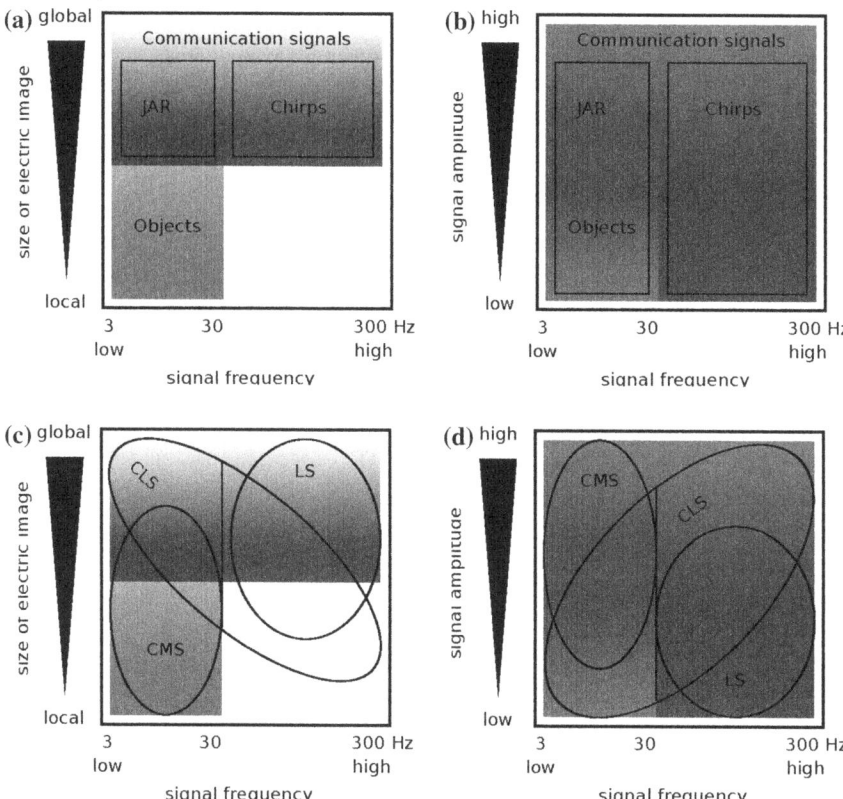

Fig. 12.11 Properties of electrosensory signals and ELL maps. Objects generate small ("local") to intermediate sized, low-frequency and medium to small amplitude signals (amplitude modulations of the EOD). In contrast, communication signals are global and of larger amplitudes. In the context of JARs, communication signals are low-frequency, in chirping contexts communications signals usually contain higher frequencies. **a** The different classes of signals shown in a plane of electric image size versus signal frequency. *Stronger colours* indicate larger amplitudes of the signals. **b** Same as in (**a**) but as a function of signal amplitude and frequency. *Stronger colours* indicate more global, spatially diffuse signals. **c** and **d** The ellipses indicate regions in the stimulus planes for which the three ELL maps are best suited. The CMS with its small receptive fields and low-frequency tuning is best for localised, low-frequency signals that are not too small in amplitude. These are objects and JAR stimuli. The LS with its large receptive fields and high-frequency tuning is great for high-frequency communication signals, in particular with small amplitudes. The CLS with intermediate receptive fields switches from low-frequency tuning under local stimulation to high-frequency tuning for global signals (indicated by the *vertical line*, Chacron et al. 2003; Krahe et al. 2008). Therefore, both distant objects and strong communication signals might be processed in this segment

the very weak signals evoked by distant objects, the pyramidal cells might need to integrate over hundreds or even thousands of electroreceptor afferents to gain the necessary sensitivity (Maler 2009b). Therefore, it seems surprising, why the LS pyramidal cells integrating over more than one thousand afferents are not suited for electrolocation, because of their high-frequency tuning. However, since the electrolocation signals are low-frequency, the information about these signals might already saturate at smaller numbers of input neurons (Fig. 12.10d). The medium-sized receptive fields of the CLS might therefore be an adequate solution for detecting distant objects. Under sufficiently local stimulation, CLS pyramidal cells are also tuned to the required low-frequency signals (Chacron et al. 2003; Krahe et al. 2008). Once an object gets closer, even less neurons are needed to compensate for the intrinsic noise of the P-units, because of the stronger signals (for example, in Fig. 12.10b a population of 100 neurons transmits about 260 bits/s at a noise intensity of 10 mV^2/Hz. The same mutual information is achieved with just 30 neurons at a smaller noise level of 1 mV^2/Hz, i.e. higher signal-to-noise ratio). Here the CMS with its small receptive fields might take over in order to match the spatial resolution of the electrosensory system to the more detailed electric image (Rasnow 1986; Lewis and Maler 2001; Maler 2009b).

The second stimulus class are the various kinds of communication signals that result from the superposition of the EODs of several fish, in particular beats, envelopes and chirps (blue regions in Fig. 12.11). These "global" stimuli affect large numbers of electroreceptors in a similar way (Kelly et al. 2008). The amplitude of these signals is given by the amplitude of the other fish's EOD at the location of the perceiving fish. Therefore, communication signals can be much larger in amplitude in comparison to electrolocation signals for nearby fish. But communication signals can also be arbitrarily small, since the far field of electric fish is that of a dipole, whose amplitude drops inversely with the squared distance. So communication signals potentially cover a huge range of amplitudes. Assuming a detection threshold of 0.5 µV/cm (Knudsen 1974) other fish can be detected up to a distance of about 1.5 m (Knudsen 1975)—two orders of magnitude more than for electrolocation.

The communication signals can be further classified according to their temporal properties. On the one hand, low-frequency beats up to about 20 Hz that elicits a jamming avoidance response (JAR) and on the other hand beats of higher frequency that elicit chirps and related behaviours (Hupé and Lewis 2008). Because of the low frequency (Fig. 12.10d) and larger amplitude (lower noise intensity in Fig. 12.10c) of JAR signals integration over a few receptor afferents should be sufficient to get close to the maximum information that can be retrieved from the noisy P-unit population. Indeed, a lesion study has shown that the CMS is necessary and sufficient for JAR behaviour (Metzner and Juranek 1997). JAR behaviour can be evoked by small amplitude beats (<0.1 % contrast) as well (Rose and Heiligenberg 1985; Kawasaki 1997). How well these small-amplitude signals are encoded in the CMS or whether the CLS with its larger receptive fields is needed for tiny jamming signals in order to compensate for the lower signal-to-noise ratio, remains to be investigated.

The higher frequencies of fast beats and chirps profit the most from the noise of the P-units if the target cell average over many neurons (light grey curves in Fig. 12.10b–d). This would fit the pyramidal cells in the CLS or even the LS with their larger receptive fields. The above mentioned lesion study showed that it is the

LS that is necessary and sufficient for chirping behaviour (Metzner and Juranek 1997). The fish are also very sensitive to low-amplitude beat signals (Knudsen 1974), which would be generated by the presence of distant fish. In this regime, the LS would play out its strength given the high number of P-unit afferents converging onto each pyramidal cell (more than one thousand, Maler 2009a). However, the fish might also want to localise the conspecific. For this, the smaller receptive fields of the CLS might be more appropriate to better evaluate the various geometries of the electric images resulting from the interaction of two nearby fish (Kelly et al. 2008).

In summary, given the large variety of electric signals the noise introduced by the electroreceptor afferents themselves seems to be appropriately used by the pyramidal cells in the different segments of the ELL. In particular, the noise is necessary to enhance encoding of high-frequency communication signals and to potentially increase the information about small amplitude signals. How exactly the number of neurons projecting onto the pyramidal cells is optimised given the fixed noise level of the P-units remains, however, to be investigated. Vice versa, one can ask whether the noise level of the P-units and their heterogeneity is optimised for the observed convergence ratios and required spatial resolutions.

Here, we have discussed only the direct input to the ELL neurons. However, as described above, the pyramidal cells also receive feedback from two distinct feedback pathways. In particular, spatially extended predictive input is subtracted from the responses of superficial pyramidal cells by means of a "negative image" mechanism, thus improving the representation of novel signals. The feedback also decorrelates the responses in the LS (Chacron and Bastian 2008) and thus potentially allows for further increases in sensitivity in higher processing stages like the Torus semicircularis by averaging over the LS cells.

Also, in addition to static snapshots of the electric images the fish could make use of motion signals (Babineau et al. 2007). For instance, during prey capture the fish scan the prey object and in a closed-loop fashion adapt their movement to the gained information about object location until capturing the prey within about one second (MacIver et al. 2001). This rapid behavioural sequence also sets tight limits for potential temporal integration mechanisms to improve prey detection. In the ELL, neurons are not particularly sensitive to motion signals, but they might pick up the signal upstrokes or downstrokes generated by moving objects (Gabbiani 1996). At the next stage of electrosensory processing, in the Torus semicircularis, some neurons indeed show direction-selective responses (Ramcharitar et al. 2005, 2006; Chacron et al. 2009).

12.5.4 Synchrony Code and Chirps

So far, we discussed how electric signals might be represented by means of populations of noisy neurons. However, sensory systems also need to process information. Relevant features of the sensory input need to be detected (Gabbiani 1996) and irrelevant signals should be discarded (Bastian 1995). In the following, we show how a neural system can make use of noisy neurons in order to extract certain features from a sensory signal. For this, we first look closer at communication signals, in particular how chirps (Fig. 12.5) are processed by the first stages of the electrosensory system.

The firing rate of P-unit afferents follows the periodic amplitude modulation of a low-frequency beat. A small chirp briefly advances the beat and thus introduces a higher frequency amplitude modulation (Fig. 12.5c). This faster signal evokes a stronger firing rate response (Fig. 12.12b) that briefly synchronises the P-unit population (Benda et al. 2005, 2006). On higher beat frequencies the situation reverses. Then the fast beat signals synchronise the receptor afferents and the small chirp briefly desynchronises the population response (Hupé et al. 2008). Similarly, large chirps that are usually emitted on high beat frequencies desynchronise the P-unit afferents as well (Benda et al. 2006).

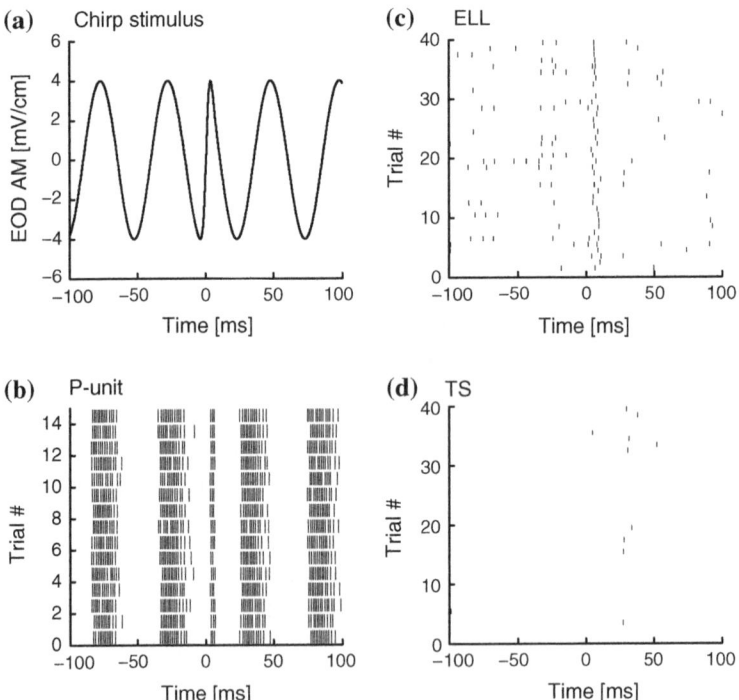

Fig. 12.12 Encoding of chirps in the electrosensory system. **a** The amplitude modulation of a 20 Hz beat with a 14 ms wide chirp with a 100 Hz frequency excursion centred at time $t = 0$ ms is the input signal. **b** A P-unit receptor afferent responds well to both the beat and the chirp. The response to the chirps is, however, usually stronger and more precisely timed than the one to the beat. **c** Pyramidal cells in the LS of the ELL weakly respond to the beat but reliably generate a burst in response to a chirp. **d** In Torus semicircularis some cells selectively respond to chirps and not to the beat any more. Data kindly provided by Henriette Walz (*panel b*) and Maurice Chacron (*panel c, d*)

What the chirps do is to change the level of synchrony in the P-unit population. This synchrony code is an additional coding dimension that is only possible in a population of spiking neurons. To differentiate between synchronous and asynchronous spikes that are fired in response to a common input is only possible because of the high noise level in the P-unit population. Without noise the activity of a neural population receiving a common input signal would be highly redundant and most spikes would be synchronous anyways.

Reading out synchronous spikes requires a non-linear operation like coincidence detection that can be achieved, for example, by linear synaptic summation and a high firing threshold. Such mechanisms potentially process the incoming information such that some aspects are tossed away. What if a pyramidal cell would only read out synchronous spikes from the population of P-unit afferents?

Middleton et al. (2009) followed this idea. They first quantified what kind of information the synchronous spikes carry about the input signal in comparison to all spikes. For this they extracted "synchronous spike trains" and "all spike" spike trains from recorded afferent responses to broad-band noise signals (Fig. 12.13a). The coherence between the input signal and one of these two spike trains is a lower bound estimate of the mutual information as a function of signal frequency. Interestingly, while all spikes code best for low signal frequencies, the synchronous spikes do not code for low frequencies but rather selectively encode higher frequency components of the input signal (Fig. 12.13b).

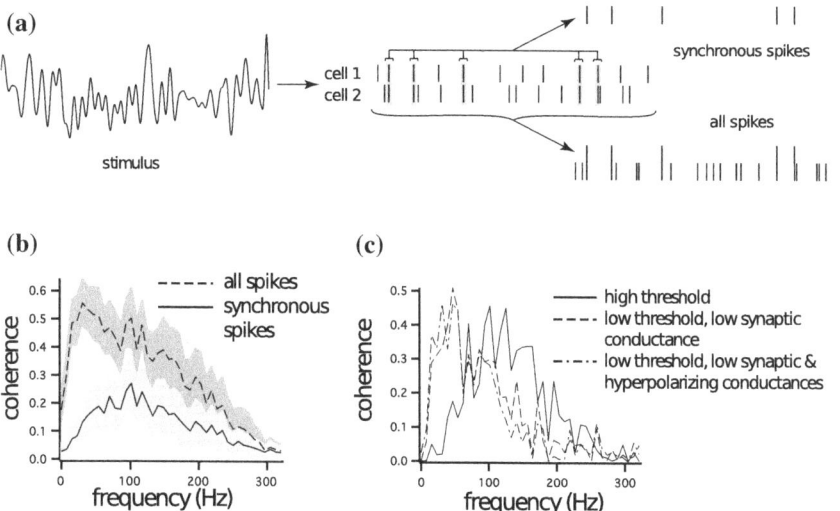

Fig. 12.13 Synchronous spikes code for high-frequency signals. **a** From the spike trains of P-unit afferents one can construct a spike train that is just the sum of all spikes and another one that extracts only spikes that simultaneously occurred in both spike trains. **b** The coherence as a spectral measure of information transmitted about the input signal is low-pass for all spikes and high-pass for synchronous spikes. **c** A model of the pyramidal cell taking its receptive field and high firing threshold into account also responds in a high-pass manner to the input signal. Figures from Middleton et al. (2009)

Naively one could suggest that there should be some pyramidal cells with low firing thresholds that decode all spikes and thus are low-pass coherent with the input signal. Other pyramidal cells with high firing thresholds should decode the synchronous spikes only and thus show a high-pass coherence with the input signal. However, if both these cells integrate over the same number of input neurons, this would mean that the mean firing rate of the synchronous-spike decoder is much lower than the one of the all-spike decoder. Simulations of the pyramidal cell responses to the recorded P-unit afferent spike trains showed that both large receptive fields and high firing thresholds are required for the synchronous-spike decoder and small receptive fields and lower firing thresholds for the all spike decoder. Only then, they have about the same firing rates and the characteristic high-pass or low-pass coherence with the input signal, respectively (Fig. 12.13c, Middleton et al. 2009). Indeed, there is experimental evidence from in vitro work that CMS pyramidal cells with their small receptive fields have indeed a lower threshold compared to the LS pyramidal cells with their large receptive fields (Mehaffey et al. 2008). This fits well with the tuning properties of the pyramidal cells measured in vivo: CMS cells are low-pass whereas LS cells are high-pass (Krahe et al. 2008), indicating that LS pyramidal cells read out synchronous spikes from the population of P-unit afferents.

For the chirps this would mean that the brief synchronisation of the P-unit population is well encoded by the pyramidal cells of the LS while the asynchronous response to low-frequency beats is suppressed. Recordings from the ELL only partly support this hypothesis: in the LS only the superficial pyramidal cells receiving the negative image through feedback loops (see above) encode a small chirp with a signal-to-noise ratio that is enhanced in comparison to the responses of the P-unit afferents (Fig. 12.12b, Marsat et al. 2009; Marsat and Maler 2010). At the next level of the electrosensory pathway, the Torus semicircularis, a sub-population of cells responds selectively to chirps (Fig. 12.12c, Vonderschen and Chacron 2011). Thus, the signal-to-noise ratio of the chirp responses over the beat is successively improved in higher sensory areas and thus facilitates detection of chirps (see Signal Detection Theory, Chap. 2). In this sense, the beat is the noise in which the chirps are embedded.

12.6 Conclusion

The exotic electrosense of weakly electric fish shares several features with auditory and visual systems. The sensory signals are amplitude modulations of periodic carriers, similar to acoustic signals. A two-dimensional array of electroreceptors provides spatial information about nearby objects and conspecifics that cast electric images on the fish, similar to visual images that are processed by the retina. The active electrosensory system processes two different classes of AM signals simultaneously: communication and electrolocation signals.

Except for the evolutionarily young anthropogenic 60 Hz electropollution there is virtually no external noise in the frequency bands used by the active electrosensory system at least of wave-type weakly electric fish. However, the highly structured and narrow-band electrocommunication signals interfere with the electrolocation signals, and thus constitute a noise source for electrolocation. Vice versa, electrolocation signals could be a relevant noise source for low amplitude communication signals.

Instead of being narrowly tuned to the fish's EOD frequency that carries the electrolocation signals, P-units are rather broadly tuned and thus open up a relatively wide frequency range that is used by conspecifics with their individual EOD frequencies and communication signals like, for example, chirps. Because of the low level or even lack of external electric noise the wide tuning of the P-units seems not to impair electrolocation.

Most strikingly, the electroreceptor afferents themselves introduce an unexpected amount of noise into the system—both as intrinsic noise and through their heterogeneity. Fundamental properties of populations of spiking neurons receiving a common input signal, however, show that such noise potentially improves the information about the signal. In particular, frequency components that are higher than the typical firing rate of the individual neurons benefit from this effect.

In this context, weakly electric fish prove to be an excellent model system to study such noisy population codes. First, the different classes of electric signals are well specified by just three parameters amplitude, frequency and spatial extent. Second, the target neurons of the electroreceptor afferents, the pyramidal cells in the ELL, are nicely separated in three distinct maps that differ in receptive field size and temporal tuning properties. In the lateral segment, pyramidal cells integrate over more than one thousand afferents and thus receive a high-quality input about high-frequency communication signals. Although of small amplitudes, the low-frequency electrolocation signals do not benefit from very large input populations. For these input signals the pyramidal cells in the centrolateral and centromedial segments with their smaller receptive fields are better suited.

Because of the noisy responses of the primary afferents, pyramidal cells could read out the level of synchrony and by this non-linear mechanism process incoming information. This is in particular relevant for encoding chirps, short duration modulations of EOD frequency used as communication signals. Depending on context, these chirps either synchronise or desynchronise the P-unit population. In vivo and in vitro data suggest that pyramidal cells in the lateral segment of the ELL indeed extract synchronous spikes from the P-unit afferents.

In this chapter, we have laid out some concepts and supporting experimental data on the role of noise in networks of spiking sensory neurons. These concepts are not limited to the electrosensory system of weakly electric fish, but we feel that this system has much to offer in terms of understanding how neural circuits use noise. The electrosensory system appears to be exposed to relatively little extrinsic noise. Instead, it appears to generate remarkably large levels of intrinsic noise. One wonders if we should start looking for a negative correlation between levels of extrinsic and intrinsic noise across sensory systems.

Acknowledgments We would like to thank the editor, Henrik Brumm and the reviewers, John Lewis und Peter McGregor, for their helpful comments to improve our manuscript. Our work was generously supported by our funding agencies, in particular by the BMBF (German Ministry of Education and Research), a Bernstein award for Computational Neuroscience to JB, and a NSERC (Natural Sciences and Engineering Research Council of Canada) Discovery grant to RK.

References

Albers HE, Karom M, Smith D (2002) Serotonin and vasopressin interact in the hypothalamus to control communicative behavior. NeuroReport 13:931–933

Allee SJ, Markham MR, Salazar VL, Stoddard PK (2008) Opposing actions of 5ht1a and 5ht2-like serotonin receptors on modulations of the electric signal waveform in the electric fish brachyhypopomus pinnicaudatus. Horm Behav 53:481–488

Arnegard ME, Carlson BA (2005) Electric organ discharge patterns during group hunting by a mormyrid fish. Proc Biol Sci 272:1305–1314

Assad C, Rasnow B, Stoddard PK, Bower JM (1998) The electric organ discharges of the gymnotiform fishes: II. *Eigenmannia*. J. Comp. Physiol. A 183:419–432

Ávila-Åkerberg O, Krahe R, Chacron M (2010) Neural heterogeneities and stimulus properties affect burst coding in vivo. Neuroscience 168:300–313

Babineau D, Lewis J, Longtin A (2007) Spatial acuity and prey detection in weakly electric fish. PLoS Comput Biol 3:e38

Babineau D, Longtin A, Lewis JE (2006) Modeling the electric field of weakly electric fish. J Exp Biol 209:3636–3651

Bacher M (1983) A new method for the simulation of electric fields, generated by electric fish, and their distortions by objects. Biol Cybern 47:51–58

Bastian J (1981) Electrolocation I. How electroreceptors of *Apteronotus albifrons* code for moving objects and other electrical stimuli. J Comp Physiol A 144:465–479

Bastian J (1982) Vision and electroreception: integration of sensory information in the optic tectum of the weakly electric fish *Apteronotus albifrons*. J Comp Physiol A 147:287–297

Bastian J (1987a) Electrolocation in the presence of jamming signals: behavior. J Comp Physiol A 161:811–824

Bastian J (1987b) Electrolocation in the presence of jamming signals: electroreceptor physiology. J Comp Physiol A 161:825–836

Bastian J (1995) Pyramidal-cell plasticity in weakly electric fish: a mechanism for attenuating responses to reafferent electrosensory inputs. J Comp Physiol A 176:63–73

Bastian J, Chacron MJ, Maler L (2002) Receptive field organization determines pyramidal cell stimulus-encoding capability and spatial stimulus selectivity. J Neurosci 22:4577–4590

Bastian J, Chacron MJ, Maler L (2004) Plastic and nonplastic pyramidal cells perform unique roles in a network capable of adaptive redundancy reduction. Neuron 41:767–779

Bastian J, Courtright J (1991) Morphological correlates of pyramidal cell adaptation rate in the electrosensory lateral line lobe of weakly electric fish. J Comp Physiol A 168:393–407

Bastian J, Schniederjan S, Nguyenkim J (2001) Arginine vasotocin modulates a sexually dimorphic communication behavior in the weakly electric fish *Apteronotus leptorhynchus*. J Exp Biol 204:1909–1923

Bell CC (2002) Evolution of cerebellum-like structures. Brain Behav Evol 59:312–326

Bell CC, Maler L (2005) Central neuroanatomy of electrosensory systems in fish. In: Bullock TH, Hopkins CD, Popper AN, Fay RR (eds) Electroreception. Springer, New York, pp 68–111

Benda J, Longtin A, Maler L (2005) Spike-frequency adaptation separates transient communication signals from background oscillations. J Neurosci 25:2312–2321

Benda J, Longtin A, Maler L (2006) A synchronization-desynchronization code for natural communication signals. Neuron 52:347–358

Berman N, Maler L (1999) Neural architecture of the electrosensory lateral line lobe: adaptations for coincidence detection, a sensory searchlight and frequency-dependent adaptive filtering. J Exp Biol 202:1243–1253

Bratton B, Bastian J (1990) Descending control of electroreception: II. Properties of nucleus praeeminentialis neurons projecting directly to the electrosensory lateral line lobe. J Neurosci 10:1241–1253

Bullock TH (1969) Species differences in effect of electroreceptor input on electric organ pacemakers and other aspects of behavior in electric fish. Brain Behav Evol 2:85–118

Bullock TH, Hamstra RH, Scheich H (1972a) The jamming avoidance response of high frequency electric fish. I. general features. J Comp Physiol A 77:1–22

Bullock TH, Hamstra RH, Scheich H (1972b) The jamming avoidance response of high frequency electric fish. II. quantitative aspects. J Comp Physiol A 77:23–48

Bulsara AR, Zador A (1996) Threshold detection of wideband signals: a noise-induced maximum in the mutual information. Phys Rev E 54:R2185–R2188

Carr CE, Maler L, Sas E (1982) Peripheral organization and central projections of the electrosensory nerves in gymnotiform fish. J Comp Neurol 211:139–153

Chacron MJ, Bastian J (2008) Population coding by electrosensory neurons. J Neurophysiol 99:1825–1835

Chacron MJ, Doiron B, Maler L, Longtin A, Bastian J (2003) Non-classical receptive field mediates switch in a sensory neuron's frequency tuning. Nature 423:77–81

Chacron MJ, Lindner B, Longtin A (2004) Noise shaping by interval correlations increases information transfer. Phys Rev Lett 92:080601

Chacron MJ, Longtin A, Maler L (2001) Negative interspike interval correlations increase the neuronal capacity for encoding time-dependent stimuli. J Neurosci 21:5328–5343

Chacron MJ, Maler L, Bastian J (2005) Electroreceptor neuron dynamics shape information transmission. Nat Neurosci 8:673–678

Chacron MJ, Toporikova N, Fortune ES (2009) Differences in the time course of short-term depression across receptive fields are correlated with directional selectivity in electrosensory neurons. J Neurophysiol 102:3270–3279

Chance FS, Abbott LF, Reyes AD (2002) Gain modulation from background synaptic input. Neuron 35:773–782

Chen L, House JL, Krahe R, Nelson ME (2005) Modeling signal and background components of electrosensory scenes. J Comp Physiol A 191:331–345

Crampton WGR, Albert JS (2006) Evolution of electric signal diversity in gymnotiform fishes. In: Ladich F, Collin SP, Moller P (eds) Communication in fishes, vol 2. Science Publishers, Enfield, pp 647–731

Crick F (1984) Function of the thalamic reticular complex: the searchlight hypothesis. Proc Natl Acad Sci U S A 81:4586–4590

Cuddy M, Aubin-North N, Krahe R (2011) Electrocommunication behaviour and non invasively-measured androgen changes following induced seasonal breeding in the weakly electric fish, *Apteronotus leptorhynchus*. J Exp Biol (in press)

Doiron B, Longtin A, Turner RW, Maler L (2001) Model of gamma frequency burst discharge generated by conditional backpropagation. J Neurophysiol 86:1523–1545

Dunlap KD (2002) Hormonal and body size correlates of electrocommunication behavior during dyadic interactions in a weakly electric fish, *Apteronotus leptorhynchus*. Horm Behav 41:187–194

Dunlap KD, Jashari D, Pappas KM (2011) Glucocorticoid receptor blockade inhibits brain cell addition and aggressive signaling in electric fish, *Apteronotus leptorhynchus*. Horm Behav 60:275–283

Dunlap KD, Pelczar PL, Knapp R (2002) Social interactions and cortisol treatment increase the production of aggressive electrocommunication signals in male electric fish, *Apteronotus leptorhynchus*. Horm Behav 42:97–108

Dunlap KD, Smith GT, Yekta A (2000) Temperature dependence of electrocommunication signals and their underlying neural rhythms in the weakly electric fish. *Apteronotus leptorhynchus.* Brain Behav Evol 55:152–162

Dunlap KD, Thomas P, Zakon HH (1998) Diversity of sexual dimorphism in electrocommunication signals and its androgen regulation in a genus of electric fish. *Apteronotus.* J Comp Physiol A 183:77–86

Ellis LD, Mehaffey WH, Harvey-Girard E, Turner RW, Maler L, Dunn RJ (2007) SK channels provide a novel mechanism for the control of frequency tuning in electrosensory neurons. J Neurosci 27:9491–9502

Enger PS, Szabo T (1968) Effect of temperature on the discharge rates of the electric organ of some gymnotids. Comp Biochem Physiol 27:625–627

Engler G, Zupanc GK (2001) Differential production of chirping behavior evoked by electrical stimulation of the weakly electric fish. *Apteronotus leptorhynchus.* J Comp Physiol A 187:747–756

Flecker A, Taphorn D, Lovell J, Feifarek B (1991) Drift of characin larvae, *Bryconamericus deuterodonoides,* during the dry season from andean piedmont streams. Environ Biol Fishes 31:197–202

Fourcaud-Trocmé N, Hansel D, van Vreeswijk C, Brunel N (2003) How spike generation mechanisms determine the neuronal response to fluctuating inputs. J Neurosci 23:11628–11640

Fugère V, Krahe R (2010) Electric signals and species recognition in the wave-type gymnotiform fish *Apteronotus leptorhynchus.* J Exp Biol 213:225–236

Fugère V, Ortega H, Krahe R (2011) Electrical signalling of dominance in a wild population of electric fish. Biol Lett 7:197–200

Gabbiani F (1996) From stimulus encoding to feature extraction in weakly electric fish. Nature 384:564–567

Goodson JL, Bass AH (2001) Social behavior functions and related anatomical characteristics of vasotocin/vasopressin systems in vertebrates. Brain Res Brain Res Rev 35:246–265

Gussin D, Benda J, Maler L (2007) Limits of linear rate coding of dynamic stimuli by electroreceptor afferents. J Neurophysiol 97:2917–2929

Gutzler SJ, Karom M, Erwin WD, Albers HE (2011) Seasonal regulation of social communication by photoperiod and testosterone: effects of arginine-vasopressin, serotonin and galanin in the medial preoptic area-anterior hypothalamus. Behav Brain Res 216:214–219

Hagedorn M, Heiligenberg W (1985) Court and spark: electric signals in the courtship and mating of gymnotid fish. Anim Behav 33:254–265

Heiligenberg W (1973) Electrolocation of objects in the electric fish *Eigenmannia* (Rhamohichthyidae, Gymnotidei). J Comp Physiol A 87:137–164

Heiligenberg W (1975) Theoretical and experimental approaches to spatial aspects of electrolocation. J Comp Physiol A 103:247–272

Heiligenberg W (1991) Neural nets in electric fish. MIT Press, Cambridge

Heiligenberg W, Dye J (1982) Labeling of electroreceptive afferents in a gymnotoid fish by intracellular injection of HRP: the mystery of multiple maps. J Comp Physiol A 148:287–296

Heiligenberg W, Metzner W, Wong CJH, Keller CH (1996) Motor control of the jamming avoidance response of *Apteronotus leptorhynchus*: evolutionary changes of a behavior and its neuronal substrates. J Comp Physiol A 179:653–674

Heiligenberg W, Rose G (1985) Phase and amplitude computations in the midbrain of an electric fish: intracellular studies of neurons participating in the Jamming Avoidance Response of *Eigenmannia.* J Neurosci 5:515–531

Hopkins CD (1972) Sex differences in electric signaling in an electric fish. Science 176:1035–1037

Hopkins CD (1973) Lightning as background noise for communication among electric fish. Nature 242:268–270

Hopkins CD (1976) Stimulus filtering and electroreception: tuberous electroreceptors in three species of gymnotoid fish. J Comp Physiol A 111:171–207

Hupé GJ, Lewis JE (2008) Electrocommunication signals in free swimming brown ghost knifefish, *Apteronotus leptorhynchus*. J Exp Biol 211:1657–1667

Hupé GJ, Lewis JE, Benda J (2008) The effect of difference frequency on electrocommunication: chirp production and encoding in a species of weakly electric fish. *Apteronotus leptorhynchus*. J Physiol Paris 102:164–172

Jaramillo F, Wiesenfeld K (1998) Mechanoelectrical transduction assisted by brownian motion: a role for noise in the auditory system. Nat Neurosci 1:384–388

Kawasaki M (1997) Sensory hyperacuity in the jamming avoidance response of weakly electric fish. Curr Opin Neurobiol 7:473–479

Kawasaki M (2005) Physiology of tuberous electrosensory systems. In: Th TB, Hopkins C, Popper A, Fay R (eds) Electroreception. Springer, New York, pp 154–194

Keller CH, Zakon HH, Sanchez DY (1986) Evidence for a direct effect of androgens upon electroreceptor tuning. J Comp Physiol A 158:301–310

Kelly M, Babineau D, Longtin A, Lewis JE (2008) Electric field interactions in pairs of electric fish: modeling and mimicking naturalistic inputs. Biol Cybern 98:479–490

Kirschbaum F (1983) Myogenic electric organ precedes the neurogenic organ in apteronotid fish. Naturwissenschaften 70:205–207

Kirschbaum F, Westby GW (1975) Development of the electric discharge in mormyrid and gymnotid fish (*Marcusenius sp.* and *Eigenmannia virescens*). Experientia 31:1290–1294

Knight BW (1972) Dynamics of encoding in a population of neurons. J Gen Physiol 59:734–766

Knill DC, Pouget A (2004) The bayesian brain: the role of uncertainty in neural coding and computation. Trends Neurosci 27:712–719

Knudsen EI (1974) Behavioral thresholds to electric signals in high frequency electric fish. J Comp Physiol 91:333–353

Knudsen EI (1975) Spatial aspects of the electric fields generated by weakly electric fish. J Comp Physiol A 99:103–118

Koch C (1984) Cable theory in neurons with active, linearized membranes. Biol Cybern 50:15–33

Krahe R, Bastian J, Chacron MJ (2008) Temporal processing across multiple topographic maps in the electrosensory system. J Neurophysiol 100:852–867

Krahe R, Gabbiani F (2004) Burst firing in sensory systems. Nat Rev Neurosci 5:13–23

Kramer B, Kirschbaum F, Markl H (1981) Species specificity of electric organ discharges is a sympatric group of gymnitoid fish from Manaus (Amazonas). In: Szabo T, Czeh G (eds) Sensory physiology of aquatic lower vertebrates. Akademia Kiado, Budapest

Kramer B, Otto B (1991) Waveform discrimination in the electric fish *Eigenmannia*: sensitivity for the phase differences between the spectral components of a stimulus wave. J Exp Biol 159:1–22

Kramer DL (1978) Reproductive seasonality in the fishes of a tropical stream. Ecology 59:976–985

Kreiman G, Krahe R, Metzner W, Koch C, Gabbiani F (2000) Robustness and variability of neuronal coding by amplitude-sensitive afferents in the weakly electric fish *Eigenmannia*. J Neurophysiol 84:189–204

Lewis JE, Maler L (2001) Neuronal population codes and the perception of object distance in weakly electric fish. J Neurosci 21:2842–2850

Lewis RS, Hudspeth AJ (1983) Voltage- and ion-dependent conductances in solitary vertebrate hair cells. Nature 304:538–541

Lissmann HW, Machin KE (1958) The mechanism of object location in *Gymnarchus niloticus* and similar fish. J Exp Biol 35:451–486

MacIver MA, Sharabash NM, Nelson ME (2001) Prey-capture behavior in gymnotid electric fish: motion analysis and effects of water conductivity. J Exp Biol 204:543–557

Maler L (2009a) Receptive field organization across multiple electrosensory maps. I. columnar organization and estimation of receptive field size. J Comp Neurol 516:376–393

Maler L (2009b) Receptive field organization across multiple electrosensory maps. II. computational analysis of the effects of receptive field size on prey localization. J Comp Neurol 516:394–422

Maler L, Sas E, Johnston S, Ellis W (1991) An atlas of the brain of the electric fish *Apteronotus leptorhynchus*. J Chem Neuroanat 4:1–38

Maler L, Sas EKB, Rogers J (1981) The cytology of the posterior lateral line lobe of high-frequency weakly electric fish (Gymnotidae): dendritic differentiation and synaptic specificity in a simple cortex. J Comp Neurol 195:87–139

Manwani A, Steinmetz PN, Koch C (2002) The impact of spike timing variability on the signal-encoding performance of neural spiking models. Neural Comput 14:347–367

Markham MR, Allee SJ, Goldina A, Stoddard PK (2009) Melanocortins regulate the electric waveforms of gymnotiform electric fish. Horm Behav 55:306–313

Marsat G, Maler L (2010) Neural heterogeneity and efficient population codes for communication signals. J Neurophysiol 104:2543–2555

Marsat G, Proville RD, Maler L (2009) Transient signals trigger synchronous bursts in an identified population of neurons. J Neurophysiol 102:714–723

Matsubara J, Heiligenberg W (1978) How well do electric fish electrolocate under jamming? J Comp Physiol A 125:285–290

McGregor PK, Westby GM (1992) Discrimination of individually characteristic electric organ discharges by a weakly electric fish. Anim Behav 43:977–986

Mehaffey WH, Maler L, Turner RW (2008) Intrinsic frequency tuning in ELL pyramidal cells varies across electrosensory maps. J Neurophysiol 99:2641–2655

Metzner W, Heiligenberg W (1991) The coding of signals in the electric communication of the gymnotiform fish *Eigenmannia*: from electroreceptors to neurons in the torus semicircularis of the midbrain. J Comp Physiol A 169:135–150

Metzner W, Juranek J (1997) A sensory brain map for each behavior? Proc Natl Acad Sci U S A 94:14798–14803

Metzner W, Koch C, Wessel R, Gabbiani F (1998) Feature extraction by burst-like spike patterns in multiple sensory maps. J Neurosci 18:2283–2300

Meyer JH, Leong M, Keller CH (1987) Hormone-induced and maturational changes in electric organ discharges and electroreceptor tuning in the weakly electric fish *Apteronotus*. J Comp Physiol A 160:385–394

Middleton J, Longtin A, Benda J, Maler L (2006) The cellular basis for parallel neural transmission of a high-frequency stimulus and its low-frequency envelope. Proc Natl Acad Sci U S A 103:14596–14601

Middleton JW, Longtin A, Benda J, Maler L (2009) Postsynaptic receptive field size and spike threshold determine encoding of high-frequency information via sensitivity to synchronous presynaptic activity. J Neurophysiol 101:1160–1170

Moller P (1995) Electric fishes: History and Behavior. Chapman and Hall, London

Moortgat KT, Keller CH, Bullock TH, Sejnowski TJ (1998) Submicrosecond pacemaker precision is behaviorally modulated: the gymnotiform electromotor pathway. Proc Natl Acad Sci U S A 95:4684–4689

Nelson ME, MacIver MA (1999) Prey capture in the weakly electric fish *Apteronotus albifrons*: sensory acquisition strategies and electrosensory consequences. J Exp Biol 202:1195–1203

Nelson ME, Xu Z, Payne JR (1997) Characterization and modeling of P-type electrosensory afferent responses to amplitude modulations in a wave-type electric fish. J Comp Physiol A 181:532–544

Partridge BL, Heiligenberg W (1980) Three's a crowd? Predicting *Eigenmannia*'s responses to multiple jamming. J Comp Physiol A 136:153–164

Pasch B, George AS, Campbell P, Phelps SM (2011) Androgen-dependent male vocal performance influences female preference in neotropical singing mice. Anim Behav 82:177–183

Pettigrew JD (1999) Electroreception in monotremes. J Exp Biol 202:1447–1454

Pollak GD, Bodenhamer RD (1981) Specialized characteristics of single units in inferior colliculus of mustache bat: frequency representation, tuning, and discharge patterns. J Neurophysiol 46:605–620

Pressley J, Troyer TW (2011) The dynamics of integrate-and-fire: mean versus variance modulations and dependence on baseline parameters. Neural Comput 23:1234–1247

Ramcharitar JU, Tan EW, Fortune ES (2005) Effects of global electrosensory signals on motion processing in the midbrain of *Eigenmannia*. J Comp Physiol A 191:865–872

Ramcharitar JU, Tan EW, Fortune ES (2006) Global electrosensory oscillations enhance directional responses of midbrain neurons in *Eigenmannia*. J Neurophysiol 96:2319–2326

Rasnow B (1986) The effects of simple objects on the electric field of *Apteronotus*. J Comp Physiol A 178:397–411

Rasnow B, Bower JM (1996) The electric organ discharges of the gymnotiform fishes: *Apteronotus leptorhynchus*. J Comp Physiol A 178:383–396

Reardon E, Parisi A, Krahe R, Chapman LJ (2011) Energetic constraints on electric signalling in wave-type eweakly electric fishes. J Exp Biol (in press)

Ricci AJ, Kennedy HJ, Crawford AC, Fettiplace R (2005) The transduction channel filter in auditory hair cells. J Neurosci 25:7831–7839

Rose G, Heiligenberg W (1985) Temporal hyperacuity in the electric sense of fish. Nature 318:178–180

Rose GR (2004) Insights into neural mechanisms and evolution of behaviour from electric fish. Nat Rev Neurosci 5:943–951

Salazar VL, Stoddard PK (2008) Sex differences in energetic costs explain sexual dimorphism in the circadian rhythm modulation of the electrocommunication signal of the gymnotiform fish brachyhypopomus pinnicaudatus. J Exp Biol 211:1012–1020

Sawtell NB, Bell CC (2008) Adaptive processing in electrosensory systems: links to cerebellar plasticity and learning. J Physiol Paris 102:223–232

Scheich H, Bullock TH, Hamstra RH (1973) Coding properties of two classes of afferent nerve fibers: high-frequency electroreceptors in the electric fish. *Eigenmannia*. J Neurophysiol 36:39–60

Schief A, von Seelen W, Stagge J, Winkler G (1971) Reception of disrupted signals by the weak electric fish *Gnathonemuspetersii*. Kybernetik 9:34–43

Schnitzler HU, Moss CF, Denzinger A (2003) From spatial orientation to food acquisition in echolocating bats. Trends Ecol Evol 18:386–394

Shumway CA (1989) Multiple electrosensory maps in the medulla of weakly electric gymnotiform fish. I. Physiological differences. J Neurosci 9:4388–4399

Smith GT, Combs N (2008) Serotonergic activation of 5HT1A and 5HT2 receptors modulates sexually dimorphic communication signals in the weakly electric fish *Apteronotus leptorhynchus*. Horm Behav 54:69–82

Stamper SA, Carrera-G E, Tan EW, Fugère V, Krahe R, Fortune ES (2010) Species differences in group size and electrosensory interference in weakly electric fishes: implications for electrosensory processing. Behav Brain Res 207:368–376

Stocks NG, Mannella R (2001) Generic noise-enhanced coding in neuronal arrays. Phys Rev E 64:030902

Stoddard PK, Zakon HH, Markham MR, McAnelly L (2006) Regulation and modulation of electric waveforms in gymnotiform electric fish. J Comp Physiol A 192:613–624

Stumpner A (2002) A species-specific frequency filter through specific inhibition, not specific excitation. J Comp Physiol A 188:239–248

Suga N (1965) Functional properties of auditory neurones in the cortex of echo-locating bats. J Physiol 181:671–700

Tallarovic SK, Zakon HH (2005) Electric organ discharge frequency jamming during social interactions in brown ghost knifefish, *Apteronotus leptorhynchus*. Anim Behav 70:1355–1365

Tan EW, Nizar JM, Carrera-G E, Fortune ES (2005) Electrosensory interference in naturally occurring aggregates of a species of weakly electric fish. *Eigenmannia virescens*. Behav Brain Res 164:83–92

Telgkamp P, Combs N, Smith GT (2007) Serotonin in a diencephalic nucleus controlling communication in an electric fish: sexual dimorphism and relationship to indicators of dominance. Dev Neurobiol 67:339–354

Triefenbach F, Zakon HH (2003) Effects of sex, sensitivity and status on cue recognition in weakly electric fish, *Apteronotus leptorhynchus*. Anim Behav 65:19–28

Triefenbach FA, Zakon HH (2008) Changes in signalling during agonistic interactions between male weakly electric knifefish, *Apteronotus leptorhynchus*. Anim Behav 75:1263–1272

Turner CR, Derylo M, de Santana CD, Alves-Gomes JA, Smith GT (2007) Phylogenetic comparative analysis of electric communication signals in ghost knifefishes (gymnotiformes: Apteronotidae). J Exp Biol 210:4104–4122

van der Sluijs I, Gray S, Amorim M, Barber I, Candolin U, Hendry A, Krahe R, Maan M, Utne-Palm A, Wagner HJ, Wong B (2011) Communication in troubled waters: responses of fish communication systems to changing environments. Evol Ecol 25:623–640

Viancour TA (1979a) Electroreceptors of a weakly electric fish. I. Characterization of tuberous receptor organ tuning. J Comp Physiol A 133:317–325

Viancour TA (1979b) Electroreceptors of a weakly electric fish. II. Individually tuned receptor oscillations. J Comp Physiol A 133:327–338

Vonderschen K, Chacron MJ (2011) Sparse and dense coding of natural stimuli by distinct midbrain neuron subpopulations in weakly electric fish. J Neurophysiol (epub)

Watanabe A, Takeda K (1963) The change of discharge frequency by A.C. stimulus in a weak electric fish. J Exp Biol 40:57–66

Wessel R, Koch C, Gabbiani F (1996) Coding of time-varying electric field amplitude modulations in a wave-type electric fish. J Neurophysiol 75:2280–2293

White JA, Rubinstein JT, Kay AR (2000) Channel noise in neurons. Trends Neurosci 23:131–137

Zakon H, Oestreich J, Tallarovic S, Triefenbach F (2002) EOD modulations of brown ghost electric fish: JARs, chirps, rises, and dips. J Physiol Paris 96:451–458

Zakon HH (1986a) The electroreceptive periphery. In: Bullock TH, Heiligenberg W (eds) Electroreception. Wiley, New York, pp 103–156

Zakon HH (1986b) The emergence of tuning in newly generated tuberous electroreceptors. J Neurosci 6:3297–3308

Zakon HH, Thomas P, Yan HY (1991) Electric organ discharge frequency and plasma sex steroid levels during gonadal recrudescence in a natural population of the weakly electric fish sternopygus macrurus. J Comp Physiol A 169:493–499

Zupanc GK, Bullock TH (2005) From electrogenesis to electroreception: an overview. In: Bullock TH, Hopkins CD, Popper AN, R. FR (eds) Electroreception. Springer, New York, pp 5–46

Zupanc GKH, SÃrbulescu RF, Nichols A, Ilies I (2006) Electric interactions through chirping behavior in the weakly electric fish. *Apteronotus leptorhynchus*. J Comp Physiol A 192:159–173

Chapter 13
Noise in Chemical Communication

Volker Nehring, Tristram D. Wyatt and Patrizia d'Ettorre

Abstract Chemical communication is ubiquitous. It is not only employed in inter-individual communication, but also used to transfer information within individuals, from cell to cell and from one organ to another within a body with a complicated network of hormones and neurotransmitters. However, how noise affects chemical communication has been largely neglected. Here, we review possible sources of noise and the effects noise has on the behaviour of receivers. We will also discuss variation in chemical cues and signals that may provide information in some contexts, but obscure messages in others. Finally, we attempt to identify strategies that senders and receivers can follow to either reduce the occurrence or mitigate the effects of noise.

13.1 Introduction

All organisms live in a chemical world, and chemical communication is likely to be so widespread because it is the most ancient form of communication (Wyatt 2014). The olfactory lives of insects are particularly well studied: female moths, for example, attract males using a volatile pheromone; males can detect and follow traces of these pheromones from large distances. Social insects organize their societies using chemical cues and signals (d'Ettorre and Moore 2008), and

V. Nehring (✉)
University Freiburg, Biology I, Hauptstrasse 1, 79104 Freiburg, Germany
e-mail: volker.nehring@biologie.uni-freiburg.de

T. D. Wyatt
Department of Zoology, University of Oxford, South Parks Road, Oxford OX1 3PS, UK
e-mail: tristram.wyatt@zoo.ox.ac.uk

P. d'Ettorre
Laboratory of Experimental and Comparative Ethology, University of Paris 13, Sorbonne Paris Cité, France
e-mail: dettorre@leec.univ-paris13.fr

H. Brumm (ed.), *Animal Communication and Noise*,
Animal Signals and Communication 2, DOI: 10.1007/978-3-642-41494-7_13,
© Springer-Verlag Berlin Heidelberg 2013

bacterial cells use chemicals for communication, for example in quorum sensing (Waters and Bassler 2005). Plants are able to detect olfactory cues emitted by other plants (Dicke et al. 2003), and chemical communication is also known from all vertebrate groups (Müller-Schwarze 2006). Even humans, although not necessarily consciously, may communicate with chemicals; for instance, odours may be important in mate choice (Havlicek and Roberts 2009) and for newborns to identify their mother (Porter and Winberg 1999).

Chemical communication is not only employed in inter-individual communication, but also used to transfer information within individuals, from cell to cell and from one organ to another within a body with a complicate network of hormones and neurotransmitters. However, noise in chemical communication has usually been a neglected topic.

13.1.1 Features of Chemical Communication

Compared to other means of information transfer, chemical communication has some distinct features that need special attention when discussing possible sources of noise (Wyatt 2014). Unlike auditory and visual signals, pheromones are not waves but actual matter that has to be transported. This particularity has some important implications: The transport of volatile pheromone molecules, for example, is much slower and more stochastic than that of sound or light waves, and, apart from short range diffusion at very small scales, relies on wind and water flow (Webster and Weissburg 2009). Furthermore, once a pheromone is released, there is no way to stop the signal. While visual displays or vocal calls disappear within fractions of a second after signalling commenced, the pheromones can continue to exist until the molecules are broken down or concentrations drop below perception levels of the receiver, which might take days and longer.

Originally, many chemicals that now serve to communicate between individuals had a physiological function within an individual and then they have evolved through the process of ritualization, which means they have been optimised for signalling purposes (Fig. 13.1). For example, many alarm pheromones of ants are related to the defensive compounds used by that species and the sex pheromones of fish include hormones or related molecules (see Wyatt 2010, 2014). In some cases, the communication function may overlap with the current functions of the molecules. In such cases there may be evidence for a trade-off between the chemical's original function and their aptitude for signalling. The cuticular hydrocarbons of insects, for example, originally evolved to protect individuals from desiccation. Their signalling function is derived, and the optimal structure of the molecules is suggested to differ between the two functions (cf. Steiger et al. 2011).

Besides these particularities in physical properties and evolution of chemical cues, there are some factors that make research in chemical communication possibly more difficult than in other modalities. The olfactory system, like the immune system, tracks a moving world of cues generated by other organisms, and must

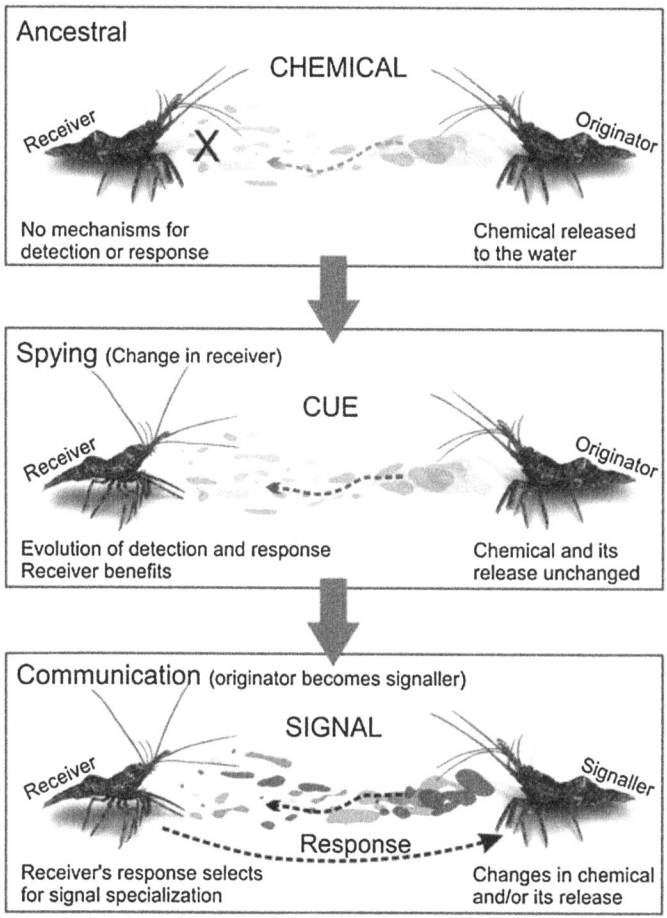

Fig. 13.1 Proposed stages in the evolution of a communication function for molecules released by an individual. Ancestrally, chemicals "leak" from the originator, but are not detected by any other individual. In the spying phase, the leaked chemicals act as cues that give some information to receivers. The transition to bilateral benefit could occur later if there is selective advantage to the sender, who then may modulate the release of the signal (adapted from Wyatt 2011)

constantly generate, test and discard receptor genes and coding strategies over evolutionary time (Bargmann 2006). This has meant that discovering the many different olfactory receptors (Fig. 13.2), unrelated in vertebrates and insects, has been a more challenging task than identifying the relatively few (and uniform) light sensitive receptor types, or the few mechanosensory receptor types needed for hearing. The processing of olfactory sensory neurons follows a non-topological model different from the higher brain processing of visual and sound inputs. This complexity explains why researchers already had a good idea about how many colours honeybees could see a hundred years ago (von Frisch 1915) while the study of the chemical world the bees live in is still revealing new insights.

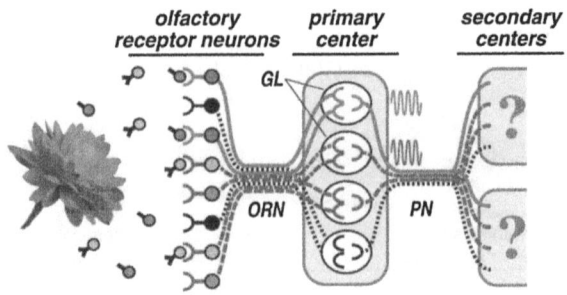

Fig. 13.2 Olfactory processing in insects and mammals: Olfactory receptor neurons (ORN) carry many receptors, often all of them are tuned to the same substances. The sensory neuron axons terminate in glomeruli (GL) in the primary centre (olfactory bulb in mammals, antennal lobe in insects); typically, glomeruli receive input from only one type of sensory neurons. In the primary centre, a first processing occurs by inhibitory interneurons between glomeruli. Each glomerulus is innervated by only one or few projection neurons (PN), which send information to the secondary centres, e.g. the cortex or the mushroom bodies (adapted from Tanaka et al. 2004)

Furthermore, investigating the perception and processing of chemical signals and cues is harder than it is for visual or auditory ones, because semiochemicals cannot simply be recorded and replayed (Wyatt 2009). Correct synthesis of pheromone molecules can be more challenging than their identification, and because of the way that olfaction works, the copies have to be exact. We can record and play back the audible calls of an animal easily enough, but we do not have a video or digital audio file equivalent for recreating chemical signals. This difficulty is also one reason Smell-O-Vision cinema never took off.

On the other hand, some aspects of chemical communication are rather easy to analyse. Many signals and their effects are hard-wired, and researchers can track the biosynthesis of pheromones, their release, the physics of the transmission, the reception, and even the behavioural responses the pheromones elicit. For some model species, all of these stages are accessible to molecular tools, which is very convenient for the study of pheromone evolution and speciation. In the European corn borer *Ostrinia nubilalis*, for example, two races have been described that differ in their chemical signals. The pheromone blend of both races consists of the same two components, (*E*)-11-tetradecenyl acetate and (*Z*)-11-tetradecenyl acetate. However, the pheromone of one strain consists of 98 % of the E-isomer, while the other one has approximate 97 % of the Z-isomer. In the field, the races are reproductively isolated, but the pheromones of laboratory-produced crosses are controlled by simple Mendelian inheritance (Lassance 2010; Löfstedt 1990). Mutations that lead to reproductive isolation have been identified in a single autosomal factor for the pheromone production, another autosomal factor for the antennal response, and a sex-linked genetic factor for the male response to the pheromone. The pheromone component ratio is controlled by a single gene for a particular fatty-acyl reductase (Lassance et al. 2010; Olsson et al. 2010).

13.1.2 What is Noise, and what is Signal?

As outlined in Chap. 2, noise in communication is anything influencing a receiver's receptors other than the signal of interest. Like in any communication system, noise is also inherent to chemical communication. For example, the signals may not always be produced and perceived accurately, which may lead to receiver errors. In addition to this biological noise, some of the variance that is observed by researchers is exclusively due to experimental error. When machines like gas chromatographs are used to measure olfactory signals, for example, the machines may not be sensitive enough to track subtle differences that have an important meaning to the receiver.

Going down to the receptor level, we could say that anything a receptor has not been selected (through natural or sexual selection) to detect is noise. If a receptor is tuned to a sex pheromone component, for example, all other substances that either bind or block the receptor would contribute to noise because they will make the receptor unavailable for the specific pheromone. It is hence important to consider the information that is of interest when deciding which variance component is noise and which is signal.

Pheromones fit the definition of signals as "any act or structure which alters the behaviour of other organisms, which evolved because of that effect, and which is effective because the receiver's response has also evolved" (Maynard Smith and Harper 2003, p. 3). At least in the world of chemical communication, where substances primarily served another purpose than communication, and have later been used by receivers to gather information, there are many examples of communication where this co-evolutionary relationship has not been established. These substances would be called "cues" as opposed to signals (in terms of Signal Detection Theory, however, this distinction is not relevant, Chap. 2). Pheromones are chemical signals used to communicate within a species. Pheromones are 'molecules that are evolved signals, in defined ratios in the case of multiple component pheromones, which are emitted by an individual and received by a second individual of the same species, in which they cause a specific reaction, for example, a stereotyped behaviour or a developmental process' (Wyatt 2010, modified after Karlson and Lüscher 1959). Chemical signals generally consist of only a few components which vary little between senders, and then the receiver's reaction to them is often hard-wired, as for example in moth sex pheromones. Chemical messages can also be highly variable "signature mixtures" consisting of multiple substances which vary not only with the genetic background of the sender, but also with factors such as the diet (Wyatt 2010). Specific signature mixtures are learned by the receivers (cf. Wyatt 2010), as it is the case, for example, of the colony-specific profiles social insects use to discriminate nestmates from non-nestmates (van Zweden and d'Ettorre 2010). Although the specific substance ratios of these signature mixtures are learned, their function for communication may have co-evolved between sender and receiver, in which case they still would be signals. In cases where we refer to both pheromones and signature mixtures, and perhaps even include allomones, which are substances for inter-specific communication, we will use the term semiochemical.

We will analyse separately the production and perception component of the chemical communication system. How is a semiochemical produced by the sender and what could be the constraints, and how is the semiochemical perceived by the receiver and what could interfere with perception? However, some noise is introduced neither by the sender nor by the receiver, hence we discuss separately the noise arising during the transmission of the information, and between the production of the semiochemical by the signaller and the receiver's perception.

13.2 Noise in Production

13.2.1 Cue Variation Between and Within Individuals

When cues or signals differ between individuals, but the inter- and intra-individual variation does not convey any information that is useful for the receiver, it is noise, since it potentially makes it harder for the receiver to identify and interpret the information.

Examples of both inter-and intra-individual variations are known. For instance, both the ratio and the absolute abundance of flower odours vary in roses (Helsper et al. 1998), and the proportion of components in the pheromone of southern green stink bug *Nezara viridula* males varies not only between populations, but also between individuals and between repeated measurements of the same individual (Miklas et al. 2000). Another example concerns female moths. When the phero-mone emission of individuals of the carnation tortrix *Cacoecimorpha pronubana* was measured repeatedly, the ratio of the different pheromone components differed between individuals (Witzgall and Frerot 1989). Between the individual females, however, there was significant variation in the ratio of some of the components. The ratio of (Z)-9-tetradecyl acetate to (E)-11-tetradecyl acetate, for example, varied between individuals from 0.04 to 1.17; each individual, however, produced the same ratio during each sampling period (high repeatability). Similar results have been obtained with other moth species; some of the inter-individual variation is heritable, and the genetics underlying the heritable variation is known (Löfstedt 1990). The odours of mice have also been shown to be at least in part based on genetic variation (Brown 1985). Although in these examples the variation in the production is quantified, nothing is known about the effect of the variation on receivers. It is possible that receivers infer the quality of senders from the semiochemical variation. This is the case in tiger moths (*Utetheisa ornatrix*), where females choose males based on how much of specific chemicals they emit, since these amounts are correlated with the amount of alkaloids the male passes to the female during mating. The female uses alkaloids to protect the eggs, and by choosing males with more pheromone the females can improve egg protection (Conner and Weller 2004).

The variance between individuals can also help avoiding inbreeding, as it occurs in many vertebrates that preferentially choose breeding partners with dissimilar smell, which usually correlates with differences in their Major Histocompatibility Complex (MHC), a gene family involved in immunity (Penn 2002). As we see, inter-individual variation in cues and signals may be noise, but can also convey relevant information. It depends on what exactly the receivers are interested in, and to estimate the noise in communication these details ought to be known.

13.2.2 Noise in Nestmate Recognition Cues

A system where both the variation in the production of chemical cues and its effect on the receiver have been studied is nestmate recognition in social insects: ants, bees, wasps and termites. Social insects bear a signature mixture of lipids on their cuticle that generally differs between colonies, to the extent that it can be used by social insect workers to discriminate between nestmates and non-nestmates. The signature mixtures vary over time (van Zweden et al. 2009) and can be influenced by the task workers are performing (Bonavita-Cougourdan et al. 1993; Greene and Gordon 2003) or by their food (cf. Lenoir et al. 1999). The influence of food on nestmate recognition labels has been studied in the laboratory to reveal recognition mechanisms. When two groups of ants from the same colony are separated and fed different diets, the signature mixtures of the groups can diverge to the point that the former nestmates begin to attack each other (Liang et al. 2001). Under field conditions, it might perhaps not frequently have negative effects on recognition. However, it is clear that food-induced signature mixture variation could at least theoretically impair nestmate recognition if no counter measures had evolved.

Besides the environmental variation, genetic variation within social insect colonies also contributes to variance in the production of the nest-specific cuticular lipid blend. The different patrilines, i.e. full sister groups in colonies where the queen has mated with more than one unrelated male, in honeybee and leaf-cutting ants differ in their signature mixtures (Arnold et al. 1996; Nehring et al. 2011), with workers from the same patriline being more chemically similar. The signature mixtures of wood ants vary more in colonies with multiply mated than with singly mated queens (Boomsma et al. 2003). The variation seems to be even higher between matrilines in colonies with multiple queens (Dani et al. 2004). It is hitherto unclear whether social insects can use the information provided by kin-informative recognition cues. If discrimination were possible, full sisters could behave more altruistically among them. This would possibly decrease the efficiency of the colony as a whole, and thus nepotistic behaviour is supposed to be selected against at the colony level. It has been demonstrated, however, that genetic variation of recognition cues in multiple-queen colonies can decrease the efficiency of nestmate recognition (Martin et al. 2009).

The case of reduced nestmate recognition efficiency in multiple-queen colonies illustrates that from a receiver perspective, variation that is generally conveying useful information, but in which none of the receivers is interested at a given time (e.g. information about the relatedness), may contribute to noise. This happens when the information is transmitted using the same set of molecules (e.g. cuticular lipids) that are used to transmit another kind of information, e.g. the nestmate recognition label.

A similar example is fertility signalling, which is also very important for regulating life in social insect colonies. Because egg laying is correlated with certain components of the cuticular lipid blend of ants, workers always have information about whether there is still a fertile queen in the colony. In general, workers do not have difficulties in assigning a significance (e.g. fertility signal) to one of the cuticular hydrocarbons among the many peaks of the background constituting the nestmate signature mixture (Holman et al. 2010; Peeters et al. 1999; Smith et al. 2009). Workers can often also detect whether other workers lay eggs, and if that happens destroy the worker-laid eggs and attack the laying workers (egg and worker policing, cf. Ratnieks 1988). Worker policing is such an important behaviour to social insect colonies that some workers even specialise on this task (Barth et al. 2010; van Zweden et al. 2007). It has been shown, though, that fertility signals on individuals of *Camponotus* ants can override the nestmate recognition system, i.e. because an individual bears the fertility signal, the workers accept the individual despite it being a non-nestmate (Moore and Liebig 2010). An example where nestmate and fertility signalling do not interfere is egg recognition in *Formica* ants. As long as there is a queen in the colony, workers destroy all eggs laid by other workers, whether nestmate or non-nestmate. They also destroy eggs laid by other queens than their own (Helanterä and Ratnieks 2008).

Whether the overlap in molecules is responsible for fertility signalling overriding nestmate recognition in *Camponotus* ants is not clear, though. It is still possible that both the nestmate label and the fertility signal are perceived correctly, but that the fertility signal overrides everything else. In other words, as soon as an individual is a fertile queen, there is generally no need to check whether it is from the same colony, since under field conditions the ants never encounter queens from other colonies. The lack of selection pressure to assure the origin of fertile individuals could therefore have allowed to increase recognition efficiency: accepting any queen saves the trouble to integrate information about the nestmate cues and the fertility signal and also precludes the ants from ever rejecting their own queen, which would be very costly in single queen colonies of *Camponotus*. In *Formica*, the situation is somehow different: these ants are exploited by numerous social parasites that deceive host ants and make them raise parasite brood (Lenoir et al. 2001; Martin et al. 2011). There is therefore more selection pressure on always watching out for social parasite queens, and thus in *Formica* ants the nestmate recognition cues override the fertility signal.

13.2.3 Inaccuracies in Cue Biosynthesis

The biosynthesis of molecules that are part of a signal might not always be precise. In the biosynthesis of peptides, for example, it can happen that the wrong amino acid is added to the peptide string (Loftfield and Vanderjagt 1963; Zaher and Green 2009). It would thus not be a surprise if enzymes synthesising other semiochemicals would make mistakes as well, so that besides large amounts of the targeted pheromone molecules, small quantities of similar molecules might always be produced. These additional molecules might not be bioactive, but they could have a different meaning to the receiver.

The relative abundance of similar hydrocarbons, for example alkanes, is often correlated between compounds of similar chain length, suggesting that the underlying biosynthesis pathways are similar for those substances (homologous series, Martin and Drijfhout 2009). Thus, the abundance of individual hydrocarbons cannot be regulated individually. It is possible that homologous series of hydrocarbons are caused by inaccuracy in biosynthesis, i.e. the insect is 'trying' to synthesise only one particular hydrocarbon, but due to errors the neighbouring hydrocarbons are produced as well.

13.3 Solutions for Coping with Noise in Signal Production

13.3.1 Acceptance Windows and Generalization

When signals are variable around a mean, detection of those signals could be enhanced by accepting not only mean signals, but also those that deviate from the mean. Receivers widen their acceptance window around the mean to include an optimal number of variants (see Chap. 2; cf. frequency tuning in auditory (Chaps. 3, 5 , 8, 10) and electroreceptors (Chap. 12)). In response to the variation in ratio of pheromone components that female moths produce, the males respond to a range of ratios as the cost of not responding to a female of their own species is high. The mean of the produced ratios and of the acceptance window are equal, but the acceptance window has a higher variance to make sure all possible female-produced ratios are included (Fig. 13.3; Löfstedt 1990). The acceptance windows of male moths to female pheromones are heritable and can evolve (Domingue et al. 2009; Lassance et al. 2010).

The response to signature mixtures, the detailed qualitative and quantitative composition of which receivers learn, is naturally more flexible than the response to pheromones, which is usually hard-wired. Often, animals learn the ratio of signature mixture components (Wyatt 2010). How this might work is revealed by studies of learning of simple odour mixtures. When honeybees are presented with two odourants in varying ratios, but only a particular ratio is associated with food, while a response to other ratios is not only unrewarded, but even punished, the

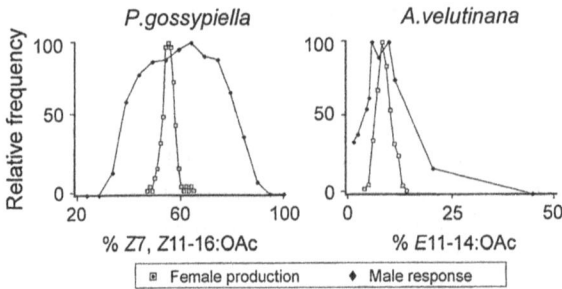

Fig. 13.3 The ratio variation for two components of female-produced sex pheromones, and the likelihood that males respond to the respective ratios (acceptance window), for two moth species (*Pectinophora gossypiella* and *Argyrotaenia velutinana*). In both cases, the male acceptance window is larger than the variation in female production (adapted from Löfstedt 1990)

bees learn to respond to the rewarded ratio and to ignore or avoid all others. When, however, responses to unrewarded ratios are never punished, the bees learn to ignore the ratio altogether and react to any ratio of two specific components, given that one of them is associated with a reward (Wright et al. 2008). The learning process involved here is termed generalisation. Generalisation is a common phenomenon that occurs in rodents as well (Cleland et al. 2009). From brain imaging it is known that mice perceive two very similar substances differently, but behaviourally respond to them in the same way. They can, however, learn to behaviourally discriminate between the substances if discrimination is rewarded (Linster et al. 2001, 2002).

13.3.2 Blending Group Level Cues

Inter-individual variation in recognition cues can reduce the efficiency of group-specific recognition, as described above for nestmate recognition. The colony members can counteract this effect by sharing semiochemicals among nestmates to reduce the intra-colonial variation. The hydrocarbons that make up the nest-specific signature mixture of an ant colony, for example, are transferred between individuals by social grooming (individuals cleaning each other's body) and from mouth to mouth, through trophallaxis (exchange of liquid food, which can also contain recognition cues). Thus the inter-individual differences are aligned, resulting in a so-called *Gestalt* odour, which is common to all colony members (Lenoir et al. 1999; Soroker et al. 1995; van Zweden et al. 2010; van Zweden and d'Ettorre 2010).

13.3.3 Separate the Transmission Channels for Different Types of Information

In some of the examples mentioned above, variation encoding some sort of information (e.g. the fertility signal in social insects) appeared to impair the interpretation of other information (the nestmate recognition cues). This happened in *Camponotus* ants, but not in *Formica* ants. One way to avoid the interference of different messages is to use different groups of molecules to transmit the information. In honeybees, for example, cuticular hydrocarbons are used for nestmate recognition (Dani et al. 2005). In contrast, the queen retinue pheromone, a fertility signal that informs workers that their queen is still alive and laying eggs, is a blend that consists mainly of fatty acids and alcohols (Slessor et al. 2005). A set of fatty acid esters is used for communication within the honeybee hive in a third context: larvae use these substances to signal their age and eventually to induce workers to cap the cell the larvae are in Slessor et al. (2005). If different kinds of information are transmitted using different sets of molecules, it is arguably less likely that the different messages interfere, than when all messages are sent using the same types of molecules.

13.4 Noise in Transmission

Between the release of a chemical message by a sender and the reception by a receiver, noise can be introduced into the communication process. Naturally, long distance-pheromones are particularly prone to this kind of noise, but even information conveyed by contact semiochemicals can degrade during transmission in the communication channel.

13.4.1 Chemicals from Other Sources

The environment is full of chemicals. Many of them may form "background noise". These can be substances that do not have any biological meaning for the receiver, or they could be the same molecules that form part of a pheromone blend the receiver is interested in, but come from a source the receiver is not looking for. It is known, for instance, that different species share at least some components of their pheromones. The most famous example is perhaps a sex pheromone component that more than a hundred species of moths share among each other, but also with elephants: the small volatile molecule (Z)-7-dodecen-1-yl acetate (cf. Kelly 1996; Wyatt 2010).

The overlap of moth and elephant pheromones may not have any relevant effect on either species. In fact, the pheromones of sympatric species are under selection to diverge, to reduce noise and error rates during mate localization (Cardé and

Haynes 2004; Smadja and Butlin 2009). In biocontrol, artificially produced pheromones disrupt the mating of moth species that are considered pests. Mating disruption is mostly thought to happen because artificial sources that emit the complete female pheromone blend in high concentration are most attractive to males (Miller et al. 2010; Minks and Cardé 1988). However, in some cases single pheromone components, which by themselves are not attractive, can disrupt mating (Ryne et al. 2001). It is possible that in these cases flooding the air with one component of a pheromone distorts the ratios of the pheromone components, which males very specifically react to. They may not recognise the pheromone blend as the one specific for their species any more when the concentration of one of the components is too high.

Interference can occur in other chemical communication apart from airborne pheromones. The same can be imagined for non-volatile semiochemicals that are deposited on substrates. Animals that place scent marks, for example to mark their territory, often counter-mark the scent of other individuals, which results in multiple overlapping marks. The scent marks are specific for different individuals, and if individual marks overlap, there is the risk that they result in a single mixed mark that would not allow receivers to identify any of the senders' identity. Research on golden hamsters and voles, however, has shown that receivers are indeed able to extract information from mixed scent marks about the different individuals that produced the marks (Johnston 2008). In hamsters and voles the top mark is preferred.

13.4.2 Deception

In the examples mentioned so far, chemicals from neutral sources entered the communication process and disturbed the information transfer, which means there is no co-evolution between the receiver and the neutral source. There are cases, however, where the purpose of emitting these chemicals is to deceive the receiver. From the receiver's perspective, deception produces noise because it limits the reliability of signals.

To attract pollinators, some flowers produce substances that are semiochemicals in a pollinator's communication system (Schiestl 2005). For instance, some orchids produce the sex pheromone of wasp females to attract males (Schiestl et al. 2003). The flower of another orchid species emits (Z)-11-eicosen-1-ol, a major component of the honeybee alarm pheromone (Brodmann et al. 2009). Hornets, which prey on bees, are attracted by the chemical and pounce on the flower as if it were a bee, in the process picking up the orchid's pollinia. Arum flowers emit a less specific attractant to lure flies into pollination: they imitate the odour of carrion (Stensmyr et al. 2002). Bolas spiders emit the sex pheromones of moths, and this attracts moth males that the spiders prey on (Stowe et al. 1987). The spiders can emit the pheromones of different prey species and even adjust the pheromone they emit to the time of the night the respective moth species is usually active (Haynes et al. 2002).

Another well-known example are social parasites that sneak into social insect colonies. Many parasites seem to be able to acquire the nestmate recognition signature of their host colony and so avoid being attacked, expelled, or killed. The parasites can change their recognition cues quickly by transferring substances from the cuticle of host individuals to their own. Furthermore, they seem to have heritable cuticular hydrocarbon profiles that are similar to that of the host species or at least support the transfer of cues (Lenoir et al. 2001). In any case, the within-colony variability of nestmate recognition cues, i.e. noise, allows parasites to deceive host workers, since the presence of some variability caused by noise broadens the range of the host individuals' labels, and a broader range is easier to match for the parasites.

In both cases, the mimicry of attractive semiochemicals as well as the acquaintance of host-specific substances, there has been significant co-evolution between host and parasite, with selection on the host to avoid exploitation, while the parasite is selected to copy the host's signals. Countermeasures by the hosts, to reduce the noise produced by parasites, are therefore different from other cases. Instead of unidirectionally increasing the precision of production, transmission, and perception of the semiochemicals, red queen dynamics may cause fluctuations in the quality of the signals (Dybdahl and Storfer 2003; Kawecki and Ebert 2004). Theoretically, when there are many bolas spiders, for example, that prey on a moth species and emit deceptive pheromone to attract the males, moth genotypes that produce and react to a slightly different ratio of sex pheromone components would be selected for, since the ratio shift would allow males to discriminate dangerous spiders from moth females. The moths' sex pheromone would change, and they would suffer less from bolas spiders. Now the pressure would be on the spiders to change their deceptive blend as well, until they are again able to attract more prey.

13.4.3 Degradation of Information During Molecule Transport

When a pheromone is carried away from an emitting animal by air or water, turbulence breaks up the air or water into pockets or filaments (Fig. 13.4). Within these pockets or filaments pheromone blends can be virtually unchanged in concentration and composition more than 100 m downstream (Murlis et al. 1992; Webster and Weissburg 2009 for water-borne chemicals). This is because the bulk flow is more important than diffusion at these scales. Still, there are several ways noise can enter the communication process and degrade the information (Fig. 13.4). This is in part also true for non-volatile pheromones that are deposited on a surface, where receivers detect them later.

Turbulent airflow can introduce noise by disrupting the pockets or separating filaments containing pheromone-laden air. As the structure of the odour filaments cannot hint to the direction of the pheromone source, this obscures the direction a

Fig. 13.4 The transport of odour molecules. **a** is not linear, but proceeds in pockets and filaments that are intermixed with "clean" medium. Inside the odour pockets, there is very little variation in concentration. Light intensity (**b**), in contrast, decreases predictably with distance to the source. Figure inspired by Webster and Weissburg 2009, picture of odour plume: courtesy of Marc Weissburg

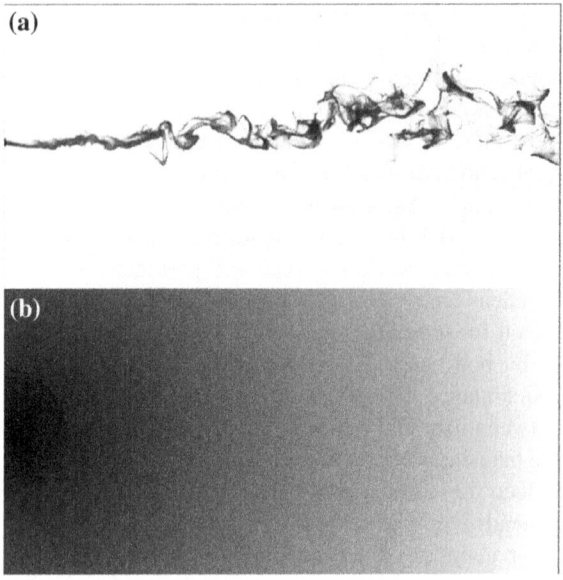

pheromone is coming from. It is therefore a common strategy to use the pheromone to identify the sender, but then to simply move upstream as long as pheromone can be detected; the direction is inferred from the airflow. The male moth *Heliothis virescens*, for example, flies upwind for a certain distance whenever the antennae make contact with filament of air containing the female pheromone. In between these upwind surges, the males fly crosswind until they make contact with the odour again (Vickers and Baker 1994).

Even low volatile hydrocarbons can be transferred between interacting individuals. In ants, for example, cuticular hydrocarbons are transferred between colony members (Soroker et al. 1995), which is thought to help maintaining a uniform colony odour (Lenoir et al. 1999). As mentioned above, in some cases the transfer may obscure information about the genetic background of the individuals, which limits the information available for individuals within the colony, but may stabilise the colony because it prevents individuals from discriminating related from less-related colony members, i.e. nepotism, the preferential treatment of kin over non-kin (van Zweden et al. 2010). Ants might also acquire hydrocarbons from prey items, which then later may lead nestmates to attack the ants because they mistake them for non-nestmates (Liang et al. 2001).

Last, molecules can degrade. Carbon chains can break, or the molecules can be otherwise chemically modified. Some gold fish steroid pheromones, for example, are rather short-lived (Sorensen et al. 2000). Also polypeptides, used by many aquatic organisms as pheromones, are quickly broken down by microorganisms. The rapid bacterial degradation of organic molecules in water may be one reason why territory scent marks have not been found in aquatic organisms (Wyatt 2011).

13.5 Solutions to Cope with Noise in Transmission

The fundamental strategy to avoid confusing semiochemicals from one's own species with molecules from other sources is to use semiochemicals that are either structurally, spatially or temporally specific, i.e. chemical signals that do not overlap with molecule blends from irrelevant sources.

13.5.1 Specific Pheromones

It is obvious that it is easier for receivers to correctly identify relevant semiochemicals when these are unique, in the sense that no irrelevant individual (e.g. the wrong species) emits the same chemical. For a moth sex pheromone, for example, the specificity would be highest—and thus noise lowest—if no other species emitted the same substance. When, instead of one single substance, two or multiple substances form the pheromone, many different pheromone blends can be constructed from the same set of pheromone components, as long as the ratio between components is species-specific. The likelihood of randomly finding a specific combination of substances in the environment is lower than that of finding one substance (Linn et al. 1987; Roelofs 1995; Wyatt 2014). Indeed, the sex pheromones of multiple moth species comprise a limited set of components, and the component ratio is the only factor securing species-specific attraction in some habitats (Fig. 13.5; Cardé and Haynes 2004).

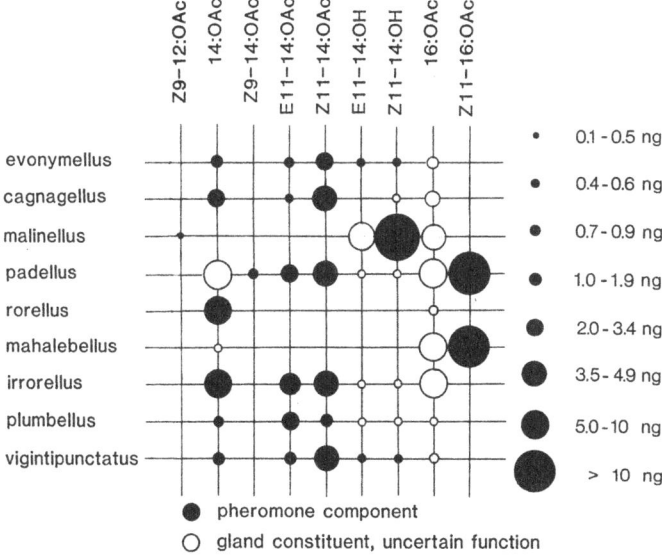

Fig. 13.5 The pheromone blends of nine congeneric moth species (*Yponomeuta sp.*) are constructed by a limited set of substances. Not presence and absence, but the relative concentration (i.e. the ratio) of substances is species-specific (adapted from Löfstedt et al. 1991)

With pheromone blends, there are two layers of specificity. First, all components need to be present for the receiver to respond; second, the ratio between the components is crucial as well. This second step (ratio specificity) allows two species to be discriminated even if they use the same components, which makes evolution of specificity relatively simple, since it is not necessary to develop new biosynthesis pathways. Single autosomal gene mutations have shown to dramatically change the ratio of sex pheromone components of cabbage looper *Trichoplusia ni* and the European corn borer (see above), to the extent that the males cannot recognise the pheromone any more, and even males of distantly related moths are attracted (Haynes and Hunt 1990; Löfstedt 1990; Roelofs 1995; Lassance et al. 2010).

Another classical example for specific pheromone blends is found in pine beetle populations that differ in the pheromone component ratios (Lanier et al. 1980). Pheromone blends are known from vertebrates as well: some mouse male pheromones, for example, are a blend and its individual components are not bioactive on their own (Novotny et al. 1985).

The signature mixtures of social insects offer an impressive example of complexity, where individuals bear complex mixtures of hydrocarbons on their cuticle that can consist of more than 60 substances. This high number of variables allows the hydrocarbons not only to be species-specific, but also colony-specific (van Zweden and d'Ettorre 2010) and in some rare cases even permits individual recognition (d'Ettorre and Heinze 2005). On top of these signature mixtures, pheromones are recruited from the cuticular hydrocarbons that allow the insects to discriminate between individuals with different functional roles within their colony (Bonavita-Cougourdan et al. 1993; Greene and Gordon 2003; Holman et al. 2010; Monnin and Peeters 1999).

An alternative to using specific blends of semiochemicals would be to use a single substance that is so unusual or complex that it is unique and thus specific. Some cockroach species have a sex pheromone comprised of a single substance that is so rare in nature that no interference from other sources is expected (for example the brown-banded cockroach *Supella longipalpa* uses supellapyrone, Fig. 13.6; Charlton et al. 1993). Peptides are also complex molecules that can be used in pheromone communication (Touhara 2008). A single point mutation can cause the exchange of an amino acid and thereby already produce a very different molecule, thus specificity can easily evolve (Fig. 13.6). The peptides used as pheromones by two different newt species, for example, differ by only two amino acids which is sufficient to make them highly species-specific (Toyoda et al. 2004).

13.5.2 Avoidance of Semiochemicals Specific to Unwanted Sources

In addition to reacting to pheromones that are specific for the target sender, receivers can also use information about unwanted senders. If, for example, more than one species is calling for mates, and the sex pheromone blends overlap but the

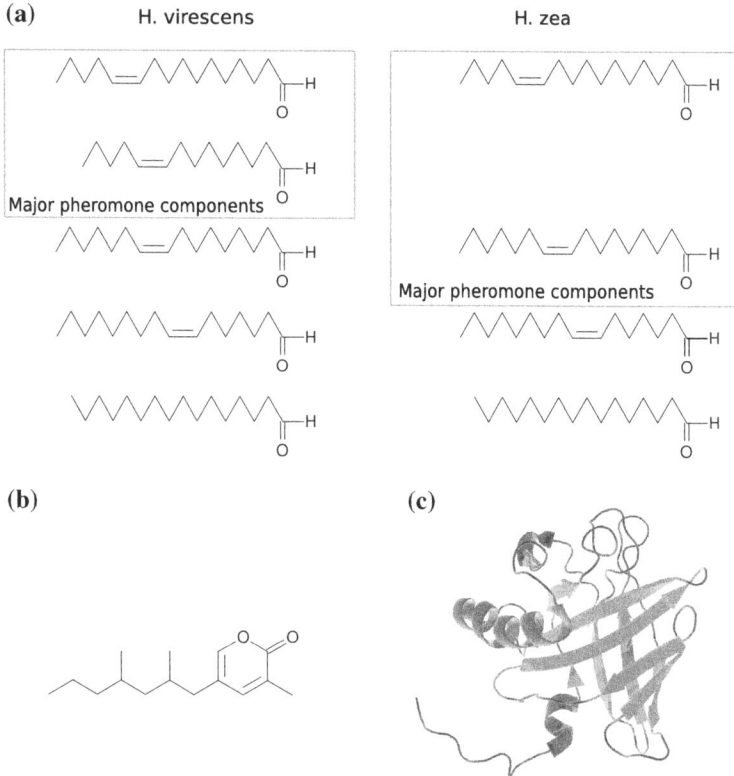

Fig. 13.6 *Pheromones and specificity* Specificity of pheromones can be achieved by using special combinations of simple molecules. **a** The compound pheromones of two moth species, *Heliothis virescens* and *Heliothis zea*. Major components make up the majority of the pheromone blend. Some of the molecules are used by both species, but the complete blend makes the specificity (Mustaparta 1997). Alternatively, single molecules can be used that are complex and rare in nature: **b** Supellapyrone, the pheromone of the brown-banded cockroach *Supella longipalpa* (a few cockroach species seem to have their own unique molecules (Gemeno and Schal 2004)). **c** Three-dimensional ribbon structure of darcin, a rodent major urinary protein. The structure of darcin has been solved by NMR and has been published with the PDFB structure file name of 29LC.PDB. In this diagram, the extended N-terminus encoding the hexahistidine purification tag has been removed (Roberts et al. 2010; Phelan et al. 2012)

blend of each species has at least one idiosyncratic component, receivers could use the other species' idiosyncratic components to avoid approaching allospecific callers. Even without ideosyncratic substances, receivers can avoid substance ratios specific for other species to optimise their own search (Baker 2008). This information is used by males so they do not waste time and energy following the pheromones of the wrong species (Cardé and Haynes 2004). Such a behaviour has been demonstrated for several lepidopterans that feed on the same maize plants (Eizaguirre et al. 2009). Interestingly, *Heliothis* moths have specific receptors innervating specific enlarged brain structures, which are dedicated to perceiving components from allospecific butterflies (Berg et al. 1998). The same seems to be true for some *Drosophila* species. Each species is repelled by odours from food that they are not specialised on (Ibba et al. 2010). Similarly, the males are repelled by a male-specific hydrocarbon, so they do not waste time with courting other males instead of females (Lacaille et al. 2007).

The use of allospecific sex pheromones for repulsion is, from an evolutionary perspective, a very easy way of reducing noise: during speciation, the pheromone, the receptors, and the neural infrastructure are already in place; only the wiring has to switch from attraction to repulsion in order to achieve reproductive isolation (Fig. 13.7).

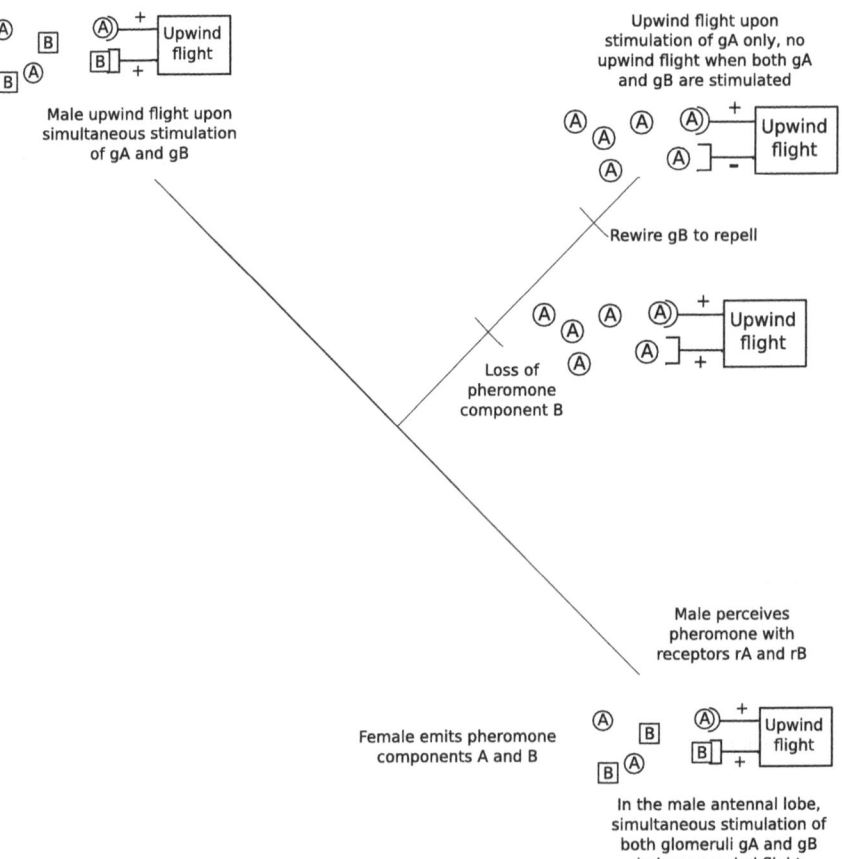

Fig. 13.7 *Speciation by rewiring on the antennal lobe* Ancestrally, there is one moth species. The females emit a pheromone composed by the substances A and B. Males have receptors for A and B, and excitation of receptor neurons that carry A- and B-receptors will trigger upwind flight. During speciation the females of one species lose one of the pheromone components, B. When both species are sympatric, cross-attraction of the derived species to the other one, and hence noise, could be reduced if males of the derived species (*right branch*) would still use their B-receptors, but rewire the olfactory processing so that A still has an attractive, but B has a repulsive effect

When more than one moth species is emitting a pheromone, the odours may be mixed by turbulent air flow, resulting in tiny intertwined odour filaments. By placing all receptors for the different components of the own pheromone blend and possible repellent substances in close proximity on the same antennal sensillum, a high spatial resolution can be achieved, allowing discrimination of the intertwined odour filaments of two different species. Fast processing allows discrimination of odourants that reach the receptors 1 ms apart, so that moths can discriminate the different filaments and find their way to the right pheromone source (de Bruyne and Baker 2008).

13.5.3 Context of Pheromone Reception

Besides a specific structure of the pheromone, a specific context for the emission of the pheromone can also reduce potential interference and thus increase the signal-to-noise ratio. The context could be for example the time of the day receivers can expect senders to emit pheromones. Moth males respond to pheromone only at times when females of their species are actually releasing pheromone (cf. Roelofs 1995). For example, the moths *Autographa californica* and *Trichoplusia ni* use the same pheromone, but signal at different times of the night.

The individual emitting a pheromone or bearing a signature mixture can also serve as a specifying context. For instance, when an ant evaluates the colony-specific signature mixture of another ant, the receiver will know that she is dealing with an ant signature mixture, since the mixture is located on the ant's cuticle. There are non-volatile cues, however, that work even if there is no direct physical interaction between the sender and the receiver. Territorial scent marks are one example. As the active space of these semiochemicals is minimal, the question arises how the senders are able to locate the scent in the first place. Receivers tend to find the marks using the same rules that the senders follow when depositing them: they prefer specific, often conspicuous, places (Johnson 1973). Anyone who has walked a domestic dog knows that lamp posts and trees are very popular among canids (Macdonald 1985). Indeed, wolves preferentially place informative scent marks in conspicuous places such as trees (Barja et al. 2008), and this behaviour is known from foxes, felids and rodents as well (Brown 1985; Macdonald 1985). Voles and mice, as another example, mark the entrance of their tunnels, and beavers even build earth mounds of up to 60 cm height, which they then mark with castoreum (a compound that is also harvested and used as a component of human perfumes). The earth mounds are located along the edges of the territory (Brown 1985). In addition, beavers mark protruding objects, a behaviour that is also known from gerenuk antelopes which preferentially mark twigs that are projecting from the general surface of shrubs and are located at a particular height (Gosling 1981).

Iguanas provide us with an example where the scent marks themselves may be visually conspicuous and may guide receivers to the mark. The femoral gland secretions, which are thought to provide information about the individuals in an area, absorb ultraviolet light. These animals are sensitive to ultraviolet light and thus a visual stimulus points receivers to where the olfactory information is, greatly increasing the efficiency of finding the marks (Alberts 1990).

The interplay between different communication modalities, for instance olfactory and visual cues, is referred to as multimodal communication. In the examples above, a visual cue points receivers to the olfactory signal; it creates the context, and thereby increases the signal-to-noise ratio. Multimodal signalling is widespread and used for communication in many taxa (see Chaps. 2, 3, 5, 7). A well-studied case involving a chemical signal is courtship in *Drosophila*. The flies use olfactory cues during courtship but in the absence of additional acoustic and optical signals the olfactory cues alone will not elicit mating (Greenspan and Ferveur 2000; Dickson 2008). In host finding, visual cues can also aid host location and detection. Tsetse flies and mosquitoes, for example, seem to use both visual and chemical odours to locate their hosts. While carbon dioxide and host odours trigger the attention, the short distance attraction to landing sites is mediated by visual cues (Harris and Foster 1994).

13.5.4 Redundancy

When semiochemicals are volatile, one factor that can limit the active space is the degradation of signals by dilution and physical degradation of molecules. If some components of a pheromone blend do not reach the receptors, the receiver might not recognise the pheromone. Redundancy can help to avoid problems due to the loss of signals or their components (Chap. 2). To circumvent signal loss due to the degradation of single components, senders could emit "backup signals", i.e. several different signals that convey the same message (Møller and Pomiankowski 1993). However, evidence for backup signals is rather scarce; different signals rarely convey the same message but rather additional information (Candolin 2003; Hebets and Papaj 2005).

The moth *Trichoplusia ni*, however, may be an example of redundancy. Females emit a pheromone blend consisting of five different substances; while the blend is still attractive when single components are left out, individual components displayed separately do not attract any males. In this example the use of a blend of substances as a pheromone may perhaps serve two purposes: on one hand it allows for high specificity, but on the other hand some of the components seem to be redundant, i.e. backups (Linn et al. 1984). Although researchers can describe these pheromone components as technically redundant, there is so far no evidence that the pheromone composition has indeed evolved this way because redundancy was selected for; the composition may have been shaped by other selection pressures, or some of the components may transmit information that has not yet been identified.

Redundancy can also be achieved by repeated or continued emission of the same signal, as is the case in many acoustic and optical signals (Chaps. 3, 5, 7, 11). A similar process can be observed in semiochemicals. If a pheromone is emitted continuously and the receivers "listen" over a period of time, they can make sure to catch the signal at least a few times; and, in addition, receiving a signal over extended periods allows averaging out random background noise that is due to a fluctuating environment (see also Chaps. 3, 6, 8). Sampling and averaging over time will also reduce the signal-to-noise ratio when measuring chemical gradients, as it has been shown in bacteria (reviewed by Bialek 1987).

13.5.5 Focussing on Ratios Instead of Concentrations

Most receivers are able to recognise specific pheromone mixtures despite variation in the absolute concentrations of the pheromone. What is more important than absolute concentrations are ratios between the components, which make the specificity of a pheromone blend. While the concentration of components can still influence the activation pattern of the glomeruli in the olfactory bulb (vertebrates) or antennal lobe (insects, cf. Fig. 13.2), the behaviour is usually only correlated with the ratios of the substances, i.e. a normalised activation pattern. The neuronal mechanisms allowing this independence from the absolute concentration are currently revealed in rats and fruit flies (Asahina et al. 2009; Cleland et al. 2007).

In other circumstances, changing ratios as different components evaporate or break down at different rates can give information. For example, mice can determine the relative age of scent marks by the loss of the more volatile molecules from their carrier proteins in urine scent marks (Hurst and Beynon 2004).

13.5.6 Sensory Adaptation

Sensory adaptation is a process during which sensory neurons "get used to" a stimulus. When an odour plume first hits the receptors, the neurons respond fully, but when the animal stays inside the odour plume for a while, the response of the olfactory receptor neurons declines. When the odourants are abundant over a period of time, intra-cell messaging changes in a way that decreases the propensity of the axon to depolarize (Todd and Baker 1999). These dynamics of receptor neurons can aid to raise the signal-to-noise ratio. Sensory adaptation is, for example, a way to blind out background noise. When exposed to certain odorants for a while, receptor neurons will only respond if the abundance of these odorants peaks over the background noise level. Everybody will have experienced this process themselves: entering a kitchen, the food's aroma will be obvious; after spending some time in the kitchen, however, the aroma will not be sensed any more unless the concentration peaks, for example when opening a pot (cf. Berglund et al. 1971). Similar processes have been described in lobsters (Borroni and Atema 1988) and many other animals: after adaptation, receptor neurons stop firing in response to the background odour levels unless there is a superimposed stimulus. Generally, sensory adaptation does not only allow to mask the background noise, but also to maintain a high local resolution for odourant concentration, while in the long term it is still possible to flexibly perceive a broad concentration range (Wark et al. 2007). The dynamics of sensory adaptation differ between receptor neuron types, suggesting that different neuron types have evolved to efficiently meet the requirements for reception of the different possible odourants (Todd and Baker 1999).

13.6 Noise in Perception and Processing

If production of semiochemicals and particularly their transmission are prone to errors, there is also the chance that noise enters the communication process during the reception and processing of chemical cues and signals.

13.6.1 Inaccuracy at the Receptor Level

Although an olfactory receptor can be very specific for a particular substance at low odourant concentrations, and does not react to anything else, most receptors lose their discriminatory power and may react to any substance if it is present at high concentration. *Drosophila* olfactory receptors are a well-studied example: when exposed to ethyl acetate in low concentration, only one receptor type responds; when the concentration of the same odourant is higher, though, two further receptor types respond (de Bruyne and Baker 2008). At very high doses, many receptors that are typically not tuned to the odourant might respond as well, which introduces noise. This characteristic, specificity at low concentrations but generality at high doses, is typical for both insect and mammal olfactory receptors (Kaupp 2010; Kay and Stopfer 2006).

13.6.2 Noise in Intracellular Signalling

Although we have listed sensory adaptation as a mean to reduce noise occurring at the receptor level, there are examples where the process may lead to the breakdown of recognition. In the turnip moth, under high sex pheromone concentrations, sensory adaptation occurs for only one out of the three pheromone components. Hence, the component ratio perceived by the central nervous system (for instance at the level of the antennal lobes) might deviate from that of the pheromone. This selective sensory adaptation may cause the effect that males do not respond to the pheromone any more when its concentration is very high (Fig. 13.8; Hansson and Baker 1991). However, it is also possible that the selective sensory adaptation to

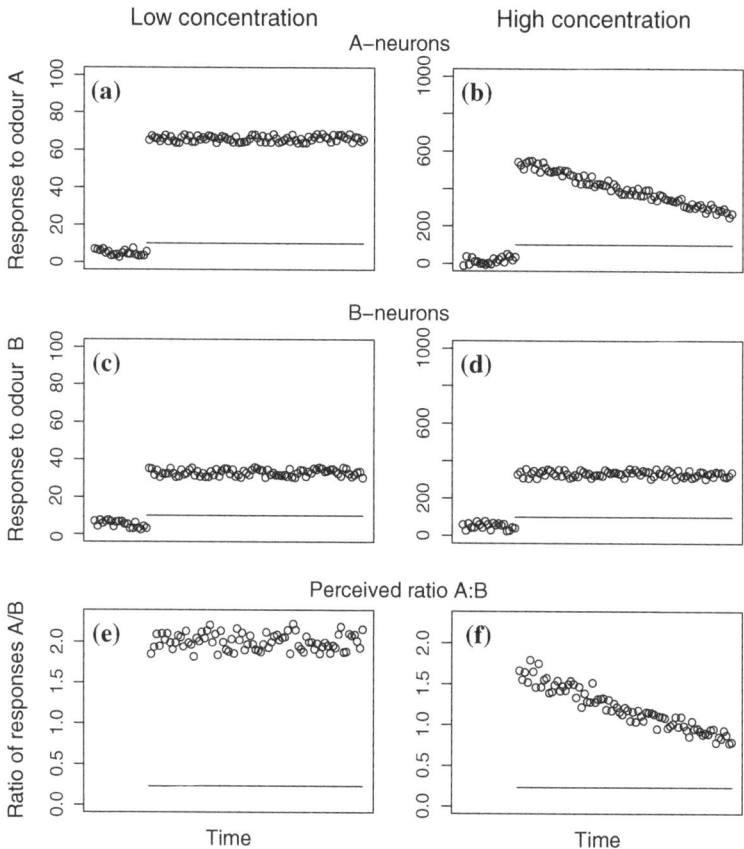

Fig. 13.8 The response of two different sensory neurons tuned to two different substances A (panel A, B) and B (panel C, D), each at two different pheromone concentrations. Note that there is sensory adaptation in the A-neurons (the response at high concentration decreases with time, panel B), but not in B-neurons (panel D). In perception, this leads to an unstable ratio of A and B at high pheromone concentrations (panel F), but not at low pheromone concentration (panel E). The horizontal line indicates the time the odour stimuli were present (adapted from Hansson and Baker 1991)

one of the pheromone components plays a yet unknown functional role in pheromone detection. As not all the tested sensory neurons for this compound adapted, the remaining cells may be sufficient to avoid a distortion of the perceived ratio due to selective adaptation.

After an odourant has bound to one of the many receptors in the membrane of a sensory neuron, an intracellular signalling cascade controls the excitation of the neuron. The cascade tends to consist of multiple steps, some of which amplify the signal. Each of these steps is susceptible to error to a certain extent. For instance, it is known that fluctuations in concentration of second messengers (molecules that relay signals from receptors on the cell surface to target molecules inside the cell) cause noise in the signal transduction of mammalian olfactory receptor neurons (Lowe and Gold 1995). Signal transduction in insect olfactory neurons does not necessarily use second messengers, and thus there are fewer steps between receptor and the action potential that could possibly introduce noise (Wyatt 2014). On the other hand, insect olfactory neurons show more baseline noise, i.e. more spontaneous firing, than those of mammals, perhaps because the second messenger steps can also act as filters that increase the signal-to-noise ratio (Kaupp 2010).

13.6.3 Concentrations Beyond the Dynamic Range

Varying concentration of odorants can also cause other problems than influencing the specificity of single olfactory receptor proteins on sensory neurons. Every sensory neuron is carrying many receptors, and whether the neuron is excited depends on the number of receptors that are activated (Fig. 13.2). In the olfactory bulb, the concentration of an odorant is then coded by the number of excited sensory neurons that are tuned to the odorant. At low odorant concentration, there will not be enough activated receptors in any of the neurons to excite it. With rising odorant concentration, more receptors and hence more sensory neurons respond. As soon as all sensory neurons are excited, increasing odorant concentration cannot lead to the activation of any more neurons, so that the message received by the central nervous system will not change either. The odorant concentration range, in which a change in the odorant concentration changes the number of the responding sensory neurons, i.e. the concentration range in which the animal can perceive changing concentration, is called the dynamic range. Note that this simple model of odourant coding ignores the temporal pattern that sensory neuron firing can evince, be it due to sensory adaptation, inhibitory neurons (in mammals), and perhaps other reasons not yet understood (Hallem and Carlson 2006; Kay and Stopfer 2006; Spors et al. 2006).

Consider a pheromone that consists of two components A and B, each of which is perceived by different sensory neuron types. The normal ratio of the two pheromone components is 2 A: 1 B (Fig. 13.9). Over the dynamic range of the sensory neurons, changing the concentration of both components equally would not change the perceived ratio between the components. However, as soon as the concentration of one component, perhaps B, falls below the perception threshold of the neurons, so that none of them responds any more, the ratio of the pheromone components is distorted. The animal will only smell A and will not recognise the pheromone any more, which is supposed to be a 2:1 blend of A and B. One may argue that it is quite expected that receivers cannot recognise the pheromone any more when its concentration is too low. However, a similar effect may occur when the pheromone concentration is very high. When all A-neurons are already excited, increasing the pheromone concentration will not change the proportion of responding A-neurons; the B-neurons, however, which are not yet all activated,

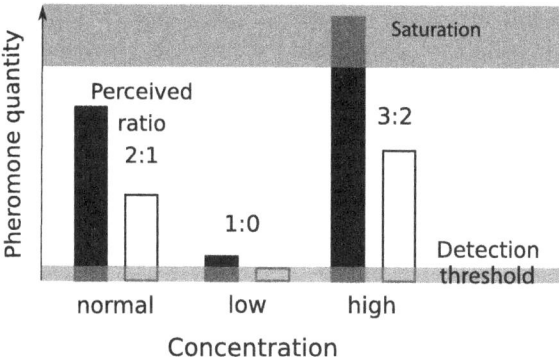

Fig. 13.9 *Threshold-saturation.* A pheromone consisting of two components (*black* and *white bars*) in a 2:1 ratio is usually also perceived in the same ratio. Ratios can get distorted at low and high pheromone concentrations when the concentration of one component is lower than the response threshold or high enough to reach saturation of all sensory neurons. In those two cases, changing the pheromone concentration changes the perceived concentration of only one of the components (the grey part is not perceived), which distorts the perceived ratio. The perceived ratio is indicated over the bars

will continue to report increasing concentration. Again, the perceived ratio of the pheromone components will be distorted; although the ratio of the components might still be 2:1, the ratio of activated sensory neurons, and thus the message received by the central nervous system, might now be 3:2, and the pheromone is not recognised any more (Fig. 13.9). In both cases, the dynamic range of the entirety of sensory neurons limits in what concentration range the animal can respond to the pheromone; when the concentration is outside of the range, noise is brought into the system.

13.7 Solutions to Reduce Noise in Perception

The binding of odourants to receptors, and also the signal transduction to the central nervous system, are probabilistic processes. The likelihood that an odourant binds a receptor increases with increasing match between receptor and odourant, and the probability that a calcium-modulated chloride channel opens increases when calcium enters the cell. However, as each step is probabilistic there is noise entering the communication during perception and processing. Noise in probabilistic processes can greatly be reduced by averaging. The greater the number of receptors, the larger the chance that an odourant is bound; the more chloride channels, the better the match of the channel opening likelihood with the strength of calcium in stream and so on. One way to cope with unwanted variation of receptor responses is to use many receptors of one type and average among them. The more receptors there are, the higher the signal-to-noise ratio will be.

More receptors will not only increase the accuracy, but also the detection threshold and the dynamic range for semiochemicals (Leon and Johnson 2003; cf. Chaps. 2, 12 for other modalities).

For very important odourants, such as sex pheromones, many insects have very high numbers of specialised receptor neurons. The high amount of information input from the receptor neurons requires adequate processing that is usually handled by large neural clusters within the antennal lobes, called macroglomeruli. These macroglomeruli are typically innervated only by the many specialised receptors that bind the sex pheromone. In moths, for example, males have enlarged antennae that house huge numbers of sex pheromone-specific receptors and sensory neurons that project to a macroglomerular complex containing a glomerulus for the olfactory sensory neurons responding to each of the pheromone components (Berg et al. 1998). The macroglomerular complex of males from a species with five components to its female pheromone would have five glomeruli. Cockroach males, but not females, have a macroglomerulus, which is the exclusive target of tens of thousands of receptors neurons that respond exclusively to one of the two components of the female pheromone, periplanone A and B. The receptors have varying sensitivity, and the convergence of many receptor neurons (36,000 for periplanone B) on a few projecting neurons (the macroglomerulus is innervated by only 20 uniglomerular projection neurons) permits a broad dose–response curve. Some of the projecting neurons respond to both components of the pheromone, others only to one of them (Boeckh and Selsam 1984; Boeckh and Tolbert 1993).

Macroglomeruli for trail pheromones in ants (Kleineidam et al. 2005), or fruit odours in *Drosophila* (Ibba et al. 2010) have also been described. In social insects they are only found in those individuals that need high sensitivity for a specific chemical or chemical blend. For example, only the foraging ant worker castes possess a macroglomerulus for trail pheromones.

There is of course a trade-off between sensitivity and accuracy for particular odourants on one hand, and the broadness and flexibility of the recognition system and discrimination ability on the other hand. Space for receptors and neurons is limited, and maintaining neural tissue is costly; thus the maximum number of receptors cannot increase over a certain level (cf. Chittka and Niven 2009). It is therefore not possible to increase the number of receptors of all types on the antenna, even if antennae are enlarged, such as in moth males. The number of receptors of each type will have to be optimised relative to each other, depending on how important sensitivity for the respective odourants is for the animal (de Bruyne and Baker 2008).

Besides averaging over a large number of modules, e.g. receptors or sensory neurons, it is also possible to average over time. If the total receptor response is averaged over a few milliseconds, the outcome of the measurement will vary less than if the measurement is made over only 1 ms. The same is true for intracellular signalling. Channel noise, for example, can be reduced when the response of channels to preceding signals is averaged over time (Kleene 1997).

13.8 Conclusion

In this chapter, we summarised the causes and consequences of noise in chemical communication, as well as the adaptations by both senders and receivers to increase the signal-to-noise ratio.

Studying noise is challenging. Often, the data do not allow the researcher to entirely partition the variation into its three components: information, system noise and experimental error. First of all, it will always be hard to take apart the experimental error and the noise inherent to the system. Second, variation can be information in one situation, but noise in another. Though it is difficult to study, we think further investigation of noise in chemical communication, using classic pheromone systems, would be very rewarding for everyone researching chemical communication.

It is important that the occurrence of noise is acknowledged. It is common practice to try to find an adaptive explanation for any observation and experimental result. However, organisms do make mistakes, and it is possible to discover that not an adaptation, but noise is the cause of the observed behaviour. Leaf-cutting ants, for example, sometimes adopt brood items from foreign colonies. This behaviour could be adaptive, since it increases the colony's worker force without any obvious cost, but it seems more likely to be a simple mistake in nestmate recognition (Fouks et al. 2011).The study of noise relies on quantifying variation in signal production, transmission and perception. Even when a study is not specifically designed to analyse noise in a communication system, it may contribute to understanding the role of noise by quantifying every aspect of the communication process. We suggest that researchers should aim at always quantifying the effects of potential pheromones instead of only testing whether or not a certain substance acts as a pheromone, since a quantitative approach is essential for estimating noise, and noise is often one of the missing elements in the discussion of pheromone communication systems.

Acknowledgments We thank Marc Weissburg for the photograph of the underwater odour plume visualised with dye and Rob Beynon for the Darcin ribbon graphic. We also thank Jan Benda, Henrik Brumm and an anonymous referee for their helpful comments on drafts of this chapter.

References

Alberts A (1990) Chemical properties of femoral gland secretions in the desert iguana, *Dipsosaurus dorsalis*. J Chem Ecol 16:13–25
Arnold G, Quenet B, Cornuet J-M et al (1996) Kin recognition in honeybees. Nature 379:498
Asahina K, Louis M, Piccinotti S, Vosshall LB (2009) A circuit supporting concentration-invariant odor perception in Drosophila. J Biol 8:9. doi:10.1186/jbiol108
Baker TC (2008) Balanced olfactory antagonism as a concept for understanding evolutionary shifts in moth sex pheromone blends. J Chem Ecol 34:971–981. doi:10.1007/s10886-008-9468-5

Bargmann CI (2006) Comparative chemosensation from receptors to ecology. Nature 444:295–301. doi:10.1038/nature05402

Barja I, Silván G, Illera JC (2008) Relationships between sex and stress hormone levels in feces and marking behavior in a wild population of Iberian wolves (*Canis lupus signatus*). J Chem Ecol 6:697–701

Barth MB, Kellner K, Heinze J (2010) The police are not the army: context-dependent aggressiveness in a clonal ant. Biol Lett 6:329–332. doi:10.1098/rsbl.2009.0849

Berg BG, Almaas TJ, Bjaalie JG, Mustaparta H (1998) The macroglomerular complex of the antennal lobe in the tobacco budworm moth *Heliothis virescens*: specified subdivision in four compartments according to information about biologically significant compounds. J Comp Physiol A 183:669–682. doi:10.1007/s003590050290

Berglund B, Berglund U, Engen T, Lindvall T (1971) The effect of adaptation on odor detection. Percept Psychophys 9:435–438

Bialek W (1987) Physical limits to sensation and perception. Annu Rev Biophys Biophys Chem 16:455–478. doi:10.1146/annurev.bb.16.060187.002323

Boeckh J, Selsam P (1984) Quantitative investigation of the odour specificity of central olfactory neurones in the American cockroach. Chem Senses 9:369–380

Boeckh J, Tolbert LP (1993) Synaptic organization and development of the antennal lobe in insects. Microsc Res Tech 24:260–280. doi:10.1002/jemt.1070240305

Bonavita-Cougourdan A, Clement J-L, Lange C (1993) Functional subcaste discrimination (foragers and brood-tenders) in the ant *Camponotus vagus* Csop: polymorphism of cuticular hydrocarbons. J Chem Ecol 19:1461–1477

Boomsma JJ, Nielsen J, Sundström L et al (2003) Informational constraints on optimal sex allocation in ants. Proc Natl Acad Sci USA 100:8799–8804. doi:10.1073/pnas.1430283100

Borroni PF, Atema J (1988) Adaptation in chemoreceptor cells. I. Self-adapting backgrounds determine threshold and cause parallel shift of response function. J Comp Physiol A 164:67–74

Brodmann J, Twele R, Francke W et al (2009) Orchid mimics honey bee alarm pheromone in order to attract hornets for pollination. Curr Biol 19:1368–1372

Brown RE (1985) The rodents II: suborder Myomorpha. In: Brown RE, Macdonald DW (ed) Social odours in mammals. Oxford University Press, Oxford, pp 345–457

de Bruyne M, Baker TC (2008) Odor detection in insects: volatile codes. J Chem Ecol 34:882–897. doi:10.1007/s10886-008-9485-4

Candolin U (2003) The use of multiple cues in mate choice. Biol Rev 78:575–595

Cardé RT, Haynes KF (2004) Structure of the pheromone communication channel in moths. In: Cardé R, Millar J (eds) Advances in insect chemical ecology. Cambridge University Press, Cambridge, pp 283–332

Charlton RE, Webster FX, Zhang A et al (1993) Sex pheromone for the brown banded cockroach is an unusual dialkyl-substituted alpha-pyrone. Proc Natl Acad Sci USA 90:10202–10205

Chittka L, Niven J (2009) Are bigger brains better? Curr Biol 19:R995–R1008. doi:10.1016/j.cub.2009.08.023

Cleland TA, Johnson BA, Leon M, Linster C (2007) Relational representation in the olfactory system. Proc Natl Acad Sci USA 104:1953–1958

Cleland TA, Narla VAA, Boudadi K (2009) Multiple learning parameters differentially regulate olfactory generalization. Behav Neurosci 123:26. doi:10.1037/a0013991.Multiple

Conner W, Weller SJ (2004) A quest for alkaloids: the curious relationship of tiger moths and plants containing pyrrolizidine alkaloids. In: Cardé RT, Millar JG (eds) Advances in insect chemical ecology. Cambridge University Press, Cambridge, pp 248–282

Dani FR, Foster KR, Zacchi F et al (2004) Can cuticular lipids provide sufficient information for within-colony nepotism in wasps? Proc R Soc Lond B 271:745–753. doi:10.1098/rspb.2003.2646

Dani FR, Jones GR, Corsi S et al (2005) Nestmate recognition cues in the honey bee: differential importance of cuticular alkanes and alkenes. Chem Senses 30:477–489. doi:10.1093/chemse/bji040

Dicke M, Agrawal AA, Bruin J (2003) Plants talk, but are they deaf? Trends Plant Sci 8:403–405

Dickson BJ (2008) Wired for sex: the neurobiology of *Drosophila* mating decisions. Science 322:904–909

Domingue MJ, Haynes KF, Todd JL, Baker TC (2009) Altered olfactory receptor neuron responsiveness is correlated with a shift in behavioral response in an evolved colony of the cabbage looper moth, *Trichoplusia ni*. J Chem Ecol 35:405–415. doi:10.1007/s10886-009-9621-9

Dybdahl MF, Storfer A (2003) Parasite local adaptation: red queen versus suicide king. Trend Ecol Evol 18:523–530

d'Ettorre P, Heinze J (2005) Individual recognition in ant queens. Curr Biol 15:1–2. doi:10.1016/j.cub.2005.10.067

d'Ettorre P, Moore AJ (2008) Chemical communication and the coordination of social interactions in insects. In: d'Ettorre P, Hughes DP (ed) Sociobiology of communication. Oxford Scholarship Online Monographs, pp 81–97

Eizaguirre M, López C, Sans A et al (2009) Response of *Mythimna unipuncta* males to components of the *Sesamia nonagrioides* pheromone. J Chem Ecol 35:779–784

Frisch K von (1915) Der Farbensinn und Formensinn der Biene. Zool Jb Physiol 35:1–182

Fouks B, d'Ettorre P, Nehring V (2011) Brood adoption in the leaf-cutting ant *Acromyrmex echinatior*: adaptation or recognition noise? Insectes Soc 58:479–485. doi: 10.1007/s00040-011-0167-9

Gemeno C, Schal C (2004) Sex pheromones of cockroaches. In: Cardé R, Millar J (eds) Advances in insect chemical ecology. Cambridge University Press, New York, pp 179–247

Gosling LM (1981) Demarkation in a gerenuk territory: an economic approach. Z Tierpsychol 56:305–322

Greene MJ, Gordon DM (2003) Social insects: cuticular hydrocarbons inform task decisions. Nature 423:32

Greenspan RJ, Ferveur J-F (2000) Courtship in *Drosophila*. Annu Rev Genet 34:205–232

Hallem EA, Carlson JR (2006) Coding of odors by a receptor repertoire. Cell 125:143–160

Hansson BS, Baker TC (1991) Differential adaptation rates in a male moth's sex pheromone receptor neurons. Naturwissenschaften 78:517–520

Harris MO, Foster SP (1994) Behavior and integration. In: Carde RT, Bell WJ (ed) Chemical ecology of insects 2. Chapman and Hall, New York, pp 3–46

Havlicek J, Roberts SC (2009) MHC-correlated mate choice in humans: a review. Psychoneuroendocrinology 34:497–512

Haynes KF, Hunt RE (1990) A mutation in pheromonal communication system of cabbage looper moth, *Trichoplusia ni*. J Chem Ecol 16:1249–1257

Haynes KF, Gemeno C, Yeargan KV et al (2002) Aggressive chemical mimicry of moth pheromones by a bolas spider: how does this specialist predator attract more than one species of prey? Chemoecology 12:99–105

Hebets EA, Papaj DR (2005) Complex signal function: developing a framework of testable hypotheses. Behav Ecol Sociobiol 57:197–214

Helanterä H, Ratnieks FLW (2008) Two independent mechanisms of egg recognition in worker *Formica fusca* ants. Behav Ecol Sociobiol 63:573–580. doi:10.1007/s00265-008-0692-3

Helsper JPFG, Davies JA, Bouwmeester HJ et al. (1998) Circadian rhythmicity in emission of volatile compounds by flowers of *Rosa hybrida* L. cv. Honesty. Planta 207:88–95. doi: 10.1007/s004250050459

Holman L, Dreier S, d'Ettorre P (2010) Selfish strategies and honest signalling: reproductive conflicts in ant queen associations. Proc R Soc Lond B 277:2007–2015. doi: 10.1098/rspb.2009.2311

Hurst J, Beynon R (2004) Scent wars: the chemobiology of competitive signalling in mice. BioEssays 26:1288–1298

Ibba I, Angioy AM, Hansson BS, Dekker T (2010) Macroglomeruli for fruit odors change blend preference in *Drosophila*. Naturwissenschaften 97:1059–1066. doi:10.1007/s00114-010-0727-2

Johnson RP (1973) Scent marking in mammals. Anim Behav 21:521–535

Johnston RE (2008) Individual odors and social communication: individual recognition, kin recognition, and scent over-marking. In: Brockmann H, Roper T, Naguib M et al. (eds) Advances in the study of behavior. Academic Press, New York, pp 439–505

Karlson P, Lüscher M (1959) "Pheromones", a new term for a class of biologically active substances. Nature 183:55–56. doi:10.1038/183055a0

Kaupp UB (2010) Olfactory signalling in vertebrates and insects: differences and commonalities. Nat Rev Neurosci 11:188–200

Kawecki TJ, Ebert D (2004) Conceptual issues in local adaptation. Ecol Lett 7:1225–1241

Kay LM, Stopfer M (2006) Information processing in the olfactory systems of insects and vertebrates. Semin Cell Dev Biol 17:433–442. doi:10.1016/j.semcdb.2006.04.012

Kelly DR (1996) When is a butterfly like an elephant? Chem Biol 3:595–602

Kleene SJ (1997) High-gain, low-noise amplification in olfactory transduction. Biophys J 73:1110–1117

Kleineidam CJ, Obermayer M, Halbich W, Rössler W (2005) A macroglomerulus in the antennal lobe of leaf-cutting ant workers and its possible functional significance. Chem Senses 30:383–392. doi:10.1093/chemse/bji033

Lacaille F, Hiroi M, Twele R et al (2007) An inhibitory sex pheromone tastes bitter for Drosophila males. Plos One 2:e661. doi:10.1371/journal.pone.0000661

Lanier GN, Classon ALF, Piston JJ et al (1980) Ips pini: the basis for interpopulational differences in pheromone biology. J Chem Ecol 6:677–687

Lassance J-M (2010) Journey in the Ostrinia world: from pest to model in chemical ecology. J Chem Ecol 36:1155–1169. doi:10.1007/s10886-010-9856-5

Lassance J-M, Groot AT, Liénard MA et al. (2010) Allelic variation in a fatty-acyl reductase gene causes divergence in moth sex pheromones. Nature 466:486–489. doi: 10.1038/nature09058

Lenoir A, d'Ettorre P, Errard C, Hefetz A (2001) Chemical ecology and social parasitism in ants. Ann Rev Entomol 46:573–599

Lenoir A, Fresneau D, Errard C, Hefetz A (1999) Individuality and colonial identity in ants. In: Detrain C, Deneubourg JL, Pasteels JM (ed) Information processing in social insects. Birkhäuser Verlag, Basel, pp 219–237

Leon M, Johnson BA (2003) Olfactory coding in the mammalian olfactory bulb. Brain Res Rev 42:23–32. doi: 10.1016/S0165-0173(03)00142-5

Liang D, Blomquist GJ, Silverman J (2001) Hydrocarbon-released nestmate aggression in the Argentine ant, Linepithema humile, following encounters with insect prey. Comp Biochem Physiol B 129:871–882

Linn CE, Campbell MG, Roelofs WL (1987) Pheromone components and active spaces: what do moths smell and where do they smell it? Science 237:650–652

Linn CEJ, Bjostad LB, Du JW, Roelofs WL (1984) Redundancy in a chemical signal; behavioral responses of male Trichoplusia ni to a 6-component sex pheromone blend. J Chem Ecol 10:1635–1658

Linster C, Johnson Morse et al (2001) Perceptual correlates of neural representations evoked by odorant enantiomers. J Neurosci 21:9837–9843

Linster C, Johnson Morse et al (2002) Spontaneous versus reinforced olfactory discriminations. J Neurosci 22:6842–6845

Loftfield RB, Vanderjagt D (1963) The frequency of errors in protein biosynthesis. Biochem J 89:82–92

Lowe G, Gold GH (1995) Olfactory transduction is intrinsically noisy. Proc Natl Acad Sci USA 92:7864–7868

Löfstedt C (1990) Population variation and genetic control of pheromone communication systems in moths. Entomol Exp Appl 54:199–218

Löfstedt C, Herrebout WM, Menken SBJ (1991) Sex pheromones and their potential role in the evolution of reproductive isolation in small ermine moths (Yponomeutidae). Chemoecology 2:20–28. doi:10.1007/BF01240662

Macdonald DW (1985) The carnivores: order Carnivora. In: Brown RE, Macdonald DW (ed) Social odours in mammals. Oxford University Press, Oxford, pp 619–722

Martin SJ, Drijfhout FP (2009) A review of ant cuticular hydrocarbons. J Chem Ecol 35:1151–1161. doi:10.1007/s10886-009-9695-4

Martin SJ, Helanterä H, Drijfhout FP (2011) Is parasite pressure a driver of chemical cue diversity in ants? Proc R Soc Lond B 278:496–503. doi:10.1098/rspb.2010.1047

Martin SJ, Helanterä H, Kiss K et al (2009) Polygyny reduces rather than increases nestmate discrimination cue diversity in Formica exsecta ants. Insectes Soc 56:375–383. doi:10.1007/s00040-009-0035-z

Maynard Smith J, Harper D (2003) Animal signals. Oxford University Press, Oxford

Miklas N, Renou M, Malosse I, Malosse C (2000) Repeatability of pheromone blend composition in individual males of the southern green stink bug, *Nezara viridula*. J Chem Ecol 26:2473–2485

Miller JR, McGhee PS, Siegert PY et al (2010) General principles of attraction and competitive attraction as revealed by large-cage studies of moths responding to sex pheromone. Proc Natl Acad Sci USA 107:22–27. doi:10.1073/pnas.0908453107

Minks AK, Cardé RT (1988) Disruption of pheromone communication in moths—is the natural blend really most efficacious. Entomol Exp Appl 49:25–36

Monnin T, Peeters C (1999) Dominance hierarchy and reproductive conflicts among subordinates in a monogynous queenless ant. Behav Ecol 10:323–332. doi:10.1093/beheco/10.3.323

Moore D, Liebig J (2010) Mixed messages: fertility signaling interferes with nestmate recognition in the monogynous ant *Camponotus floridanus*. Behav Ecol Sociobiol 64:1011–1018. doi:10.1007/s00265-010-0916-1

Murlis J, Elkinton JS, Cardé RT (1992) Odor plumes and how insects use them. Annu Rev Entomol 37:505–532. doi:10.1146/annurev.ento.37.1.505

Mustaparta H (1997) Olfactory coding mechanisms for pheromone and interspecific signal information in related moth species. In: Carde RT, Minks AK (eds) Insect Pheromone Research: New Directions, Springer

Møller AP, Pomiankowski A (1993) Why have birds got multiple sexual ornaments? Behav Ecol Sociobiol 32:167–176. doi:10.1007/BF00173774

Müller-Schwarze D (2006) Chemical ecology of vertebrates. Cambridge University Press, Cambridge

Nehring V, Evison SEF, Santorelli LA, d'Ettorre P, Hughes WOH (2011) Kin-informative recognition cues in ants. Proc R Soc Lond B 278:1942–1948. doi:10.1098/rspb.2010.2295

Novotny M, Harvey S, Jemiolo B, Alberts J (1985) Synthetic pheromones that promote inter-male aggression in mice. Proc Natl Acad Sci USA 82:2059–2061

Olsson SB, Kesevan S, Groot AT et al (2010) Ostrinia revisited: evidence for sex linkage in European Corn Borer Ostrinia nubilalis (Hubner) pheromone reception. BMC Evol Biol 10:285. doi:10.1186/1471-2148-10-285

Peeters C, Monnin T, Malosse C (1999) Cuticular hydrocarbons correlated with reproductive status in a queenless ant. Proc R Soc Lond B 266:1323–1327. doi:10.1098/rspb.1999.0782

Phelan MM, Mclean L, Beynon RJ, Hurst JL, Lian L (2012) Structural insights into the specificity of darcin, an atypical major urinary protein. PDB 2L9C, doi:10.2210/pdb2l9c/pdb

Penn DJ (2002) The scent of genetic compatibility: sexual selection and the major histocompatibility complex. Ethology 108:1–21

Porter RH, Winberg J (1999) Unique salience of maternal breast odors for newborn infants. Neurosci Biobehav Rev 23:439–449

Ratnieks FLW (1988) Reproductive harmony via mutual policing by workers in eusocial Hymenoptera. Am Nat 132:217–236

Roberts SA, Simpson DM, Armstrong SD, Davidson AJ, Robertson DH, McLean L, Beynon RJ, Hurst JL (2010) Darcin: a male pheromone that stimulates female memory and sexual attraction to an individual male's odour. BMC Biol 8:75

Roelofs WL (1995) Chemistry of sex attraction. Proc Natl Acad Sci USA 92:44–49

Ryne C, Svensson GP, Löfstedt C (2001) Mating disruption of *Plodia interpunctella* in small-scale plots: effects of pheromone blend, emission rates, and population density. J Chem Ecol 27:2109–2124

Schiestl FP (2005) On the success of a swindle: pollination by deception in orchids. Naturwissenschaften 92:255–264

Schiestl FP, Peakall R, Mant JG et al (2003) The chemistry of sexual deception in an orchid-wasp pollination system. Science 302:437–438. doi:10.1126/science.1087835

Slessor KN, Winston ML, Le Conte Y (2005) Pheromone communication in the honeybee (Apis mellifera L). J Chem Ecol 31:2731–2745

Smadja C, Butlin RK (2009) On the scent of speciation: the chemosensory system and its role in premating isolation. Heredity 102:77–97. doi:10.1038/hdy.2008.55

Smith AA, Hölldobler B, Liebig J (2009) Cuticular hydrocarbons reliably identify cheaters and allow enforcement of altruism in a social insect. Curr Biol 19:78–81. doi:10.1016/j.cub.2008.11.059

Sorensen PW, Scott AP, Kihslinger RL (2000) How common hormonal metabolites function as relatively specific pheromonal signals in the goldfish. In: Norberg B, Kjesbu OS, Taranger GL et al. (eds) Proceedings of the sixth international symposium on the reproductive physiology of fish. Institute of Marine Research and University of Bergen, Bergen, pp 125–128

Soroker V, Vienne C, Hefetz A (1995) Hydrocarbon dynamics within and between nestmates in *Cataglyphis niger* (Hymenoptera: Formicidae). J Chem Ecol 21:365–378

Spors H, Wachowiak M, Cohen LB, Friedrich RW (2006) Temporal dynamics and latency patterns of receptor neuron input to the olfactory bulb. J Neurosci 26:1247–1259. doi:10.1523/JNEUROSCI.3100-05.2006

Steiger S, Schmitt T, Schaefer HM (2011) The origin and dynamic evolution of chemical information transfer. Proc R Soc Lond B 278:970–979. doi:10.1098/rspb.2010.2285

Stensmyr MC, Urru I, Collu I et al (2002) Pollination: rotting smell of dead-horse arum florets. Nature 420:625–626

Stowe MK, Tumlinson JH, Heath RR (1987) Chemical mimicry: bolas spiders emit components of moth prey species sex pheromones. Science (80-) 236:964–967

Tanaka N, Awasaki T, Shimada T (2004) Integration of chemosensory pathways in the *Drosophila* second-order olfactory centers. Curr Biol 14:449–457. doi:10.1016/j

Todd JL, Baker TC (1999) Function of peripheral olfactory organs. In: Hansson BS (ed) Insect olfaction. Springer, Berlin, pp 67–96

Touhara K (2008) Sexual communication via peptide and protein pheromones. Curr Opin Pharmacol 8:759–764

Toyoda F, Yamamoto K, Iwata T et al (2004) Peptide pheromones in newts. Peptides 25:1531–1536

van Zweden JS, Fürst MA, Heinze J, D'Ettorre P (2007) Specialization in policing behaviour among workers of the ant *Pachycondyla inversa*. Proc R Soc Lond B 274:1421–1428. doi:10.1098/rspb.2007.0113

van Zweden JS, Dreier S, d'Ettorre P (2009) Disentangling environmental and heritable nestmate recognition cues in a carpenter ant. J Insect Physiol 55:159–164. doi: DOI: 10.1016/j.jinsphys.2008.11.001

van Zweden JS, d'Ettorre P (2010) Nestmate recognition in social insects and the role of hydrocarbons. In: Blomquist GJ, Bagneres AG (eds) Insect hydrocarbons: biology, biochemistry, and chemical ecology. Cambridge University Press, Cambridge, pp 222–243

van Zweden JS, Brask JB, Christensen JH et al (2010) Blending of heritable recognition cues among ant nestmates creates distinct colony gestalt odours but prevents within-colony nepotism. J Evol Biol 23:1498–1508. doi:10.1111/j.1420-9101.2010.02020.x

Vickers NJ, Baker TC (1994) Reiterative responses to single strands of odor promote sustained upwind flight and odor source location by moths. Proc Natl Acad Sci USA 91:5756–5760

Wark B, Lundstrom BN, Fairhall A (2007) Sensory adaptation. Curr Opin Neurobiol 17:423–429

Waters CM, Bassler BL (2005) Quorum sensing: cell-to-cell communication in bacteria. Annu Rev Cell Dev Biol 21:319–46. doi: 10.1146/annurev.cellbio.21.012704.131001

Webster DR, Weissburg MJ (2009) The hydrodynamics of chemical cues among aquatic organisms. Annu Rev Fluid Mech 41:73–90

Witzgall P, Frerot B (1989) Pheromone emission by individual females of carnation tortrix, Cacoecimorpha pronubana. J Chem Ecol 15:707–717

Wright GA, Kottcamp SM, Thomson MGA (2008) Generalization mediates sensitivity to complex odor features in the honeybee. Plos One 3:e1704

Wyatt TD (2009) Fifty years of pheromones. Nature 457:262–263

Wyatt TD (2010) Pheromones and signature mixtures: defining species-wide signals and variable cues for identity in both invertebrates and vertebrates. J Comp Physiol A 196:685–700. doi:10.1007/s00359-010-0564-y

Wyatt TD (2011) Pheromones and behavior. In: Breithaupt T, Thiel M (eds) Chemical communication in crustaceans. Springer, New York, pp 23–38

Wyatt TD (2014) Pheromones and animal behavior: chemical signals and signature mixtures, 2nd edn. Cambridge University Press, Cambridge

Zaher HS, Green R (2009) Quality control by the ribosome following peptide bond formation. Nature 457:161–166. doi:10.1038/nature07582

Part IV
Impacts of Anthropogenic Noise

Chapter 14
Anthropogenic Noise and Conservation

Peter K. McGregor, Andrew G. Horn, Marty L. Leonard
and Frank Thomsen

Abstract Anthropogenic noise is a common but evolutionarily recent influence on communicating animals and evidence is accumulating of its adverse impacts on human health, therefore it has potential relevance to conservation. However, demonstrating that this potential is realised is not straightforward. A particular issue is the difficulty of assessing likely impacts from the limited evidence on the main factors influencing impacts—from the hearing abilities of animals of conservation concern through to the characteristics of emitted sound fields in natural environments. Further issues include the likely underestimation of behavioural effects, and a lack of knowledge of how animals trade off costs and benefits. In this chapter, we aim to highlight the main themes emerging from the growing interest in the effects of anthropogenic noise on conservation. We predominantly consider the marine environment (with examples drawn mainly from marine mammals) and the terrestrial environment (with bird examples). An important consideration that emerges from the increasing levels of anthropogenic noise and difficulties in assessing specific impacts is the need to develop interim guidance, while more detailed information is gathered and assessed.

P. K. McGregor (✉)
Centre for Applied Zoology, Cornwall College Newquay, Trenance Gardens, Newquay TR7 2LZ, UK
e-mail: peter.mcgregor@cornwall.ac.uk

A. G. Horn · M. L. Leonard
Department of Biology, Life Science Centre, Dalhousie University, 1355 Oxford Street, Halifax, NS B3H 4J1, Canada
e-mail: aghorn@dal.ca

M. L. Leonard
e-mail: mleonard@dal.ca

F. Thomsen
DHI, Agern Alle 5, DK-2970 Hørsholm, Denmark
e-mail: frth@dhigroup.com

H. Brumm (ed.), *Animal Communication and Noise*,
Animal Signals and Communication 2, DOI: 10.1007/978-3-642-41494-7_14,
© Springer-Verlag Berlin Heidelberg 2013

14.1 Introduction

Post-industrial revolution humans produce more and louder noise than any other species on the planet. Given that we are also ubiquitous and numerous, anthropogenic noise has become a dominating feature of most animals' environments. Animals have evolved to communicate in the presence of natural biological and physical sources of noise, including other animals of the same and different species (see other chapters in this volume). However, most anthropogenic noise differs from natural noise in features including intensity, distribution, persistence and timescale that are likely to make an adaptive response by most species problematic. Therefore anthropogenic noise has the potential to impact conservation.

Another reason to consider that anthropogenic noise will have conservation impacts in addition to being common and new (on an evolutionary time scale) is its documented adverse effects on human health. For example, a recent study in Western Europe indicated that at least one million healthy life years are lost every year from traffic-related noise. These losses are principally through stress-related effects linked to sleep disturbance and annoyance but also include ischaemic heart disease, cognitive impairment of children, tinnitus (WHO 2011) and incident diabetes (Sørensen et al. 2013). These studies and similar demonstrations in humans of the role of noise as a stressor, suggest that comparable effects could occur in other vertebrates. The WHO study (2011) also pointed out that exposure to noise in Europe is increasing whereas other stressors such as exposure to dioxins and benzene are declining. It is not clear whether this relative difference in noise v. other stressors also applies to animal populations, although it is clear that anthropogenic noise is increasing (see Sect. 14.3).

Our chapter is different in content and scope from the others in this book. We aim to appraise the significance of anthropogenic noise for issues related to conservation. Mitigation of, and adaptation to, noise are fundamental processes in communication and signal detection (well demonstrated by the other chapters in this volume). However, anthropogenic noise can increase errors by signal receivers (see Chap. 2) and such errors can reduce individual fitness. Reductions in individual fitness can translate into effects at a population level and therefore become relevant to conservation. However, demonstrating that the potential impact of noise on conservation is realised, particularly through effects on communication, is not straightforward.

We will discuss anthropogenic noise in terrestrial and marine environments with a taxonomic coverage largely limited to birds and marine mammals because these are the groups with which we are most familiar. This chapter is not intended as an exhaustive review of the importance of noise to animal communication and conservation, rather we highlight what we see as the main themes emerging from the growth in this field. Several recent reviews provide more details and other emphases (e.g. Pepper et al. 2003; Warren et al. 2006; Nowacek et al. 2007; Slabbekoorn and Ripmeester 2007; Southall et al. 2007; Barber et al. 2009a, b; Popper and Hastings 2009b; OSPAR 2009b; Goodwin and Shriver 2010; Tasker et al. 2010; Kociolek 2011; Ortega 2012; Slabbekoorn 2013).

As this book amply demonstrates, noise is a common problem for all modalities of animal communication—acoustic, visual, chemical, tactile or electrical (for acoustic see Chaps. 3–10, visual see Chap. 11, electrical see Chap. 12 and chemical see Chap. 13) and includes intrinsic noise in the reception system of receivers (e.g. noise of receptor cells, see Chaps. 3 and 12). We will focus on extrinsic acoustic noise because this is likely to be the only context in which noise will be familiar to those with conservation interests. However, we believe that it would be productive to apply the approach we develop in this chapter to other communication modalities. Consider, for example, noise in the chemical modality. Anthropogenic chemicals have long been included in legal definitions of pollutants and many have, or mimic, biological signalling functions. Therefore, pollution by such chemicals can also be considered as noise in chemical communication systems. A specific example in freshwater habitats is the widespread presence of anthropogenic sex hormone mimicking chemicals and other endocrine disrupters, which have demonstrable behavioural and physiological effects with potential conservation implications (e.g. Tyler and Jobling 2008).

We begin this chapter by characterising terrestrial and marine environments with respect to the potential for anthropogenic noise to have consequences for conservation through communication effects, including differences in sources of noise. We then consider how noise impacts in general have been assessed. The Sect. 14.2 discusses potential and demonstrated conservation impacts of noise through effects on communication. It is subdivided into evidence for proximate costs (with effects at population level inferred) and evidence for population level effects where proximate causes are inferred. In Sect. 14.2 we look at management of anthropogenic noise and mitigation measures; dealing with terrestrial and marine environments separately because we believe that, unlike previous sections, an integrated approach yields fewer additional insights. We conclude by identifying where further work is necessary and interim approaches that can be applied now.

14.2 Characteristics of Terrestrial and Marine Environments

There are several characteristics of marine and terrestrial environments that affect both the potential for anthropogenic noise to impact communication and the implications for conservation. These range from the physics of sound transmission to the ease of observing impacts and will be considered individually before looking at their combined effects.

14.2.1 Sound Transmission

The speed of sound in salt water is approximately 1,500 ms^{-1} whereas in air it is about four and a half times slower at approximately 330 ms^{-1}. Sound is also attenuated less in seawater, especially at lower frequencies and can thus travel considerable distances (Urick 1983; Richardson et al. 1995; Ainslie 2010). In consequence, the active space of acoustic signals is much larger underwater than it is in air. For example, bottlenose dolphin *Tursiops truncatus* whistles have an estimated active space of up to 25 km (Janik 2000); and fin whales *Balaenoptera physalus* could communicate over ranges of up to 100 km, depending on conditions (Stafford et al. 2007). By contrast, the active space of a bird's song would be measured in tens of metres (Lohr et al. 2003; Nemeth and Brumm 2010). As a result, the area over which a given noise might be of concern is much larger underwater than on land; noise from a shipping lane may interfere with communication across a wide area of sea, whereas noise from a highway is likely to interfere with signalling only in the bird territories within a few road-widths of the road.

14.2.2 Frequencies Used in Communication

The wide range of frequencies emitted by animals is illustrated in this section by marine taxa. At the lower end of the frequency scale are calls in the region of 20 Hz by baleen whales such as fin whales which are presumed to be reproductive displays (Watkins et al. 1987). The higher end of the scale are clicks of more than 300 kHz produced by odontocetes such as whitebeaked dolphins *Lagenorhynchus albirostris* (Mitson and Morris 1988, see also Rasmussen and Miller 2002) which are used for navigation (echolocation). Consequently, the hearing of most marine mammals investigated to date spans a very wide bandwidth (see Chap. 10). Southall et al. (2007) divided marine mammals into four functional hearing groups. The three families of pinnipeds were placed in one category with a designated hearing range of 75 Hz–75 kHz. Cetaceans were placed in three functional groups (1) low-frequency cetaceans, e.g. fin whale (7 Hz–22 kHz); (2) mid-frequency cetaceans, e.g. bottlenose dolphin (150 Hz–160 kHz); (3) high-frequency cetaceans, e.g. harbour porpoise *Phocoena phocoena* 200 Hz–180 kHz. This designation of species into functional groups is preliminary as hearing studies with published audiograms are available for ~20 of the 128 species and subspecies of marine mammals. For the species in which hearing has yet to be measured (and this includes all species of baleen whales), hearing range has been derived from the acoustic properties of the emitted signals and anatomical features (see Ketten 1997).

Fish show a more restricted bandwidth of emitted sounds than marine mammals. Most fish signals are well below 1 kHz, albeit with exceptions (Zelick et al. 1999; Popper et al. 2003; Ladich 2008, see Chap. 4). Hearing ability is diverse and dependent on anatomical features. Taxa with no swim bladder, for example sharks and flatfish, are only sensitive to particle motion. Species such as cod *Gadus morrhua* have swim bladders but no apparent connection between swim bladder and ear. Such species are sensitive to particle motion and pressure. Species such as herring *Clupea harengus* have tight connections between pressure receptors and inner ear and exhibit high sensitivity and a wide bandwidth extending to frequencies well above 1 kHz (see Popper and Fay 2011). Hearing has been investigated in fewer than 200 of the 30,000 species of fish, so our knowledge of the fish hearing spectrum (see review by Popper and Hastings 2009a) is more limited than of cetaceans (see previous paragraph).

In addition to fish and marine mammals, invertebrates such as decapod crustaceans have been described as being sensitive to sound, i.e. the particle motion component (Popper et al. 2001) and the shore crab *Carcinus maenas* responds physiologically to playback of ship noise (Wale et al. 2013). Cephalopods are sensitive to frequencies below 20 Hz (Packard et al. 1990; Mooney et al. 2012). Sea turtles have shown hearing capabilities in the lower frequency band (Bartol et al. 1999; Dow Piniak et al. 2012; Lavender et al. 2012). If and to what extent underwater sound is used by marine birds and how sensitive they are to sound is unknown (Dooling and Therrien 2012) although attempts are underway to document underwater hearing in some species (Johansen et al. 2013).

14.2.3 Use of Sound

Animals use sound for a range of activities including detecting predators and prey, communication, navigation and foraging. Echolocation is well characterised in marine mammals (e.g. Au 1993) and bats (e.g. Jones and Teeling 2006). The use of sound for navigation and orientation is less well characterised in other groups, although it is possible that fish use the surrounding acoustic environment (acoustic scene information) for orientation (Fay and Popper 2000; Montgomery et al. 2006) and infrasound may provide navigation cues for some birds (e.g. Bingman and Cheng 2005). As a general rule, however, on land sound is primarily a tool for communication, while in marine environments it serves a broader range of functions.

14.2.4 Habitat Biases

Terrestrial habitats differ from underwater habitats in the visibility of effects. One consequence of this difference is that more is known about the immediate effects of noise on land animals than those living underwater, because it is easier to observe (and conduct) experiments with most terrestrial animals, including assessing their hearing ability. A second consequence is that the habitat destruction associated with noise production is more visible on land than underwater and this has contributed to a difference in the perceived relative importance of noise and habitat destruction in the two habitats. On land it is considered that the conservation consequences of habitat destruction around noise sources (e.g. the cleared area around a gas well) are more important than the effects of noise *per se* (e.g. interference with communication). Underwater, the reverse is often the case as the difficulty of observing habitat destruction associated with noise production may result in such potentially significant conservation effects being overlooked and attention being concentrated on noise alone. Anthropocentric biases are different; terrestrial noise is readily appreciated to affect humans, whereas marine noise is viewed mainly in terms of its effects on animals. This effect is enhanced if the marine species have iconographic status (e.g. humpback *Megaptera novaeangliae*, blue *Balaenoptera musculus* and killer whales *Orcinus orca*), with the result that the well-being of such species is widely considered.

14.2.5 Intentional v. Incidental Noise production

Most anthropogenic noise is a by-product of activities such as travel (road, shipping and aircraft noise), construction (e.g. pile driving), extraction (e.g. blasting), industrial activity and wind farms (Blickley and Patricelli 2010). Noise resulting from sound that is intentionally introduced is much more common in the marine environment through sonar and geophysical surveys (e.g. airguns), with terrestrial examples limited to alarm and warning sounds. Clearly, the scope for mitigation is greater when anthropogenic noise is an incidental by-product than when it is vital for the outcome of the activity.

14.2.6 Summary

It will be clear from the rest of this chapter that anthropogenic noise has received far less attention in relation to its impacts on, and conservation implications for, terrestrial animals than marine animals. This is likely a combination of a failure to consider the impact of noise on terrestrial animals due to anthropocentric bias and the presumed greater effects of visible habitat destruction. This bias is also despite the relative ease of observation and measurement of impacts on land. However, as much terrestrial noise is an incidental by-product of our activities (cf. for example the essential role of sound in marine seismic surveys) there may be more scope for mitigation on land.

14.3 Sources of Noise

At first consideration, anthropogenic noise would seem to differ in several characteristics from natural sources of noise, such as wind, other species, waterfalls, waves and thermal energy (marine environment reviewed by Hildebrand 2009; Ainslie 2010). The first difference is that anthropogenic sounds are often more intense than natural noises (exceptions include large waterfalls, storms, undersea earthquakes, sea floor volcanic eruptions and sperm whale *Physeter macrocephalus* echolocation clicks, all of which are relatively localised in space and time.) A second difference is that most anthropogenic sounds contain more low frequencies than natural noises (exceptions are high-frequency sounds produced by some machinery and the hiss of tyres on road surfaces). A third difference is the relative commonness of high-intensity impulse sounds produced by anthropogenic sources (naturally occurring exceptions are lightning strikes and echolocation clicks of most odontocetes). Intense impulse sounds such as airgun firing, blasting charges, pile strikes and sonar pings are more likely to have acute impacts including temporary or permanent injury to auditory systems. By contrast, continuous noise including road, ship and aircraft traffic noise, drilling, construction, industrial activities, low- and mid-frequency sonar systems (see Southall et al. 2007; Tasker et al. 2010) and acoustic harassment/deterrent devices are more likely to produce chronic effects such as masking and stress.

Some of the key acoustic characteristics of marine anthropogenic noise are summarised in Table 14.1. The source levels of sounds can provide a first impression of their potential impacts; however, inferring impact from source levels is complicated by two things. First, source levels are usually determined by measuring sound levels in the acoustic far-field and extrapolating back to determine the level at 1 m from the source (see Ainslie 2010). In many cases a simple Xlog (R/1 m) scaling is used and not an actual propagation loss correction. The resulting source level is therefore not independent of the environment in which the measurements were taken and it is difficult to compare results obtained in different studies. Second, effects on living animals are dependent on many other acoustic characteristics in addition to the sound level at the receiver (for a discussion of these in the marine environment, see Southall et al. 2007). Finally, as there is at least one biological source of naturally occurring high-intensity impulse sounds (odontocete echolocation clicks, see previous paragraph), it is possible that marine animals may be adapted to deal with high-intensity impulse sounds.

Anthropogenic sources of noise are increasing in their distribution and abundance. In the US, for example, road traffic nearly tripled between 1970 and 2007 and aircraft traffic, by some measures, more than tripled between 1980 and 2007 (Barber et al. 2009a). Unfortunately, this increase significantly offsets the reduction in intensity of many sound sources (e.g. sound levels from US aircraft engines dropped 20 dB(A) in the past three decades, Bronzaft and Hagler 2010) that resulted from a growing awareness of noise pollution and consequent regulations (discussed below). In the seas, ambient noise levels have increased in several regions over the past decades due to increased ship traffic (e.g. Ross 1993; Andrew et al. 2011).

14.4 Assessing Noise Impacts

Anthropogenic noise can have many different impacts on individual fitness that can translate into conservation consequences, such as permanent or temporary threshold shifts, flight reactions and disruption of activities such as foraging and

Table 14.1 Acoustic characteristics of some marine anthropogenic sounds (Adapted from Hildebrand 2009; OSPAR 2009a; CEDA 2011; Thomsen et al. 2011)

Sound source	Source level (dB re 1μPa-1 m)	Bandwidth (Hz)	Major amplitude (Hz)	Duration (ms)	Source direction
TNT (1–100 lbs)	272–287$_{Peak}$	2–1000	6–21	0.001–0.01	Omni
Pile driving (4–4.7 m ø pile)	243–257$_{Peak-to-Peak}$	20–100,000	100–500	0.05	Omni
Airgun array	260–262 $_{Peak-to-Peak}$	50–100,000	10–120	0.03–0.06	Vertical
Echo sounders	230–245	10,500–100,000	Various	0.00001–0.002	Vertical
Military sonar mid-frequency	223–235$_{Peak}$	2,600–8,200	Various	0.5–2	Horizontal
Military sonar low- frequency	214–240$_{Peak}$	100–500	Various	6–100	Horizontal
Large vessels	180–190	6 to > 30,000	<200	Continuous	Omni
Trailer Suction Hopper Dredger	168–186	30 to > 20,000	100–500	Continuous	Omni
Small boats and ships	160–180	20 to > 10,000	<1000	Continuous	Omni
Drilling (Drillship and drilling operation)	145–190	10–10,000	<100	Continuous	Omni
Wind turbine in operation	142–151	16–20,000	30–200	Continuous	Omni
Acoustic deterrent/harassment devices	132–200$_{Peak}$	1,800–30,000	Various	0.015–0.6	Omni

migration. We detail three approaches that have been formalised to assess effects of noise on animal populations. These have mainly been applied in the marine environment, but we discuss their actual and potential application to terrestrial environments.

14.4.1 Zone of Influence Model

This approach to assessing noise impacts is based, at least partly, on the distance between the source and the receiver; the rationale is that sound intensity falls with increasing distance from the source and therefore impacts are likely to lessen, or at least to change, with distance. Richardson et al. (1995) defined a nested series of zones of influence centred on the source (Fig. 14.1):

- The zone of audibility is the most extensive and is defined by the receiver's ability to detect noise.
- The zone of responsiveness is the area within which the receiver reacts behaviourally or physiologically to the sound. (For examples of behavioural disruption in a terrestrial environment see Kaseloo and Tyson 2004).
- The zone of masking is the area where noise interferes with the detection of biologically relevant signals such as echolocation clicks or social signals. It is highly variable.
- The zone closest to the source is where the received sound level is high enough to cause hearing loss, discomfort or injury. In air, continuous noise >110 dB(A) causes permanent threshold shifts in birds, noise >93 dB(A) causes temporary threshold shifts (Dooling and Popper 2007). The physiological effects of noise

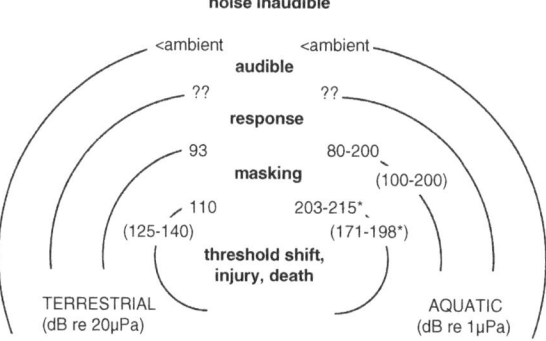

Fig. 14.1 An illustration of the zones of influence model after Richardson et al. (1995). Bold text shows names of zones. The source is at the centre of the concentric circles. Indicative threshold values in dB (*e5 indicates re 1μPa^2·s in water) for the boundaries between zones are taken from Dooling and Popper (2007) (terrestrial) and Southall et al. (2007) (aquatic, for pinnipeds and cetaceans). Values in brackets indicate thresholds for impulsive sounds

exposure on marine mammals and fish are reviewed by Southall et al. (2007) and Popper and Hastings (2009a, b) respectively.

The zones of influence model has been applied in various marine impact studies (e.g. Erbe and Farmer 2000; Madsen et al. 2006a; Thomsen et al. 2006) and formalised for terrestrial habitats (with birds in particular) by Dooling and Popper (2007). However, we have to bear in mind that the relationship between the type of effect elicited and distance to the sound source is not straightforward. One reason is that the complexity of sound transmission (particularly underwater, but also in complex built environments such as cities) inevitably leads to sound fields that are more complicated than the concentric circles of the Richardson et al. (1995) model. A second reason is that while distance between source and receiver might adequately relate to some of the properties of a sound wave (e.g. received sound pressure level and duration), other sound characteristics do not. For example, kurtosis ('peakedness'; see Southall et al. 2007), rise time and overall pattern of occurrence can also define sound effects and these features do not relate simply to distance to source. Furthermore, studies have shown that physiological effects are related to the dose of exposure, which involves the duration of the exposure (see Southall et al. 2007; Kastelein et al. 2012). This means that physiological effects can potentially occur at sound pressure levels that do not cause a behavioural response when the animals are exposed for a long period. Thus, the influence zone for physiological effects can be larger than the zone of responsiveness (see also WODA 2013). Finally, although zones of noise influence are a very useful starting point in classifying impacts, they can mislead. For example, behavioural reactions might lead to severe consequences such as stranding (see Cox et al. 2006) so that a zone where initial responsiveness occurs might well become the zone of injury or even death.

14.4.2 Population Consequence of Acoustic Disturbance Model

A second approach addressing how acoustic disturbance could lead to population level consequences of relevance to conservation is the Population Consequence of Acoustic Disturbance model (PCAD model, Fig. 14.2) developed for marine mammals (NRC 2005). The model involves several steps, from a characterisation of the sound source to population effects, but most of the transfer functions are not well understood. For example, acoustic disturbance can lead to disruption in feeding behaviour in cetaceans such as killer whales, but the effects on variables such as survival, maturation and reproduction are largely unknown (see Williams et al. 2002, 2006 for estimated costs of behavioural reactions).

Similarly, in terrestrial environments the causal link between the immediate effects of noise on signalling and population level effects is indirect, compared to the more extreme effects of noise. Deafness or repeated interruption of foraging is

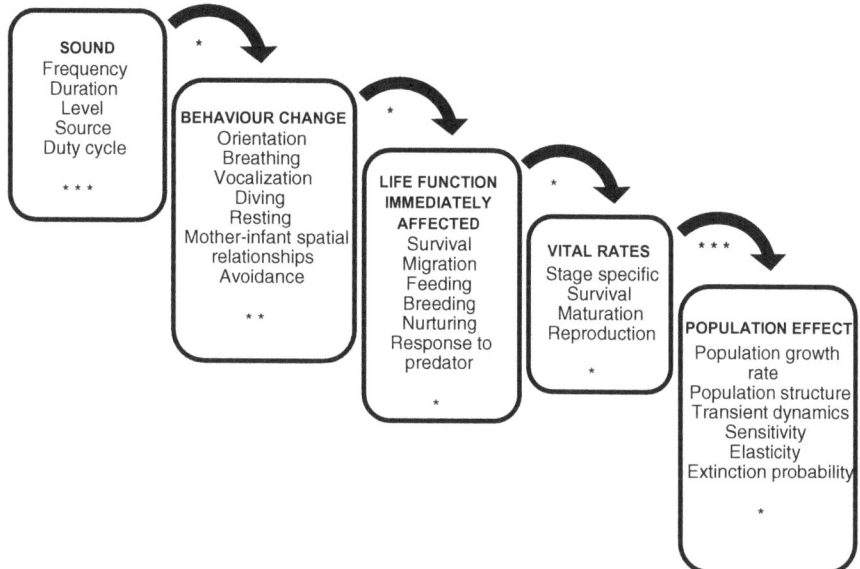

Fig. 14.2 Overview of the PCAD Model (NRC 2005). The number of * within the boxes indicate how well the features of the model can be measured. The number of * under the transfer arrows indicate how well the transfer functions are known

clearly detrimental, and likely to reduce survival or reproductive success. However, it is less clear that slight changes in song structure or difficulties in signal reception associated with effects of noise on communication will cause appreciable harm, once all the other factors that may affect an animal's fitness are factored in. Demonstrating effects of noise can be hampered by poorly documented study design, both with regards to a proper characterisation of the source signal and the adequate sampling of behaviour (reviewed by Nowacek et al. 2007 and OSPAR 2009a). Even if the causal links between an effect of noise on communication and a decrement in survival and reproductive success can be demonstrated, this stress is only one of a number that can affect population viability, all of which must be weighed to evaluate whether the effect of noise specifically on animal communication should be a conservation concern. Unless these effects can be shown to be as detrimental as the more obviously extreme effects of noise (e.g. flight responses) the effects of noise on communication are less likely to have priority in conservation efforts, particularly given that the most immediate threats to population viability are habitat destruction and fragmentation. In the marine environment there are similar difficulties in weighing effects of noise relative to more immediate pressures such as fishing (including bycatch effects) and physiological reactions to contaminant loads (see Thomsen et al. 2011).

In both terrestrial and underwater environments it is likely that a combination of impacts will produce population level responses, yet methods for assessing cumulative impacts are still in their infancy (e.g. Wright 2009). A further factor to bear in mind is that population census estimates can be highly variable, making it very difficult to detect effects even in areas of high anthropogenic impacts of several sorts (see Thomsen et al. 2011). Variability in difficult to census species (e.g. many cetacean stock assessments) is understandable and compounded by the resolution with which such census estimates can be made. The usual result is an inability to detect change even when a considerable percentage of the population has been lost.

14.4.3 Risk Assessment Framework

The third approach is a risk assessment framework, which Boyd et al. (2008) have suggested would result in a more systematic approach to noise impact studies. The risk assessment framework involves a stepwise procedure including:

1. hazard identification (characterisation of the potential threats of a source);
2. dose-response assessment (assessment of the quantitative relation between received sound and the effect);
3. exposure assessment (specifying the number of individuals that might be exposed to the hazard);
4. overall characterisation of the risk, leading to risk management with appropriate mitigation measures (details in Boyd et al. 2008).

It looks as though step 1 might be relatively straightforward, although measurements of sound sources need to be standardised much more and some of the more complex issues related to source characteristics (e.g. vertical differences in emitted sound levels) and transmission of sound (e.g. water column vs. sediment transport) need to be explored more thoroughly (see recommendations in IACMST 2006; OSPAR 2009b; Southall et al. 2009; TNO 2011).

The dose–response assessment and exposure assessment (steps 2 and 3) will be more difficult to apply in the marine environment as the distribution of receivers is highly variable and areas of high importance are therefore quite difficult to identify (e.g. Coull et al. 1998; Hammond 2006). The most challenging step might be to assess the relationship between dose (e.g. properties of the received sound) and response, as results from studies investigating the effects of sound on marine mammals, fish and other marine life are, to date, highly equivocal.

14.4.4 Summary

Assessing conservation impacts of noise is complicated by the need to translate noise impacts on individual (or small group) communication behaviour into effects on individual fitness that will have population level consequences of interest to conservation. We are some way from a robust impact assessment of noise that can be applied to contexts of potential conservation concern; however, the approaches discussed above are a step forward.

14.5 Noise Impacts: Potential and Documented

Most reviews of the impact of noise list a variety of effects. For example, noise that causes death or injury (e.g. death in herring due to pile driving noise, Caltrans 2001) is clearly of potential conservation concern. However, overviews of impacts of noise (e.g. Table 14.2, marine impacts and effects for fish and marine mammals) can be difficult to interpret. Two important caveats apply to many studies of noise impacts (illustrated here with marine examples, but which apply equally strongly to terrestrial examples). The first caveat is that results can be equivocal, with both documented presence and absence of effects. For example, Nowacek et al. (2007) and Popper and Hastings (2009b) provide examples of well controlled studies which elicited no apparent behavioural or physiological response, even though some studies involved very high received sound levels. The second caveat is that some studies reporting effects have methodological problems which make it difficult to assess their validity (see Popper and Hastings 2009b for examples of fish injured by pulsed sounds of pile driving). This caveat also applies to some behavioural studies where responses were not documented properly and/or received sound levels were unknown (for a discussion see OSPAR 2009a). This emphasises that research on noise-related impacts is still in its infancy even though attempts to standardise methodologies have been undertaken (e.g. Tyack et al. 2004; ANSI/ASA 2009; TNO 2011).

In this section, we review the empirical evidence for the effects of noise on communication and, ultimately, population viability (rather than direct effects with severe conservation implications such as death and injury). As noted above, establishing the link between anthropogenic noise, communication and conservation is difficult, because the effects are indirect and because many other factors are involved (e.g. Kight and Swaddle 2011). Currently studies have shown two types of effect. First, studies have shown an apparent effect of noise on communication, but the link between the demonstrated proximate cost and an ultimate cost in survival or reproductive success is inferred rather than demonstrated. Second, studies have shown a decrease in population density or diversity in relation to noise, but the relationship is usually a correlation, so factors other than noise or its effect on communication might account for the relationship.

Table 14.2 Overview of documented effects of underwater noise on aquatic life and example studies (adapted from OSPAR 2009a; see also Richardson et al. 1995; Würsig and Richardson 2002; Popper and Hastings 2009b)

Impact	Documented effect	Example study
Physiological—non-auditory	Mortality and physical injury	Mortality in herring (*Clupea harengus*) and injury in herring and five other species—pile driving—Caltrans (2001)
Physiological—auditory	Permanent damage to hair cells	Pink snapper (*Pagrus auratus*)—seismic airgun—McCauley et al. (2003)
	Temporary threshold shift	Northern pike (*Esox lucius*) and lake chub (*Couesius plumbeus*)—seismic airgun—Popper et al. (2005)
		Harbour porpoise (*Phocoena phocoena*)—pile driving—Lucke et al. (2009)
Perceptual	Masking	White whale (*Delphinapterus leucas*)—ice breaker—Erbe and Farmer (2000)
		Goldfish (*Carassius auratus*), lined Raphael catfish (*Platydoras costatus*), pumpkinseed sunfish (*Lepomis gibbosus*)—white noise—Wysocki and Ladich (2005)
		Drum (*Sciaena umbra*), damselfish (*Chromis chromis*), Lusitanian toadfish *Halobatrachus didactylus*—boat/ship noise—Codarin et al. 2009, Vasconcelos et al. 2007
Behavioural	Stranding and beaching	Cuvier's beaked whales (*Ziphius cavirostris*) and other species—mid-frequency active sonar—Cox et al. (2006)
	Interruption of behavioural pattern	Sole (*Sola solea*) and cod (*Gadus morrhua*)—pile driving—Mueller-Blenkle et al. (2010)
		Humpback whales (*Megaptera novaengliae*)—seismic surveys—McCauley et al. (2000)
	Adaptive shifting of vocalisation intensity and/or frequency	Humpback whales-sonar—Miller et al. (2000)
	Displacement from area (short or long term)	Cod and haddock (*Melanogrammus aeglefinus*)—seismic surveys—Engås et al. (1996)
		Killer whales (*Orcinus orca*) acoustic harassment devices—Morton and Symonds (2002)
	Attention shifts and reduction in foraging performance	Threespined stickleback (*Gasterosteus aculeatus*)—white noise—Purser and Radford 2011

14.5.1 Evidence of Proximate Costs, with Population Level Effects Inferred

14.5.1.1 Costs of Threshold Shift

Threshold shifts (i.e. reduced sensitivity to sounds) resulting from exposure to noise are a likely cost of communicating in noise. Permanent damage to hearing (permanent threshold shifts, PTS, see Chaps. 8 and 10) often through damage to hair cells (e.g. pink snapper *Pagrus auratus* from seismic airgun sounds, McCauley et al. 2003) is considered a form of injury; however, as part of its effect will be through communication, we include it here. Whereas PTS is considered an auditory injury, temporary threshold shift (TTS) represents auditory fatigue (with effects on hair cells, variation in middle ear muscular activity and blood flow that are recoverable; Southall et al. 2007, see Chaps. 4, 8 and 10). Nevertheless, part of the effect of TTS will be through communication. TTS has been documented in a variety of fish and marine mammals (overview in Southall et al. 2007 and Popper and Hastings 2009a, b, Ladich this volume; examples in Table 14.2; for detailed discussion see Chaps. 4 and 10). Both threshold effects are likely to have similar costs to masking discussed in the Sect. 14.5.1.2. It is worth pointing out that in marine mammals TTS can occur at frequencies that are very different from the main frequency of the received sound (see for example Lucke et al. 2009) so conclusions on effects based solely on the frequency spectrum of the emitted sound are problematic. We have presented aquatic examples in this section because most examples on threshold effects come from the marine environment. Also, they are arguably of more concern because of better transmission of sound in water compared to air, and because in terrestrial environments human health concerns might be expected to keep levels below TTS effects in mammals.

14.5.1.2 Costs of Masking

The most obvious purported cost of communicating in noise is a decrease in survival or reproductive success because of signal masking. As this volume shows (Chaps. 3–10), both signallers and receivers have a range of adaptations to reduce these costs, but these strategies are presumably adaptations to the conditions that prevail in nature, and thus might well fail in the face of anthropogenic noise (but see Cunnington and Fahrig 2013).

In terrestrial animals, signals that are imperfectly detected or discriminated might result in poor predator detection (e.g. Francis et al. 2009), lower mating success (e.g. Bee and Swanson 2007; Samarra et al. 2009; Gross et al. 2010; Gordon and Uetz 2012), smaller territories (e.g. Parris et al. 2009), poorer flock cohesion (e.g. Lohr et al. 2003) or reduced parental care (e.g. Leonard and Horn 2005, 2012; Schroeder et al. 2012; but see Leonard and Horn 2008; Naguib et al. 2013).

In the marine environment, cetaceans are known to use vocalisations to coordinate movements (Ford 1989; Janik and Slater 1998; Miller 2006), in reproductive behaviour (e.g. Payne and Webb 1971; Oleson et al. 2007), and to maintain contact between group members (e.g. Ford et al. 1989) and mothers and their calves (e.g. Smolker et al. 1993). Cetaceans also use passive sonar when hunting, detecting acoustical cues from potential prey (e.g. Barrett-Lennard et al. 1996). Noise could adversely affect all of these uses of sound by masking. However, the role of masking will remain speculative until there is evidence that marine mammals communicate or orientate over the large distances indicated by the enormous active space of their signals. In fish, close range signals such as reproductive calls of some species (e.g. cod Brawn 1961; Hawkins and Rasmussen 1978; Lusitanian toadfish *Halobatrachus didactylus* Vasconcelos et al. 2007; damselfish *Chromis chromis* and drums *Sciaena umbra* Codarin et al. 2009) can be masked by continuous sound and at least potentially disrupt mating and spawning.

In both environments animals have been shown to gather information by eavesdropping on signals of more distant individuals (McGregor 2005), therefore noise has the potential to adversely affect such interactions by reducing the extent of the communication network (see Janik 2005 for marine mammals).

14.5.1.3 Production Costs of Attempts to Overcome Masking

For signallers, another possible cost of communicating in noise is the cost of changing the signal so that it is less likely to be masked. Such changes include increases in intensity, rate, duration or frequency, all of which might increase the usual costs of signal production, such as energy expenditure or predator attraction (e.g. Gil and Gahr 2002; Parris et al. 2009) as well as increased social aggression (Brumm and Ritschard 2011). It should be remembered, however, that evidence for such signalling costs in terrestrial environments is still scant and controversial for most signalling systems (Searcy and Nowicki 2005; Zollinger et al. 2011). In the marine environment there is evidence of changes in vocalisations in the presence of noise (change in intensity: beluga *Delphinapteras leucus* Scheifele et al. 2005; killer whale: Holt et al. 2009; and/or change in frequency: right whale *Eubalaena glacialis* Parks et al. 2011). In the presence of sonar other species change signal duration (humpback whale Miller et al. 2000) and/or frequency (humpback whale Miller et al. 2000). However, the costs for the individuals (Bejder et al. 2009) and their fitness consequences are currently the subjects of discussion.

14.5.1.4 Increased Stress and Impaired Decision Making

Receivers, too, might bear costs in trying to detect and discriminate signals in noise. Given the many options they have to better perceive masked signals (e.g. Dooling and Popper 2007), the costs they might incur are varied. One cost that is likely to be universal, however, is that the extra effort required to perceive masked signals might cause receivers to miss critical stimuli, such as alarm calls or acoustic cues from predators (Quinn et al. 2006; Rabin et al. 2006). Similarly, the extra cognitive effort needed to process masked signals might impair decision-making more generally, resulting in poor behavioural choices and perhaps physiological stress (Bateson 2007; Kight and Swaddle 2011; Owens et al. 2012; but see Zheng 2012; Crino et al. 2013). Similarly stranding or beaching events that have occurred in species such as Cuvier's beaked whales *Ziphius cavirostris* and other species in response to mid-frequency active sonar (Cox et al. 2006) are most likely related to stress and impaired decision making. Interruption of normal behaviour patterns due to playback of pile driving noise to sole *Sola solea* and cod (Mueller-Blenkle et al. 2010) and in humpback whales due to seismic surveys (McCauley et al. 2000) could induce stress.

14.5.1.5 Changes in the Location or Timing of Signalling

Both senders and receivers may incur costs when they change the location or timing of communication to avoid interference from noise. If a bird has to sing from higher perches to overcome traffic noise, for example, it may increase its exposure to predators (e.g. Díaz et al. 2011; Halfwerk et al. 2012; see also McLaughlin and Kunc 2013). Similarly, if birds have to shift the timing of vocal behaviour to avoid noisy periods (Bergen and Abs 1997; Fuller et al. 2007; Kaiser et al. 2011; Arroyo-Solís et al. 2013; see Chap. 7) or have to sing more to compensate for masking (Díaz et al. 2011), the change in their overall time budget will almost certainly entail trade-offs with other behaviours (e.g. Conomy et al. 1998; Díaz et al. 2011). Displacement from locations for short and long durations has been found for fish in response to seismic surveys (cod and haddock *Melanogrammus aeglefinus* Engås et al. 1996) and for cetaceans in response to acoustic harassment devices (e.g. killer whales, Morton and Symonds 2002), marine construction activities (gray whales *Eschrichtius robustus* Bryant et al. 1984) and pile driving (harbour porpoises, Brandt et al. 2011).

14.5.1.6 Evolutionary Changes

Noise-induced changes in signal structure, especially long-term learned or evolved changes, as seen in nestling tree swallows *Tachycineta bicolor* (Leonard and Horn 2008) and, possibly, adult European blackbirds *Turdus merula* (Slabbekoorn and Ripmeester 2007, but see Mendes et al. 2011), may have evolutionary consequences, for example shifting the preference of females for particular songs or the ability of males to sing preferred songs, thus potentially reducing gene flow between urban and rural populations, for example (Slabbekoorn and Peet 2003; Montague et al. 2012; see also Luther and Derryberry 2012). Growing evidence suggests learned changes in songs may lead to genetic differentiation, presumably via mate choice (e.g. Leader et al. 2005, see more extensive discussion therein and in Slabbekoorn and Ripmeester 2007; Slabbekoorn 2013). Nonetheless, the possibility of anthropogenic noise having evolutionary consequences still remains speculative (for a detailed discussion of urban song divergence in birds, see Chap. 7).

In summary, several effects of noise on animal signalling appear to be costly, but none have been directly linked to survival, reproductive success or any other more direct measure of population viability, despite the theoretical possibilities.

14.5.2 Evidence of Population Level Effects, with Proximate Cause Inferred

An alternative approach to establishing an immediate effect of noise on communication, and then inferring its possible effects on population viability, is to correlate potential masking noise with measures of population viability, while attempting to rule out other habitat effects. For instance, some studies have shown that reductions in abundance (van der Zande et al. 1980; Eigenbrod et al. 2009; Kaiser et al. 2011) or diversity (Stone 2000; Francis et al. 2009) extend farther from noise sources than one would expect if the reductions were caused by habitat degradation or disturbance. In some particularly convincing studies of the effects of road noise, declines in density or diversity correlated with traffic load, but at distances beyond where direct mortality and disturbance could be an issue, suggesting that signal interference is the most likely cause (Reijnen et al. 1995, 1997; Forman and Deblinger 2002; Peris and Pescador 2004; Jaeger et al. 2005; Eigenbrod et al. 2009; but see Fahrig et al. 1995; Benítez-López et al. 2010). Most convincing of all are studies in which the breeding density (Bayne et al. 2008; Francis et al. 2011; Blickley et al. 2012), diversity (Francis et al. 2009) or reproductive success (Habib et al. 2007; Schroeder et al. 2012) of birds is lower near noisy compared to silent machinery, or near playback of traffic or machinery noise compared to silent controls (see also Barrass 1985, cited in Eigenbrod et al. 2009). One study further shows how such effects of noise on birds can have far-reaching effects on how ecosystems function; Francis et al. (2012) found that key pollinating and seed dispersing bird species change where they forage in response to noise.

Whether the above patterns are caused by the effect of noise on signalling *per se*, as opposed to some other behavioural effect, is unclear. More direct evidence implicating signal masking comes when the declines are stronger for species that have lower frequency vocalisations and thus are more likely to be masked by anthropogenic noise. Several studies have, indeed, shown this pattern for declines in abundance or diversity with traffic noise (Rheindt 2003; Francis et al. 2009, 2010; Parris and Schneider 2009; Goodwin and Shriver 2010; Herrera-Montes and Aide 2011; Proppe et al. 2013; see also Hu and Cardoso 2009; Hoskin and Goosem 2010), or have shown that declines in reproductive success are best explained by the noise levels in the frequency band that would mask songs, rather than in other frequencies (Halfwerk et al. 2011).

14.6 Anthropogenic Noise and Environmental Management

There is a stark contrast between marine and terrestrial environments in relation to anthropogenic noise and conservation impacts—noise in the marine environment is now a major issue for both the public and regulators, whereas in terrestrial environments noise receives much less attention in general, and particularly so with regard to conservation. For example, when a new road is proposed, habitat destruction, particularly wetland crossings, is tightly regulated, but, as a rule, the impact of noise for non-human animals associated with the presence of the road is only addressed in special cases, such as when species at risk are known to be highly sensitive to disturbance. Noise concerns as they relate to humans, on the other hand, are often considered to be fairly tightly regulated (but see abstract for emerging human health concerns). Such regulations only contribute to animal conservation in areas where humans are also likely to be affected.

In this section we discuss management of marine noise and then terrestrial noise, in part because management approaches have different histories and patterns of application that would make an integrated approach unwieldy. We include all anthropogenic noise, whether or not it is likely to have an effect through communication. We shall then consider whether communication effects need separate additional management and legislative treatment.

14.6.1 Management of Marine Noise

14.6.1.1 Background and Assessments

Since concerns about the potential impacts of underwater noise on marine life were raised in the early 1970s (e.g. Payne and Webb 1971), the issue has been debated (with some accompanying controversy) between scientists, the public, industry and other stakeholders including non-governmental organisations. This is especially during the past decade (e.g. OGP/IAGC 2007; Weilgart 2007; Parson et al. 2008). A milestone in scientifically driven debate was the formation of national fora such as the UK Working Group on Underwater Sound (now Underwater Sound Forum, see Defra 2010, working group report see IACMST 2006) and the US Joint Subcommittee on Ocean Science and Technology (JSOST, Southall et al. 2009). Also worth mentioning are information and guidance papers compiled by large industry platforms such as CEDA (2011) and WODA 2013. Internationally, the Intersessional

Correspondence Group on Underwater Sound within OSPAR (Convention for the Protection of the Marine Environment of the North East Atlantic former Oslo-Paris Commission) was formed. This group has published two reports, one providing a background on impacts of man-made sound in the environment (OSPAR 2009a), the other assessing current (as of 2009) pressures due to underwater noise in the North East Atlantic (OSPAR 2009b). Both reports have been accepted by all 14 OSPAR member states and therefore carry some weight in informing policy. In this context, the OSPAR quality status report (QSR) 2010 (OSPAR 2010) is of particular relevance. The QSR provides a holistic assessment of the status of the North East Atlantic including for the first time underwater noise impacts. Although OSPAR (2010) notes the scarcity of information on noise-related effects it points out that OSPAR regions II (Greater North Sea) and III (Celtic Seas) seem to be most affected by noise and also calls for developing guidance on options for mitigation of noise and its effects. The importance of these sometimes time consuming and complex assessments should not be underestimated; they significantly inform policy makers on current status and, even more importantly, future research and management needs.

14.6.1.2 Legal Instruments

There are existing regulatory frameworks such as the US Marine Mammal Protection Act (1972) and the EU Habitats Directive (1992) which protect a variety of marine mammals and fish species that are sensitive to sound. Similarly, the EU-EIA Directive requires member states to perform an Environmental Impact Assessment (EIA) for projects likely to have significant impacts on the environment. EIAs can involve methods for assessing impacts of sound and can lead to mitigation measures such as 'soft-start' procedures during pile driving for offshore wind farms (e.g. Cefas 2004; JNCC 2009b). EIAs are also undertaken in many other parts of the world (for further information, see http://www.iaia.org).

The legal instruments mentioned above all go some way towards managing noisy activities. Yet until recently there was no regulatory framework specifically addressing underwater noise. This changed in Europe with the publication in June 2008 of the EU Marine Strategy Framework Directive (MSFD). The purpose of the MSFD is 'establishing a framework for community action in the field of marine environmental policy'. The MSFD aim is to protect, conserve, and where possible, restore the marine environment in order to maintain biodiversity and provide diverse and dynamic oceans and seas which are clean, healthy and productive. The Directive requires Member States to achieve 'Good Environmental Status' (GES) in their marine environment by 2020 at the latest. Annex one of the MSFD lists the 11 qualitative descriptors for GES, one of which states that 'the introduction of energy, including underwater noise, is at levels that do not adversely affect the marine environment'. Based on advice from an expert group (see Tasker et al. 2010) the EU has decided on two indicators that further specify GES. Indicator one addresses the distribution in time and place of loud, low- and mid-frequency impulsive sounds. The second indicator deals with continuous low-frequency sound (details in EU 2010). Whereas indicator one will perhaps require only an annual desk based assessment of activities generating low-frequency pulses, such as pile driving and seismic surveys, indicator two will most likely involve measuring ambient noise, perhaps at a regional level which would represent huge progress in identifying trends in existing pressures such as those from shipping (see Tasker et al. 2010; van der Graaf et al. 2012). Details of requirements for such monitoring are currently being investigated by an EU expert group and were planned to emerge in 2013 to keep up to speed with the very ambitious timeline of the Directive. The issue of ship noise has also been picked up by a corresponding group of the International Maritime Organisation (IMO, see IMO 2009)

14.6.1.3 Marine Mitigation Measures

The EU Marine Strategy Framework Directive will most likely lead to management measures that have to be undertaken in order to achieve Good Environmental Status. In this section, we shall discuss some of the existing and emerging measures to mitigate effects of underwater noise. However, before we do this we shall make a more general remark. It should have been clear from the previous section that our knowledge on sound-related effects has made huge progress in recent decades; yet, the overall picture remains incomplete, especially looking at the population level consequences of noise exposure. This calls for more controlled and replicable impact studies, especially looking at behavioural disturbance due to the potentially large impact ranges. Such studies are important because the information provided can be used to assess the cost to society of resulting mitigation measures. This is even more important as some of the structures now being considered by regulators, such as offshore wind farms, result from efforts to reduce other adverse impacts on the environment.

Geographical and seasonal restrictions Noise impacts can be mitigated effectively through geographical and seasonal restrictions on sound production, thereby protecting times and locations critical to mating, breeding, feeding or migration. An example is the moratorium that the Spanish Ministry of Defence has maintained since 2004 on the use of sonar within 50 nautical miles of the Canary Islands, following stranding events involving beaked whales *Ziphiidae* (see also Weilgart 2007; OSPAR 2009a). Protection zones could be designated under the EU Habitat Directive (Natura 2000 sites). Yet we have to remember that marine species are highly mobile and that distributional shifts (see for example harbour porpoises in the North Sea between 1994 and 2005; Hammond 2006) might lead to incongruity between protected areas and their originally postulated conservation objective. It is also likely that noise produced in the vicinity of a protected area can impact receivers therein; in Europe this has to be addressed in specialised Appropriate Assessments.

Another form of spatial restriction is the application of safety zones to avoid ensonification of receivers at distances thought to be critical, e.g. causing injury.

For example, the Joint Nature Conservation Committee of the UK (JNCC) advises an exclusion zone of 500 m for seismic survey operations (JNCC 2009a). Specially trained marine mammal observers are required to detect marine mammals within the safety zone and passive acoustic monitoring (PAM) can additionally be used to detect marine mammals at night or during averse sighting conditions (see JNCC 2009a). PAM technology will very likely become more advanced in identifying senders and also in supporting real-time monitoring; however, it can only monitor safety zones if the species of interest produces sound most of the time. This may not be the case (see for example Barrett-Lennard et al. 1996; see Weilgart 2007; Compton et al. 2008 for other issues with regards to soft-start).

Noise exposure criteria Criteria for noise exposure were set by the US National Marine Fisheries Service in 2003 at 180 dB re 1 µPa (rms) for cetaceans and 190 dB re 1µPa for pinnipeds. More recently a US group of experts have suggested modified criteria for three functional hearing groups of cetaceans (low-, mid- and high-frequency; see Sect. 14.2.2) and pinnipeds in air and underwater both for injury (PTS) and behavioural response (using a TTS criterion). Both pulses (single and multiple) and non-pulses were considered (Southall et al. 2007). For fish, Popper et al. (2006) and Carlson et al. (2007) proposed interim criteria for injury and TTS for pile driving for different hearing groups. We have to bear in mind that all of these criteria are provisional as no hearing studies have been undertaken for most species (i.e. they are based on extrapolation from one species in which hearing abilities have been measured). Recently, Lucke et al. (2009) found TTS in a harbour porpoise at much lower received levels than those postulated by Southall et al. (2007) (see also Kastelein et al. 2012).

The exposure criteria are set for received sound pressure levels; however, these will be very difficult to establish as measurements might not be feasible in every case and thus modelling has to address site-specific transmission loss characteristics (see Madsen et al. 2006a for some values in the North Sea and Baltic). To avoid this issue Tasker et al. (2010) proposed criteria based on source rather than received levels. Most of the criteria address a limited set of sound types (e.g. one category 'pulse' or activity 'pile driving') and extrapolating criteria across sound types might not be appropriate due to differences in sound characteristics. We should also bear in mind that for many fish species particle motion rather than pressure is the appropriate stimulus (see Popper and Fay 2011). Multiple exposures to sounds below threshold can lead to injuries or TTS depending on the duty cycle and the overall dose received over time (Kastelein et al. 2012). Cumulative exposure criteria have to be considered that are very different from those for single strikes (Carlson et al. 2007; Southall et al. 2007). It is therefore evident that current suggestions will have to be revised when new data becomes available.

'**Soft-start**' methods (in which strike amplitude is slowly increased over several strikes in order to provide receivers with an opportunity to leave the area before adverse levels are reached) are applied before seismic surveys and pile driving operations. It is not yet known if this method achieves the desired effect (see Madsen et al. 2006b; Miller et al. 2009). Tools which emit an aversive signal so that receivers move out of the potential injury zone (**acoustic management**

devices or '**pingers**') have been used inter alia during offshore wind farm construction activities in Denmark (Tougaard et al. 2006). In general, these devices are quite effective in displacing receivers out of the immediate zone of danger (e.g. Culik et al. 2001). However, this of course raises the issue that one 'evil' (the effects of exposure to pile driving and other sounds) is replaced by another (the effects of exposure to pingers).

Technical mitigation measures Engineering solutions focus on the reduction of sound at the source. Examples are cofferdams, bubble curtains or plastic sleeves around pile drivers (see for example Nehls et al. 2007; Lucke et al. 2011). There are also attempts to develop alternative ship propeller designs, and methods to improve wake flows into ship propellers (Renilson 2009).

14.6.2 Management of Terrestrial Noise

14.6.2.1 Legal Instruments

The protection that exists for non-human animals generally relies on regulations that protect habitat and prevent disturbance, rather than regulations that protect against noise *per se*. These regulations fall under a wide range of legislation and policies, such as those applying to protected areas, critical habitat and environmental assessments. Extending these tools to incorporate the effects of noise requires broader interpretations of habitat and disturbance than are usually applied. Even then, such broader interpretations usually come to the fore only for the most intense effects of noise that result in obvious disturbance, such as escape behaviour or interruptions of foraging bouts (e.g. Pepper et al. 2003). For example, the Migratory Bird Convention Act, one of the most powerful tools for wildlife protection in North America, prohibits the disturbance of birds or their nests. Thus the prohibitions of the act can be applied to regulate noise that startles or disturbs birds, and can result in recommended set-backs based on flushing distances or the point where activities are interrupted. Even these blunt tools for measuring disturbance are generally applied only to species that are considered to be particularly sensitive to noise, such as waders and raptors (e.g. Rodgers and Smith 1995; Bautista et al. 2004; Wright et al. 2010). The Act offers no protection for the subtler, but spatially and temporally more extensive, effects of noise such as those that could affect communication, unless one takes a very broad view of the meaning of "disturbance". Similarly, Canada's Species at Risk Act, like much such legislation around the world, protects the critical habitat of endangered species, i.e. the habitat essential for the species' survival or recovery. Again, in a broad interpretation, preventing disturbance may include establishing buffer areas around nest sites (Pepper et al. 2003), but would not apply to noise at the levels that are high enough to mask signals but too low to interrupt other activities. Even these gains in protection against more extreme noise levels are hard-won, because the sound environment that animals depend on is, somewhat understandably, not considered as integral to an animal's habitat as are the more localised and tangible resources of food and shelter.

Given the lack of regulations and policies that target effects on non-human animals, it is worthwhile considering whether the many measures protecting humans from harmful noise are useful for protecting other animals. Setting aside regulations of extreme levels that can cause hearing impairment, anthropogenic noise is regulated in most countries based on two criteria: whether it interrupts human activities, such as conversation, and whether it is annoying to a certain percentage of the population (OECD 1995). Although annoyance seems to be a subjective way to gauge impacts on humans, it has been adopted because it is a readily obtained metric of objective impacts that underlie it, which include a wide range of cognitive and health impairments (Guski et al. 1999; Ouis 2001; Moudon 2009). Where these regulations apply, such as in residential areas, they might well be adequate for protecting most animals from deleterious effects of signal masking. Regulations protecting people from bothersome noise vary, but generally apply at levels around 60 dB(A) or less. Lab studies show signal masking at these levels (Chaps. 3, 6 and 8), but in the field receivers can readily overcome such masking (e.g. by turning their heads or moving slightly—Dooling and Popper, 2007). This result suggests that higher levels of noise are needed to mask signals. Of course, there is considerable doubt about this conclusion, especially because masking effects should vary considerably across species. Also, the regulations in place for human health and well-being offer no protection in areas where people are absent, which is often where species of conservation concern occur (Blickley and Patricelli 2010). As a remedy to these shortcomings, Blickley and Patricelli (2010) suggest that, because the effects of noise vary among species, a list of standards be developed that is specific to given species or groups of species. While it seems far too early to come up with such a list, given all the unknowns raised in this chapter and throughout this book, some such attempts have been made, mostly for local applications (e.g. Reijnen et al. 1995, 1997; Barber et al. 2011; Patón et al. 2012). As noted above, the measures in place for terrestrial species lag far behind those that are now routine for marine species, where much less is known but the potential harm is so severe that regulators have been more inclined to take a precautionary approach (see below).

14.6.2.2 Terrestrial Mitigation Measures

What can be done to mitigate the effects of noise on terrestrial animal communication? Of course, the most effective mitigation is to simply keep noise sources away from animals, a measure taken frequently for humans but rarely for non-human animals. Simply changing the timing of noise-related activities would be effective for species that signal preferentially at a certain time of day. This is analogous to the geographical and time restrictions discussed for the marine environment.

Other options for reducing noise are now available. For example, better noise reduction for car engines and tyres, quieter asphalt for roads, improved noise barriers and road siting will all contribute to reduce traffic noise (Makarewicz and Kokowski 2007; van Langevelde and Jaarsma 2009). However, most of these measures are directed at human complaints about interference with conversation (<3000 Hz) or disturbance from low-frequency rumbles (<100 Hz) and as such are more likely to fit the frequency range of large terrestrial mammals rather than birds or insects. (An exception might be high-pitched whines that humans also find annoying.) There are several engineering measures that could be employed to reduce noise that could affect animal signals, including traffic speed and flow control, quieter aircraft engines and more effective silencing of construction machinery and plant. We know of no instances in which any such engineering measures have been deployed to address conservation concerns. As Slabbekoorn and Ripmeester (2007) point out, however, small changes to current mitigation methods could have a disproportionately beneficial effect for the frequency range of most bird song (see also Halfwerk et al. 2011). Slight increases in barrier height can block road noise from the tree canopy where birds communicate, and angles or absorbent material on barrier top edges can absorb traffic noise (Slabbekoorn and Ripmeester 2007). One of the most effective noise absorption barriers is vegetation, which also provides additional habitat for the birds (Slabbekoorn and Ripmeester 2007) and an aesthetically pleasing driving experience for people.

14.7 Conclusion

We consider that the potential for anthropogenic noise to adversely impact conservation is amply demonstrated in both marine and terrestrial environments. The key challenges are to establish whether this potential is realised and if so, to assess its effects relative to more commonly considered conservation issues such as habitat loss. A further challenge (in terms of the topic of this book) is to establish which effects of anthropogenic noise have relevance to conservation through impacts on communication.

There are several reasons why establishing the link between anthropogenic noise and conservation impacts will be difficult:

- The evidence base that feeds into the management process is limited. Although arguably there is more information for marine than terrestrial animals, it is apparent that knowledge of hearing abilities is still limited. Similarly, although we have some good information on amplitude and frequency characteristics of marine sound sources, we have far less information on the complexity of the emitted sound field. This leads to huge uncertainties when calculating sound transmission and ultimately the levels and characteristics of sound at the receiver which determine noise impacts.

- Behavioural effects are likely to be underestimated. Although traditionally seen as less significant impacts in risk management approaches, behavioural effects can be pervasive and linked to fitness consequences. The potential for adverse behavioural reactions to have significant conservation impacts can be seen in certain circumstances such as the strandings observed in beaked whales in response to active sonar (see Sect. 14.6.1.3).
- A relatively poor understanding of how animals trade off costs and benefits. For example the benefits of staying in an area to gain access to food or mates might outweigh the costs of exposure to noise such as signal masking and threshold shifts. This approach has been more fully developed in the welfare literature where it is termed adaptive cost gauging (Barnard and Hurst 1996; Barnard 2007) and it has been suggested as a complicating factor when interpreting response to playback (McGregor 2008).
- There is little information on adverse fitness consequences of noise through effects on communication networks, information networks and soundscapes. However, such networks are likely to be as important as resource networks such as food webs.

The case for gathering more information on effects of noise on conservation is clear. We require more controlled studies looking at the nature, extent and persistence of behavioural responses to noise to assess likely population level consequences of acoustic disturbance. We also require more studies of masking and the efficacy of mitigation measures. However, the increase in anthropogenic noise means that advice is needed now, before such extra information has been gathered. One response to this need is to develop rules of thumb guidelines to be used in the absence of behavioural studies of the species of concern and in the absence of masking studies for the site at risk. A response to the absence of acoustic mapping tools (e.g. Barber et al. 2011) for non-human animals is to apply, as an interim measure, tools for acoustic mapping in humans (e.g. residential layouts that optimise disturbance zones, Theobald et al. 1997; and distance-based identification of open country quiet areas, Votsi et al. 2012; see Nega et al. 2013 for an application to urban park planning).

A continuing challenge will be that regulators rarely have the psychological or behavioural ecology background to understand the effects noise has on fitness through communication. Similarly, scientists who understand how signal masking can incur psychological costs such as distraction and physiological costs such as elevated heart rate that can translate into fitness costs, are unlikely to understand issues of legislation implementation and enforcement.

To conclude, the current perception of anthropogenic noise in biological conservation is similar to that of infectious diseases (Smith et al. 2009). In particular, neither noise nor disease has been cited as the principle cause of species extinction, but both can contribute to the effects of other drivers such as habitat loss. Also our current state of knowledge makes their relative contribution to species extinction hard to assess. It is encouraging that the role of infectious disease in conservation is beginning to be recognised. The same may be true for anthropogenic noise, where a recent report of the impact of low-frequency sound on cephalopods (André et al. 2011) led to an editorial in *New Scientist* (2011). We hope that this chapter further raises the profile of anthropogenic noise in conservation.

Acknowledgments Peter McGregor would like to thank former members of the Communication Crew at Copenhagen University, COWRIE project co-workers and members of the Marine Noise Network for discussion and information on noise. Particular thanks to my co-authors for agreeing to collaborate on the chapter and to Henrik Brumm, Friedrich Ladich and an anonymous reviewer for helpful suggestions on earlier drafts.

Andy Horn and Marty Leonard would like to thank the Natural Sciences and Engineering Research Council of Canada for funding, the many students who have helped with the research in various ways, and the Coldwell, Hynes, and Minor families for continued use of their land.

Frank Thomsen would like to thank the members of the noise expert groups set up by OSPAR and the EU/ICES for the stimulating discussions that informed many thoughts in this paper. Particular thanks go to Mark Tasker and René Dekeling for their effort holding things together.

References

Ainslie MA (2010) Principles of sonar performance modelling. Springer in association with Praxis Publishing Chichester, UK

André M, Solé M, Lenoir M, Durfort M, Quero C, Mas A, Lombarte A, van der Schaar M, López-Bejar M, Morell M, Zaugg S, Houégnigan L (2011) Low-frequency sounds induce acoustic trauma in cephalopods. Front Ecol Enviro 9:489–493

Andrew RK, Howe BM, Mercer JA (2011) Long-time trends in ship traffic noise for four sites off the North American West Coast. J Acoust Soc Am 129:642–651

ANSI/ASA (2009) American national standard quantities and procedures for description and measurement of underwater sound from ships—Part 1: general requirements. Acoustical Society of America, Melville

Arroyo-Solís A, Castillo JM, Figueroa E, López-Sánchez JL, Slabbekoorn H (2013) Experimental evidence for an impact of anthropogenic noise on dawn chorus timing in urban birds. J Avian Biol 44:288–296

Au WWL (1993) The sonar of dolphins. Springer, New York

Barber JR, Crooks KR, Fristrup KM (2009a) The costs of chronic noise exposure for terrestrial organisms. Trends Ecol Evol 1176:1–10

Barber JR, Fristrup KM, Brown CL, Hardy AR, Angeloni LM, Crooks KR (2009b) Conserving the wild life therein–Protecting park fauna from anthropogenic noise. Parks Sci 26(3):11

Barber JR, Burdett CL, Reed SE, Warner KA, Formichella C, Crooks KR, Theobald DM, Fristrup KM (2011) Anthropogenic noise exposure in protected natural areas: estimating the scale of ecological consequences. Landscape Ecol 26:1281–1295

Barnard CJ (2007) Ethical regulation and animal science: why animal behaviour is special. Anim Behav 74:5–13

Barnard CJ, Hurst JL (1996) Welfare by design: the natural selection of welfare criteria. Anim Welfare 5:405–433

Barrass A (1985) The effects of highway traffic noise on the phonotactic and associated reproductive behaviour of selected anurans. Vanderbilt University, Nashville

Barrett-Lennard LG, Ford JKB, Heise KA (1996) The mixed blessing of echolocation: differences in sonar use by fish-eating and mammal-eating killer whales. Anim Behav 51:553–565

Bartol SM, Musick JA, Lenhardt M (1999) Auditory evoked potentials of the loggerhead sea turtle (*Caretta caretta*). Copeia 3:836–840

Bateson M (2007) Environmental noise and decision making possible implications of increases in anthropogenic noise for information processing in marine mammals. Int J Comp Psychol 20:169–178

Bautista L, García JT, Calmaestra RG, Palacín C, Martín CA, Morales MB, Bonal R, Viñuela J (2004) Effect of weekend road traffic on the use of space by raptors. Conserv Biol 18:726–732

Bayne E, Habib L, Boutin S (2008) Impacts of chronic anthropogenic noise from energy-sector activity on abundance of songbirds in the boreal forest. Conserv Biol 22:1186–1193

Bee M, Swanson E (2007) Auditory masking of anuran advertisement calls by road traffic noise. Anim Behav 74:1765–1776

Bejder L, Samuels A, Whitehead H, Finn H, Allen S (2009) Impact assessment research: use and misuse of habituation, sensitisation and tolerance in describing wildlife responses to anthropogenic stimuli. Mar Ecol-Prog Ser 395:177–185

Benítez-López A, Alkemade R, Verweij PA (2010) The impacts of roads and other infrastructure on mammal and bird populations: a meta-analysis. Biol Conserv 143:1307–1316

Bergen F, Abs M (1997) Etho-ecological study of the singing activity of the blue tit *Parus caeruleus*, great tit *Parus major* and chaffinch *Fringilla coelebs*. J f Orn 138:451–467

Bingman VP, Cheng K (2005) Mechanisms of animal global navigation: comparative perspectives and enduring challenges. Ethol Ecol Evol 17:295–318

Blickley J, Patricelli G (2010) Impacts of anthropogenic noise on wildlife: research priorities for the development of standards and mitigation. J Int Wildl Law Policy 13:274–292

Blickley JL, Blackwood D, Patricelli GL (2012) Experimental evidence for the effects of chronic anthropogenic noise on abundance of Greater Sage-Grouse at leks. Conserv Biol 26:461–471

Boyd I, Brownell B, Cato D, Clarke C, Costa D, Evans PGH, Gedamke J, Genrty R, Gisiner B, Gordon J, Jepson P, Miller P, Rendell L, Tasker M, Tyack P, Vos E, Whitehead H, Wartzok D, Zimmer W (2008) The effects of anthropogenic sound on marine mammals—a draft research strategy. European Science Foundation and Marine Board, Oostende

Brandt MJ, Diederichs A, Betke K, Nehls G (2011) Responses of harbour porpoises to pile driving at the Horns Rev II offshore wind farm in the Danish North Sea. Mar Ecol-Prog Ser 421:205–216

Brawn VM (1961) Sound production by the cod (*Gadus callarias*). Behaviour 18:239–255

Bronzaft AL, Hagler L (2010) Noise: the invisible pollutant that cannot be ignored. Emerging Environmental Technologies 2:75–96

Brumm H, Ritschard M (2011) Song amplitude affects territorial aggression of male receivers in chaffinches. Behav Ecol 22:310–316

Bryant PJ, Lafferty CM, Lafferty SK (1984) Reoccupation of Laguna Guerrero Negro, Baja California, Mexico by gray whales. In: Jones ML, Swartz SL, Leatherwood S (eds) The gray whale *Eschrichtius robustus*. Academic Press, Orlando, pp 375–387

Caltrans (2001) Fisheries Impact Assessment, Pile Installation Demonstration Project for the San Francisco—Oakland Bay Bridge. East Span Seismic Safety Project PIDP EA 01208. Caltrans Contract 04A0148, Task Order 205.10.90, PIPD 04-ALA-80.0.0/0.5

Carlson TJ, Hastings MC, Popper AN (2007) Update on recommendation for revised interim sound exposure criteria for fish during pile driving activities. California Department of Transportation, San Diego, p. 8

CEDA (2011) CEDA position paper: underwater sound in relation to dredging. Terra et Aqua 125:23–28

Cefas (2004) Offshore wind farms—Guidance note for environmental impact assessment in respect of FEPA and CPA Requirements, Version 2

Codarin A, Wysocki LE, Ladich F, Picciulin, M (2009) Effects of ambient and boat noise on hearing and communication in three fish species living in a marine protected area (Miramare, Italy). Mar Pollut Bull 58:1880–1887

Compton R, Goodwin L, Handy R, Abbott V (2008) A critical examination of worldwide guidelines for minimising the disturbance to marine mammals during seismic surveys. Mar Policy 32:255–262

Conomy JT, Dubovsky JA, Collazo JA, Fleming WJ (1998) Do black ducks and wood ducks habituate to aircraft disturbance? J Wildl Manage 62:1135–1142

Coull KA, Johnstone R, Rogers SI (1998) Fisheries sensitivity maps in British waters. UKOOA Ltd, 9 Albyn Terrace, Aberdeen

Cox TM, Ragen TJ, Read AJ, Vos E, Baird RW, Balcomb K, Barlow J, Caldwell J, Cranford T, Crum L, D'Amico A, D'Spain G, Fernández A, Finneran J, Gentry R, Gerth W, Gulland F, Hildebrand J, Houser D, Hullar T, Jepson PD, Ketten D, MacLeod CD, Miller P, Moore S, Mountain D, Ponganis P, Rommel S, Rowles T, Taylor B, Tyack P, Wartzok D, Gisiner R, Mead J, Lowry L, Benner L (2006) Understanding the impacts of anthropogenic sound on beaked whales. J Cetacean Res Manag 7:177–187

Crino OL, Johnson EE, Blickley JL, Patricelli GL, Breuner CW (2013) The effects of experimentally elevated traffic noise on nestling white-crowned sparrow stress physiology, immune function, and life-history. J Exp Biol 216:2055–2062

Culik BM, Koschinski S, Tregenza N, Ellis GM (2001) Reactions of harbour porpoises *Phocoena phocoena* and herring *Clupea harengus* to acoustic alarms. Mar Ecol-Prog Ser 211:255–260

Cunnington GM, Fahrig L (2013) Mate attraction by male anurans in the presence of traffic noise. Anim Cons 16:275–285

Defra (2010) UK marine science strategy. Department for Environment Food and Rural Affairs, London

Díaz M, Parra A, Gallardo C (2011) Serins respond to anthropogenic noise by increasing vocal activity. Behav Ecol 22:332–336

Dooling R, Popper A (2007) The effects of highway noise on birds. California Department of Transportation, Sacramento

Dooling RJ, Therrien SC (2012) Hearing in birds: what changes from air to water. In: Popper AN, Hawkins A (eds) The effects of noise on aquatic life. Advances in experimental medicine and biology, vol 730. Springer Verlag, New York, pp 77–82

Dow Piniak WE, Mann DA, Eckert SA, Harms CA (2012) Amphibious hearing in sea turtles. In: Popper AN, Hawkins A (eds) The effects of noise on aquatic life. Advances in experimental medicine and biology, vol 730. Springer Verlag, New York, pp 83–87

Eigenbrod F, Hecnar SJ, Fahrig L (2009) Quantifying the road-effect zone: threshold effects of a motorway on anuran populations in Ontario, Canada. Ecol Soc 14:24 [online] URL: http://www.ecologyandsociety.org/vol14/iss1/art24/

Engås A, Løkkeborg S, Ona E, Soldal A (1996) Effects of seismic shooting on local abundance and catch rates of cod (*Gadus morhua*) and haddock (*Melanogrammus aeglefinus*). Can J Fish Aquat Sci 53:2238–2249

Erbe C, Farmer DM (2000) Zones of impact around icebreakers affecting beluga whales in the Beaufort Sea. J Acoust Soc Am 108:1332–1340

EU (2010) Commission Decision of 1 September 2010 on criteria and methodological standards on good environmental status of marine waters, European Commission 2010/477/EU, Brussels

Fahrig L, Pedlar JH, Pope SE, Taylor PD, Wegner JF (1995) Effect of road traffic on amphibian density. Biol Conserv 73:177–182

Fay R, Popper AN (2000) Evolution of hearing in vertebrates: the inner ears and processing. Hearing Res 149:1–10

Ford JKB (1989) Acoustic behaviour of resident killer whales (*Orcinus orca*) off Vancouver Island, British Columbia. Can J Zool 67:727–745

Forman R, Deblinger R (2002) The ecological road-effect zone of a Massachusetts (U.S.A.) suburban highway. Conserv Biol 14:36–46

Francis C, Ortega C, Cruz A (2009) Cumulative consequences of noise pollution: noise changes avian communities and species interactions. Curr Biol 19:1415–1419

Francis CD, Kleist NJ, Ortega CP, Cruz A (2010) Vocal frequency change reflects different responses to anthropogenic noise in two suboscine tyrant flycatchers. Proc Roy Soc B 279:2727–2735

Francis CD, Paritsis J, Ortega CP, Cruz A (2011) Landscape patterns of avian habitat use and nest success are affected by chronic gas well compressor noise. Landscape Ecol 26:1269–1280

Francis CD, Kleist NJ, Ortega CP, Cruz A (2012) Noise pollution alters ecological services: enhanced pollination and disrupted seed dispersal. Proc Roy Soc B Biol Sci 279:2727–2735

Fuller RA, Warren PH, Gaston KJ (2007) Daytime noise predicts nocturnal singing in urban robins. Biol Lett 3:368–370

Gil D, Gahr M (2002) The honesty of bird song: multiple constraints for multiple traits. Trends Ecol Evol 17:133–141

Goodwin SE, Shriver WG (2010) Effects of traffic noise on occupancy patterns of forest birds. Conserv Biol 25:406–411

Gordon SD, Uetz GW (2012) Environmental interference: impact of acoustic noise on seismic communication and mating success. Behav Ecol 23:707–714

Gross K, Pasinelli G, Kunc HP (2010) Behavioral plasticity allows short-term adjustment to a novel environment Am Natur 176:456–464

Guski R (1999) Personal and social variables as co-determinants of noise annoyance. Noise Health 3:45–56

Habib L, Bayne E, Boutin S (2007) Chronic industrial noise affects pairing success and age structure of ovenbirds *Seiurus aurocapilla*. J Applied Ecol 44:176–184

Halfwerk W, Holleman LJM, Lessells CM, Slabbekoorn H (2011) Negative impact of traffic noise on avian reproductive success. J Applied Ecol 48:210–219

Halfwerk W, Bot S, Slabbekoorn H (2012) Male great tit song perch selection in response to noise-dependent female feedback. Funct Ecol 26:1339–1347

Hammond PS (2006) Small cetaceans in the European Atlantic and North Sea (SCANS II) LIFE 04 NAT/GB/000245

Hawkins AD, Rasmussen KJ (1978) The calls of gadoid fish. J Marine Biol Assoc 58:891–911

Herrera-Montes MI, Aide TM (2011) Impacts of traffic noise on anuran and bird communities. Urban Ecosyst 14:415–427

Hildebrand J (2009) Anthropogenic and natural sources of ambient noise in the ocean. Mar Ecol Prog Ser 395:5–20

Holt MM, Noren DP, Veirs V, Emmons CK, Veirs S (2009) Speaking up: killer whales (*Orcinus orca*) increase their call amplitude in response to vessel noise. J Acoustic Soc Am 125:EL27–EL32

Hoskin C, Goosem M (2010) Road impacts on abundance call traits, and body size of rainforest frogs in Northeast Australia. Ecol Soc 15:art15

Hu Y, Cardoso GC (2009) Are bird species that vocalize at higher frequencies preadapted to inhabit noisy urban areas? Behav Ecol 20:1268–1273

IACMST (2006) Underwater sound and marine life—IACMST Working Group Report No.6, Inter-Agency Committee on Marine Science and Technology—National Oceanographic Centre, Southampton

IMO—MEPC (2009) Shipping noise and marine mammals—Submitted by the United States. MEPC 57/INF.4 m 17 December 2007 (available under: http://www.gc.noaa.gov/documents/mepc_57_20_inf_4.pdf)

Jaeger JAG, Bowman J, Brennan J, Fahrig L, Bert D, Bouchard J, Charbonneau N, Frank K, Gruber B, von Toschanowitz KT (2005) Predicting when animal populations are at risk from roads: an interactive model of road avoidance behavior. Ecol Model 185:329–348

Janik VM (2000) Source levels and the estimated active space of bottlenose dolphin (*Tursiops truncatus*) whistles in the Moray Firth, Scotland. J Comp Physiol A 186:673–680

Janik VM (2005) Underwater acoustic communication networks in marine mammals. In: McGregor PK (ed) Animal Communication Networks. Cambridge University Press, Cambridge, pp 390–415

Janik VM, Slater PJB (1998) Context-specific use suggests that bottlenose dolphin signature whistles are cohesion calls. Anim Behav 56:829–838

JNCC (2009a) Guidelines for minimising acoustic disturbance to marine mammals from seismic surveys. Joint Nature Conservation Committee, Peterborough

JNCC (2009b) ANNEX B—Statutory nature conservation agency protocol for minimising the risk of disturbance and injury to marine mammals from piling noise. Joint Nature Conservation Committee, Peterborough

Johansen S, Larsen ON, Dalsgaard JC, Seidelin L, Wahlberg M (2013) In-air and underwater hearing in the great cormorant (*Phalacrocorax carbo*). Paper presented at the 3rd International Conference on the Effects of Sound on Aquatic Life, Budapest

Jones G, Teeling EC (2006) The evolution of echolocation in bats. Trends Ecol Evol 21:149–156

Kaiser K, Scofield DG, Alloush M, Jones RM, Marczak S, Martineau K, Oliva MA, Narins PM (2011) When sounds collide: the effect of anthropogenic noise on a breeding assemblage of frogs in Belize, Central America. Behaviour 148:215–232

Kaseloo P, Tyson K (2004) Synthesis of noise effects on wildlife populations. US Department of Transportation, Federal Highway Administration, McLean

Kastelein RA, Gransier R, Hoek L, Olthuis J (2012) Temporary threshold shifts and recovery in a harbor porpoise (*Phocoena phocoena*) after octave-band noise at 4 kHz. J Acoust Soc Am 132:3525–3537

Ketten D (1997) Structure and function in whale ears. Bioacoustics 8:103–135

Kight CR, Swaddle JP (2011) How and why environmental noise impacts animals: an integrative, mechanistic review. Ecol Lett 14:1052–1061

Kociolek AV, Clevenger AP, St Clair CC, Proppe DS (2011) effects of road networks on bird populations. Conserv Biol 25:241–249

Ladich F (2008) Sound communication in fishes and the influence of ambient and anthropogenic noise. Bioacoustics 17:35–37

Lavender AL, Bartol SM, Bartol IK (2012) Hearing capabilities of loggerhead sea turtles (*Caretta caretta*) throughout ontogeny. In: Popper AN, Hawkins A (eds) The effects of noise on aquatic life. Advances in experimental medicine and biology, vol 730. Springer Verlag, New York, pp 89–92

Leader N, Wright J, Yom-Tov Y (2005) Acoustic properties of two urban song dialects in the orange-tufted sunbird (*Nectarinia osea*). Auk 122:231–245

Leonard M, Horn A (2005) Ambient noise and the design of begging signals. Proc Roy Soc B Biol Sci 272:651–656

Leonard M, Horn A (2008) Does ambient noise affect growth and begging call structure in nestling birds? Behav Ecol 19:502–507

Leonard ML, Horn AG (2012) Ambient noise increases missed detections in nestling birds. Biol Lett 8:530–532

Lohr B, Wright T, Dooling RJ (2003) Detection and discrimination of natural calls in masking noise by birds: estimating the active space of a signal. Anim Behav 65:763–777

Lucke K, Lepper PA, Blanchet M-A, Siebert U (2011) The use of an air bubble curtain to reduce the received sound levels for harbor porpoises (*Phocoena phocoena*). J Acoust Soc Am 130:3406–3412

Lucke K, Siebert U, Lepper PA, Blanchet MA (2009) Temporary shift in masked hearing thresholds in a harbor porpoise (*Phocoena phocoena*) after exposure to seismic airgun stimuli. J Acoust Soc Am 125:4060–4070

Luther DA, Derryberry EP (2012) Birdsongs keep pace with city life: changes in song over time in an urban songbird affects communication. Anim Behav 83:1059–1066

Madsen PT, Johnson M, Miller PJO, Aguilar Soto N, Lynch J, Tyack P (2006a) Quantitative measures of air gun pulses recorded on sperm whales (*Physeter macrocephalus*) using acoustic tags during controlled exposure experiments. J Acoust Soc Am 120:2366–2379

Madsen PT, Wahlberg M, Tougaard J, Lucke K, Tyack P (2006b) Wind turbine underwater noise and marine mammals: implications of current knowledge and data needs. Mar Ecol Prog Ser 309:279–295

Makarewicz R, Kokowski P (2007) Prediction of noise changes due to traffic speed control. J Acoust Soc Am 122:2074–2081

McCauley RD, Fewtrell J, Duncan AJ, Jenner C, Jenner MN, Penrose JD, Prince RIT, Adhitya A, Murdoch J, McCabe K (2000) Marine seismic surveys: a study of environmental implications. APPEA Journal 2000:692–708

McCauley RD, Fewtrell J, Popper AN (2003) high intensity anthropogenic sound damages fish ears. J Acoust Soc Am 113:631–642

McGregor PK (2005) Animal Communication Networks. Cambridge University Press, Cambridge

McGregor PK (2008) Designing experiments to test for behavioural effects of sound. Bioacoustics 17:336–338

McLaughlin KE, Kunc HP (2013) Experimentally increased noise levels change spatial and singing behaviour. Biol Lett 9:20120771

Mendes S, Colino-Rabanal VJ, Peris SJ (2011) Bird song variations along an urban gradient: the case of the European blackbird (*Turdus merula*). Landsc Urban Plan 99:51–57

Miller PJO (2006) Diversity in sound pressure levels and estimated active space of resident killer whale vocalizations. J Comp Physiol A 192:449–459

Miller PJO, Biassoni N, Samuels A, Tyack PL (2000) Whale songs lengthen in response to sonar. Nature 405:903

Miller PJO, Johnson MP, Madsen PT, Biassoni N, Querob M, Tyack PL (2009) Using at-sea experiments to study the effects of airguns on the foraging behavior of sperm whales in the Gulf of Mexico. Deep Sea Res Pt I Oceanogr Res Pap 56:1168–1181

Mitson RB, Morris R (1988) Evidence of high frequency acoustic emissions from the white beaked dolphin (*Lagenorhynchus albirostris*). J Acoust Soc Am 83:825–826

Mooney TA, Hanlon R, Madsen PT, Christensen-Dalsgaard J (2012) Potential for sound sensitivity in cephalopods. In: Popper AN, Hawkins A (eds) The effects of noise on aquatic life. Advances in experimental medicine and biology, vol 730. Springer Verlag, New York, pp 125–128

Montague MJ, Danek-Gontard M, Kunc HP (2012) Phenotypic plasticity affects the response of a sexually selected trait to anthropogenic noise. Behav Ecol 24:343–348

Montgomery JC, Jeffs A, Simpson SD, Meekan M, Tindle C (2006) Sound as an orientation cue for the pelagic larvae of reef fishes and decapod crustaceans. Adv Mar Biol 51:143–196

Morton AB, Symonds HK (2002) Displacement of *Orcinus orca* (L.) by high amplitude sound in British Columbia, Canada. ICES J Mar Sci 59:71–80

Moudon AV (2009) Real noise from the urban environment: how ambient community noise affects health and what can be done about it. Am J Prev Med 37:167–171

Mueller-Blenkle C, McGregor P, Gill A, Andersson M, Metcalfe J, Bendall V, Sigray P, Wood D, Thomsen F (2010) Effects of pile-driving noise on the behaviour of marine fish. COWRIE Report Fish 06–08

Naguib M, van Oers K, Braakhuis A, Griffioen M, de Goede P, Waas JR (2013) Noise annoys: effects of noise on breeding great tits depend on personality but not on noise characteristics. Anim Behav 85:949–956

Nega T, Yaffe N, Stewart N, Fu W-H (2013) The impact of road traffic noise on urban protected areas: a landscape modeling approach. Transport Res D Tr E 23:98–104

Nehls G, Betke K, Eckelmann S, Ros M (2007) Assessment and costs of potential engineering solutions for the mitigation of the impacts of underwater noise arising from the construction of offshore windfarms. COWRIE Ltd. (Collaborative Offshore Wind Research Into The Environment), London

Nemeth E, Brumm H (2010) Birds and anthropogenic noise: are urban songs adaptive? Am Nat 176:465–475

New Scientist editorial (2011) A noisy noise annoys a cephalopod. New Sci 210(2808):5

Nowacek DP, Thorne LH, Johnston DW, Tyack PL (2007) Responses of cetaceans to anthropogenic noise. Mamm Rev 37:81–115

NRC (2005) Marine mammal populations and ocean noise—determining when noise causes biologically significant effects. The National Academics Press, Washington

OECD (1995) Roadside Noise Abatement. Organisation for Economic Co-operation and Development, Paris

OGP/IAGC (2007) Seismic surveys and marine mammals. International Association of Oil and Gas Producers/International Association of Geophysical Contractors, London

Oleson EM, Calambokidis J, Burgess WC, McDonald MA, LeDuc CA, Hildebrand JA (2007) Behavioral context of call production by eastern North Pacific blue whales. Mar Ecol Prog Ser 330:269–284

Ortega CP (2012) Effects of noise pollution on birds: a brief review of our knowledge. Ornithol Monogr 74:6–22

OSPAR (2009a) JAMP assessment of the impacts of anthropogenic underwater sound in the marine environment. OSPAR Convention for the Protection of the Marine Environment of the North-East Atlantic (www.ospar.org)

OSPAR (2009b) Overview of the impacts of anthropogenic underwater sound in the marine environment. OSPAR Convention for the Protection of the Marine Environment of the North-East Atlantic (www.ospar.org)

OSPAR (2010) Quality status report 2010. OSPAR Commission, London

Ouis D (2001) Annoyance from road traffic noise: a review. J Environ Psych 21:101–120

Owens JL, Stec CL, O'Hatnick A (2012) The effects of extended exposure to traffic noise on parid social and risk-taking behavior. Behav Process doi:10.1016/j.beproc.2012.05.010

Packard A, Karlsen HE, Sand O (1990) Low frequency hearing in cephalopods. J Comp Physiol A 166:501–505

Parks SE, Johnson M, Nowacek D, Tyack PL (2011) Individual right whales call louder in increased environmental noise. Biol Lett 7:33–35

Parris K, Schneider A (2009) Impacts of traffic noise and traffic volume on birds of roadside habitats. Ecol Soc 14:art10

Parris K, Velik-Lord M, North JMA (2009) Frogs call at higher pitch in traffic noise. Ecol Soc 14:25. http://www.ecologyandsociety.org/vol14/iss1/art25/

Parson ECM, Dolman SJ, Wright AJ, Rose NA, Burns WCG (2008) Navy sonar and cetaceans: just how much does the gun need to smoke before we act? Mar Pollut Bull 56:1248–1257

Patón D, Romero F, Cuenca J, Escudero JC (2012) Tolerance to noise in 91 bird species from 27 urban gardens of Iberian Peninsula. Landscape Urban Plan 104:1–8

Payne R, Webb D (1971) Orientation by means of long range acoustic signalling in baleen whales. Ann NY Acad Sci 188:110–141

Pepper C, Nascarella M, Kendall RJ (2003) A review of the effects of aircraft noise on wildlife and humans, current control mechanisms, and the need for further study. Environ Manag 32:418–432

Peris S, Pescador M (2004) Effects of traffic noise on passerine populations in Mediterranean wooded pastures. App Acoust 65:357–366

Popper AN, Carlson TJ, Hawkins AD, Southall BL (2006) Interim criteria for injury of fish exposed to pile driving operations: a white paper. (available at: http://www.wsdot.wa.gov/NR/rdonlyres/84A6313A-9297-42C9-BFA6-750A691E1DB3/0/BA_PileDrivingInterimCriteria.pdf)

Popper AN, Fay RR (2011) Rethinking sound detection by fishes. Hear Res 273:25–36

Popper AN, Fay RR, Platt C, Sand O (2003) Sound detection mechanisms and capabilities of teleost fishes. In: Collin SP, Marshall NJ (eds) Sensory processing in aquatic environments. Springer Verlag, New York, pp 3–38

Popper AN, Hastings MC (2009a) The effects of anthropogenic sources of sound on fishes. J Fish Biol 75:455–489

Popper AN, Hastings MC (2009b) The effects of human-generated sound on fish. Integr Zool 4:43–52

Popper AN, Salmon M, Horch KW (2001) Acoustic detection and communication by decapod crustaceans. J Comp Physiol A 187:83–89

Popper AN, Smith ME, Cott PA, Hanna BW, MacGillivary AO, Austin M, Mann DA (2005) Effects of exposure to seismic airgun use on hearing of three fish species. J Acoust Soc Am 117:3958–3971

Proppe DS, Sturdy CB, Cassady St Clair C (2013) Anthropogenic noise decreases urban songbird diversity and may contribute to homogenization. Glob Change Biol 19:1075–1084

Purser J, Radford AN (2011) Acoustic noise induces attention shifts and reduces foraging performance in three-spined sticklebacks (Gasterosteus aculeatus). PLoS ONE 6(2):e17478. doi:10.1371/journal.pone.0017478

Quinn J, Whittingham M, Butler SJ, Cresswell W (2006) Noise, predation risk compensation and vigilance in the chaffinch Fringilla coelebs. J Avian Biol 37:601–608

Rabin LA, Coss RG, Owings DH (2006) The effects of wind turbines on antipredator behavior in California ground squirrels (Spermophilus beecheyi). Biol Conserv 131:410–420

Rasmussen MH, Miller LA (2002) Whistles and clicks from white-beaked dolphins, Lagenorhynchus albirostris, recorded in Faxaflói Bay, Iceland. Aquat Mamm 28:78–89

Reijnen R, Foppen R, ter Braak C, Thissen J (1995) The effects of car traffic on breeding bird populations in woodland. 3. Reduction of density in relation to the proximity of main roads. J Applied Ecol 32:187–202

Reijnen R, Foppen R, Veenbaas G (1997) Disturbance by traffic of breeding birds. Evaluation of the effect and considerations in planning and managing road corridors. Biodivers Conserv 6:567–581

Renilson M (2009) Reducing underwater noise pollution from large commercial vessels. International Fund for Animal Welfare, London

Rheindt F (2003) The impact of roads on birds: does song frequency play a role in determining susceptibility to noise pollution? J f Orn 144:295–306

Richardson WJ, Malme CI, Green CR Jr, Thomson DH (1995) Marine mammals and noise, vol 1. Academic Press, San Diego

Rodgers J Jr, Smith H (1995) Set-back distances to protect nesting bird colonies from human disturbance in Florida. Conserv Biol 9:89–99

Ross DG (1993) On ocean underwater ambient noise. Acoust Bull 18:5–8

Samarra FIP, Klappert K, Brumm H, Miller P (2009) Background noise constrains communication: acoustic masking of courtship song in the fruit fly Drosophila montana. Behav 146:1635–1648

Scheifele PM, Andrew S, Cooper RA, Darre M, Musiek FE, Max L (2005) Indication of a Lombard vocal response in the St. Lawrence river beluga. J Acoust Soc Am 117:1486–1492

Searcy W, Nowicki S (2005) The evolution of animal communication: reliability and deception in signaling systems. Princeton University Press, Princeton

Schroeder J, Nakagawa S, Cleasby IR, Burke T (2012) Passerine birds breeding under chronic noise experience reduced fitness. PLoS ONE 7:e39200. doi:10.1371/journal.pone.0039200

Slabbekoorn H (2013) Songs of the city: noise-dependent spectral plasticity in the acoustic phenotype of urban birds. Anim Behav 85:1089–1099

Slabbekoorn H, Peet M (2003) Birds sing at a higher pitch in urban noise. Nature 424:267

Slabbekoorn H, Ripmeester E (2007) Birdsong and anthropogenic noise: implications and applications for conservation. Mol Ecol 17:72–83

Smith KF, Acevedo-Whitehouse K, Pedersen AB (2009) The role of infectious diseases in biological conservation. Anim Conserv 12:1–12

Smolker RA, Mann J, Smuts BB (1993) Use of signature whistles during separations and reunions by wild bottlenose dolphin mothers and infants. Behav Ecol Sociobiol 33:393–402

Southall B, Berkson J, Bowen D, Brake R, Eckman J, Field J, Gisiner R, Gregerson S, Lang W, Lewandoski J, Wilson J, Winokur R (2009) Addressing the effects of human-generated sound on marine life: an integrated research plan for U.S. federal agencies. Interagency task force on anthropogenic sound and the marine environment of the joint subcommittee on ocean science and technology, Washington

Southall BL, Bowles AE, Ellison WT, Finneran JJ, Gentry RL, Greene CRJ, Kastak D, Ketten DR, Miller JH, Nachtigall PE, Richardson WJ, Thomas JA, Tyack P (2007) Marine mammal noise exposure criteria: initial scientific recommendations. Aquat Mamm 33:411–521

Sørensen M, Andersen ZJ, Nordsborg RB, Becker T, Tjønneland A, Overvad K, Raaschou-Nielsen A (2013) Long-term exposure to road traffic noise and incident diabetes: a cohort study. Environ Health Persp 121:217–222

Stafford KM, Mellinger DK, Moore SE, Fox CG (2007) Seasonal variability and detection range modeling of baleen whale calls in the Gulf of Alaska, 1999–2002. J Acoust Soc Am 122:3378–3390

Stone E (2000) Separating the noise from the noise: a finding in support of the Niche Hypothesis that birds are influenced by human induced noise in natural habitats. Anthrozoos 13:25–231

Tasker ML, Amundin M, Andre M, Hawkins T, Lang I, Merck T, Scholik-Schlomer A, Teilmann J, Thomsen F, Werner S, Zakharia M (2010) Marine strategy framework directive—task group 11 report—underwater noise and other forms of energy. European Commission Joint Research Centre and International Council for the Exploration of the Sea, Luxembourg

Theobald DM, Miller JR, Hobbs NT (1997) Estimating the cumulative effects of development on wildlife habitat. Landsc Urban Plan 39:25–36

Thomsen F, Lüdemann K, Kafemann R, Piper W (2006) Effects of offshore wind farm noise on marine mammals and fish. Biola, Hamburg, Germany on behalf of COWRIE Ltd, Newbury

Thomsen F, McCully SR, Weiss L, Wood D, Warr K, Barry J, Law R (2011) Cetacean stock assessment in relation to exploration and production industry activity and other human pressures: review and data needs. Aquat Mamm 37:1–92

TNO (2011) Ainslie MA (eds) Standard for measurement and monitoring of underwater noise, part 1: physical quantities and their units. TNO, Den Haag

Tougaard J, Carstensen J, Wisz MS, Jespersen M, Teilmann J, Ilsted Bech N, Skov H (2006) Harbour porpoises on Horns Reef—Effects of the Horns Reef wind farm. Final Report to Vattenfall A/S. NERI, Roskilde

Tyack P, Gordon J, Thompson D (2004) Controlled-exposure experiments to determine the effects of noise on marine mammals. Mar Technol Soc J 37:39–51

Tyler CR, Jobling S (2008) Roach, sex, and gender-bending chemicals: the feminization of wild fish in English rivers. Bioscience 58:1051–1059

Urick R (1983) Principles of underwater sound. McGraw Hill, New York

van der Graaf S, Ainslie MA, Andre M, Brensing K, Dalen J, Dekeling RPA, Robinson S, Tasker ML, Thomsen F, Werner S (2012) European marine strategy framework directive—good environmental status (MSFD GES): report of the technical subgroup on underwater noise and other forms of energy. Milieu Ltd, Belgium

van der Zande A, ter Keurs W, van der Weijden WJ (1980) The impact of roads on the densities of four bird species in an open field habitat: evidence of a long-distance effect. Biol Conserv 18:299–321

van Langevelde F, Jaarsma C (2009) Modeling the effect of traffic calming on local animal population persistence. Ecol Soc 14:art39

Vasconcelos RO, Amorim MCP, Ladich F (2007) Effects of ship noise on the detectability of communication signals in the Lusitanian toadfish. J Exp Biol 210:2104–2112

Votsi NEP, Drakou EG, Mazaris AD, Kallimanis AS, Pantis JD (2012) Distance-based assessment of open country Quiet Areas in Greece. Landsc Urban Plan 104:279–288

Wale MA, Simpson SD, Radford AN (2013) Size-dependent physiological responses of shore crabs to single and repeated playback of ship noise. Biol Lett 9:20121194. http://dx.doi.org/10.1098/rsbl.2012.1194

Warren P, Katti M, Ermann M, Brazel A (2006) Urban bioacoustics: it's not just noise. Anim Behav 71:491–502

Watkins WA, Tyack PL, Morre KE, Bird JE (1987) The 20-Hz signals of finback whales (*Balaenoptera physalus*). J Acoust Soc Am 82:1901–1912

Weilgart L (2007) The impacts of anthropogenic ocean noise on cetaceans and implications for management. Can J Zool 85:1091–1116

WHO (2011) Burden of disease from environmental noise: quantification of healthy life years lost in Europe. World Health Organization, Geneva www.euro.who.int/__data/assets/pdf_file/0008/136466/e94888.pdf accessed 31/3/11

Williams R, Bain DE, Ford JKB, Trites AW (2002) Behavioural responses of male killer whales to a 'leapfrogging' vessel. J Cetacean Res Manage 4:305–310

Williams R, Lusseau D, Hammond PS (2006) Estimating relative energetic costs of human disturbance to killer whales (*Orcinus orca*). Biol Conserv 133:301–311

WODA (2013) Technical Guidance on: Underwater Sound in Relation to Dredging. World Organisation of Dredging Associations, Delft

Wright AJ (2009) Report of the workshop on assessing the cumulative impacts of underwater noise with other anthropogenic stressors on marine mammals: from ideas to action Monterey, California, 26–29th August 2009, Okeanos—Foundation for the Sea—http://www.sound-in-the-sea.org/download/CIA2009_en.pdf, Darmstadt

Wright MD, Goodman P, Cameron TC (2010) Exploring behavioural responses of shorebirds to impulsive noise. Wildfowl 60:150–167

Würsig B, Richardson WJ (2002). Effects of noise. In: Perrin WF, Würsig B, Thewissen JGM (eds) Encyclopedia of marine mammals. Academic Press, New York, pp 794–802

Wysocki LE, Ladich F (2005) Hearing in fish under noise conditions. J Assoc Res Otolaryngol 6:28–36

Zelick R, Mann DA, Popper AN (1999) Acoustic communication in fishes and frogs. In: Fay RR, Popper AN (eds) Comparative hearing: Fish and amphibians. Springer, New York, pp 363–411

Zheng W (2012) Auditory map reorganization and pitch discrimination in adult rats chronically exposed to low-level ambient noise. Front Syst Neurosci 6:65. doi:10.3389/fnsys.2012.00065

Zollinger SA, Goller F, Brumm H (2011) Metabolic and respiratory costs of increasing song amplitude in zebra finches. PLoS ONE 6:e23198

Index

A

Abiotic noise, 67, 112–118, 189, 190, 196, 218, 312
Abroscopus albogularis, 196, 197
Absorption, 252, 253
Acoustic adaptation hypothesis, 195, 196, 198
Acoustic niche partitioning, 96, 97, 137, 197, 198
Acoustic orientation, 65
Acridotheres tristis, 213
Acris crepitans, 111, 160
Action potential, 351
Active space, 42, 93, 117, 122, 142, 230, 412, 424
Adaptive fastidiousness, 20
Adaptive growth hypothesis, 321
Adaptive gullability, 19
AEP. *See* Auditory evoked potential
Agelaius phoeniceus, 206, 214
Aggressive calls, 98, 99, 107
Air gun, 300
Air sacs, 254
Alarm call, 188, 235
Alarm pheromone, 374
Alerting component, 15
Alerting signal, 15
Allobates femoralis, 97, 151, 152
Altered environments, 321
Alternative signals, 43
Ambient noise, 69, 233, 252, 253, 258–262, 264–266
Ameerega trivittata, 15, 45, 97
American gray flycatcher. *See Empidonax wrightii*
American robin. *See Turdus migratorius*
American toad. *See Bufo americanus*
Ammodramus aurifrons, 215
Amolops tormotus. See Odorrana tormota

Amphibolurus muricatus, 314
Amplitude modulation, 343
Ampullary electroreceptors, 343
Andropadus virens, 197
Animation technology, 325
Anolis, 314, 327
Antennal lobes, 394, 398
Anthochaera carunculata, 210
Anthropogenic noise, 3, 65, 77, 83, 84, 118–123, 139, 144, 192, 194, 199, 208, 218, 220, 230, 239, 409–411, 414, 415, 421, 423, 426, 427, 432–434
Ants, 374, 379, 380, 383, 386, 398
Apenodytes patagonicus, 208, 209
Apis mellifera, 379, 381, 383, 384
Apteronotus leptorhynchus, 334
Arbitrary preferences, 22, 25
Ash-throated flycatcher. *See Myiarchus cinerascens*
ASSR. *See* Auditory steady-state response
Asymptotic Threshold Shift, 296
ATS. *See* Asymptotic threshold shift
Audiogram, 141
Auditory evoked potential (AEP), 74, 291
Auditory feedback, 194
Auditory filter, 275
Auditory grouping, 160–167
Auditory induction, 169, 172
Auditory masking, 274
Auditory objects, 244, 246
Auditory perception, 273
Auditory scene analysis, 160–167, 244, 273
Auditory sensitivity, 68
Auditory steady-state response, 292
Auditory stream segregation, 161, 163
Auditory weighting functions, 299
Australian magpie. *See Cracticus tibicen*
Australian quacking frog. *See Crinia georgiana*

H. Brumm (ed.), *Animal Communication and Noise*,
Animal Signals and Communication 2, DOI: 10.1007/978-3-642-41494-7,
© Springer-Verlag Berlin Heidelberg 2013

Lightning Source UK Ltd.
Milton Keynes UK
UKOW06f0338121115

262514UK00001B/8/P

9 783642 414954